AF352934

Non-Destructive Testing of Structures

Non-Destructive Testing of Structures

Editor

Magdalena Rucka

MDPI • Basel • Beijing • Wuhan • Barcelona • Belgrade • Manchester • Tokyo • Cluj • Tianjin

Editor
Magdalena Rucka
Gdansk University of
Technology
Poland

Editorial Office
MDPI
St. Alban-Anlage 66
4052 Basel, Switzerland

This is a reprint of articles from the Special Issue published online in the open access journal *Materials* (ISSN 1996-1944) (available at: https://www.mdpi.com/journal/materials/special_issues/ structures_NDT).

For citation purposes, cite each article independently as indicated on the article page online and as indicated below:

LastName, A.A.; LastName, B.B.; LastName, C.C. Article Title. *Journal Name* **Year**, *Volume Number*, Page Range.

ISBN 978-3-03943-965-2 (Hbk)
ISBN 978-3-03943-966-9 (PDF)

Contents

About the Editor

Magdalena Rucka is a graduate of the Faculty of Civil and Environmental Engineering at the Gdańsk University of Technology (2002), and is currently the head of the Department of Mechanics of Materials and Structures. She received her Ph.D. in 2005, her habilitation in 2011 and the title of professor in 2020. She has five patents and over 65 articles indexed in the database Web of Science. These publications have >750 citations in Web of Science. She serves as the editorial board member of *Materials*. She is a specialist in structural dynamics and non-destructive testing. Her research interests centre on damage detection and imaging, mainly with the use of ultrasonic and electromagnetic wave propagation methods.

Editorial

Special Issue: "Non-Destructive Testing of Structures"

Magdalena Rucka

Department of Mechanics of Materials and Structures, Faculty of Civil and Environmental Engineering, Gdańsk University of Technology, Narutowicza 11/12, 80-233 Gdańsk, Poland; magdalena.rucka@pg.edu.pl or mrucka@pg.edu.pl; Tel.: +48-58-347-1891

Received: 31 October 2020; Accepted: 3 November 2020; Published: 6 November 2020

Abstract: The Special Issue "Non-Destructive Testing of Structures" has been proposed to present recent developments in the field of diagnostics of structural materials and components in civil and mechanical engineering. The papers highlighted in this editorial concern various aspects of non-invasive diagnostics, including such topics as condition assessments of civil and mechanical structures and connections of structural elements, the inspection of cultural heritage monuments, the testing of structural materials, structural health monitoring systems, the integration of non-destructive testing methods, advanced signal processing for the non-destructive testing of structures (NDT), damage detection and damage imaging, as well as modeling and numerical analyses for supporting structural health monitoring (SHM) systems.

Keywords: non-destructive testing; structural health monitoring; civil engineering structures; mechanical structures; damage detection and visualization; modeling and simulations

---

Engineering structures are gradually destroyed over time due to the influence of atmospheric conditions, excessive loads, and processes of natural aging. Since damage in a structural element may lead to improper operation of the whole object, various damage detection and structural health monitoring (SHM) methods have thus been investigated and developed to improve reliability and safety and to solve the maintenance problems of infrastructural and mechanical structures.

The term non-destructive testing (NDT) covers a wide group of measurement and analysis techniques used in the process of assessing the current state of structural materials or elements. It is a quick and effective approach, the main advantage of which is the ability to examine the material in a non-invasive manner, without damaging or changing the composition or shape of the inspected object. In recent years, there has been growing interest in the development of non-invasive methods, including in the field of civil and mechanical engineering.

The Special Issue "Non-Destructive Testing of Structures" has been proposed to gather the experience of civil and mechanical research communities in relation to the latest advances and trends in the field of diagnostics of structures and their components. A total of 22 papers were published in the Special Issue to touch different aspects connected with novel NDT approaches, with particular emphasis on the development of single and integrated measurement techniques, damage imaging procedures, advanced signal processing as well as modeling and numerical analyses for supporting SHM systems. A brief description of the Special Issue papers is presented below.

A wide range of experimental and numerical studies is reported in the papers in this Special Issue [1–8]. Nazarko and Ziemiański [1] monitored the axial bolt forces by means of elastic wave propagation signals. A series of laboratory tests were carried out on flange connections with six bolts, subjected to static tensile tests. Some bolts were equipped with washer load cells for the precise measurement of axial force. Moreover, selected bolts were equipped with piezoelectric transducers (actuator and sensor working in a pitch-catch configuration) in order to register the elastic wave signals. The results of ultrasonic testing were next integrated with the artificial neural network (ANN) for both signal compression and as an interface tool. The results showed that ANNs were able to predict the axial forces in bolts with relatively good accuracy. The proposed approach revealed the significant potential for real-life NDT inspections [1]. The effect of cladding stiffening on steel structures was studied by Korcz-Konkol and Iwicki [2]. Their work was undertaken to analyze what kind of non-destructive approach can be used for the condition assessment of existing buildings. Experimental investigations on small-scale shear panels made of trapezoidal sheeting were conducted to observe the behavior of the diaphragm under increasing and repeated load. The force–displacement relation and strains in selected areas of the sheeting were recorded during tests in a testing machine. Additionally, numerical calculations by the finite element method (FEM) in Abaqus software were conducted. The results obtained revealed the nonlinear, hysteretic force–displacement behavior of the panel and the occurrence of persistent deflections and stresses remaining even after the unloading. The authors concluded that indicators like the stresses, force–displacement paths and modes of failure could be potentially used for structural health monitoring of existing buildings in terms of parasitic stressed-skin action [2]. Palacz et al. [3] presented some aspects of numerical analyses for supporting SHM systems. They concentrated on FEM-based wave propagation modeling. The purpose of the study was to indicate the problems of numerical nature, which are immanent to modelling, as well as practical aspects that may lead to significant errors and the misinterpretation of results. The authors carried out numerical studies on one-dimensional structures. They analyzed the dynamic response covering natural frequencies and modes of natural vibrations and the accuracy of their representations [3]. Li et al. [4] presented a method dedicated to detection, localization, and quantitative analysis of local damage to beam structures based on vibration signal analysis. A novel algorithm based on the unscented Kalman filter was designed. The effectiveness of the method was demonstrated on experimental data acquired for a four-layer frame model composed of plexiglass plates and aluminum columns. The non-destructive testing algorithm was used to evaluate and monitor the strength, stiffness, damping of the frame. The results showed that the proposed damage detection algorithm is much more efficient than conventional methods [4]. Moonens et al. [5] described a promising technology patented under the name effective structural health monitoring (eSHM). In the paper, they addressed the possibility of assessing the fatigue life of a straight lug component by capillary-based structural health monitoring. The eSHM assumes the integrating structurally small and pressurized capillaries into the component. The concept is as follows: a fatigue crack breaches the capillary network and causes a leak to flow to the open atmosphere; next, the loss of pressure in the galleries is detected by a pressure sensor. The concept was verified numerically using the finite element and extended finite element methods. Various capillary sizes and shapes were analyzed. The authors revealed the great potential of the eSHM, especially with the use of additive manufacturing for part production [5]. Liu et al. [6] investigated the influence of a crack size on the stress evaluation of low alloy steel with metal magnetic memory (MMM) technology. A sample made of ferromagnetic material with rectangular grooves of different depths and widths was tested. The metal magnetic memory detector was used to store the $H_p(y)$ signal of the sample under different stresses. The fracture morphology was observed by using scanning electron microscopy (SEM). The results showed how different types of grooves influenced the magnetic intensity gradient changes [6]. The paper by Roskosz et al. [7] was focused on the evaluation of the hardness of ferromagnetic steel. Measurements of the Barkhausen noise and hardness were carried out on specimens subjected to plastic strain and thermochemical treatment. A new methodology was developed for the determination of correlations between the Barkhausen noise number of events and hardness [7]. Maciusowicz and Psuj [8] presented a new approach to the non-destructive evaluation of the easy magnetization axis in grain-oriented steel based on the Barkhausen phenomenon and its time–frequency characteristics. In the experiment, a sample of the conventional cold-rolled electrical steel sheet was used. Measurements were made for several angular settings towards the rolling and transverse directions. A procedure of signal transformation into the time–frequency domain by short-time Fourier transform was presented. The proposed method allowed detailed observation of the relationship between magnetic Barkhausen noise properties expressed in time and frequency for subsequent time moments [8].

The second group of papers has been devoted to the condition assessment of concrete and cement-based materials and structures [9–17]. Szeląg [9] presented a paper in which he reviewed the achievements in the field of the diagnostics of the cracking in cement composites. The four most important aspects related to the evaluation of a cracking pattern were addressed, namely: (i) the process of crack formation and their evolution into a branched system of cracks; (ii) the diagnostic techniques of digital extraction of the cracking patterns based on image analysis; (iii) the quantitative parameters used to describe the development degree and morphology of the cracks system; (iv) the influence of a cracking pattern on selected properties of cement composites. From the presented current state of the art in the field of the analysis of the cracking patterns in cement composites, it can be concluded that current works carried out in the field of crack assessment significantly contribute to the development of non-destructive testing methods in the field of cement composite technology [9]. The paper by Kuryłowicz-Cudowska [10] deals with the concept of a practical computation method to simulate the temperature distribution in a concrete structure. A model for the hardening of concrete, consistent with in situ measurement capabilities, was developed. Investigations consisted of three complementary parts: laboratory tests of high-performance concrete, measurements of temperature evolution in the bridge deck, and numerical simulations of temperature field in a concrete box bridge girder using the finite difference method. A new approach for the identification of the model parameters and heat transfer coefficients has been proposed. The results showed that the temperature history of concrete hardening, supplemented with maturity method equations, made it possible to estimate an early-age compressive strength of the cast-in-place concrete. It was also revealed that the proposed solution has great potential in the monitoring systems for concrete objects [10]. Słoński and Tekieli [11] presented a study on the monitoring and evaluation of surface cracks in concrete structural elements. The main goal was to integrate two computer vision methods, i.e., the digital image correlation (DIC) technique and region-based convolutional neural network (CNN) for the automatic detection and localization of multiple cracks. The proposed methodology was verified experimentally on post-tensioned, precast crane runway beams after more than 50 years of exploitation, subjected to three-point bending tests. At first, the DIC method was used for monitoring the deformation fields of the beam side surface. Next, the trained convolutional neural network was applied for the detection and localization of cracks. The results showed that the developed computer vision system was highly effective in automatically identifying the number of cracks and precisely pointing to their localization [11]. Sadowski et al. [12] proposed a complex methodology for the effective control of the entire process of executing floors made of cement-based materials. The methodology, developed based on many years of the authors' experience, has three control stages: (i) control before starting the floor; (ii) ongoing control during the execution of the floor; (iii) control tests after finishing the execution of the floor. The authors indicated the need to use necessary semi-destructive and non-destructive methods for the examination of the technical condition of floors in various construction objects [12]. Perkowski and Tatara [13] investigated the detection of brittle damage in concrete beams using transmission ultrasonic tomography. Tomographic experiments were carried out on three reinforced concrete beams on a laboratory scale and one prefabricated beam on a natural scale. A method of reducing errors in the tomographic reconstruction of longitudinal wave velocity maps caused by the simplifying assumptions of the straightness of the fastest wave propagation paths was developed. The method consisted of the appropriate extension of the measured propagation times of the wave transmitted between opposite sending–receiving transducers if the actual propagation paths deviate from straight lines. The conducted research has shown that the accuracy of the tomography imaging of brittle damage in concrete could be effectively supported by the graph theory and Dijkstra's algorithm [13]. Logoń and Schabowicz [14] used the acoustic emission (AE) method to recognize micro-events in quasi-brittle cement composites and to identify the destruction process. Experimental tests were conducted on a quasi-brittle composite of a cement paste reinforced with a high volume of dispersed polypropylene fibers. The main innovation of this research was the detection of AE micro-events using the sound spectrum in the area preceding the occurrence of critical cracks that initiate the destruction process in cement composites. It was confirmed that the application of the 3D spectrum enabled a better recognition of micro- and macro-changes in the structure of the samples based on the analysis of sound intensity, amplitudes, and frequencies [14]. Jasiński [15] evaluated changes in the stress state of masonry specimens made of autoclaved aerated concrete using the acoustoelastic method. The experimental studies were divided into two parts. At first, the empirical relationships between compressive stress and the velocity of the longitudinal ultrasonic wave, including humidity, were determined. Next, nine masonry walls were tested in axial compression. The obtained results showed that precise values of mean stress in the wall were determined on the basis of the measured velocity of ultrasonic wave propagation at a high number of measuring points, and the reduced number of measuring points resulted in a significant underestimation of mean stress [15]. The next two articles concerned bonding between a concrete substrate and cladding. Topolář et al. [16] experimentally examined the influence of the bonding system between a concrete substrate and large-format tiles. The bond strength was observed during mechanical loading under conditions that corresponded to real-life applications. Non-destructive diagnostics were performed using the following techniques: ultrasonic pulse velocity test, acoustic emission method, strain measurement, and acoustic tracing. The experiment showed that the loading caused no damage to the ceramic tile and all the detected failures took place in the adhesive layer or in the concrete slab [16]. Wojtczak et al. [17] applied the guided wave propagation technique for the condition assessment of concrete beams strengthened with steel plates. The novelty was in the comprehensive theoretical–numerical–experimental analysis of wave propagation in steel–concrete adhesively bonded specimens and the possibility of debonding imaging using weighted root mean square-based ultrasonic diagnostics. The research aimed at the damage identification and imaging in adhesive joints of composite beams with different levels of debonding. The experimental wave field was measured by scanning laser Doppler vibrometry. Next, the wave signals were processed by the weighted root mean square calculations. The obtained results indicated that the quality of damage maps strongly depended on the location of excitation [17].

Another group of articles concerned non-destructive testing of bridge structures [18–20]. Kwiatkowski et al. [18] presented a comparison of various measuring methods applied to investigate the behavior of a suspension bridge under different load scenarios. Three techniques, namely terrestrial laser scanning, tachymetry and photogrammetry, were examined on the bridge with a 165 m span. The testing range consisted of conducting the non-contact measuring of the bridge and cable displacements under dynamic and static loads and static loads. The obtained results enabled an assessment of the usefulness of the used measuring techniques. Some particular conclusions about the integration of the used methods have been drawn, especially in the context of measurements of structures under static and dynamic loading [18]. Mac et al. [19] focused on the detection of delamination in a concrete bridge deck. The main goal of the study was to develop a comprehensive system for bridge inspection using passive infrared thermography. Experimental tests were conducted on laboratory-scale concrete specimens with embedded artificial delamination of different width-to-depth ratios. Infrared cameras were employed to capture the surface temperature of the structure. The relations between the delamination size and depth and the heat energy were described. The optimal time to inspect the concrete bridge deck was determined [19]. Miśkiewicz et al. [20] described interdisciplinary and comprehensive non-destructive diagnostic tests of a soil–steel bridge made of corrugated sheets. A non-standard program of bridge loading was used during the inspection, including static and moving loads. The load test design was based on the finite element method simulations. In situ measurements were done by means of inductive sensors, an optical total station, and terrestrial laser scanner. The results obtained by terrestrial laser scanning were used to build a precise image of structural deformation. The accuracy of laser mapping was significantly increased using the information coming from the total station and the inductive sensors. A combination of testing methods was revealed as an effective tool in the non-destructive diagnostics of structures and an interesting alternative for the standard approach, in which the measurements are done in a limited number of points [20].

The diagnostics of cultural heritage structures were the subjects of interest in the last two papers from this Special Issue [21,22]. Tomaszewska et al. [21] investigated the problem of the modal identification of a masonry lighthouse. In the experiment, operational modal analysis was applied to the structure. Three identification techniques (peak picking, eigensystem realization algorithm and natural excitation technique with eigensystem realization algorithm) were applied to obtain modal parameters. The exact structural model was built by means of the finite element method. In the diagnostic process, the material properties were determined via numerical model validation applied to the first pair of natural frequencies and their related mode shapes, determined experimentally. As a result, the elastic modulus, Poisson's ratio and material density of brick, sandstone and granite masonry were determined [21]. Rucka et al. [22] presented the results of the integrated ground penetrating radar (GPR) and ultrasonic testing (UT) inspection conducted on a historical floor. The aim of the study was to present the practical aspects of the application of both techniques in detecting and imaging the underfloor inclusions, such as air voids, brick walls, pipes, rubble and human remains. In order to better understand the phenomena of electromagnetic and ultrasonic wave propagation within the air voids and concentrated inclusions, laboratory tests were conducted on two concrete slabs stacked on top of each other and gradually moved apart to simulate a slot of varying thickness. The numerical simulations of electromagnetic waves were performed to support the interpretation of the GPR results. The obtained results showed that the integration of the GPR and UT methods provided effective imaging of the floor and the area under it. Ultrasonic testing proved to be a good technique for identifying air voids, while the GPR method allowed for the detection of concentrated anomalies and determining the degree of ground homogeneity under a floor. The presented research revealed the possibilities and limitations of both methods, indicating their complementarity in the context of the non-destructive diagnostics of historical buildings [22].

In summary, the papers presented in this Special Issue concern various aspects of non-invasive diagnostics and present recent developments in the field of the diagnostics of structural materials and components in civil and mechanical engineering. Both interesting practical solutions and a significant contribution to the new approaches to NDT were achieved.

Conflicts of Interest: The author declares no conflicts of interest.

References

1. Nazarko, P.; Ziemiański, L. Application of elastic waves and neural networks for the prediction of forces in bolts of flange connections subjected to static tension tests. *Materials* **2020**, *13*, 3607. [CrossRef] [PubMed]
2. Korcz-Konkol, N.; Iwicki, P. Corrugated sheeting as a member of a shear panel under repeated load-experimental test. *Materials* **2020**, *13*, 4032. [CrossRef] [PubMed]
3. Palacz, M.; Zak, A.; Krawczuk, M. FEM-based wave propagation modelling for SHM: Certain numerical issues in 1D structures. *Materials* **2020**, *13*, 2051. [CrossRef] [PubMed]
4. Li, X.; Shi, D.; Yu, Z. Nondestructive damage testing of beam structure based on vibration response signal analysis. *Materials* **2020**, *13*, 3301. [CrossRef]
5. Moonens, M.; Wyart, E.; De Baere, D.; Hinderdael, M.; Ertveldt, J.; Jardon, Z.; Arroud, G.; Guillaume, P. On the influence of capillary-based structural health monitoring on fatigue crack initiation and propagation in straight lugs. *Materials* **2019**, *12*, 2965. [CrossRef]
6. Liu, B.; Fu, P.; Li, R.; He, P.; Dong, S. Influence of crack size on stress evaluation of ferromagnetic low alloy steel with metal magnetic memory technology. *Materials* **2019**, *12*, 4028. [CrossRef]
7. Roskosz, M.; Fryczowski, K.; Schabowicz, K. Evaluation of ferromagnetic steel hardness based on an analysis of the Barkhausen noise number of events. *Materials* **2020**, *13*, 2059. [CrossRef]
8. Maciusowicz, M.; Psuj, G. Use of Time-Frequency Representation of Magnetic Barkhausen Noise for Evaluation of Easy Magnetization Axis of Grain-Oriented Steel. *Materials* **2020**, *13*, 3390. [CrossRef] [PubMed]

Materials **2020**, *13*, 4996

9. Szelag, M. Evaluation of cracking patterns in cement composites-from basics to advances: A review. *Materials* **2020**, *13*, 2490. [CrossRef]

10. Kurylowicz-Cudowska, A. Determination of thermophysical parameters involved in the numerical model to predict the temperature field of cast-in-place concrete bridge deck. *Materials* **2019**, *12*, 3089. [CrossRef]

11. Słoński, M.; Tekieli, M. 2D digital image correlation and region-based convolutional neural network in monitoring and evaluation of surface cracks in concrete structural elements. *Materials* **2020**, *13*, 3527. [CrossRef] [PubMed]

12. Sadowski, L.; Hola, A.; Hola, J. Methodology for controlling the technological process of executing floors made of cement-based materials. *Materials* **2020**, *13*, 948. [CrossRef] [PubMed]

13. Perkowski, Z.; Tatara, K. The use of Dijkstra's algorithm in assessing the correctness of imaging brittle damage in concrete beams by means of ultrasonic transmission tomography. *Materials* **2020**, *13*, 551. [CrossRef]

14. Logoń, D.; Schabowicz, K. The recognition of the micro-events in cement composites and the identification of the destruction process using acoustic emission and sound spectrum. *Materials* **2020**, *13*, 2988. [CrossRef] [PubMed]

15. Jasiński, R. Identification of Stress States in Compressed Masonry Walls Using a Non-Destructive Technique (NDT). *Materials* **2020**, *13*, 2852. [CrossRef]

16. Topolář, L.; Kocáb, D.; Šlanhof, J.; Schmid, P.; Daněk, P.; Nováček, J. Testing the influence of the material bonding system on the bond strength of large-format tiles installed on concrete substrate under mechanical loading. *Materials* **2020**, *13*, 3200. [CrossRef]

17. Wojtczak, E.; Rucka, M.; Knak, M. Detection and imaging of debonding in adhesive joints of concrete beams strengthened with steel plates using guided waves and weighted root mean square. *Materials* **2020**, *13*, 2167. [CrossRef]

18. Kwiatkowski, J.; Anigacz, W.; Beben, D. Comparison of Non-Destructive Techniques for Technological Bridge Deflection Testing. *Materials* **2020**, *13*, 1908. [CrossRef]

19. Mac, V.H.; Tran, Q.H.; Huh, J.; Doan, N.S.; Kang, C.; Han, D. Detection of delamination with various width-to-depth ratios in concrete bridge deck using passive IRT: Limits and applicability. *Materials* **2019**, *12*, 3996. [CrossRef] [PubMed]

20. Miśkiewicz, M.; Sobczyk, B.; Tysiąc, P. Non-Destructive Testing of the Longest Span Soil-Steel Bridge in Europe—Field Measurements and FEM Calculations. *Materials* **2020**, *13*, 3652. [CrossRef] [PubMed]

21. Tomaszewska, A.; Drozdowska, M.; Szafrański, M. Material parameters identification of historic lighthouse based on operational modal analysis. *Materials* **2020**, *13*, 3814. [CrossRef]

22. Rucka, M.; Wojtczak, E.; Zielińska, M. Integrated application of GPR and ultrasonic testing in the diagnostics of a historical floor. *Materials* **2020**, *13*, 2547. [CrossRef]

Publisher's Note: MDPI stays neutral with regard to jurisdictional claims in published maps and institutional affiliations.

Article

Corrugated Sheeting as a Member of a Shear Panel Under Repeated Load—Experimental Test

Natalia Korcz-Konkol * and Piotr Iwicki

Department of Metal Structures, Faculty of Civil and Environmental Engineering, Gdańsk University of Technology, ul. G. Narutowicza 11/12, 80-233 Gdańsk, Poland; piotr.iwicki@pg.edu.pl
* Correspondence: natalia.korcz@pg.edu.pl

Received: 30 June 2020; Accepted: 8 September 2020; Published: 11 September 2020

Abstract: In stressed-skin design, the cladding stiffening effect on structures is taken into account. However, the "traditional" design is more usual, wherein this effect is neglected. Even if the diaphragm actions are not regarded, in particular cases such as big sheds (and others), the parasitic (unwanted) stressed-skin action may occur with the result of leakage or even failure. The structures of this kind have already been built. Thus, an important question arises: How can one assess them if there is a need to correct or redesign them? What kind of non-destructive approach can be used to achieve that? Experimental tests of small-scale shear panels made of trapezoidal sheeting were designed in order to observe the behaviour of the diaphragm under increasing and repeated load. The tests were oriented toward force–displacement relations and strains in selected areas of the sheeting. The results revealed nonlinear, hysteretic force–displacement behaviour of the panel and the occurrence of the persistent deflections and stresses which remain even after the unloading. The relation among the stresses, force–displacement paths and modes of failure can be potentially used in monitoring systems of existing buildings in terms of parasitic stressed-skin action.

Keywords: steel structures; stressed-skin effect; diaphragm design; trapezoidal sheeting; condition assessment of steel structure; structural health monitoring

---

1. Introduction

Cladding of steel structures affects the stiffness and spatial character of structural performance, and consequently, deflections and forces in particular structural members. This action is called stressed-skin or diaphragm action. The idea of stressed-skin design was born in the 1960s, and came to be widely known throughout the decades. The intention is to take advantage of the interaction between structural elements and panels (roofing, wall cladding or even flooring). According to [1], roof and floor panels may be treated as the web and edge members (elements along the structure, e.g., purlins) as the flanges of the deep plate girder. Similarly, wall diaphragms may be treated as bracing. That is typical response of the stressed-skin structure on the horizontal forces (such as wind loads); however, in the case of a pitched-roof, vertical forces (such as snow loads) may be resisted in a similar way as well.

Recent European Recommendations for stressed-skin design [2] were formulated in 1995. Since then, significant changes were noted in steel construction: nowadays the structures are bigger, taller, made of more slender members (e.g., cold-formed profiles), have new types of cladding and fasteners, etc. What is more, the development of hardware and numerical software dedicated to structures has brought in more and more accessible tools to consider more complex cases. These are probably some of the reasons for the increasing interest of the researchers in stressed-skin actions. The improvements to the analytical procedures are investigated, e.g., new factors affecting stiffness of the diaphragms but neglected in the procedures and analyses of schemes of the fasteners, e.g., the lack of seam fasteners [3–6], and schemes known from practice but not included in recommendations [7,8].

Some of the necessary improvements of design codes have been proposed earlier [9]. Other studies focus on numerical aspects of taking into account the diaphragm actions, such as [8,10–12].

Simultaneously, in "traditional design" stabilisation of particular elements, e.g., purlins, by sheeting is much more eagerly taken into account (this problem was analysed, e.g., in [13]), while the diaphragm effect is neglected; thus, extra global stiffening of the structure by sheeting is considered beneficial. However, it has to be remembered that regardless of possible considering the stressed-skin actions of the cladding, it brings extra structural stiffness to a certain extent. The significance of this aspect is controlled by many parameters. In [14,15] the attention was drawn to selected roof failures attributed to the parasitic (undesired) stressed-skin action. Further discussion and explanations are presented also in [16]. The unwanted diaphragm actions may trigger significant force increases in the edge members (e.g., purlins), purlin to cleat connections, purlin to cladding connections and cladding itself. These situations arise in big sheds, but can be also important in small buildings, e.g., with cold-formed structure of the frames. The problem is even more complex because of the fact that buildings may be susceptible to fatigue accumulation deficits, reaching failure conditions many years after their construction, possibly causing leakage of the cladding, and even fatigue failure. These alarming reports should lead to much more cautious design of steel structures in the traditional approach.

The problem of the underestimation of the importance of the diaphragms is known and actual not only in steel structures. Another example may be the infilled reinforced concrete (RC) structures when subjected to earthquakes. According to [17], the infill masonry walls have a significant contribution to the global seismic response of RC structure. Results of the blind test prediction carried out in a scaled RC structure confirmed that the strength and stiffness of the infill (diaphragm) should definitely be taken into account in design processes as a very important factor.

A list of problems to solve thusly arose. How can one evaluate the condition of the structure in the existing buildings designed without the consideration of stressed-skin effect, currently considered as being in the group of risk of parasitic stressed-skin action? How can one assess them if there is a need to correct or redesign the structure before the leakage or failure occurred? What kind of non-destructive approach can be used to achieve this?

This paper attempts to address the issues above. Experimental tests of small-scale shear panels made of trapezoidal sheeting were designed in order to observe the behaviour of diaphragms under increasing and repeated load. The measurement results were the force–displacement relation (in a testing machine) and strains in selected areas of the sheeting using strain gauges (SG). The results revealed the hysteresical character of panel work and persistent displacements and strains which remain even after unloading the structure, which can be the first step to leakage, even failure. The outcomes showed the relation between the stresses, force–displacement paths and the mechanism of failure, to be potentially applied in the system of monitoring existing buildings in terms of parasitic stressed-skin action.

2. Materials and Methods

2.1. Experimental Model—General Description

Experimental research was intended to investigate a wide range of panel behaviour at shear. The experimental set-up was designed for the corrugated sheet investigation; however, it can also be used for other types of panels (e.g., sandwich, wood or textile panels or parts of roofs consisting of purlins and trapezoidal sheeting). A static model of the permanent set-up is presented in Figure 1, the main elements are shown in Figure 2. Four pin-ended rectangular hollow section members form a square frame with four hinged nodes in the frame plane. The axial dimension of the frame was 850 mm (width and length: *a* and *b* in Figure 1). The dimensions of the set-up were assessed according to the technological condition that the largest dimension in the direction of square frame diagonal must be less than 1350 mm, i.e., the maximum allowable spacing between the fastening elements of the testing machine.

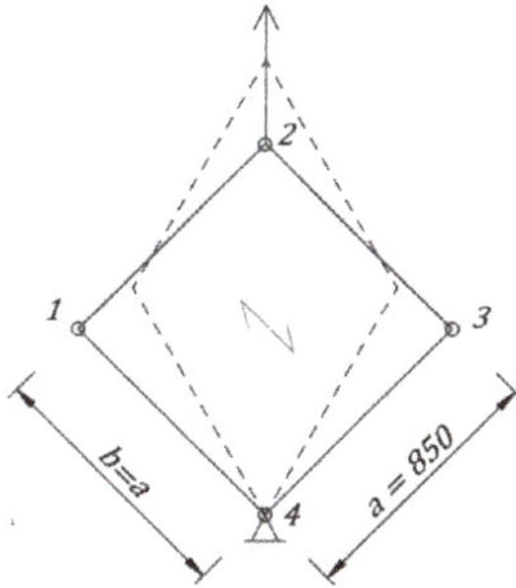

Figure 1. Static scheme of the experimental set-up [mm].

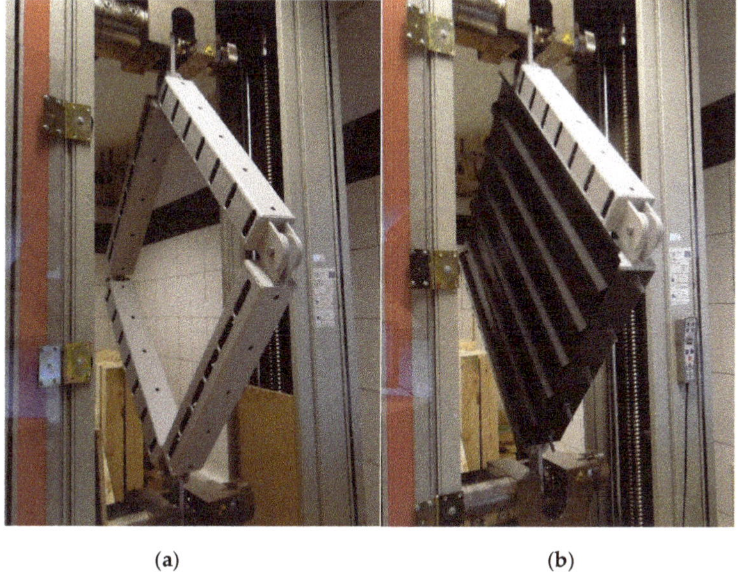

(a)   (b)

Figure 2. Experimental set-up: (**a**) permanent elements, (**b**) set-up with the analysed sheeting.

In nodes 2 and 4 (see Figure 1) extra elements were provided to allow fastening the set-up to the testing machine. These extra elements were totally fixed in both top and bottom fastening element of the testing machine. Next, every extra element was pin-joint in the hinges of the square frame with possible rotation about the axis perpendicular to the plane of the frame; see Figures 3b and 4a. Similar hinges have been constructed in nodes 1 and 3 in Figure 1, as presented in Figure 2. The bottom fastening element was fixed in the initial point during the experiment while the testing machine was inducing displacement of the top fastening element in the vertical direction. Due to pin connection of members, the permanent set-up (square frame) was a mechanism not to resist shear (comparing to the diaphragm stiffness). It allowed us to investigate the resistance and stiffness of shear panels made of corrugated sheeting installed to the frame. Sheet/purlin (or sheet/rafter) fasteners were applied, so flexibility, location and influence of the connection on corrugated sheeting were taken into account. Flexibility of the support elements (purlins, rafters) was included with the use of U-shaped cold-formed plates. The plates were fixed to rigid rectangular hollow section (RHS) profiles in order to fasten the sheeting. Specifications of the applied elements are widely discussed in Section 2.2.

2.2. Experimental Model—Specification of the Elements

The model of the experimental set-up is presented in Figure 1; the set-up fastened to the testing machine is presented in Figure 2. The main parts of the permanent experimental set-up were four pin-ended rectangular hollow section profiles (RHS 100 × 4), which formed the mechanism of a square frame (see Figures 1 and 2a). The RHS 100 × 4 profile was assumed rigid enough not to affect the final results. Next the U-shaped cold-formed plates were fastened to the RHS members using bolts as presented in Figures 2b and 3b. The thickness of the U-shaped plate was 0.5 mm; however, in the fastening plane the thickness was increased to 1.5 mm by adding extra plates. The corrugated sheeting is typically fastened to the support structure (e.g., purlin or girder) using self-drilling screws. In order to allow for fastening the corrugated sheeting to the U-shaped plates and for the re-use of the permanent set-up, long holes were designed in the RHS profiles; see Figure 2a. The case of trapezoidal sheeting was investigated, with the height of 18 mm, the length in the direction of the corrugation of 950 mm and the width in the direction parallel to the corrugation of 830 mm. The geometry of the trapezoidal sheeting was chosen according to Eurocode procedures [1,18–20], based on the comparison of the value of the ultimate limit state (ULS) combination of loading perpendicular to the plane of the sheeting typical for Poland and the allowable loading for the specimen trapezoidal sheeting. The comparison proved that the space between the supports of the sheeting should not exceed 1 m, while the spacing between support elements in the tests was equal to 0.85 m due to machine limitations. Taking this into consideration, the reduction scale is not such a significant issue. On the other hand, in the experiment only one part of the panel was taken into account; in real structures there are more panels building one roof, which means more complex behaviour. However, in this paper the concentration is primarily on the local effects (support, fasteners) which seem to be crucial in case of parasitic stressed-skin action. In this case, even the reduced scale specimen seems to fulfil its function. The trapezoidal sheeting geometry and the model of the sheet/purlin or sheet/rafter fasteners (self-drilling screws of a diameter 5.5 mm and ethylene propylene diene monomer (EPDM) washer is presented in Figure 3. The screws were used in the centreline of every corrugation.

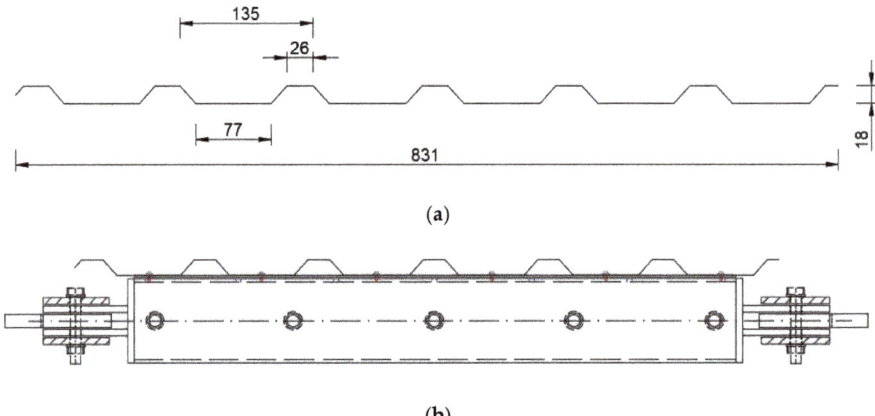

(a)

(b)

Figure 3. Specimens' details: (**a**) trapezoidal sheeting geometry [mm]; (**b**) trapezoidal sheeting fastened to the permanent frame (scheme of the fasteners).

2.3. Experimental Model—Specifications of the Measurements

Experimental analyses were conducted in the testing machine as presented in Figures 1 and 2. The displacement increment over time of the top fastening element of the machine (node 2) was assumed constant. Displacements (node 2), forces (node 4) and time were measured. Additionally, the changes of strains in selected points and directions of the trapezoidal sheeting in relation to the time were registered. The layout of the strain gauges used in the tests is presented in Figure 4. SG 1 and 5 registered the changes of the strains in the direction perpendicular to the corrugations in the side of the middle corrugation, 50 mm from the frame edge (on both sides of the sheeting). SG 2, 3 and 4 formed a rectangular rosette (on one—bottom side of the sheeting) which allowed us to assess principle stresses (assuming elastic modulus $E = 210$ GPa).

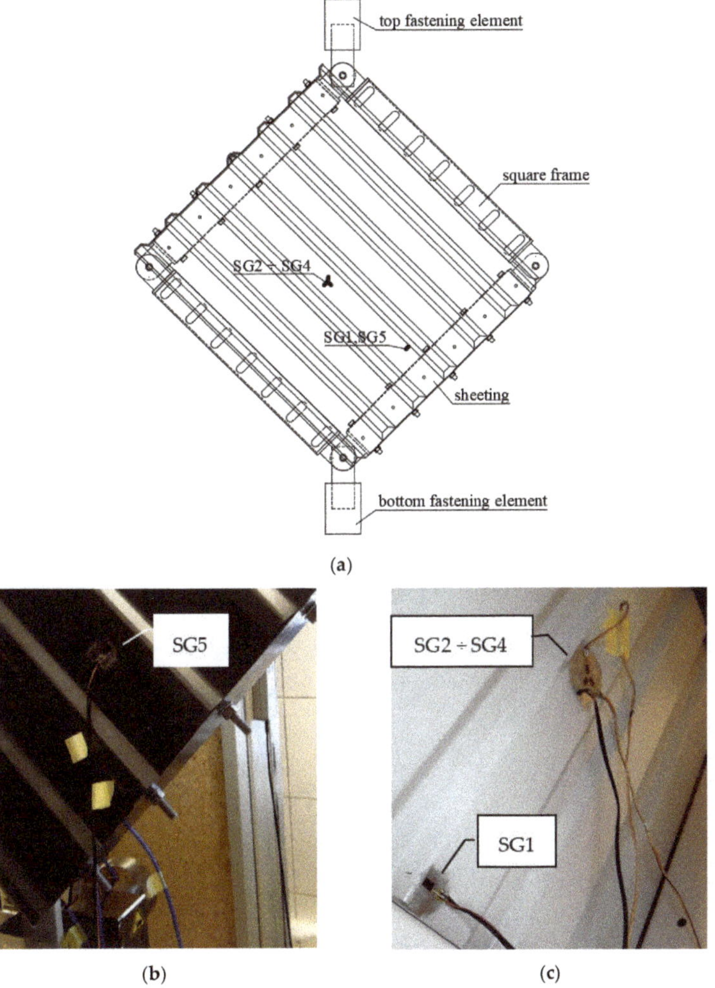

Figure 4. Strain gauge (SG) measurements: (**a**) scheme of the SG location; (**b**) SG 5 during the test; (**c**) SG 1, 2, 3 and 4 during the test.

Four tests of trapezoidal sheeting panels were conducted. The test 1 assumed the displacement was increasing constantly until the panel failure. In tests 2, 3 and 4 the repeated displacement increase/decrease was implemented. The loading protocol in tests 2, 3 and 4 was as follows:

- Initial force 50 N.

- Increase of the force (displacement increase = constant) until the value of 1000 N—force = constant for 30 s—decrease of the force to the value of 300 N (displacement decrease = constant)—force = constant for 30 s.
- Increase of the force (displacement increase = constant) until the value of 2000 N—force = constant for 30 s—decrease of the force to the value of 300 N (displacement decrease = constant)—force = constant for 30 s.
- Increase of the force (displacement increase = constant) until the value of 3000 N—force = constant for 30 s—decrease of the force to the value of 300 N (displacement decrease = constant)—force = constant for 30 s.
- Increase of the force (displacement increase = constant) until the value of 4300 N—force = constant for 30 s—decrease of the force to the value of 300 N (displacement decrease = constant)—force = constant for 30 s.
- Increase of the force (displacement increase = constant) until the value of 6300 N—force = constant for 30 s—decrease of the force to the value of 300 N (displacement decrease = constant)—force = constant for 30 s.
- Increase of the force (displacement increase = constant) until the failure of the panel.

The initial force in the testing machine was 50 N. Nevertheless, in the initial stage of loading, the loose spaces of the set-up disturbed the measurements (about 3,5 mm and circa 300 N). In the paper diaphragms without this initial stage are analysed.

2.4. Numerical Model—General Description

A 3D numerical model of the experimental research was built in ABAQUS software [21]. Geometrically and materially non-linear static analysis was performed. An elastic-plastic model of the material (steel S250GD + Z) was reflected by the following properties: elastic modulus $E = 210$ GPa, Poisson's ratio $\nu = 0.3$ and yield stress $f_y = 250$ MPa.

The frame members and corrugated sheets were modelled by 20,100 shell elements with four nodes and four integration points (S4) and the size of 8–12 mm. Profiles were built with sharp corners (radius of curved elements equal to zero). The meshed structure is presented in Figure 5a.

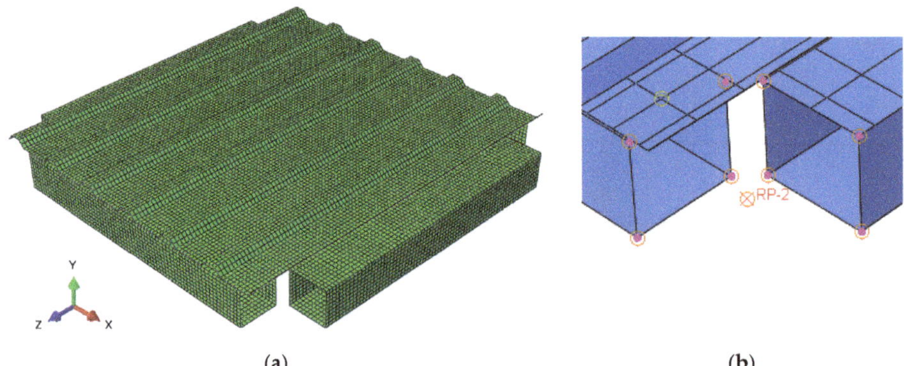

(a)   (b)

Figure 5. Numerical model: (a) view of the whole model with global coordinates; (b) detail of the frame hinge.

The frame members were modelled in a simplified way. Each of the RHS profiles, which in tests were the supports for the sheeting, was built as one element together with the U-shaped cold-formed plate. The thickness of the U-shaped plate (0.5 mm on the sides, 1.5 mm on the top) was assigned, so that the stiffness of the profile was mapped. Hinge connections in the frame were included using four reference points (RP in Figure 5b) in four axial nodes of the frame, which were tied (six degrees of freedom fixed) with four corners of the shell frame element. This way allowed us to model the hinge without the necessity of building the details of the connection. The connection is presented in Figure 5b.

Fasteners (self-drilling screws) were mapped using tie connections between nodes (six degrees of freedom fixed). The propping effect was included by modelling the contact between the sheeting and the frame (with the separation allowed after contact).

Boundary conditions were assigned using references points. In nodes 1 and 3 the displacement in y direction was fixed; in node 4—displacements in x, y and z directions were fixed (for numbers of nodes see Figure 1). In node 2 the displacement increase was applied in the direction of the frame diagonal until the failure of the panel. Stress maps were obtained. The numerical analysis results are presented in Section 3.2.

3. Results and Discussion

3.1. Laboratory Tests

In the first step of the laboratory analysis, the four-pinned frame of the experimental set-up (see Figure 2a) was tested without the sheeting. The force–displacement outcomes confirmed the assumption that the frame alone, not including trapezoidal sheeting, has a mechanism of negligible stiffness, with regard to shear stiffness of the analysed panel; see Figure 6.

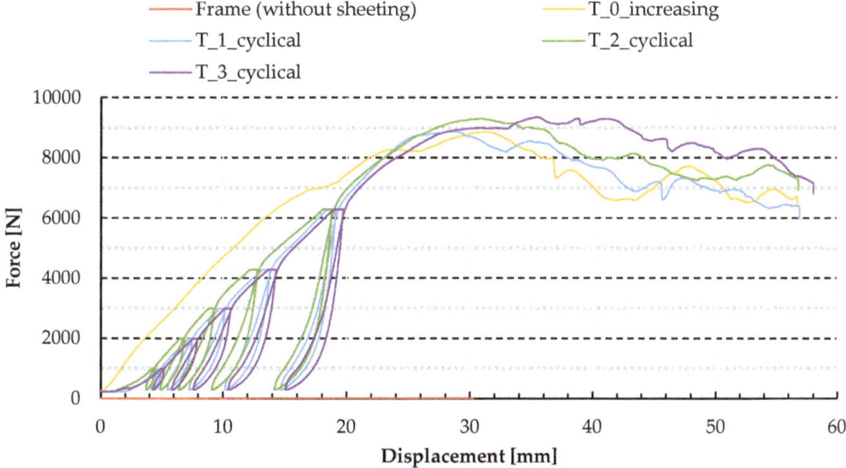

Figure 6. Force–displacement paths obtained in experimental tests.

Next, major tests of trapezoidal sheeting panels were conducted. In the first step, the constant displacement increase was applied leading to panel failure. The force–displacement and force–time path of the main nodes were registered; see "T_0_increasing" in Figures 6 and 7. Moreover, strains were measured by SG (for the location of the SG; see point 2.3), leading to the corresponding stresses: SG1 and SG5—stresses in the direction perpendicular to the corrugation (top and bottom side of the sheeting respectively) and SIG_1 and SIG_2—the principal stresses linked with the outcomes of strain gauge rosette; see Figure 7. Note that the stresses related to profile distortion (in the location of strain gauges SG1 and SG5) start to increase earlier and become much greater than the stresses related to global shear of the panel (in the location of strain gauges SG2, 3, 4). It confirms the prediction that in this variant of panel geometry (relatively high trapezoidal sheeting compared to the panel planar dimensions), the profile distortion dominates the shear strain. What is more, the absolute values of SIG_2 principal stresses were about 1.5–2 times greater than the values of SIG_1 principal stresses.

The values were affected by the orthotropy of the trapezoidal sheeting and the location of the SGs on the bottom flange.

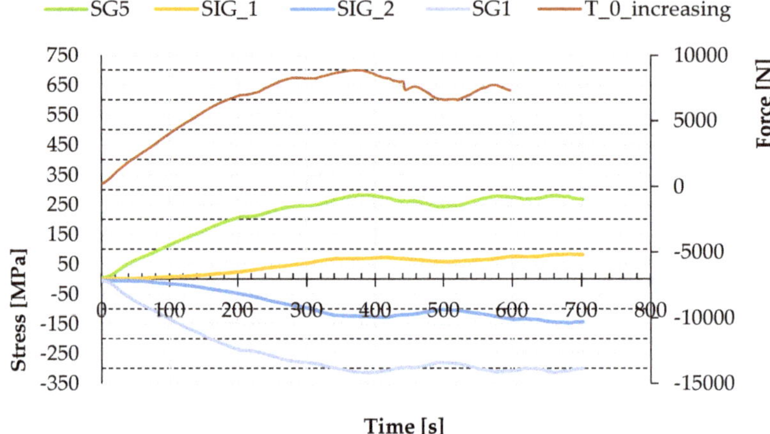

Figure 7. Force–time and stress–time paths obtained in experimental tests for increasing loading.

In the next step, three cyclic loading tests were conducted on trapezoidal sheeting panels. The diaphragms were loaded six times in every test; each step brought an increasing force level. The last increment led to failure of the panel. The force–displacement path of the main nodes registered by the testing machine is presented in Figures 6 and 8; see "T_1_cyclical," "T_2_cyclical" and "T_3_cyclical".

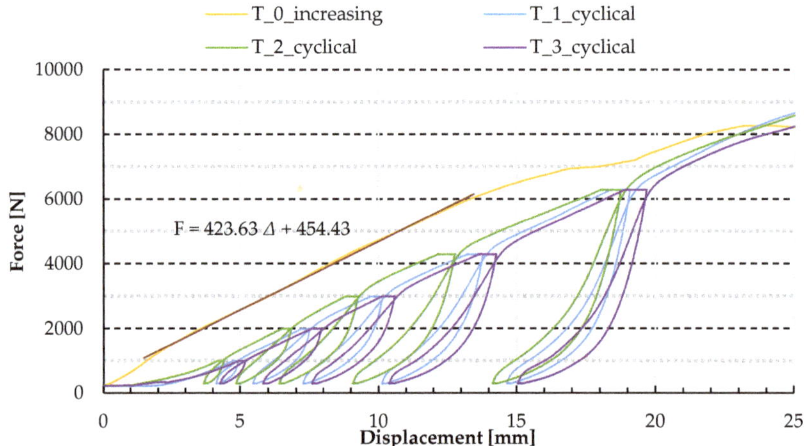

Figure 8. Force–displacement paths obtained in experimental tests—the extract of the outcomes.

The force–displacement relations corresponding to the repeated loading revealed a wider characteristic of the panel compared to the permanently increasing load (compare T_1-3_cyclical with the T_0_increasing in Figure 8). Hence the diaphragm displays two types of stiffnesses: primary and secondary. When the panel bears a particular force level for the first time, its stiffness is lower (primary stiffness); when the force level is achieved again, after unloading, the stiffness is greater (secondary stiffness). In the course of primary loading, the displacements between elements (e.g., loose space, clearances) are successively removed, some of them permanently. As a result, during reloading the displacement does not reach the level prior to the primary loading. Moreover, stiffness depends on the direction of change which is represented graphically by the hysteresis curves. The area in the centres of hysteresis loops can be interpreted as the energy dissipation mainly due to friction between elements of the panel. The scale of the hysteresis can rely on the rapidity of the loading/reloading.

The trend line for the test with increasing load is shown in Figure 8, where: *F* is force and Δ is displacement. According to the trend line equation, the primary stiffness of the panel is 0.424 kN/mm. The force–displacement relations for increasing and repeated loading make us conclude that both primary and secondary stiffness of the panels in four tests are comparable. Simultaneously, translation of the diagrams is observed. It can arise from different initial clearances in particular panels triggered by assembly imperfections, thus means it can also occur in real structure situation.

Similarly to the increasing loading case, in the case of repeated loading strains were also measured by strain gauges (the location of the strain gauges is addressed in point 2.3), leading to stresses: SG1 and SG5—the stresses in the direction perpendicular to the corrugation (top and bottom side of the sheeting respectively); and SIG_1 and SIG_2—the principal stresses in the location of strain gauges rosette, as presented in Figure 9.

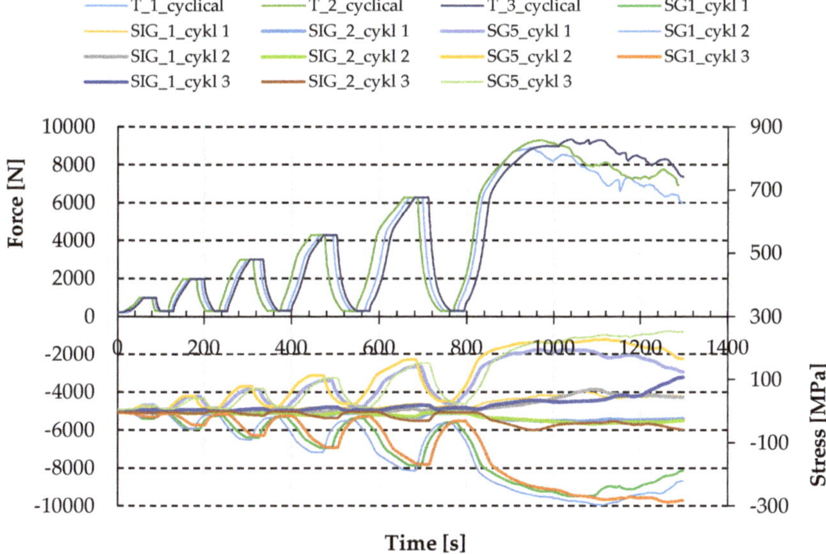

Figure 9. Force–time and stress–time paths obtained in experimental tests for repeated loading.

After reaching a particular load level, the force was fixed for the time equal to 30 s. During that time, the displacements were increasing while the force and the stresses in the sheeting remained constant; see Figures 8 and 9. It can signalise the local plasticity of the trapezoidal sheeting, e.g., in the area of screws and the longitudinal fold lines of the sheeting—even in the case of load increments when the stresses (indirectly) measured by strain gauges SG1 and SG5 in the area close to the support were below the yield stress.

Permanent displacements increased at every loading; the stresses followed; see Figure 9. That confirms the theory of local plasticity of the trapezoidal sheeting, which can lead to leakage or even failure of the sheeting.

Figure 10 shows deflection modes observed during tests. First of all, profile distortion occurs; next comes hole elongation, leading to the rotation of the sheeting. Hole elongation due to plasticity of the sheeting turned out to be a crucial failure mechanism in this series of tests.

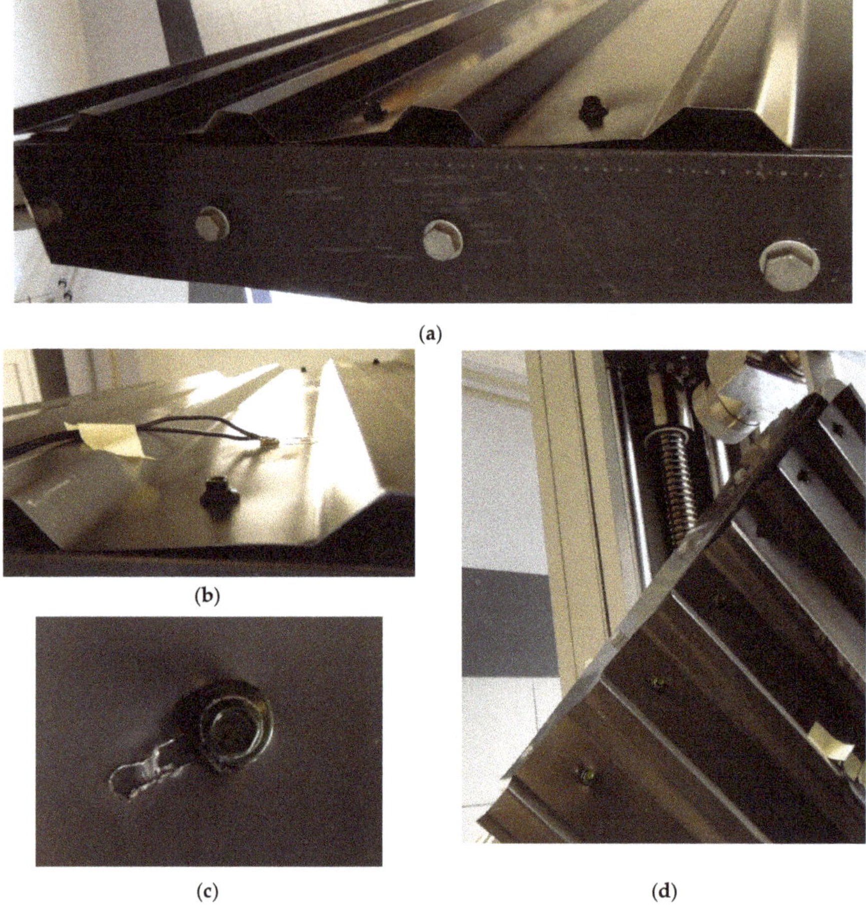

Figure 10. Deflections of trapezoidal sheeting observed during laboratory tests: (**a**) profile distortion—global view; (**b**) profile distortion—local view; (**c**) hole elongation; (**d**) rotation of the sheeting due to hole elongation.

At the first stage of the unloading, both displacements and stresses decrease rapidly (small decrease of the displacements means that the structure is rigid at the beginning of the uploading). Visual observations confirm that the profile distortion decreases significantly. It is interpreted as the fact that in first unloading stage the traction between elements supresses the displacements between members and the stiffness is mostly influenced by taking back the profile distortion.

3.2. Numerical Analysis

The 3D numerical model of the experiment was run in ABAQUS software [21]. In numerical estimations, the structure was built in simplified way, as was described in detail in Section 2.4 (screws as tie connections between two nodes, no holes, profiles with sharp corners, only increasing force considered). As a result, the numerical stiffness was greater than experimental stiffness and the effect of primary/secondary stiffness (which arises from, among others, hole elongation, natural spacing between assembled element—parameters not mapped in the numerical simulations) was not captured. However, other selected results—displacements in the direction perpendicular to the panel and stress map in the direction perpendicular to the corrugation (the direction of the strain gauges SG1 and SG5)—are presented in Figure 11. The results confirmed the dominant character of profile distortion (shape of the profile deformation, stress distribution in the area of the support, opposite signs of the stresses on both sides of the sheeting). The shapes of profile deformation are compared in Figure 12. The results in the area of screw fasteners are compared in Figure 13.

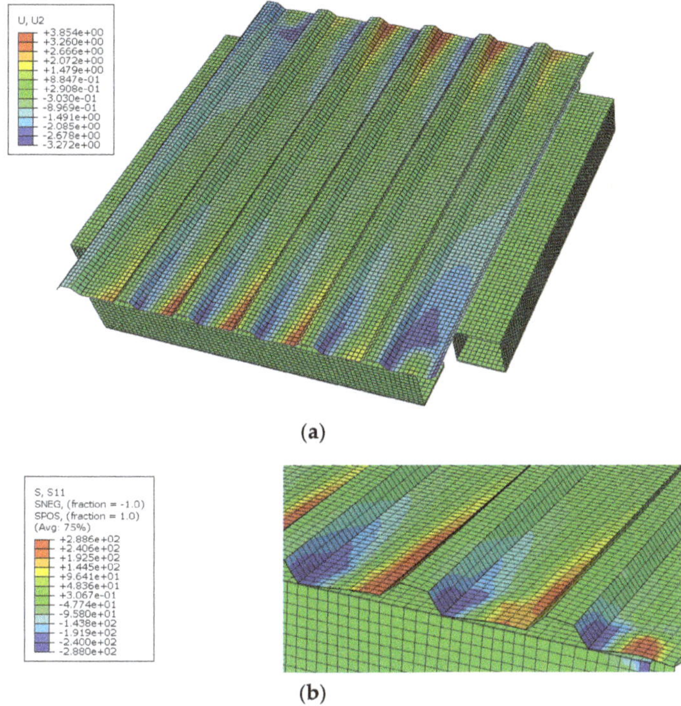

(a)

(b)

Figure 11. Selected results of the numerical analysis: (**a**) displacements in the direction perpendicular to the panel [mm]; (**b**) stresses in the direction perpendicular to the corrugation [MPa].

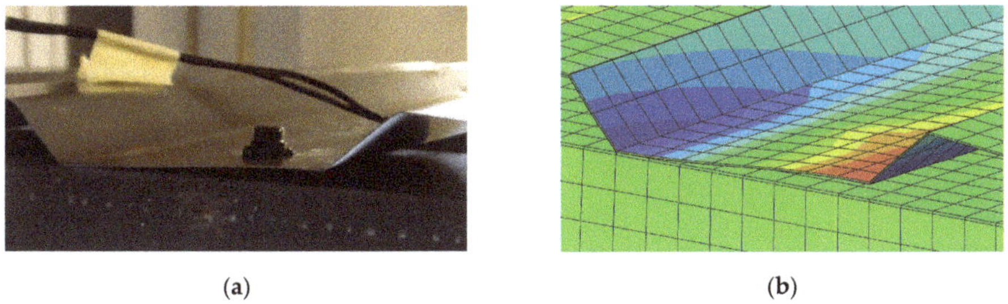

(a) **(b)**

Figure 12. The comparison of the shape of the profile deformation: (**a**) test, (**b**) numerical model.

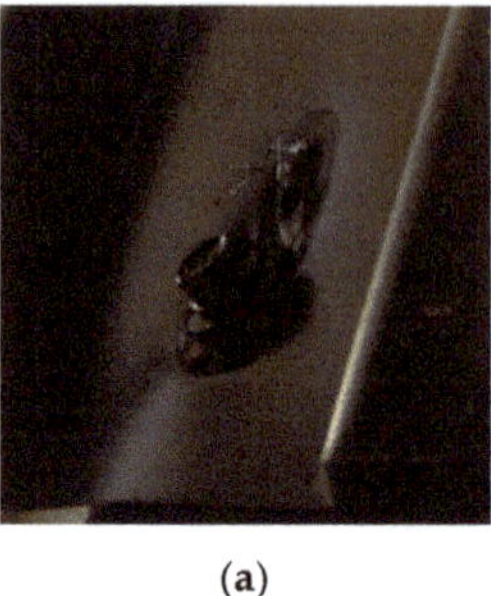 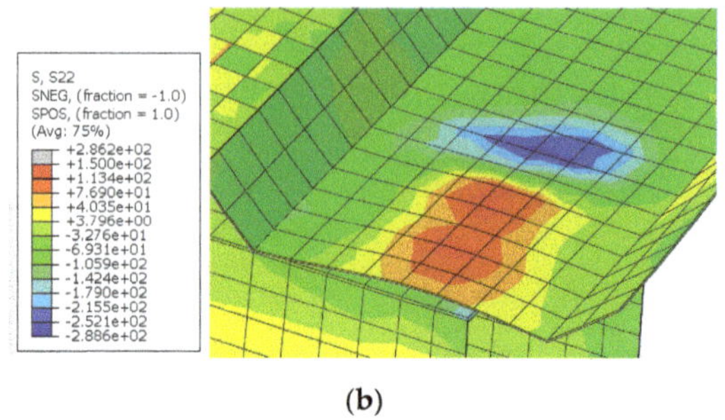

(a) (b)

Figure 13. The comparison of the results in the area of self-drilling screw: (**a**) test, (**b**) numerical model (stresses in the direction parallel to the corrugation [MPa]).

4. Conclusions

Experimental tests on a small-scale shear panel made of trapezoidal sheeting were conducted in order to observe the behaviour of the diaphragm under increasing and repeated load. The force–displacement relation (in a testing machine) and the strains in selected areas of the sheeting (using strain gauges) were measured. The results revealed the hysteresical character of panel work and occurrence of persistent deflections and stresses, which remain even after the unloading and suggest invisible plastic strains in the panel. In the case of fatigue accumulation, the latter may lead to the leakage, even failure.

The experimental results showed the character of work of the cladding in case of stressed-skin effect under repeated load (primary/secondary stiffness), the important phenomena and the failure mechanism, especially in the support area of the panels. The outcomes show the relation between the stresses, force–displacement paths and the mechanism of failure, which potentially can be used in the monitoring of existing buildings in order to assess parasitic stressed-skin action in a non-destructive way. What is more, roof rigidity determined by measurements may be incorporated in metal sheet design to determine the flexibility and stiffness of connections.

The results give a basis to the non-destructive testing of the existing structures. By knowing the displacement ranges of the existing structure panels (e.g., using inductive sensors), it seems that the condition of the structure can be assessed (also for structures which were designed without consideration of the stressed-skin effect) by the comparison of the results with the data obtained in corresponding laboratory tests. What is more, measurements which record the loading–unloading cycles and even cycles of loading in opposite directions gives the possibility of obtaining displacement paths in order to access the hysteresis character of work of the existing structure. If the hysteresis is significant even in the range of the typical loading conditions, it suggests that the structure adjusts to the loading in non-linear way, which means potentially dangerous situation—correction or redesigning of the structure should be considered. Registering of the displacement of the panels in a real structure can be complemented by registering the strains using the strain gauges in the areas predicted as crucial, e.g., near the support of the panel, which would help to access the condition of the sheeting and even the support elements.

There are still some factors which were not included in this study and are planned for the next test series or require further analysis:

- Further experimental series, especially with other schemes of the screws, with two sheets of covering with/without seam fasteners, with thermal insulation between sheeting and the support element, etc.
- Improvement of the numerical model in terms of repeated loading and stiffness estimations.
- The analysis of the influence of the test speed on the hysteresis loop.
- The reduction scale in cases of higher profiles of the sheeting.

- The modification of the experimental set-up so that another failure mechanism could be achieved.

Author Contributions: Conceptualisation and methodology, P.I.; methodology and validation, N.K.-K.; investigation, N.K.-K.; formal analysis, N.K.-K and P.I.; visualisation, N.K.-K.; writing—original draft preparation, N.K.-K.; writing—review and editing, P.I.; supervision, P.I. All authors have read and agreed to the published version of the manuscript.

Funding: This research received no external funding.

Acknowledgments: The numerical calculations were performed using the computing resources of CI TASK at Gdańsk University of Technology.

Conflicts of Interest: The authors declare no conflict of interest.

References

1. EN 1993-1-3:2006. *Eurocode 3. Design of Steel Structures. Part 1-3: General Rules. Supplementary Rules for Cold-Formed Members and Sheeting*; European Committee for Standardization: Brussels, Belgium, 2006.
2. European Convention for Constructional Steelwork—TC7, TWG 7.5. In *European Recommendations for the Application of Metal Sheeting Acting as a Diaphragm. Stressed Skin Design*; ECCS: Brussels, Belgium, 1995.
3. Lendvai, A.; Joó, A.; Dunai, L. Experimental full-scale tests for development of diaphragm action—Part I.—Experimental results. *Thin-Walled Struct.* **2018**, *132*, 729–739. [CrossRef]
4. Lendvai, A.; Joó, A.; Dunai, L. Experimental full-scale tests for development of diaphragm action—Part II.—Effect of structural components on shear flexibility. *Thin-Walled Struct.* **2018**, *132*, 740–758. [CrossRef]
5. Lendvai, A.; Joó, A. Improvement of stressed skin design procedure based on experimental and numerical simulations. *J. Constr. Steel Res.* **2019**, 105874. [CrossRef]
6. Wrzesien, A.M.; Lim, J.B.P.; Xu, Y.; MacLeod, I.A.; Lawson, R.M. Effect of stressed skin action on the behaviour of cold-formed steel portal frames. *Eng. Struct.* **2015**, *105*, 123–136. [CrossRef]
7. Korcz, N.; Urbańska-Galewska, E. Influence of sheet/purlin fasteners spacing on shear flexibility of the diaphragm. *MATEC Web Conf.* **2018**, *219*, 1–8. [CrossRef]
8. Korcz, N.; Urbańska-Galewska, E. Influence of fasteners and connections flexibility on deflections of steel building including the stressed skin effect. *Tech. Sci.* **2018**, *2*, 131–148. [CrossRef]
9. Davies, J.M. Developments in stressed skin design. *Thin-Walled Struct.* **2006**, *44*, 1250–1260. [CrossRef]
10. Nagy, Z.; Pop, A.; Mois, I.; Ballok, R. Stressed skin effect on the elastic buckling of pitched roof portal frames. *Structures* **2016**, *8*, 227–244. [CrossRef]
11. Gryniewicz, M. Metoda Modelowania Konstrukcji Hal Stalowych Obudowanych Blachą Trapezową. Ph.D. Thesis, Bialystok University of Technology, Białystok, Poland, 28 September 2018.
12. Bakhti, F.; Tremblay, R.; Rogers, C.A. Revisiting the SDI and ECCS methods for in-plane shear flexibility of metal roof deck diaphragms using 3D non-linear finite element analysis. In Proceedings of the 15th World Conference on Earthquake Engineering—15WCEE, Lisbon, Portugal, 24–28 September 2012.
13. Rzeszut, K.; Gastecki, A.; Czajkowski, A. Parameter identification in FEM models of thin-walled purlins restrained by sheeting, Recent advances in computational mechanics. In Proceedings of the 20th International Conference on Computer Methods in Mechanics (CMM 2013), Poznań, Poland, 27–31 August 2013.
14. Davies, J.M.; Roberts, M.J.; Wang, Y.C. Recent developments in stressed skin theory. In Proceedings of the 8th International Conference on Thin-Walled Structures—ICTWS, Lisbon, Portugal, 24–27 July 2018.
15. Davies, J.M.; Roberts, M.J.; Wang, Y.C. The testing and analysis of stressed skin diaphragms. In Proceedings of the 8th International Conference on Thin-Walled Structures—ICTWS, Lisbon, Portugal, 24–27 July 2018.
16. Davies, J.M.; Roberts, M.J.; Wang, Y.C. Stressed-skin action in sandwich panel roofs. In Proceedings of the 9th International conference on Steel and Aluminium Structures—ICSAS19, Bradford, UK, 3–5 July 2019.
17. Furtado, A.; Rodrigues, H.; Arêde, A.; Varum, H.; Grubišić, M. Prediction of the earthquake response of a three-storey infilled RC structure. *Eng. Struct.* **2018**, *171*, 214–235. [CrossRef]
18. Polish Committee for Standarization. *PN-EN 1990:2004 Eurocode 0. Basis of Structural Design*; Polish Committee for Standarization: Warsaw, Poland, 2004.
19. Polish Committee for Standarization. *PN-EN 1991-1-3:2005 Eurocode 1. Actions on Structures. Part 1-3: General Actions. Snow Loads*; Polish Committee for Standarization: Warsaw, Poland, 2008.

20. Polish Committee for Standarization. *PN-EN 1991-1-4:2008 Eurocode 1. Actions on Structures. Part 1-4: General Actions. Wind Actions*; Polish Committee for Standarization: Warsaw, Poland, 2005.
21. ABAQUS. *Theory Manual*; Version 6.8; Hibbit, Karlsson & Sorenses Inc.: Providence, RI, USA, 2008.

Article

Material Parameters Identification of Historic Lighthouse Based on Operational Modal Analysis

Agnieszka Tomaszewska [1],*, Milena Drozdowska [1],* and Marek Szafrański [2],*

[1] Department of Structural Mechanics, Faculty of Civil and Environmental Engineering, Gdańsk University of Technology, Narutowicza 11/12, 80-233 Gdańsk, Poland
[2] Department of Rail Transportation and Bridges, Faculty of Civil and Environmental Engineering, Gdańsk University of Technology, Narutowicza 11/12, 80-233 Gdańsk, Poland
* Correspondence: atomas@pg.edu.pl (A.T.); mildrozd@pg.edu.pl (M.D.); mszafran@pg.edu.pl (M.S.)

Received: 16 July 2020; Accepted: 27 August 2020; Published: 28 August 2020

Abstract: In the present paper, the identification of the material parameters of a masonry lighthouse is discussed. A fully non-invasive method was selected, in which the material properties were determined via numerical model validation applied to the first pair of natural frequencies and their related mode shapes, determined experimentally. The exact structural model was built by means of the finite element method. To obtain experimental data for the inverse analysis, operational modal analysis was applied to the structure. Three methods were considered: peak picking (PP), eigensystem realization algorithm (ERA) and natural excitation technique with ERA (NExT-ERA). The acceleration's responses to environmental excitations, enhanced in some periods of time by sheet piling hammering or by sudden interruptions like wind stroke, were assumed within the analysis input. Different combinations of the input were considered in the PP and NExT-ERA analysis to find the most reasonable modal forms. A number of time periods of a free-decay character were considered in the ERA technique to finally calculate the averaged modal forms. Finally, the elastic modulus, Poisson's ratio and material density of brick, sandstone and granite masonry were determined. The obtained values supplement the state of the art database concerning historic building materials. In addition, the numerical model obtained in the analysis may be used in further cases of structural analysis.

Keywords: material parameters identification; vibration measurements; testing; operational modal analysis; lighthouse; historic tower

---

1. Introduction

The parameter identification of the materials used in historic structures demands the application of non-destructive means. The need to preserve structural safety and integrity is usually superior to any goal of structural testing. However, recognizing the properties of materials used in historic structures helps to select modern materials for some restoration works. The properties may be identified in the inverse analysis, in which the structural parameters are determined based on the measured structural response to a known, or unknown, action. The present paper discusses the identification of the material parameters of a slender masonry lighthouse. A fully non-invasive method was selected, in which the material properties were determined by the numerical model validation applied to the first pair of natural frequencies and the related mode shapes that were experimentally determined.

To obtain experimental data for the inverse analysis, operational modal analysis (OMA) was applied to the structure. OMA is widely discussed in the literature concerning structural health monitoring (SHM). Numerous OMA methods have been developed [1,2], and they are dedicated to certain vibration cases, i.e., free-decay or ambient. Their efficiency depends on the quality of the vibration signals. Subsequently, the quality of the OMA results determine the possibility of its further usage. Natural frequencies or modal forms, when identified with a considerable error, will induce untrue results or conclusions. Therefore, the quality of the OMA results must be treated with care.

In the present paper, the modal identification (MID) of the lighthouse is discussed considering three kinds of structural excitation. The following MID techniques are applied herein: the eigensystem realization algorithm (ERA), the peak picking technique based on correlation analysis (PP-CA), and the natural excitation technique with ERA (NExT-ERA). The application of these three techniques increases the chance of true results being obtained.

The considered tower is a historic lighthouse situated in Gdańsk (Poland). The brick construction is massive and rigid due to the spiral granite staircase filling the whole tower interior. The modal identification of similar structures is described in the literature [3]. Examples for masonry bell towers are presented in [4,5]. The SHM systems built for such structures are described in [6,7]. The systems monitor natural frequencies, and correlate them with temperature changes or with a day's anthropogenic activity in the city. Ref. [8] concerns the identification of the foundation stiffness of the masonry tower, based on experimentally identified modal data.

The aim of this paper is to determine reasonable MID results for the lighthouse, in order to use these in the material parameter identification of the structural FEM model. The outcomes of three OMA methods under various excitations were compared, whereby the excitation caused by sheet piling hammering was a unique one. The opportunity to measure the lighthouse's vibrations under such conditions arose during the modernization of the port quay in Gdańsk. The basic excitation was the environmental impact most affected by wind and river influences. The study presents the results of the PP-CA technique in the three following excitation cases: ambient, ambient enhanced by the ground vibrations enforced by the hammer action, and the two combined, i.e., ambient enhanced by the hammer action in selected time periods. The NExT-ERA results relate to the first and the third above-mentioned excitation kinds. The ERA technique was applied to the short-term free-decay parts of the ambient signals, as well as to the steady-state vibrations generated by the sheet piling hammering. In the first case, natural frequencies and related mode shapes and damping ratios were identified. In the second case, forced vibration frequencies and related response shapes were obtained.

The provided results prove that in the case of a broad spectrum of excitations acting on a real structure, the OMA results obtained using different techniques may not be repeatable. To approach the true results, the analysis should be performed with care and skepticism. Furthermore, the application of a few OMA methods should be considered to ensure more objective inferences. Such an approach accepted in the described study provided reasonable MID results, which were further applied in the numerical model of validation for the lighthouse so as to finally determine the mechanical properties of the materials used in the construction, as well as their elastic support.

2. Materials and Methods

2.1. The Lighthouse

2.1.1. Description of the Structure

The construction of the lighthouse (Figure 1a) was finished in October 1894. It is located approximately 50 m off the Dead Vistula River channel in the Port of Gdańsk across Westerplatte Peninsula.

The cross-section of the lighthouse is octagonal. It is 27.3 m high, with a masonry-build, load-bearing corpus of 22.5 m and a light room placed above it. There are two granite terraces at a height of 20.2 m and 22.5 m. The external diameter of the masonry part varies with height, from 6.8 m with a wall thickness of 1.04 m at the bottom to 4.2 m with a wall thickness of 0.68 m at the corpus top. A brick column in the tower's center supports the spiral granite stairs. The external stairs, leading to the entrance, are also made of granite. The walls are mainly made of brick. Only the lower wall is made of sandstone, up to the height of 4.2 m. The light room is of a lighter construction than the tower corpus as it is built of steel beams, posts and sheeting with wood paneling.

Figure 1. (**a**) the lighthouse; (**b**) sheet piling hammering next to the lighthouse.

During World War II, an external wall of the upper part of the construction and the deck were damaged. The tower was mostly repaired between the 1940s and 1950s. The most recent renovation took place in 2003. The building's inventory was made including the foundation. The tower stands on a concrete slab and oaken piles. The ground conditions under the slab were not specified. The technical condition of the structure was recognized as good. There are no visible cracks in the structure, suggesting that the tower is somehow damaged.

2.1.2. Dynamic Measurements

The measurements were performed using a 16-bit HBM QuantumX acquisition system (Hottinger Baldwin Messtechnik GmbH, Darmstadt, Germany) together with one-dimensional piezoelectric accelerometers Isotron Endevco 7752-1000 (PCB Piezotronics Inc., Depew, NY, USA) with a voltage sensitivity of 1 V/g (±20%) and an amplitude response of 0.02–500 Hz (±5%). The sensors were located on the stairs along two vertical axes (points P1–P5 and points P6–P10 in the Figures 2 and 3). The sensors locations were chosen based on the lighthouse's geometry. The construction is nearly axisymmetric—the axial symmetry is disrupted by the windows and the entrance. In such structures, bending mode shapes appear as a pair of orthogonal vectors with very close natural frequencies. The vectors are aligned with 'principal axes' related to the details that disrupt the axial symmetry. It is reasonable to assume, then, that the first pair of (bending) modes is present in the directions weakened by openings in the walls.

Accelerations in two horizontal directions were measured at each point, i.e., perpendicular and parallel to the river's axis (designated as directions x and y in Figure 2, respectively). Due to the equipment limitations, four series of measurements were carried out with the reference points P4 and P9.

The three following types of data were collected for further processing: SA—ambient vibration signals, SH—ambient vibrations amplified by vibratory hammer, and SC—combination of SA and SH signals. An explanation is given in Figure 4. The equal lengths of the signals in each group were considered as follows: SA and SC—1200 s, and SH—900 s, with a sampling frequency of 200 Hz. The SH signals were the shortest because of the limited time of hammering in subsequent measurement series. The location of the vibratory hammer in relation to the tower is shown in Figure 3. It operated at a setting of about 34 Hz.

2.2. Modal Identification Techniques

2.2.1. Peak Picking Method Based on the Correlation Analysis (PP-CA)

Ambient vibration responses to the (assumed) white noise signal are considered in the PP-CA method. The method is based on the correlation analysis of response signals in the frequency domain. This was proposed in [9] as the first OMA technique. Auto- and cross-correlation functions, determined for different measuring points together with the coherence functions, allow us to identify natural frequencies, as briefly described in [10]. Additional study of the phase shifts allows us to determine mode shapes. The damping ratios cannot be identified, which is the method's limitation. The practical applications of the PP-CA technique are addressed in, e.g., [11–13].

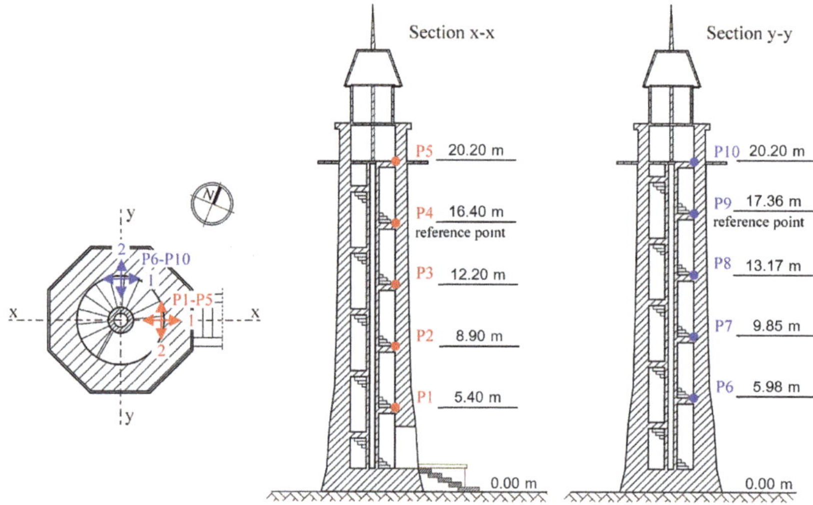

Figure 2. Measuring points' locations (P1–P10) and directions *x* and *y* of measurements (sensors 1 and 2).

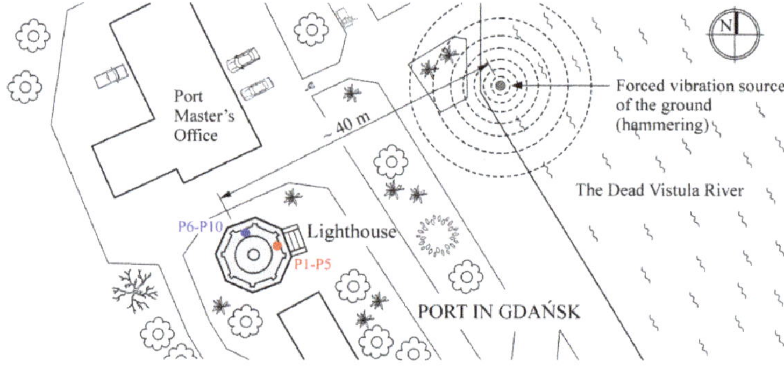

Figure 3. Situation plan.

2.2.2. Eigensystem Realization Algorithm (ERA)

The eigensystem realization algorithm belongs to the group of direct, multi-input/multi-output system identification methods in the time domain [14]. This method evolved from the Ho-Kalmann minimum realization problem, and estimates the modal parameters, i.e., frequencies, damping ratios and mode shapes, based on free-decay, finite-time and noisy experimental data [15,16]. The method has been extensively discussed and used in various mechanical and engineering problems, e.g., machine construction optimization (see, e.g., [17]), vehicle-bridge dynamic interaction (see, e.g., [18,19]), and the damage detection and technical condition assessment of structures (see, e.g., [20,21]). The input is a free-decay structural response to a certain action or initial condition. In the present study, the environmental impacts are considered as the excitation source. Directly measured, short-time free-decay signals caused by blasts of wind or river undulations are considered as the input. A similar approach was applied in the paper [11], in which the ERA efficacy in the case of a massive masonry tower was studied.

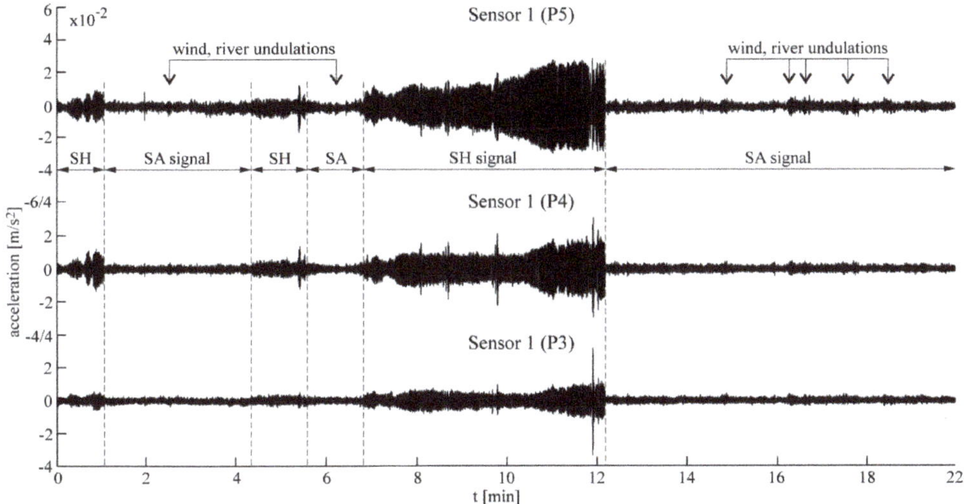

Figure 4. Selected acceleration time series with different types of signals—sensor 1 (P3–P5). Full length record is the SC signal.

2.2.3. Natural Excitation Technique and ERA (NExT-ERA)

Free-decay data can also be obtained by applying a correlation analysis to the measured ambient vibration signals (assumed as stochastic). The correlation analysis transforms stochastic data into the deterministic domain, that is, into the correlation functions of decay character, which can be used as an input for the ERA algorithm. The natural excitation technique is this transformation [22]. The NExT-ERA methodology and applications were widely studied in the literature, e.g., [23–26], and are also used in the presented study.

2.3. Numerical Model of the Lighthouse

Based on the technical documentation and the tower inventory, a detailed numerical model of the lighthouse was created using the SIMULIA Abaqus FEA 2019 software [27] and the finite elements method (FEM) (Figure 5). The model contains brick walls, an inner granite staircase, brick columns and a sandstone base. The light room part was intentionally omitted, with it having a negligible influence on the modal parameters of the structure's corpus (its mass accounts for 0.42% of the entire structural mass). A 10-node volume element (quadratic tetrahedron) with three translational degrees of freedom at each node (element C3D10) was chosen to generate a mesh. Overall, the model consists of 223,458 nodes and 146,358 elements in its entirety.

The homogeneous linear elastic material model is accepted for all materials. Mass density values were assumed on the grounds of the Polish design standards PN-EN ISO 12524:2003, PN-EN ISO 69446:1999 and PN-91/B-02020. The initial elastic modulus of brick masonry was chosen according to PN-EN 1996-1-1. The elastic moduli of sandstone and granite were selected based on Ref. [28]. It was suspected that the granite stairs were rigid and played the major role in lighthouse dynamic behavior, so the initial value was high. The sandstone base was also predicted to be stiff, however there were no cracks in the construction, suggesting little difference between the brick and sandstone parts, thus the value chosen was of a lower range. Initially, the model was fixed at the tower's footing, but the support conditions were modified during the model's calibration, and vertical linear spring elements were included to model the elastic support of the lighthouse. Horizontal support conditions were assumed to be rigid.

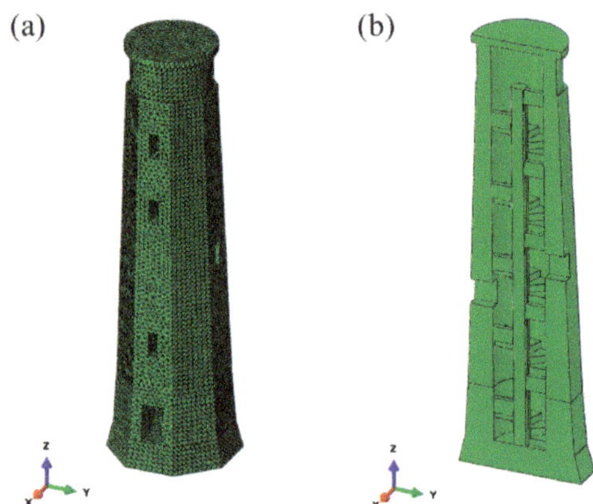

Figure 5. The finite element model of the lighthouse (Abaqus): (**a**) side view, (**b**) vertical section.

3. Results

3.1. Modal Identification

The following figures present the obtained results. Selected response spectra are presented in Figure 6. Subsequently, the mode shapes obtained from the PP-CA method for the three kinds of the input signals are plotted in Figure 7. Similar results of the NExT-ERA realization for the SA and SC signals are presented in Figure 8. The accepted modal amplitude coherence (MAC) criterion (95%) allowed us to determine only two mode shape vectors. Selected ERA solutions for different time intervals of ambient responses (SA signals) excited by blasts of wind or river undulations (see Figure 5) are presented in Figure 9, together with the average solution. The MAC criterion at a level of 98% is accepted. In the case of direction 1 (perpendicular to the river), five different parts of visible amplitudes and decay character were selected, while in the case of direction 2 (parallel to the river) three parts of signals could be selected. Signals between 4 s and 8 s in length, depending on the visible amplitude level, were processed by ERA. The considered lighthouse has a cantilever-type structure, so the prediction of the modal forms is intuitive. Thus, the most reasonable results of PP-CA and NExT-ERA were selected and put together with the theoretical forms and the average ERA results. Juxtaposition is presented in Figure 10. The summary of the identified natural frequencies and damping coefficients is presented in Table 1. All the presented results refer to the first pair of natural frequencies, visible in the spectra in Figure 6 (2.26 Hz, 2.78 Hz). However, a reasonable peak is also observed for the frequency 32.46 Hz, which corresponds to the forced vibrations caused by the vibratory hammer action. The related response shape was identified by the ERA and the NExT-ERA techniques. The results are presented in Figure 11, with a similar numerical mode shape.

Table 1. Natural frequencies and damping coefficient identified by different methods.

Method	Natural Frequency—First Mode (Hz)	Natural Frequency—Second Mode (Hz)	Damping Coefficient
Peak Picking	2.24	2.80	-
NExT-ERA	2.25	2.74	0.0226
ERA	2.26	2.78	0.0244
Numerical	2.23	2.81	-

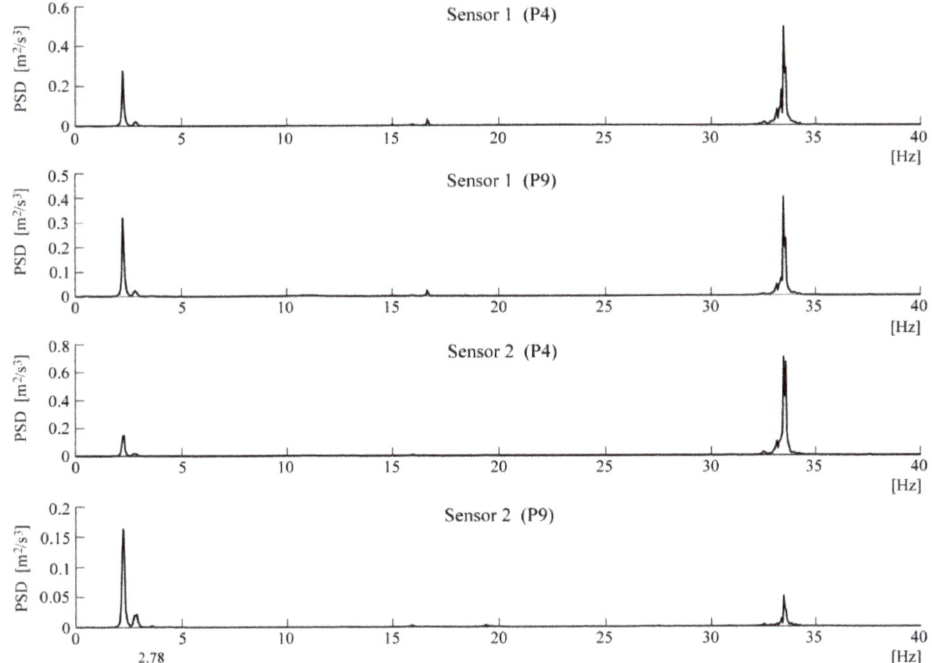

Figure 6. Selected spectra of acceleration records of the combined signals (SC).

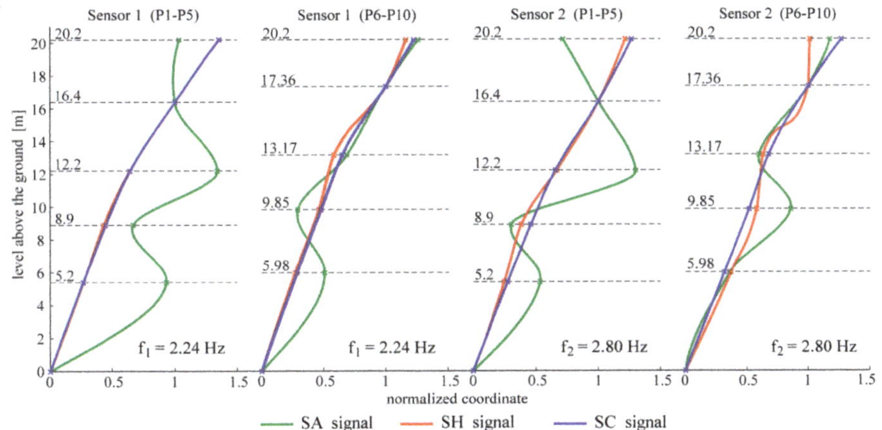

Figure 7. Mode shapes obtained via the PP-CA method for the three kinds of input signals.

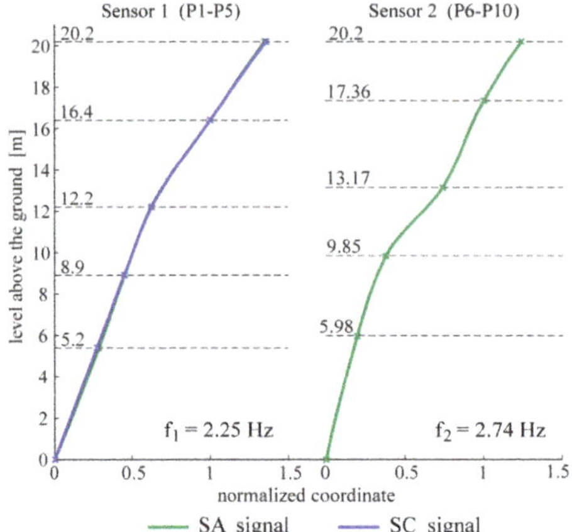

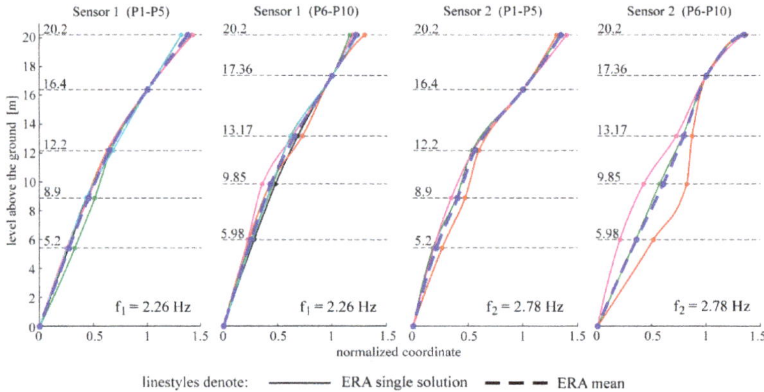

Figure 8. Mode shapes obtained via the NExT-ERA algorithm for the two kinds of input signals.

Figure 9. Mode shapes obtained via the ERA for different free-decay parts of the SA signals and the average result.

3.2. FEM Model Validation

The primary FEM model, including a fixed support and the initial material parameters, produced the first pair of natural frequencies with the values of 6.96 Hz and 7.01 Hz, in the x and y directions, respectively. Such values were approximately three-fold higher than the measured ones. Thus, the boundary conditions were modified by adding vertical springs, and the model was manually calibrated to the natural frequencies and mode shapes obtained with the PP-CA technique in the combined signal variant. The following parameters of the model were validated: material parameters (elastic modulus and mass density) and the supporting spring stiffness. Four supporting springs were considered. The bottom surface of the base was divided into four areas, in which nodes were kinematically bounded in the z direction to the spring situated in this area (see Figure 11).

The initial values were updated to obtain a sufficiently high compliance between numerical and experimental frequencies and mode shapes. The initial and final material parameters are summarized in Table 2. The initial stiffness of the supporting springs was $k_x = 1 \times 10^{10}$ N/m and $k_y = 2 \times 10^{10}$ N/m. The final stiffness of the springs was $k_x = 3.75 \times 10^9$ N/m and $k_y = 3.75 \times 10^{10}$ N/m. Table 1 compares the numerical and experimental frequencies of the lighthouse.

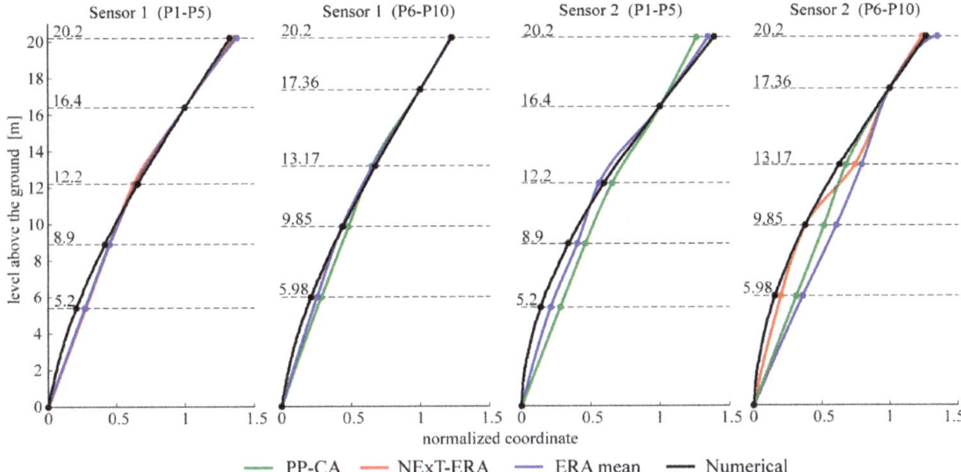

Figure 10. The comparison of the mode shapes obtained using different methods together with numerical results. PP-CA—the results based on SC signals, NExT-ERA—the results based on SA signals.

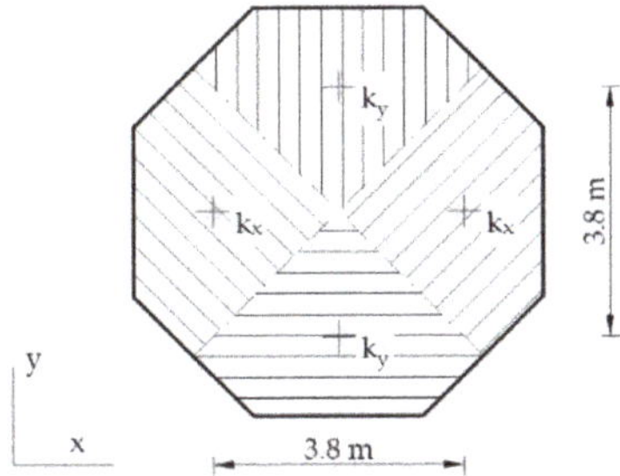

Figure 11. Supporting springs layout.

Table 2. Initial material parameters and parameters obtained by FEM model updating.

Material	Initial Elastic Modulus (GPa)	Final Elastic Modulus (GPa)	Poisson's Ratio	Initial Density (kg/m³)	Final Density (kg/m³)
Brick masonry	1	2.4	0.167	2200	2100
Sandstone	10	2.4	0.2	2400	2100
Granite	80	26	0.3	2600	2000

4. Discussion

The identification of the material parameters of a real historic structure, based on modal identification results, is discussed in this paper. The efficacy of the modal identification is crucial in obtaining true material parameters. The efficacy appeared to be different for the natural frequencies and the mode shapes. The three different methods considered in this study provide convergent results for the natural frequencies. The values are between 2.24 and 2.26 Hz, and 2.74 and 2.80 Hz, for the first pair of natural frequencies (see Table 1). Both modes are the first two bending mode shapes in two orthogonal directions. The obtained values of the fundamental frequencies are similar to the characteristics of other masonry towers. In the paper [29], the values for 30 masonry towers with heights of 16–46 m are summarized. They are between 0.61 Hz and 5.28 Hz. The damping coefficient corresponding to the first natural frequency varies between 0.0226 and 0.0244. The values are included in the most common range of damping coefficient determined for masonry towers, which is 0.02–0.03, according to [30]. The mode shapes identified by the application of the three OMA techniques to the three kinds of input signals are not always repeatable. First of all, the PP-CA method, which is dedicated to ambient excitations, does not produce a reliable mode shape for the SA signals (see Figure 7). The application of the ambient signals enhanced by the sheet peel hammering significantly improves the result, however untypical changes in the mode shapes curvatures are visible in the upper parts of the modes. The most reasonable results were obtained for the combined signals, examples of which are presented in Figure 4. In this case, the mode shapes show the curvature of one sign along the whole length; moreover, they show the typical shapes of the modes of a cantilever supported by a rotational spring—such a simplified model can approximate the considered tower. It is worth mentioning that the efficacy of the mode shapes' identification does not relate to the signal amplitudes in general. The average root mean square (RMS) of all signals in subsequent groups equals 0.0014 for SA, 0.0039 for SH and 0.0027 for SC signals. The SH signals have the biggest amplitudes, related to the hammering frequency. The identification of modes related to other frequencies is enhanced somehow, but the best solutions are obtained when the combined signals are considered with moderate RMS and with different dynamic influences. Limited results are obtained by the NExT-ERA technique. Only two modal vectors are obtained based on the SA signals, and both of them are in directions perpendicular to the wall planes (Figure 8). The SH signals produced no results, and the SC signals provide only one modal vector in the x axis of the tower (Sensors 1, P1–P5 line). Contrary to the PP-CA case, the application of SC signals does not improve the results obtained for the SA signals with the NExT-ERA algorithm. The different efficiencies of various OMA techniques in the practical cases of six British lighthouses are also reported in the paper [3].

The application of the ERA technique to the short-term free-decay parts of SA signals leads to a set of modal forms. Then, the averaged mode shape vector is found as the final ERA solution. The most considerable results obtained using three identification methods are compared in Figure 10, together with the numerical results. The PP-CA results were obtained for the case of the SC signals, while the NExT-ERA results were obtained for the SA signals. The comparison of the results shows the most repeatable forms in the line P1–P5, measured by sensor 1. This suggests the best quality of the signals collected by these sensors. The line is situated along the tower wall on the river side, and sensor 1 measures the accelerations in the direction perpendicular to the river (direction x, see Figure 2). This line and these sensors receive the highest excitation from the river's undulations. This may be the factor affecting the signal's quality. Besides, the structure looks integral, and no damages are visible. However, maybe the renovations after World War II play some role in the signal deteriorations. It is worth mentioning that the different lengths of signal considerations in the NExT-ERA and the ERA techniques did not improve the presented results.

The enhanced OMA analysis validated the data for the FEM model and the material parameters' identification via the validation process. The results obtained and presented in Table 2 are included in the range of material parameters reported for other masonry structures. In the papers [4,31,32], the investigations of masonry bell towers are described. The elastic modulus of the masonry tower described in [4] was assessed by a simplified global analysis based on the first natural frequencies obtained through experimental analysis. The results were, depending on direction, 3.579 and 4.746 GPa, with frequencies of 1.294 and 1.489 Hz. However, in the FE model the elastic modulus value of 2.819 GPa was implemented. A similar tower is the object of another study [31], where the elastic modulus is stated as 1.8 GPa with a first frequency of 0.67 Hz. In the paper [32], the elastic modulus of masonry varied from 1.8 GPa to 2.5 GPa, relative to the wall height. The first frequencies were assumed as 0.585 Hz and 0.709 Hz. In every case, stairs were unimportant from a structural point of view, and were thus not included in any numerical model. According to [28], the elastic modulus of granite varies from 2.59 GPa to 88.79 GPa, and in the case of sandstone from 3.4 GPa to 71.7 GPa. In another paper [33], 48.8 GPa is the identified value of the elastic modulus of stone masonry. The values of the elastic moduli of granite masonry applied in three different historic structures are presented in [34]. They belong in the range of 20.8–39.2 GPa, with a unit weight in the range of 24.1–26.4 kN/m^3. All those values are similar to the material parameters identified in the present study.

An interesting and rare result concerns the shape response identification, which is induced by the sheet peel hammering (see Figure 12). This shape could be identified for the x direction only (perpendicular to the river), which corresponds to the direction of wave propagation in the ground generated by the hammer action. This particular response could be identified using the ERA and the NExT-ERA methods. This shape is similar to the eigenmode of the numerical model (third bending one) (see Figure 12). The related numerical and experimental frequencies differ considerably. The distant values of the frequencies prove the safety of the peel hammering works in the vicinity of the lighthouse.

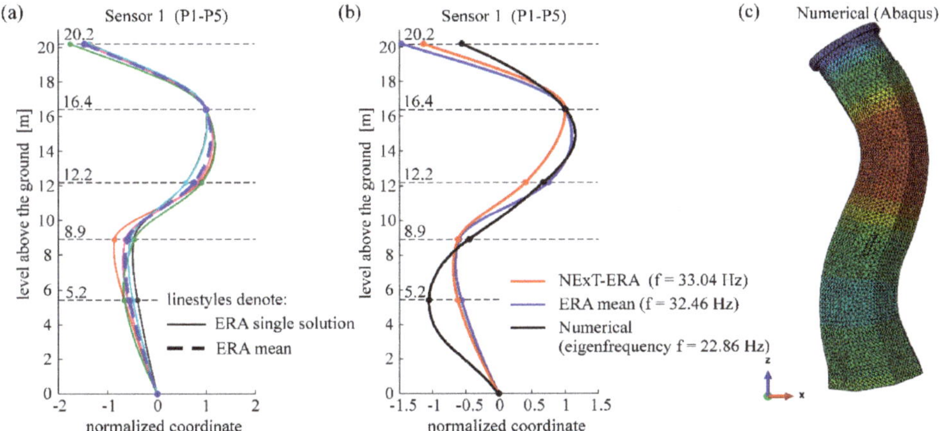

Figure 12. The response shape induced by the peel hammering and similar numerical mode shape: (**a**) ERA results; (**b**) experimental response shape (NExT-ERA and ERA mean results) in comparison with numerical third mode shape (**c**) the third numerical bending mode shape (Abaqus).

5. Conclusions

The presented study proves that in the case of a real structure with a limited possibility of dynamic excitation, reliable mode shape identification is a complex issue. Such a task demands the engagement of a few OMA techniques to authenticate the identified modal forms. The importance of the careful OMA inference is emphasized when the results are applied in the mathematical model of structural identification. The analysis provided the information concerning the mechanical parameters of the materials used in the considered lighthouse construction. The results complement the knowledge concerning such parameters for historic building materials.

Author Contributions: Conceptualization, A.T. and M.S.; Investigation, A.T., M.D. and M.S.; Methodology, A.T. and M.S.; Software, M.D. and M.S.; Validation, A.T. and M.D.; Visualization, M.D. and M.S.; Writing—original draft, A.T. and M.D.; Writing—review and editing, A.T., M.D. and M.S. All authors have read and agreed to the published version of the manuscript.

Funding: This research received no external funding.

Acknowledgments: The authors express a great gratitude to Stefan Jacek Michalak, the owner of the lighthouse, for the consent to measurements and providing the technical documentation of the structure. Calculations have been partially made at the Academic Computer Centre in Gdansk TASK.

Conflicts of Interest: The authors declare no conflict of interest.

References

1. Rainieri, C.; Fabbrocino, G. *Operational Modal Analysis of Civil Engineering Structures*; Springer: New York, NY, USA, 2014.
2. Brincker, R.; Ventura, C. *Introduction to Operational Modal Analysis*; John Wiley & Sons: Hoboken, NJ, USA, 2015.
3. Brownjohn, J.M.W.; Raby, A.; Bassitt, J.; Antonini, A.; Hudson, E.; Dobson, P. Experimental modal analysis of British rock lighthouses. *Mar. Struct.* **2018**, *62*, 1–22. [CrossRef]
4. Ivorra, S.; Pallarés, F.J. Dynamic investigations on a masonry bell tower. *Eng. Struct.* **2006**, *28*, 660–667. [CrossRef]
5. Diaferio, M.; Foti, D.; Giannoccaro, N.I.; Ivorra, S. Model updating based on the dynamic identification of a baroque bell tower. *Int. J. Saf. Secur. Eng.* **2017**, *7*, 519–531. [CrossRef]
6. Ubertini, F.; Comanducci, G.; Cavalagli, N. Vibration-based structural health monitoring of a historic bell-tower using output-only measurements and multivariate statistical analysis. *Struct. Heal. Monit.* **2016**, *15*, 438–457. [CrossRef]
7. Azzara, R.M.; Girardi, M.; Iafolla, V.; Lucchesi, D.M.; Padovani, C.; Pellegrini, D. Ambient Vibrations of Age-old Masonry Towers: Results of Long-term Dynamic Monitoring in the Historic Centre of Lucca. *Int. J. Archit. Herit.* **2019**, 1–17. [CrossRef]
8. Tomaszewska, A.; Szymczak, C. Identification of the Vistula Mounting tower model using measured modal data. *Eng. Struct.* **2012**, *42*, 342–348. [CrossRef]
9. Bendat, J.; Piersol, A. *Engineering Applications of Correlation and Spectral Analysis*; John Wiley & Sons: Hoboken, NJ, USA, 1980.
10. Tomaszewska, A. Influence of statistical errors on damage detection based on structural flexibility and mode shape curvature. *Comput. Struct.* **2010**, *88*, 154–164. [CrossRef]
11. Tomaszewska, A.; Szafranski, M. Study on applicability of two modal identification techniques in irrelevant cases. *Arch. Civ. Mech. Eng.* **2020**, *20*, 1–11. [CrossRef]
12. Brownjohn, J.M.W. Ambient vibration studies for system identification of tall buildings. *Earthq. Eng. Struct. Dyn.* **2003**, *32*, 71–95. [CrossRef]
13. Chen, G.; Omenzetter, P.; Beskhyroun, S. Operational modal analysis of an eleven-span concrete bridge subjected to weak ambient excitations. *Eng. Struct.* **2017**, *151*, 839–860. [CrossRef]
14. Maia, N.M.M.; Montalvao Silva, J.M.M. *Theoretical and Experimental Modal Analysis*, 1st ed.; Research Studies Press Ltd.: Baldock, UK, 1998.
15. Juang, J.-N.; Pappa, R.S. An eigensystem realization algorithm for modal parameter identification and model reduction. *J. Guid. Control Dyn.* **1985**, *8*, 620–627. [CrossRef]
16. Juang, J.-N. *Applied System Identification*; Prentice-Hall PTR: Englewood Clifs, NJ, USA, 1994.
17. Rusinski, E.; Dragan, S.; Moczko, P.; Pietrusiak, D. Implementation of experimental method of determining modal characteristics of surface mining machinery in the modernization of the excavating unit. *Arch. Civ. Mech. Eng.* **2012**, *12*, 471–476. [CrossRef]
18. Szafranski, M. Dynamics of the small-span railway bridge under moving loads. *MATEC Web. Conf.* **2019**, *262*, 1–8. [CrossRef]
19. Szafrański, M. A dynamic vehicle-bridge model based on the modal identification results of an existing EN57 train and bridge spans with non-ballasted tracks. *Mech. Syst. Signal Pr.* **2021**, *146*, 107039–107056. [CrossRef]
20. Bernagozzi, G.; Mukhopadhyay, S.; Betti, R.; Landi, L.; Diotallevi, P.P. Output-only damage detection in buildings using proportional modal flexibility-based deflections in unknown mass scenarios. *Eng. Struct.* **2018**, *167*, 549–566. [CrossRef]
21. Binczyk, M.; Kalitowski, P.; Szulwic, J.; Tysiac, P. Nondestructive testing of the miter gates using various measurement methods. *Sensors (Switzerland)* **2020**, *20*, 1749. [CrossRef]
22. James, G.; Carne, T.; Lauffer, J. The natural excitation technique (NExT) for modal parameter extraction from operating structures. *Modal Anal. Int. J. Anal. Exp. Modal Anal.* **1995**, *10*, 260.
23. Siringoringo, D.M.; Fujino, Y. System identification of suspension bridge from ambient vibration response. *Eng. Struct.* **2008**, *30*, 462–477. [CrossRef]

24. Nayeri, R.D.; Tasbihgoo, F.; Wahbeh, M.; Caffrey, J.P.; Masri, S.F.; Conte, J.P.; Elgamal, A. Study of time-domain techniques for modal parameter identification of a long suspension bridge with dense sensor arrays. *J. Eng. Mech.* **2009**, *135*, 669–683. [CrossRef]

25. Caicedo, J.M. Practical guidelines for the natural excitation technique (NExT) and the eigensystem realization algorithm (ERA) for modal identification using ambient vibration. *Exp. Tech.* **2011**, *35*, 52–58. [CrossRef]

26. Hosseini Kordkheili, S.A.; Momeni Massouleh, S.H.; Hajirezayi, S.; Bahai, H. Experimental identification of closely spaced modes using NExT-ERA. *J. Sound Vib.* **2018**, *412*, 116–129. [CrossRef]

27. Habbit, D.; Karlsson, B.; Sorensen, P. *ABAQUS Analysis User's Manual*; Hibbit, Karlsson, Sorensen Inc.: Providence, RI, USA, 1997.

28. Karagianni, A.; Karoutzos, G.; Ktena, S.; Vagenas, N.; Vlachopoulos, I.; Sabatakakis, N.; Koukis, G. Elastic properties of rocks. *Bull. Geol. Soc. Greece* **2010**, *43*, 1539–1548. [CrossRef]

29. Zanotti Fragonara, L.; Boscato, G.; Ceravolo, R.; Russo, S.; Ientile, S.; Pecorelli, M.L.; Quattrone, A. Dynamic investigation on the Mirandola bell tower in post-earthquake scenarios. *Bull Earthq. Eng.* **2017**, *15*, 313–337. [CrossRef]

30. Bull, J.W. *Computational Modelling of Masonry, Brickwork and Blockwork Structures*; Saxe-Coburg Publications: Stirling, UK, 2001.

31. Russo, G.; Bergamo, O.; Damiani, L.; Lugato, D. Experimental analysis of the "Saint Andrea" Masonry Bell Tower in Venice. A new method for the determination of "Tower Global Young's Modulus E". *Eng. Struct.* **2010**, *32*, 353–360. [CrossRef]

32. Gentile, C.; Saisi, A. Ambient vibration testing of historic masonry towers for structural identification and damage assessment. *Constr. Build. Mater.* **2007**, *21*, 1311–1321. [CrossRef]

33. Smoljanović, H.; Nikolić, Ž.; Živaljić, N. A finite-discrete element model for dry stone masonry structures strengthened with steel clamps and bolts. *Eng. Struct.* **2015**, *90*, 117–129. [CrossRef]

34. Arêde, A.; Almeida, C.; Costa, C.; Costa, A. In-situ and lab tests for mechanical characterization of stone masonry historical structures. *Constr. Build. Mater.* **2019**, *220*, 503–515. [CrossRef]

Article

Non-Destructive Testing of the Longest Span Soil-Steel Bridge in Europe—Field Measurements and FEM Calculations

Mikołaj Miśkiewicz [1], Bartosz Sobczyk [1,*] and Pawel Tysiac [2]

[1] Department of Mechanics of Materials and Structures, Faculty of Civil and Environmental Engineering, Gdansk University of Technology, 80-233 Gdańsk, Poland; mikolaj.miskiewicz@pg.edu.pl

[2] Department of Geodesy, Faculty of Civil and Environmental Engineering, Gdansk University of Technology, 80-233 Gdańsk, Poland; pawel.tysiac@pg.edu.pl

* Correspondence: bartosz.sobczyk@pg.edu.pl; Tel.: +48-58-347-17-55

Received: 30 June 2020; Accepted: 11 August 2020; Published: 18 August 2020

Abstract: The article describes interdisciplinary and comprehensive non-destructive diagnostic tests of final bridge inspection and acceptance proposed for a soil-steel bridge made of corrugated sheets, being the European span length record holder (25.74 m). As an effect of an original concept a detailed and precise information about the structure short-term response was collected. Periodic diagnostics of bridge deformations was done one year after it was built. Load test design was based on numerical simulations performed by means of finite element method (FEM). In situ measurements were done with the aid of: inductive sensors, optical total station, and terrestrial laser scanner. The results produced by terrestrial laser scanning were used to build a precise image of structure deformation in 3D space during the tests. The accuracy of laser mapping was significantly increased using the information coming from total station and inductive sensors. These have higher accuracy and therefore can be used as reference. Thus, new quality in measurements is introduced. Good correspondence between in situ values and FEM estimations was achieved. Therefore, such a combination of testing methods can be used in non-destructive diagnostics of structures and is an interesting alternative for the standard approach, in which the measurements are done in limited number of points.

Keywords: non-destructive testing; soil-steel bridge; terrestrial laser scanning; finite element method (FEM); modelling and simulations

---

1. Introduction and Research Background

Engineering, mechanical, aerospace, offshore structures, etc., or their parts, elements, or connections are subjected to various and complex loading conditions during their life cycle. Sometimes these conditions are not properly identified during their design or, basically, could not have been predicted because of some accidental circumstances that emerge at the site. Consequently, failures and damages of the aforesaid objects occur. Because of that, great attention is focused on understanding of the structures response, enabling appropriate failure prevention, determination of real properties of the built-in material, identification and detection of damage, and failure mechanisms or monitoring of structure health (SHM). A lot of review papers have been written that gather knowledge about non-destructive testing of different structures, see for example [1–6]. In this article, we deal with non-destructive testing issues related to: detailed, automated, high-accuracy geodetic measurements, especially using terrestrial laser scanning (TLS); determination of structural response of a bridge; structural identification (St-Id) processes; and structural health monitoring (SHM). Therefore, recent achievements in this field are shown below.

In situ, field tests and measurements are commonly utilized for the purpose of non-destructive testing. In [7] TLS automatic inspection system, enabling detection and measurement of damage together with the verification of the quality and durability of surface repairs, as required by industry standards, is presented. A new approach for assessing the dimensional accuracy and structural performance of spatial structure elements, using three dimensional (3D) laser scanning, has been shown in [8]. In the paper [9], a hybrid digital reconstruction of a structure geometry using digital close range photogrammetry and laser scanning techniques is described. A comparison between the TLS measurements and the close-range photogrammetry done for deforming concrete beams, using the Structure from Motion algorithm, has been discussed in [10]. The article [11] reviews the abilities and drawbacks of a variety of remote sensing technologies, including laser scanning, and their applications for the purpose of automated measurements on a building site. The problems of structural identification by means of TLS and finite element method (FEM) are also considered. Some case studies in this field has been done for example for a historical minaret in [12] or mitre gates in [13]. Deformations of arched structures has been analyzed in [14] based on TLS measurement and FEM estimations, which were calibrated according to TLS results. Some other interesting case studies, in which TLS was employed and FEM calculations were done are available in [15–17]. Automated geodetic measurements are also used for the purpose of SHM. Some possibilities and case studies for timber structures are discussed in the review [1]. Laser scanning, refer to [18], has been used to investigate the health of historic masonry tower in Italy. Application of global position system (GPS)-based measurements of a cable stayed bridge in Romania, done for the purpose of SHM, is presented in the paper [19]. Dynamic responses of a suspension bridge in China have been monitored by means of high sampling-rate robotic total station, as it is described in [19]. Application of different sensors and geodetic measurements for the purpose of short and long term measurements of pedestrian bridge response, as well as a structural health monitoring (SHM) system to assess the behavior of a composite footbridge made of fiber-reinforced plastics (FRP) is discussed in [20]. Other example of composite bridge field tests is shown in [21]. An example of the use of interferometric radar to measure bridge deflections and evaluate its health is presented in [22]. Non-destructive testing of structures can also be done with the help of theoretical and computational models, numerical modelling, or simulations. These are used either independently, for example [23–25], or in combination with other testing methods, like the ones presented in preceding paragraphs. Some other examples of non-destructive testing supported by FEM are presented in the papers [26–29].

The above mentioned articles show that detailed, automated, high-accuracy geodetic measurements are important and are often employed for the purpose of non-destructive testing. They also point out high relevance of numerical models and computational techniques that support the testing. Nevertheless, the study of recent literature revealed a research gap related to the abilities of automated deformation measurements. The accuracy of TLS surveying, even when it is improved by the use of other devices or algorithms, is not better than 1 mm (refer for example to papers cited in the preceding paragraph or to [30–33]). There are structures experiencing very small displacements in typical conditions and therefore, automated geodetic measurements cannot be used in every case. An attempt to deal with this issue is discussed here.

In this article, we present interdisciplinary and comprehensive diagnostic, non-destructive tests of final inspection and acceptance of soil-steel single span bridge. Soil-steel bridges, viaducts, or culverts are structures built of flexible steel corrugated sheets or stiff concrete shells, which are covered with compacted soil having large angle of internal friction—backfill. Typically, span length of bridges of this type ranges from a few to several meters, rarely exceeding 25 m, while the backfill depth at the mid-span ranges usually from 0.5 m to a few meters (refer for example to [34]). The shell, buried in ground, is either a vault or a pipe [35]. Soil-steel bridges are designed in such a way that the constructional elements interact with each other to sustain their self-weights, weight of additional equipment, and traffic loads. The final effect of this interaction is beneficial. The bridges of this kind have the following advantages. They have specific architectural style. Their cost of construction is relatively low compared to the construction cost of steel or concrete bridges with standard abutments. This is because in the case of soil-steel bridges, the foundations are smaller, there are no abutments and the superstructure is made in the majority of the backfill. The time of the construction is short. Maintenance works for the soil-steel bridges are almost unnecessary. More information about this particular structural solution can be found in the handbook [35].

The analyzed bridge is a part of the national road number 16 linking the cities of Grudziądz and Olsztyn in Poland and is located near Ostróda city. To the best knowledge of the authors of this paper, this is the European span length record holder (25.74 m) in the category of soil-steel bridges. Because of that, it was very important to collect a lot of detailed information about the structure condition and its response under loads and when the bridge is unloaded. Therefore, standard programme of final inspection and acceptance, including mainly measurements of the bridge displacements in limited number of points under short-term static loads and its modal properties, was extended. In consequence, inter alia, detailed laser scanning of the bridge deformations was done, as well as a check of stresses in the steel shell.

TLS method was chosen to gather information about the bridge displacements, because it allows to capture image of the whole deformed shell in a short period of time. Nevertheless, preliminary FEM simulations of the short-term bridge response revealed that its elastic deformations under traffic loads are very small and do not exceed 2 mm. Therefore, in this case, in order to get reliable measurement results, the accuracy of TLS needed to be additionally improved. As reported in the literature review, currently, maximum resolution of TLS, even when it is supported by the use of other devices or algorithms is 1 mm and would have been insufficient.

In order to overcome this problem a new approach for high-accuracy scanning is proposed. We claim that TLS measurements can be enhanced using the information coming from total station and inductive sensors, that have much higher accuracy. Its application for the purpose of short-term static test surveying is the main goal of this research.

Additionally, deformations of the bridge, after one-year from the date the structure was built, were also checked by means of laser scanning. Moreover, computational model of the structure is created in the FEM environment in order to evaluate appropriateness of the bridge behavior and the measured values are compared with the estimated, calculated ones. The FEM model is calibrated using the measured data. Therefore, this paper is concentrated on the selection and presentation of appropriate non-destructive testing tools and methods, including FEM computational models, that enable to gather comprehensive and detailed information about the state and response of the analyzed soil-steel bridge structure. The main attention is focused on comprehensive, high-accuracy, geodetic measurements of bridge deformations and their comparisons with FEM response predictions.

2. The Soil-Steel Bridge

The considered soil-steel bridge is a new one, built in 2017. It is a single-span arch structure, as shown in Figures 1 and 2. The steel shell is made from UltraCor, 9.5 mm thick, corrugated sheets, produced by ViaCon Polska Ltd. company ([36]) (Rydzyna near Leszno, Poland), which are shown in Figure 3. The corrugated sheets are connected to each other with bolts. S315MC steel was used to produce the shell. The bridge foundations are built of Franki piles. The UltraCor structure is covered by soil compacted in a way that its relative density is not less than 98% of the maximum density determined in the laboratory. In the close vicinity of steel sheets, the compaction is not less than 95%. The steel arc span length is 25.74 m, whereas the total length of the bridge equals 95.70 m. The arc sagitta equals 9.0 m. These basic dimension, characterizing the structure, are depicted in Figure 1.

The bridge carries the national road number 16 and crosses a path for animals and a road built for the purpose of road maintenance (refer to Figure 1).

It is worth to mention that due to the aforesaid dimensions (large total length) it may not be so obvious if this structure should be named as a bridge or a tunnel. Nevertheless, we claim that this is a bridge from the following reasons. The soil-steel structure is made of flexible corrugated sheets and there is no base slab below the service road. Moreover, according to the polish law, appropriate road signs need to be put at the tunnel entrance and its exit. As seen in Figures 1 and 2, there are no such signs. In addition, under the terms of the construction contract the structure was named as a bridge, which meets the definitions provided in the regulations.

Figure 1. Soil-steel bridge side-view taken in October 2019.

Figure 2. Close up of one of the bridge sides taken in October 2019.

Figure 3. The corrugated steel sheets and their bolt connections.

3. Non-Destructive Testing Program

Each bridge, according to the Polish Regulations and instructions (refer to [37]) has to be tested before it is accepted for exploitation, to decide whether it is built appropriately or not. The final acceptance and inspection of the considered one took place on 18 July 2017. It was carried out by Aspekt Laboratorium Ltd. (Jaworzno, Poland) and a Research team from Gdańsk University of Technology (GUT) Department of Mechanics of Materials and Structures and Department of Geodesy. In consequence, basic parameters were checked for the bridge following the instruction [37]. The stiffness of the bridge was established based on the measurements of vertical deflections of the structure under static loads. This was done by means of inductive sensors and total station surveying (see Figure 4). The stability of supports was checked through the foundation settlements measure, carried out using precise geodetic levelling. Modal properties were determined using impact hammer, inductive displacement sensors and accelerometers. Visual inspection of the bridge was done after series of tests, in order to check if the structure sustained any damage, cracking, breakage, etc. The bridge passed all the standard tests.

In addition to standard measurements described in the previous paragraph, extended test programme was launched and accepted by the Road Authorities, Contractor, and Designer, because of large dimensions: length, width, and height of the steel shell, which make the bridge the current European span-length record holder in the category of soil-steel road bridges. Therefore, additional dynamic tests were done to identify the eigenmodes and eigenfrequencies of the structure with higher precision. Also terrestrial laser scanning of steel shell deformations was done (see Figure 4) in order to capture short-term deformations of the bridge and perform its periodic diagnostics.

a)  b)  c) 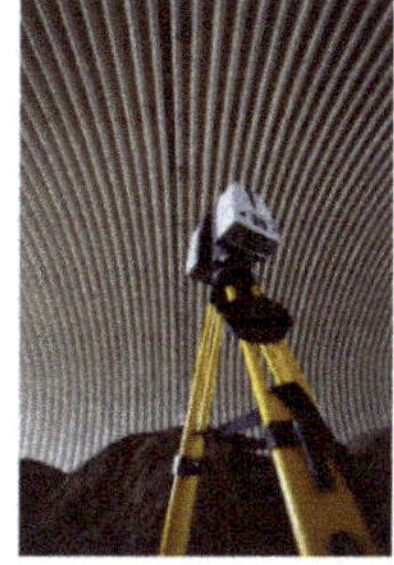

Figure 4. Devices and sensors used to measure basic bridge properties: (**a**) vertical translation inductive sensor, (**b**) total station, (**c**) laser scanner.

In this case a novel approach to increase the quality and accuracy of the scan has been proposed by the authors of this paper and it was successfully achieved during the tests. It is described in details in the next chapter. Finally, strains of the steel corrugated shell were monitored at the site during the tests using electrical resistance strain gauges and corresponding stresses were calculated. It needs to be emphasized that such an approach to non-destructive testing during final acceptance and inspection of a bridge is rarely seen in diagnostics of the structure. It enabled to collect a very detailed information about the bridge. Location of the network of sensor used to collect the data during the static tests is depicted in Figures 5 and 6. In Figure 6 only sections 2-2 and 4-4 are shown. This is because of the fact that the loads were applied to the bridge on the road lanes above these sections and in sections 1-1 and 3-3 the deformations were rather small. What is more terrestrial laser scanning of the short-term bridge deformations, being one of the most important aspect of this research, was done in sections 2-2 and 4-4.

To load the bridge and study its response during the tests 4-axle trucks were used. The total mass of each of the truck was 32 tons. The trucks were positioned in a way to maximize the loading effects in the measurement points. Two configurations of trucks positions were studied during the classical static tests. These were denoted as S1 and S2. The trucks were put on the bridge one by one, steadily in columns. Schematic drawing of trucks locations during the S1 test configuration is presented in Figure 7. Views from three different perspectives that document trucks positions during the S1 test configuration are shown in Figure 8.

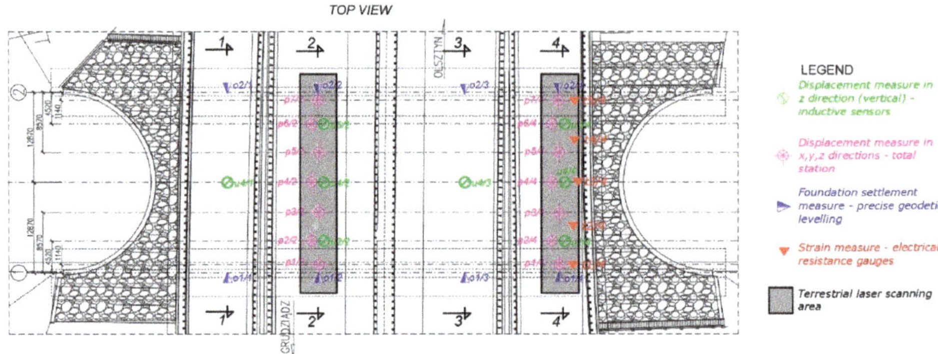

Figure 5. Location of the network of sensor used to collect the data during the static tests. Top view of the bridge.

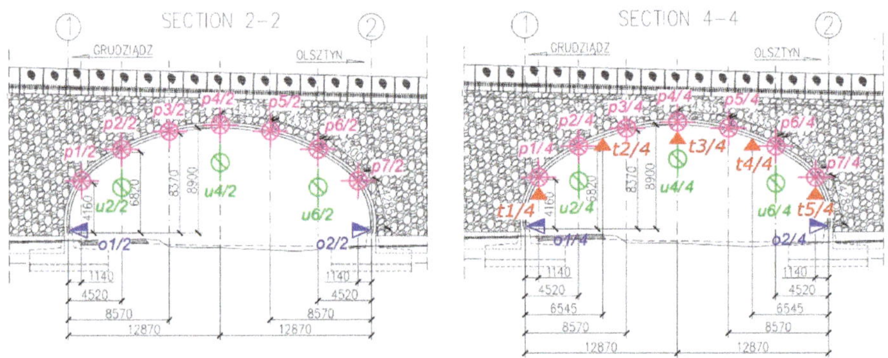

Figure 6. Location of the network of sensor used to collect the data during the static tests. sections 2-2 and 4-4 (dimensions in (mm)).

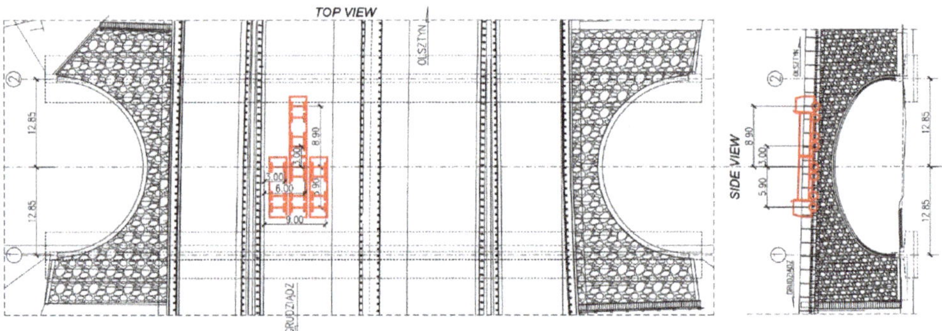

Figure 7. Schematic drawing of trucks locations during the S1 test configuration (dimensions in (m)).

Figure 8. The trucks, positioned during the S1 test configuration. Views from three different perspectives.

Schematic drawing of trucks locations during the S2 test configuration is presented in Figure 9. Views from three different perspectives that present trucks positions during the S2 test are depicted in Figure 10.

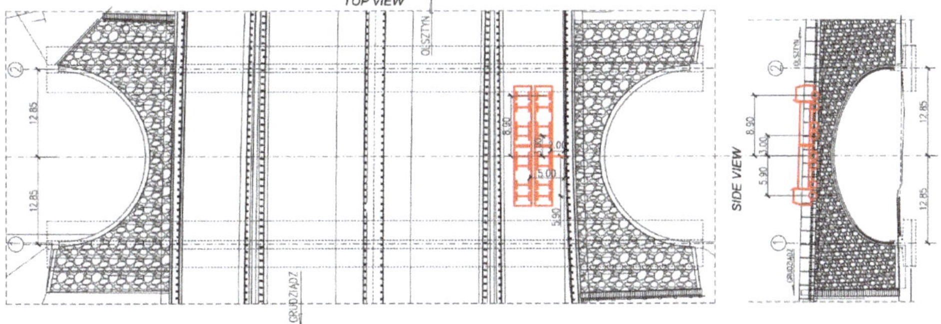

Figure 9. Schematic drawing of trucks locations during the S2 test configuration (dimensions in (m)).

Figure 10. The trucks, positioned during the S2 test configuration. Views from three different perspectives.

The bridge response was also checked under moving ballast of two trucks positioned side by side. These test configurations were denoted as M11 to M16 for the trucks standing on the road lane above 2-2 section and M21 to M26 for the trucks located above the 4-4 section. The trucks stopped in 6 equally spaced locations over the bridge. In Figure 11 schematic drawing of trucks locations at M24 stop is shown, whereas trucks moving between M23 and M24 testing positions are depicted in Figure 12.

More details about the final acceptance tests, carried out in Poland, can be also found in [38–40]. It is worth to mention that the concept of final acceptance tests based on moving loads is often considered (refer for example to [41,42]). Soil-steel bridges are structures, which response strongly depend on the behavior of the backfill. When the bridge is loaded the backfill and the shell deform together and thus its shape can slightly change. This deformation will be stronger when a flexible shell is used to cross the obstacle. When moving load concept testing is done and a bridge is loaded in a sequence of steps, the change of its shape should be much smaller, as the whole structure can easily accommodate the loads and the permanent deformations do not occur. In such a case the interpretation of measured results is easier. On the other hand, a moving load is smaller than the extreme load that can be applied to the structure. From this reason both types of load application, namely in the classic static way and with moving trucks, are important in the case of this bridge.

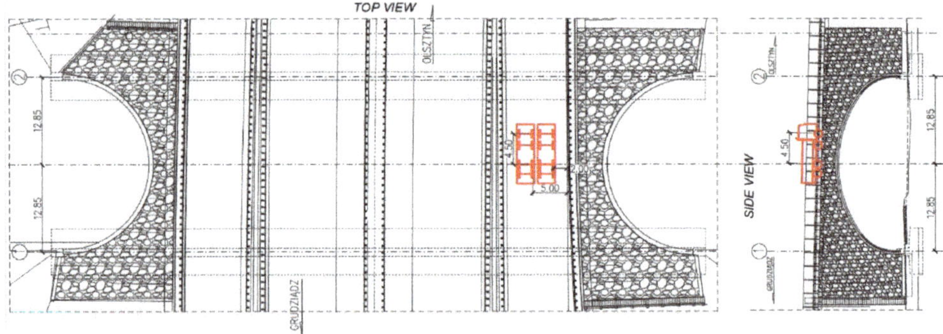

Figure 11. Schematic drawing of trucks locations at M24 stop (dimensions in (m)).

Figure 12. The trucks moving between M23 and M24 testing positions.

4. High Quality Laser Scanning—Methodology and Short-Term Static Tests Results

The preliminary numerical simulations of the structure response revealed that the shell deformations under standard loading conditions are very small—a few millimeters only. Therefore, we tried to find a method allowing us to gather as much data, about the bridge behavior, as it is possible in a relatively short period of time, having at the same time very high quality and accuracy. Owing to the research gap, identified in the introduction, we came up with an idea to perform TLS surveying of the bridge cross sections. Then, adjust and transform the scanning results appropriately, based on total station and displacement inductive sensor measurements, to significantly increase its accuracy. It is worth mentioning that soil-steel or concrete-soil composite bridges are typically subjected to in situ tests in a limited number of points and using some basic techniques of measures (refer to papers [43,44]). Some more advanced studies can be found in [41,42,45–48].

If the problem is defined as stated above, the in situ measurements methodology is divided into two stages. A properly prepared testing programme is launched at first, including laser scanning, total station and inductive sensors displacement measurements. During the second stage, proper post-processing of the data is done. Our experience in laser scanning measurements post-processing and aggregation of the results is available also in the following publications [49–52]. Our recent use of laser scanning during the final acceptance test is described in [53]. The measurements were done for the new European record holder in span length among extradosed type bridges.

The procedure and methodology for obtaining precise and accurate high quality results of terrestrial laser scanning is now presented. Leica P30 LiDAR (Light Detection and Ranging) laser scanning technology and robotic Leica TM50 total station were used for the purpose of geodetic surveying. It is worth mentioning that in this case the accuracy of the point cloud is the main research problem. According to the manufacturer's specifications, the accuracy of a point model, created during a single scan operation in one position, is about 1 mm and is going to be improved later on.

At first, appropriate preparation of the test field was done. The laser scanner was positioned in the line of the axis of each measured bridge cross sections 2-2 and 4-4 (as shown in Figure 5) and remained in the same place during each testing. Hence, the alignment of scan positions was not an issue. The total station was placed in a way that all the target prisms in the measurement points were visible. The positions of prisms (total station measure) and targets (laser scanning) were in accordance with the testing project. It has to be mentioned here, that during the tests a decision was made to do the total station survey in 9 instead of 7 points. Additional measures were done in o1 and o2 points. Thus, for each cross section under consideration, regarding the total station surveying, data were gathered from o1, o2, p1, p2, p3, p4, p5, p6, and p7. Moreover, two reference points, which are not marked in Figure 5, were established on the steel shell in the region where the deformations under the loads were very small, but still in the vicinity of the measured section. These were used to check the correctness of deformation registration. Additionally, the appropriateness of the total station surveying was checked using the data gathered by inductive displacement sensors. In Table 1 vertical displacements registered by total station (p2/4, p4/4, p6/4) and inductive sensors (u2/4, u4/4, u6/4), during S2 test, when maximum load was put on the road lane right above 4-4 section, are compared.

Table 1. Comparison of vertical displacements measured by means of total station and inductive sensors during S2 tests in the section 4-4.

Type of Measurement	Vertical Displacement (mm)		
	Point 2	Point 4	Point 6
Total station (p measuring point)	0.2	1.7	0.4
Inductive sensor (u measuring points)	0.07	1.59	0.28

On the basis of the results presented in Table 1 and owing to the precision of the used devices (inductive sensor ±0.01 mm, total station ±0.1 mm), it can be concluded that the total station measurements were appropriately done. It is worth mentioning that when a total station measurement is uncertain, additional correction of the vertical displacement can be done with regard to the inductive sensor measure. Consequently, such an updated (adjusted) result is used later on, during the process of TLS shape determination of the deformed bridge. However, in this case it was not necessary to do any additional adjustments.

The maximum distance of the targets from the scanner was approximately 20 m and the time of measurement of 9 points by the total station was about 3 min. Therefore, the laser scanning resolution was set to 3 mm per 10 m. In consequence, each scan of the whole bridge cross-section deformations lasted 3 min and 30 s. This allowed for obtaining the best measurement resolution and relatively many points of measure that were used at a later stage to create a curve representing the deformed shape of the structure. A large number of data, which was saved in such a case was not an issue, because the used instrument has the ability to reduce the noise at the level of 0.4 mm per 10 m. The surveying was done for the unloaded and loaded structure, according to the final acceptance testing programme.

After the survey ended, the post-processing of the geodetic data began. The data projection was used to compare the results from laser scanner with the ones from the total stations. In order to do so and because the total station measurements were taken in 9 points only, interpolation of the total station results between these points was done using the polynomial function. Similar considerations can be found in [54]. The chosen polynomial was of 9th degree. Thus, approximated shape of the tested cross section based on total station survey has been determined. The fitting root mean squared error (RMSE) came out at about 2 mm in all cases. This is compatible with measuring capabilities of the scanner. Therefore, we can say that the interpolation is accurate.

After that, approximation was done by means of the least squares methods of the laser scanning point cloud in order to find its best fitting with regard to the polynomial function resulting from the total station measurements. In effect a precise image of the bridge cross section in the reference state was obtained. It should be mentioned here that appropriate transformation of coordinates obtained from the total station into the coordinate system of laser scanner was carried out as well.

At that point, data were collected, containing images of reference and deformed states of the structure. Therefore the displacements of the steel shell were easily calculated for all the cross sections under consideration. The resulting deformations of the bridge cross section 2-2 in the Y-Z and X-Y planes under loading schemes S1 and M11–M16 are shown in Figure 13. Similarly in Figure 14, shell deformations are shown for the 4-4 cross section for the following loading conditions S2 and M21–M26.

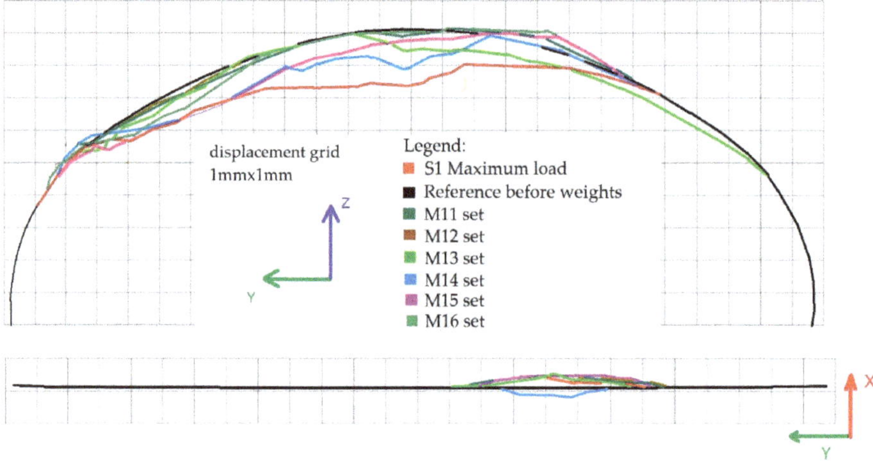

Figure 13. Deformations of the bridge cross section 2-2 in the Y-Z (top) and X-Y (bottom) planes under loading schemes S1 and M11–M16.

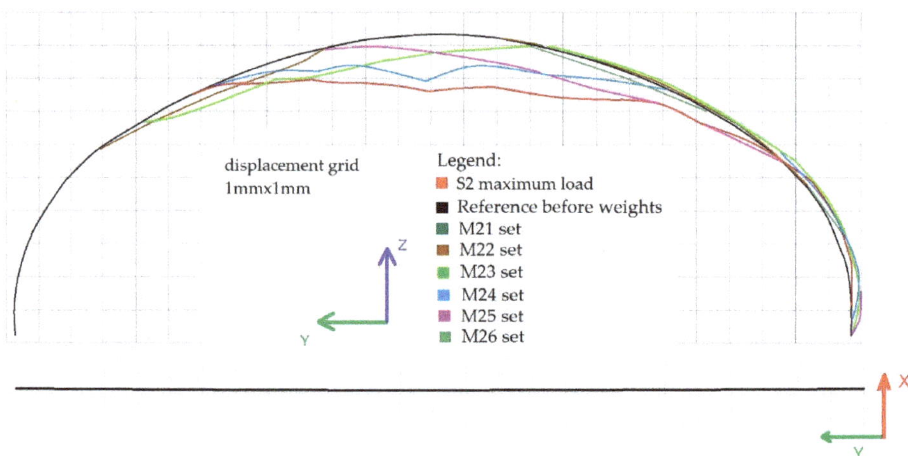

Figure 14. Deformations of the bridge cross section 4-4 in the Y-Z (top) and X-Y (bottom) planes under loading schemes S2 and M21–M26.

It should be noted that temperature changes may influence response of the bridge. However, the weather, on the day of tests, was cloudy or partly cloudy (see Figures 8, 10 and 11) and therefore the bridge was not strongly exposed to the sun and maintained constant temperature. For this reason discussion about deformations of the bridge due to temperature changes can be omitted.

In the end, we did manage to adjust the resolution of the laser scanning to the level of ~0.1 mm in lateral X and Y directions (in the location of total station prisms) and ~0.01 mm in the vertical Z direction (where inductive sensors were located) for the 2 mm × 2 mm grid of measurement points. The resulting accuracy of the laser scanning was ~0.3 mm, while the standard one obtained without any improvements is ~1 mm, as mentioned before. It is worth reminding, that in order to attain the accuracy, we adjusted results coming from laser scanner in the studied cross sections based on the information provided by the remaining sensors. Consequently, every point, registered by the scanning was influenced by this operation and therefore, accuracy of the whole scan was improved. Finally, it was possible to benefit from one of the advantages of the scanner which is its high scan rate in comparison to the other methods used, still maintaining accuracy of the gathered data. Therefore, we proved that the laser scanning technology can be used in measurements of displacements, being less than 1 mm. We consider our solution as original and innovative because such an aggregation of data coming out from different surveying devices and appropriate interpretation of results has not been done before in the field of static measurements of soil-steel bridges. Finally, we can say that we found a method to control and maintain very high resolution and accuracy of terrestrial laser scanning. Thus, very detailed images of the bridge cross sectional deformations in 3D were captured. It is also important, that the image of the unloaded bridge satisfies the conditions of the reference state in the case of structure monitoring and possible periodic diagnostics during its lifecycle. From the aforesaid reasons, we propose to include scanning of this type as a possible method of measurements during final acceptance and testing of bridge structures.

5. Periodic Diagnostics of the Bridge One Year after It Was Built

The next part of the research was focused on the periodic diagnostics of the bridge. In 2017 a scan of the whole unloaded steel shell had been made, which was repeated one year later. This enabled to check not only short-term elastic displacements, caused by real traffic conditions during final-acceptance and inspection, but also the eventual permanent displacement of the analyzed soil-steel structure. The periodic diagnostics is done to learn about the order of the bridge permanent displacements, whose occurrence is typical for this type of structures, during their exploitation. Here, great accuracy of measurements is not so important, as we expect that the displacements, that could have occurred, can be as big as a dozen or so mm. Therefore, additional improvement of scanning accuracy, as done for the purpose of short-term tests, presented in the Section 4, is not done.

In 2017, after short-term static tests were completed, the bridge was unloaded and scanning control points were stabilized in order to prepare for additional measurements. After that, full image of the bridge was registered. Because of the large dimensions of the structure, position of the scanner during the measurement process was changed 12 times. This process was repeated in the year 2018.

It needs to be emphasized that the surveying in 2018 was done at weather conditions similar to the ones observed the year before. Therefore, the results of periodic diagnostics are not affected by temperature variations.

As new scanning required new alignment of the device, the TLS accuracy could have been additionally checked. The Iterative Closest Points (ICP) algorithm was used to align scan positions correctly, besides of reference points that were stabilized outside of the structure. The positions of elements, which were used for the algorithm processing, were not changed during this time period and they have not been influenced by any external factors. The results of this alignment are presented in the form of histogram in Figure 15.

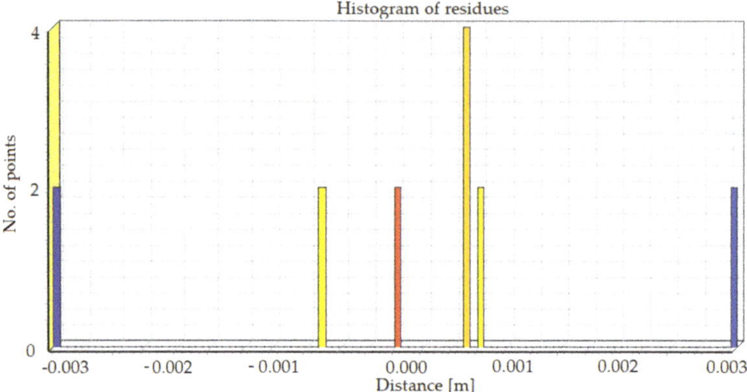

Figure 15. Histogram of residues which shows the precision of scan alignment.

The standard deviation resulting from the alignment was $\sigma = 1.6$ mm. Therefore, in order to reliably estimate the accuracy of the measurement, in accordance with the principles of normal distribution, we assumed that it is $3 \times \sigma = \sim 5$ mm. In consequence, the accuracy of computed deformation differences (registered in the period between 2017–2018) should have not exceeded this value.

As usual, the noise reduction was done during the process of point cloud data filtration. The noise reduction algorithm described in [55] was used here, which locally fits a surface into the point cloud and then remove points that are too far from that surface. In the end, it was possible to exclude points lying outside the possible deformation shape that had a negative impact on the final results and to retain only the ones maintaining reasonable accuracy.

After the noise reduction had been done, it was possible to calculate the periodic diagnostics displacements of the steel shell that occurred between the years 2017 and 2018. The method described in paper [56] (available in CloudCompare software) was used for this purpose. This particular approach was chosen because most of the available comparative methods are based on the determination of the closest distance from a point or require 3D modelling of the compared surface. Thereby, they eliminate the problem of roughness, which cannot be eliminated here, as the corrugation of the steel sheets is an important feature of the whole structure and cannot be omitted. To find the distances between two point clouds, two main steps were done. First, normal surface and its orientation in 3D was estimated on a scale consistent with local surface roughness. Then, the mean surface change was measured along the surface normal direction and the local confidence interval was determined. This, in our opinion, allowed to obtain reliable information about the structure, which has varying roughness, as evidenced by the observed noise in the cloud of points and terrain diversity. The roughness is also reflected by the average change along the surface normal direction, as it was done in the second major calculation step of this procedure. In effect the distances between two point clouds were obtained and the deformations contours (magnitudes of displacements) are shown in Figure 16.

The results of the calculations presented in Figure 16 reveal that the upper part of the bridge has moved to the south up to 7 mm, while the shell in the vicinity of foundation remained undeformed. To check the accuracy of the obtained deformations a series of statistical tests, based on the Gaussian normal distribution, were run. Figure 17 presents the statistics of the Gauss function corresponding to the calculated distance uncertainties. On this basis it can be stated that the results were correctly determined in relation to the reference shown in Figure 15.

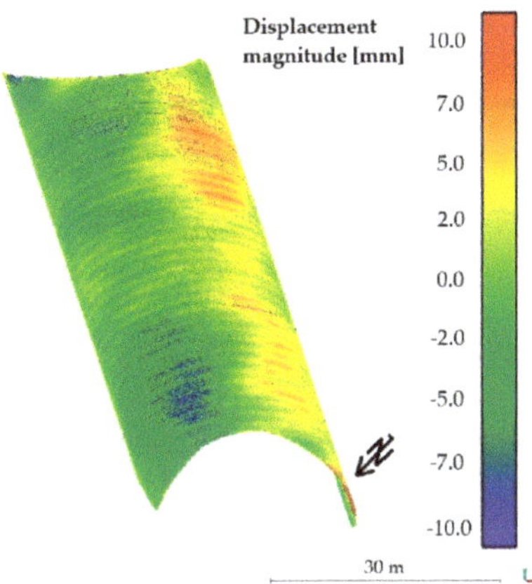

Figure 16. Periodic diagnostics of bridge deformations (displacement magnitude) measured one year after the final acceptance and inspection.

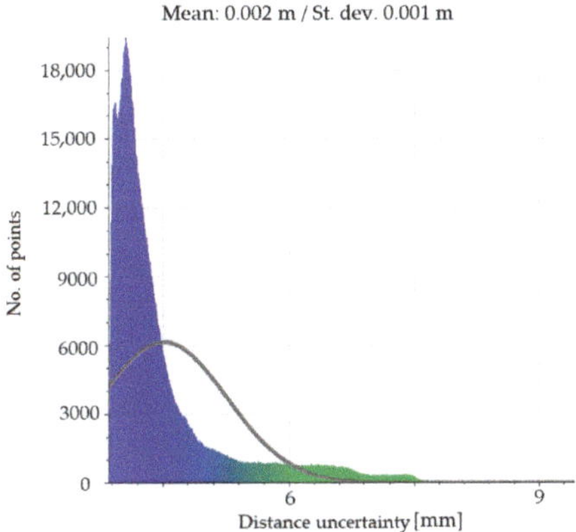

Figure 17. Distance uncertainty statistic.

Owing to the above, the uncertainty of the measured deformations (Figure 16) equals approximately 2 mm (within the computed σ from alignment). This means that the laser scanning enabled to register all of the points characterizing the largest geometric changes. Nevertheless, this uncertainty suggests that some discrepancies of the measurements occurred during the scanning. These resulted possibly from the standard issues connected with stabilization of points, their indication, and establishment. However, because of the fact that the measurements were repeated one year after the final acceptance tests and during this period of time the construction works were still ongoing, such an accuracy of measurements is to a large extent satisfactory. For example in paper [51] laser scans of an external system were done using other techniques and the resulting precision and accuracy was not as good as here.

Periodic diagnostic deformation images were created also in sections 2-2 and 4-4, that were previously tested during static tests in 2017. This enabled to additionally cross-check calculations of deformations of the whole bridge, presented in Figure 16. To create them it was required to perform a number of computational operations, enabling interpretation of deformation changes of the bridge. The results obtained from scanning during short-term tests, were quite helpful in achieving this goal. Displacements of the unloaded steel shell one year after the final acceptance and inspection tests is shown in Figure 18 for the 2-2 section and in Figure 19 for the 4-4 section. The directions of the extreme values of displacements are also shown in Figures 18 and 19.

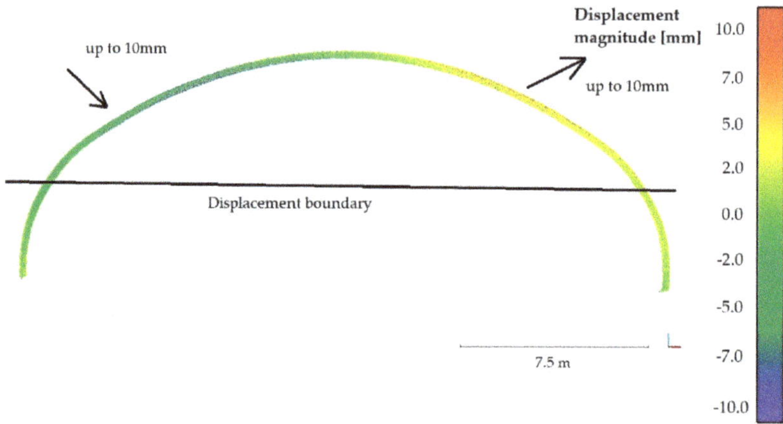

Figure 18. Displacements of the unloaded steel shell 2-2 section one year after the final acceptance and inspection tests.

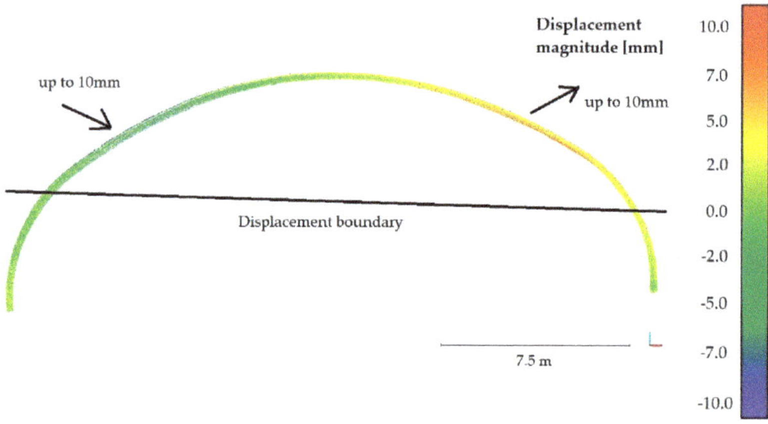

Figure 19. Displacements of the unloaded steel shell 4-4 section one year after the final acceptance and inspection tests.

The aforesaid results reveal, that during one-year period of time, the bridge did not exhibit deformations below the black border line, marked in the figures and named as "displacement boundary". In view of that, we noticed a certain negative feature of the computed periodic diagnostics deformations (see Figures 16, 18 and 19). Despite attaining satisfactory result of calculations, the displacements were not obtained in the lower parts of the steel shell. However, because of the measurement accuracy, their exact values are not so certain and may rather indicate the displacement tendency. Similarly, this could have resulted from the properties of the algorithm used. On the other hand, what also is interesting, the "displacement boundary" intersects the shell approximately at the same height at which bolt connections were done (refer to Figure 3). As one year passed since the date of final acceptance tests, slippage of this connections could have occurred and some rotations of the shell have become possible in this areas. This may explain why displacements of the upper part of the shell are only observed. Hence, a research gap has been identified here. These aspects will be analyzed and addressed in further studies.

Finally, it can be concluded that the periodic diagnostics displacements measured one year after the final acceptance tests are relatively small, 7 mm ± 2 mm. Here, we would like to emphasize that the extreme values of permanent displacements are observed above the aforesaid "displacement boundary."

Therefore, maximum and certain displacement of the bridge did not exceed 9 mm. This is a very important information as the order of the permanent displacement was identified. Occurrence of deformation of this order is typical for soil-steel structures. Thus, the bridge experienced a change of its shape, which however does not affect its response and load-bearing capacity.

6. Numerical Analyses and Comparison of Results

6.1. Computational Model

In order to design the final acceptance test programme, the bridge response needs to be estimated. Therefore, numerical simulations of its behavior under static loads and natural frequency extraction were performed by means of FEM (refer to [57]) using Abaqus 6.14-2 code. The computational model was created according to the bridge technical drawings and it is a particular one. It is created to describe the global response of the soil-steel bridge and to check whether the assumptions and approach to design were appropriate. Thus, the computational domain needs to remain in correspondence with the precision required by the design process. From this reason, we employ basic modelling techniques and some simplifications. The model should not be overdetailed, but it still has to allow for obtaining reliable results.

Its size (distance from the bridge to the outer boundaries of the domain) was selected in such a way that it did not affect the results in the area of the steel shell. A structured mesh of finite elements was created. First-order solid C3D8I elements (refer to [58]) were used to model the soil continuum. These elements are enhanced by incompatible modes. This formulation enables very strong reduction of the locking effect, as stiffening in bending, being a consequence of the parasitic shear stresses or Poisson's effect, is almost eliminated. The steel parts were represented by shell S4 elements (see for example papers [34,59–61] for other applications of shell elements) with linear shape functions and full 2×2 in-plane Gauss-Legendre integration scheme and some additional procedures preventing the locking effect. In the central part of the structure, namely in the region where the in situ measurements were taken, the corrugated sheets were modelled in detail having regard to the real geometry. A simplification was made close to the structure ends, where the steel sheets were treated as a shell endowed with equivalent orthotropic properties and thickness. It is worth mentioning that, often, the whole steel shell is treated as the one having some equivalent properties and uniform thickness (see for example papers [42,62]). Nevertheless, such a simplification cannot be done here, because we are interested in accurate estimation of stresses in the shell-corrugated sheets. The whole computational domain of the soil-steel composite structure is shown in Figure 20 and the detail depicting modelling approach of steel shell in Figure 21.

Figure 20. Finite element method (FEM) computational domain of the bridge.

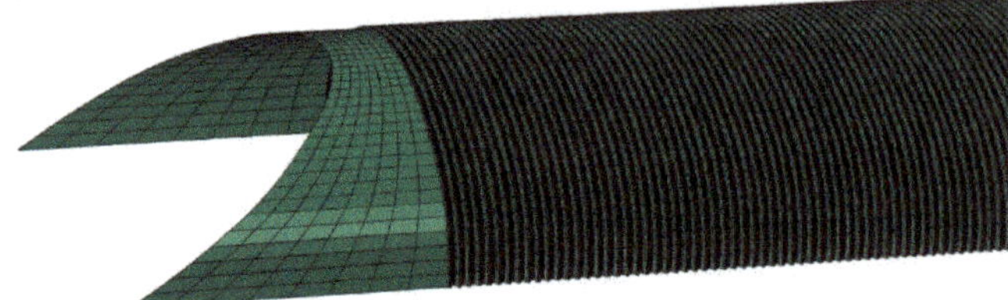

Figure 21. Steel shell in the numerical model: corrugated sheets in the center, equivalent shell close to the bridge ends.

The soil and steel were treated in the analyses as isotropic homogenous materials. The elastic constants of the soil were determined on the basis of the requirement of backfill compaction—minimal relative density of the material should be not less than 98% of the maximum density of the soil achieved in the laboratory. This corresponds, according to the polish design standards still used by site engineers [63], to the elastic modulus being 170 MPa. However, often, at the site, the properties of the backfill are better than minimums. So it was also in this case. Aggregate with very good parameters was used to construct the backfill. The reports on the compaction acceptance of the backfill layers showed that the compaction was ranging from 100% to 103%. Therefore, other possible soil properties are also considered, with elastic modulus being 200 MPa (corresponding to 100% compaction) or 230 MPa (corresponding to 103% compaction). Thus, we try to calibrate the numerical model, based on the obtained measurements and assign soil properties enabling best fitting of the FEM estimations. The material properties of steel and soil used in the computational analyses are collected in Table 2. An evident simplification of the material law for the backfill is done. Nevertheless, it is justified by the following reasons. A global model is created and we focus mainly on the appropriate estimation of displacements and stresses in the steel shell. Therefore, detailed description of soil material law is not so important. The backfill main properties are calculated based on well-established engineering procedures, which should enable reasonable description of its response. Second, behavior of the bridge under design loads is recreated and by assumption this loading condition should not lead to any failure. What is more, if failure occur in the backfill, the real response of the bridge would be much different from the FEM estimation, which would also clearly indicate that the simplified approach is not a proper one. This will be verified in the next chapter. It is also worth mentioning that the backfill is treated in a similar way by others and good results are obtained, as presented for example in [64].

Table 2. Yielding of steel rafter in all the considered local models.

Material	Elastic Modulus (MPa)	Poisson's Ratio
Steel corrugated sheets	210,000	0.3
Soil (backfill)	170–230	0.2

It also needs to be noted that the bridge was assembled in summer and its temperature during the tests was similar to the one at which it was mounted. There was only a couple of degrees difference. What is more, soil elastic modulus is only slightly affected by the temperature changes, as reported in [65]. Therefore, the material properties of the backfill does not need to be adjusted.

The following boundary conditions were assigned to the model. The steel shell is fixed in the foundations. Vertical displacements are restrained at the lateral lower surface of the embankment. The displacements perpendicular to the vertical side surfaces of the embankment are constrained as well. A contact was defined between steel shell and the surrounding backfill.

The loads were applied to the model in accordance with trucks positions, described in the Section 3 and their axle loads. In consequence, uniformly distributed loads were defined at truck tires-road contact area. The shapes of tire contact patches were simplified and treated as rectangular. No additional application of temperature loading was necessary to compensate eventual change of the bridge geometry due to the change of temperature during different testing configurations. This is due to the fact, described earlier as well, that the weather during the whole testing day (cloudy or partly cloudy) was basically the same, the bridge was not exposed to the sun, therefore variation of its temperature was negligible.

Finally, static analyses were executed to estimate bridge response during the in situ tests.

6.2. Comparison of FEM Estimations and In Situ Measured Values

Now comparisons of selected, representative results of the in situ measurements and corresponding FEM estimations are presented. Total deformations of the steel shell, measured by terrestrial laser scanning and total station are compared with the results coming from FEM calculations, in which different soil properties were assigned (refer to Table 2). These are shown in Figures 22–24, correspondingly, in 2-2 section under maximum load during S1 test, in 4-4 section under maximum load during S2 test and in 4-4 section during M24 moving load test configuration. It is worth mentioning that typically elastic displacements are analyzed during final acceptance tests. However, this is a specific case of a soil-steel structure. When the bridge was unloaded after S1 and S2 tests a slight change of shell shape was observed. Since point-cloud is created as a result of laser scanning, it is very hard to calculate elastic deformations coming from TLS. This is the reason why total deformations are compared in this chapter. Nevertheless in the case of moving load tests (refer for instance to Figure 24) the total deformation is nearly the same as the elastic deformation of the structure, because such a testing approach allows to significantly reduce this effect, as it was described in the Section 3. It has to be additionally noted that total station surveying was done in 9 points, as shown in Figure 6, in each cross section under consideration. In Figures 22–24, spline function was used to connect these 9 points and in effect deformed shape resulting from total station measurements was approximated.

It is seen in Figures 22–24 that the FEM estimations, even for different variants of the backfill elastic modulus, are generally in accordance with the measurements with regard to their quality and quantity. The backfill elastic modulus is an important parameter that affect the bridge response. It is a bit underestimated, when E = 170 MPa. Better quality results are obtained for E = 200 MPa and E = 230 MPa. In order to investigate displacements of the steel shell in detail, now, the extreme vertical displacements, that were obtained in the p4/2 point for the test S1, p4/4 for the test S2 and the corresponding FEM values (for different backfill material properties) are analyzed in Table 3. In Table 3 both total ($U_{in\text{-}situ,tot}$) and elastic ($U_{in\text{-}situ,el}$) in situ measured displacements are presented to highlight the issue of the shell shape change when the bridge was unloaded.

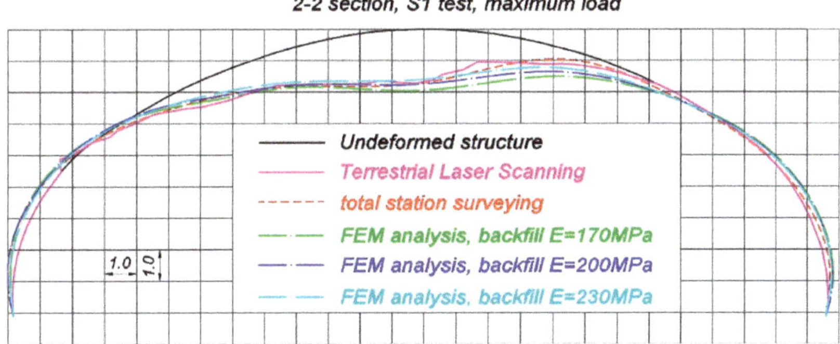

Figure 22. Comparison of the deformed shapes of the bridge in 2-2 section during the S1 test, when maximum load was applied, measured using terrestrial laser scanning, total station, and estimated by means of FEM (deformations in (mm) are 1000 times scaled).

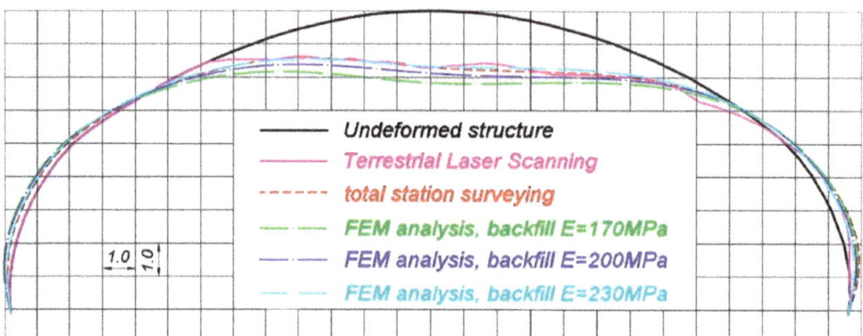

Figure 23. Comparison of the deformed shapes of the bridge in 4-4 section during the S2 test, when maximum load was applied, measured using terrestrial laser scanning, total station, and estimated by means of FEM (deformations in (mm) are 1000 times scaled).

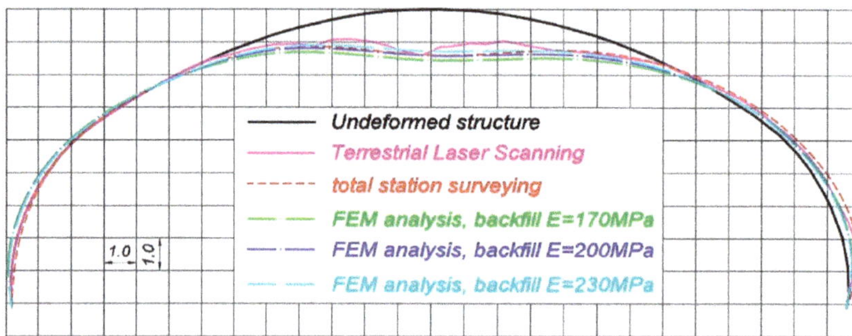

Figure 24. Comparison of the deformed shapes of the bridge in 4-4 section during the moving load M24 configuration, measured using terrestrial laser scanning, total station, and estimated by means of FEM (deformations in (mm) are 1000 times scaled).

In Table 4 vertical in situ displacements in the p4/4 point during the moving load tests M21–M26 are compared with the ones resulting from FEM calculations. Total displacement and the elastic values are not distinguished in Table 4, since they are almost the same.

The results from Tables 3 and 4 reveal that the response of the soil-steel composite bridge is underestimated, when the backfill elastic modulus is E = 170 MPa. In this case, the elastic displacement ratios $U_{in\text{-}situ,el}/U_{FEM}^{E\,=\,170}$ for p4/2 and p4/4 locations, during S1 and S2 tests, are correspondingly 79% and 77%. The ratio between the biggest in situ vertical displacement during moving load test (M24 test configuration) and the corresponding FEM estimation $U_{in\text{-}situ,M24}/U_{FEM,M24}^{E\,=\,170}$ equals 84%. Similarly, for the backfill elastic modulus equaling 200 MPa, the ratios $U_{in\text{-}situ,el}/U_{FEM}^{E\,=\,200}$ in p4/2 and p4/4 points, during S1 and S2 tests are correspondingly 86% in both locations, while the ratio $U_{in\text{-}situ,M24}/U_{FEM,M24}^{E\,=\,200}$ for the M24 test configuration is 94%. Finally the results are compared for the backfill having assigned E = 230 MPa. In this situation the ratios $U_{in\text{-}situ,el}/U_{FEM}^{E\,=\,230}$ are 98% for p4/2 and 95% for p4/4 measurement point, whereas for the case of the moving load tests the ratio $U_{in\text{-}situ,M24}/U_{FEM,M24}^{E\,=\,230}$ equals 104%. Based on these comparisons, it can be concluded, that from an engineering point of view, the response of the bridge is well recreated for the backfill with E = 200 MPa and E = 230 MPa. As the moving load test is probably the most effective way to examine and measure the bridge response (this issue has been already described in Section 3), it can be stated the real properties of the backfill regarding its elastic modulus are somewhere between 200 MPa and 230 MPa.

Table 3. Comparison of the in situ measured and estimated by means of FEM extreme displacements.

Displacement	p4/2 (mm)	p4/4 (mm)
FEM value, backfill E = 170 MPa $U_{FEM}^{E=170}$	1.91	2.09
FEM prediction, backfill E = 200 MPa $U_{FEM}^{E=200}$	1.74	1.86
FEM prediction, backfill E = 230 MPa $U_{FEM}^{E=230}$	1.53	1.68
In-situ, elastic value $U_{in\text{-}situ,el}$	1.5	1.6
In-situ, total value $U_{in\text{-}situ,tot}$	1.7	1.7

Table 4. Comparison of the in situ measured and estimated by means of FEM displacements in p4/4 measuring point during the moving load tests (M21–M26).

Test Denotation	FEM Value $U_{FEM}^{E=170}$, Backfill E = 170 MPa (mm)	FEM Value $U_{FEM}^{E=200}$, Backfill E = 200 MPa (mm)	FEM Value $U_{FEM}^{E=230}$, Backfill E = 230 MPa (mm)	In-Situ Value $U_{in\text{-}situ}$ (mm)
M21	−0.03	0.02	−0.02	0.10
M22	0.04	−0.04	0.05	0.20
M23	0.69	0.63	0.58	0.70
M24	**1.54**	**1.38**	**1.25**	**1.30**
M25	0.36	0.34	0.32	0.40
M26	−0.02	0.02	0.01	0.10

Nevertheless, the geometry of the shell at the beginning of the tests was slightly different than it was designed, as it had adjusted its shape during backfilling. Moreover, it continued to adjust its shape during final acceptance tests under different loading conditions and the whole bridge experienced some small permanent deformations. This is typical for soil-steel structures. In consequence, during static and moving loads tests, the loads caused by trucks were applied to the bridge having a bit different geometry. Ideal geometry of the structure, based on the technical drawings of the bridge, was defined in the numerical model. This could have caused the differences between the compared in situ and FEM values shown in Figures 22–24 and Tables 3 and 4. It is also worth to mention that the bridge shell is built of small corrugated steel sheets which are connected to each other with bolts (as shown in Figure 3). Such a connection technique requires the steel sheets to overlap each other. In effect, the whole shell is somehow locally stiffened, which may contribute also to the global stiffness of the whole system. A continuous shell without connections is created in the computational model. Therefore, one may expect that the structure has an internal margin of safety. Owing to the aforesaid considerations, we claim that the backfill properties are close to the ones represented by elastic modulus which is 200 MPa. In effect, we can say that the computational model has been calibrated with aid of the measurements. Thus, it reflects the real response of the analyzed structure. In effect it can be used for the purpose of the bridge diagnostics or support interpretation of SHM data.

Finally, the stresses calculated on the basis of strains, registered during the static tests S2 in t2/4 and t4/4 points, are shown together with corresponding FEM predictions in Table 5. The results presented in Table 5 were calculated assuming that the elastic modulus of the backfill is E = 200 MPa, which has been calibrated based on the displacement measurements, being the most reliable ones in this research. The points t2/4 and t4/4 were located approximately in the quarter-spans of the steel arch and extreme values were obtained there. Additionally, stresses in t3/4 point are also compared in Table 5. In Table 6 stresses calculated from strains in t2/4 point during the moving load tests M21–M26 are compared with the ones resulting from FEM calculations. Total and elastic values are not distinguished in Table 6, since they are almost the same.

Table 5. Comparison of the in-situ measured and estimated by means of FEM (for the backfill, having E = 200 MPa) extreme stresses.

Stress	t2/4 (MPa)	t3/4 (MPa)	t4/4 (MPa)
FEM prediction S_{FEM}, backfill E = 200 MPa	−6.21	−1.31	−6.52
In-situ, elastic value $S_{in\text{-}situ,el}$	−5.74	−2.13	−6.24
In-situ, total value $S_{in\text{-}situ,tot}$	−6.61	−2.50	−6.49

Table 6. Comparison of the in situ measured and estimated by means of FEM (for the backfill, having E = 200 MPa) stresses in t2/4 measuring point during the moving load tests (M21–M26).

Test Configuration	FEM Prediction S_{FEM} (MPa)	In-Situ Measure $S_{in\text{-}situ}$ (MPa)
M21	−0.17	0.12
M22	−0.11	0.25
M23	−1.41	−1.12
M24	**−4.55**	**−3.99**
M25	−1.44	−1.12
M26	−0.14	0.37

Similarly, as in case of displacements analysis, the stress ratios $S_{in\text{-}situ,el}/S_{FEM}$ for t2/4 and t4/4 from Table 5 are calculated and they are correspondingly 92% and 96%. The ratio between the biggest in situ stress during the moving load test (M24 test configuration) and the corresponding FEM estimation $S_{in\text{-}situ,M24}/S_{FEM,M24}$ is 88%. It has to be emphasized here, that because of characteristics of the used devices (strain gauges) the accuracy of the calculated in situ stresses is ±1.2 MPa. The accuracy of displacements measure was way better than for the stresses, thus the displacements comparisons, presented in the preceding paragraph, seem to be more reliable and because of that were used for the purpose of the backfill stiffness properties calibration. From this reason the discussion about the cause of differences between measurements and numerical calculations has been done based on the registered displacements. Although the accuracy of stress measurements is not so good in relation to the values that were registered, which were relatively small and close to the device accuracy, it can be still concluded that there definitely is a correspondence between the in situ measurements and FEM estimations in the field of stress comparisons.

7. Conclusions and Final Remarks

Interdisciplinary and comprehensive non-destructive diagnostic tests of final bridge inspection and acceptance of a soil-steel bridge made of corrugated sheets, being the European span length record holder (25.74 m), done in the year 2017 and 2018, were described. Non-standard program of bridge loading was used during the inspection. On the one hand, classically, static loads were applied to the bridge by appropriate arrangement of trucks. On the other hand, the bridge response was studied under moving loads, to reduce permanent deformations when it was unloaded.

Because of the large dimension of the structure and also because its deformations under standard traffic loads are very small (a few mm only), sophisticated terrestrial laser scanning was done in order to measure short-term response of the bridge. Its accuracy was significantly increased by additional correction of results based on total station surveying and inductive sensors vertical displacement control. The resulting 3D deformation images of the steel shell sections, with the achieved precision (X, Y ~0.1 mm; Z ~0.01 mm; for the 2 mm × 2 mm grid of measurement points) and accuracy of the laser scanning ~0.3 mm, can be considered as a unique achievement. Such a high precision and accuracy of the geodetic measurement allowed to reliably identify bridge displacements, which under maximum loads were not more than 1.7 mm. Moreover, this image satisfies the conditions of the reference state in the case of structure monitoring, and possible periodic diagnostics during its life cycle.

A control of the structure geometry shape was done one year after the final inspection. In effect, maximum total displacement of the bridge, which occurred in one-year period, determined in terrestrial laser scanning equaled 7 mm. The scan accuracy in this case was ~2 mm. This is typical for structures of this type. Thus, we can say that the bridge experienced a change of its shape, which however does not affect its response and load bearing capacity.

A computational model of the bridge was built in order to verify the appropriateness of the design process. In situ short-term response measurements were compared with the numerical calculations resulting from FEM analyses. Calibration of the backfill elastic modulus based on the measured response of the bridge was done and the resulting accuracy of the displacements predictions was at the level of 95%. The model reflects the real response of the structure in the field of both quality and quantity of displacements and its strain. This confirms that the approach to modelling proposed in this paper was appropriate and that the bridge was correctly designed. In effect it can be used for the purpose of the bridge diagnostics or support interpretation of SHM data.

The bridge underwent all the required final acceptance tests and inspections and currently there are no doubts about its behavior during the designed lifetime. Moreover, its accurate periodic diagnostics is also possible.

Author Contributions: Conceptualization, M.M.; data curation, M.M., B.S. and P.T.; formal analysis, M.M., B.S. and P.T.; investigation, M.M., B.S. and P.T.; methodology, M.M., B.S. and P.T.; resources, M.M.; supervision M.M.; visualization, B.S., P.T.; writing—original draft, B.S., P.T.; writing—review and editing, M.M., B.S. and P.T. All authors have read and agreed to the published version of the manuscript.

Funding: This research received no external funding.

Acknowledgments: The authors acknowledge the support provided by: the main contractor Budimex S.A., subcontractor ViaCon Polska Ltd. and research teams from Aspekt Laboratorium Ltd. and Gdańsk University of Technology (Department of Mechanics of Materials and Structures, Department of Geodesy) in conducting presented studies. Abaqus calculations were carried out at the Academic Computer Centre in Gdańsk.

Conflicts of Interest: The authors declare no conflict of interest.

References

1. Palma, P.; Steiger, R. Structural health monitoring of timber structures—Review of available methods and case studies. *Constr. Build. Mater.* **2020**, *248*, 118528. [CrossRef]
2. Towsyfyan, H.; Biguri, A.; Boardman, R.; Blumensath, T. Successes and challenges in non-destructive testing of aircraft composite structures. *Chin. J. Aeronaut.* **2020**, *33*, 771–791. [CrossRef]
3. Luo, D.; Li, Y.; Li, J.; Lim, K.S.; Nazal, N.A.M.; Ahmad, H. A Recent progress of steel bar corrosion diagnostic techniques in RC structures. *Sensors* **2018**, *19*, 34. [CrossRef] [PubMed]
4. Bień, J.; Kużawa, M.; Kamiński, T. Strategies and tools for the monitoring of concrete bridges. *Struct. Concr.* **2020**. [CrossRef]
5. Schabowicz, K. Non-destructive testing of materials in civil engineering. *Materials* **2019**, *12*, 3237. [CrossRef]
6. Bien, J.; Kamiński, T.; Kużawa, M. Taxonomy of non-destructive field tests of bridge materials and structures. *Arch. Civ. Mech. Eng.* **2019**, *19*, 1353–1367. [CrossRef]
7. Makuch, M.; Gawronek, P. 3D point cloud analysis for damage detection on hyperboloid cooling tower shells. *Remote Sens.* **2020**, *12*, 1542. [CrossRef]
8. Liu, J.; Zhang, Q.; Wu, J.; Zhao, Y. Dimensional accuracy and structural performance assessment of spatial structure components using 3D laser scanning. *Autom. Constr.* **2018**, *96*, 324–336. [CrossRef]
9. Fawzy, H.E.D. 3D laser scanning and close-range photogrammetry for buildings documentation: A hybrid technique towards a better accuracy. *Alex. Eng. J.* **2019**, *58*, 1191–1204. [CrossRef]
10. Mistretta, F.; Sanna, G.; Stochino, F.; Vacca, G. Structure from motion point clouds for structural monitoring. *Remote Sens.* **2019**, *11*, 1940. [CrossRef]
11. Moselhi, O.; Bardareh, H.; Zhu, Z. Automated data acquisition in construction with remote sensing technologies. *Appl. Sci.* **2020**, *10*, 2846. [CrossRef]

12. Korumaz, M.; Betti, M.; Conti, A.; Tucci, G.; Bartoli, G.; Bonora, V.; Korumaz, A.G.; Fiorini, L. An integrated Terrestrial Laser Scanner (TLS), Deviation Analysis (DA) and Finite Element (FE) approach for health assessment of historical structures. A minaret case study. *Eng. Struct.* **2017**, *153*, 224–238. [CrossRef]

13. Binczyk, M.; Kalitowski, P.; Szulwic, J.; Tysiąc, P. Nondestructive testing of the miter gates using various measurement methods. *Sensors* **2020**, *20*, 1749. [CrossRef] [PubMed]

14. Yang, H.; Xu, X.; Neumann, I. An automatic finite element modelling for deformation analysis of composite structures. *Compos. Struct.* **2019**, *212*, 434–438. [CrossRef]

15. Laefer, D.F.; Truong-Hong, L.; Carr, H.; Singh, M. Crack detection limits in unit based masonry with terrestrial laser scanning. *NDT E Int.* **2014**, *62*, 66–76. [CrossRef]

16. Lubowiecka, I.; Armesto, J.; Arias, P.; Lorenzo, H. Historic bridge modelling using laser scanning, ground penetrating radar and finite element methods in the context of structural dynamics. *Eng. Struct.* **2009**, *31*, 2667–2676. [CrossRef]

17. Xu, W.; Xu, X.; Yang, H.; Neumann, I. Optimized finite element analysis model based on terrestrial laser scanning data. *Compos. Struct.* **2019**, *207*, 62–71. [CrossRef]

18. Pieraccini, M.; Dei, D.; Betti, M.; Bartoli, G.; Tucci, G.; Guardini, N. Dynamic identification of historic masonry towers through an expeditious and no-contact approach: Application to the "Torre del Mangia" in Siena (Italy). *J. Cult. Herit.* **2014**, *15*, 275–282. [CrossRef]

19. Yu, J.; Zhu, P.; Xu, B.; Meng, X. Experimental assessment of high sampling-rate robotic total station for monitoring bridge dynamic responses. *Measurement* **2017**, *104*, 60–69. [CrossRef]

20. Miśkiewicz, M.; Pyrzowski, L.; Sobczyk, B. Short and long term measurements in assessment of FRP composite footbridge behavior. *Materials* **2020**, *13*, 525. [CrossRef]

21. Siwowski, T.; Rajchel, M.; Kulpa, M. Design and field evaluation of a hybrid FRP composite—Lightweight concrete road bridge. *Compos. Struct.* **2019**, *230*, 111504. [CrossRef]

22. Oktavianus, Y.; Sofi, M.; Lumantarna, E.; Kusuma, G.; Duffield, C. Long-term performance of trestle bridges: Case study of an Indonesian marine port structure. *J. Mar. Sci. Eng.* **2020**, *8*, 358. [CrossRef]

23. Bezza, A.; Destuynder, P.; Fabre, C.; Wilk, O. Energy methods for non-destructive testing in a two dimensional damaged structure. *Appl. Math. Model.* **2018**, *60*, 300–319. [CrossRef]

24. Sabik, A. In-plane shear nonlinearity in failure behavior of angle-ply laminated shells. *Compos. Struct.* **2019**, *225*, 111164. [CrossRef]

25. Miśkiewicz, M.; Bruski, D.; Chróścielewski, J.; Wilde, K. Safety assessment of a concrete viaduct damaged by vehicle impact and an evaluation of the repair. *Eng. Fail. Anal.* **2019**, *106*, 104147. [CrossRef]

26. Fan, H. Critical buckling load prediction of axially compressed cylindrical shell based on non-destructive probing method. *Thin Walled Struct.* **2019**, *139*, 91–104. [CrossRef]

27. Fedorko, G.; Molnar, V.; Ferková, Ž.; Peterka, P.; Krešák, J.; Tomaskova, M. Possibilities of failure analysis for steel cord conveyor belts using knowledge obtained from non-destructive testing of steel ropes. *Eng. Fail. Anal.* **2016**, *67*, 33–45. [CrossRef]

28. Shen, Y.; Giurgiutiu, V. Combined analytical FEM approach for efficient simulation of Lamb wave damage detection. *Ultrasonics* **2016**, *69*, 116–128. [CrossRef]

29. Grębowski, K.; Rucka, M.; Wilde, K. Non-destructive testing of a sport tribune under synchronized crowd-induced excitation using vibration analysis. *Materials* **2019**, *12*, 2148. [CrossRef]

30. Janowski, A. The method of measuring the membrane cover geometry using laser scanning and synchronous photogrammetry. In Proceedings of the 15th International Multidisciplinary Scientific GeoConference SGEM 2015, Albena, Bulgaria, 18–24 June 2015; pp. 1175–1185.

31. Huang, P.; Wu, M.; Shen, B.; Pang, Y.; Wu, F.; Lan, X.; Luo, H.; Zheng, W. Ultrastrong, flexible and lightweight anisotropic polypropylene foams with superior flame retardancy. *Compos. Part. A Appl. Sci. Manuf.* **2019**, *116*, 180–186. [CrossRef]

32. Gawronek, P.; Makuch, M.; Mitka, B.; Gargula, T. Measurements of the vertical displacements of a railway bridge using TLS technology in the context of the upgrade of the polish railway transport. *Sensors* **2019**, *19*, 4275. [CrossRef] [PubMed]

33. Artese, S.; Zinno, R. TLS for dynamic measurement of the elastic line of bridges. *Appl. Sci.* **2020**, *10*, 1182. [CrossRef]

34. Maleska, T.; Beben, D. Numerical analysis of a soil-steel bridge during backfilling using various shell models. *Eng. Struct.* **2019**, *196*, 109358. [CrossRef]

35. Beben, D. *Soil-Steel Bridges; Geotechnical, Geological and Earthquake Engineering;* Springer International Publishing: Cham, Switzerland, 2020; Volume 49, ISBN 978-3-030-34787-1.

36. UltraCor® Structures. Available online: https://viacon.pl/en/konstrukcje-ultracor (accessed on 13 August 2020).

37. Ordinance Number 35 of the General Director for National Roads and Motorways, Recommendations on the Final Acceptance and Inspection of Road Bridges Issued on 12th September 2008. Available online: https://www.gddkia.gov.pl/userfiles/articles/z/zarzadzenia-generalnego-dyrektor_6335//documents/35z.pdf (accessed on 13 August 2020).

38. Miśkiewicz, M.; Pyrzowski, L.; Wilde, K.; Chróścielewski, J. Numerical analysis and in situ tests of the Grot-Rowecki Bridge in Warsaw. In *Advances in Mechanics: Theoretical, Computational and Interdisciplinary Issues*; CRC Press: London, UK, 2016; pp. 405–408.

39. Olaszek, P.; Łagoda, M.; Casas, J. Diagnostic load testing and assessment of existing bridges: Examples of application. *Struct. Infrastruct. Eng.* **2013**, *10*, 834–842. [CrossRef]

40. Miśkiewicz, M.; Pyrzowski, L. Load tests of the movable footbridge over the port canal in Ustka. *Balt. Geod. Congress* **2017**, 242–246. [CrossRef]

41. Machelski, C. Soil-steel structure shell displacement functions based on tensometric measurements. *Stud. Geotech. Mech.* **2018**, *40*, 170–179. [CrossRef]

42. Mellat, P.; Andersson, A.; Pettersson, L.; Karoumi, R. Dynamic behaviour of a short span soil–steel composite bridge for high-speed railways—Field measurements and FE-analysis. *Eng. Struct.* **2014**, *69*, 49–61. [CrossRef]

43. Machelski, C.; Korusiewicz, L. Testing the load capacity of a soil-steel box structure. *Roads Bridg. Drog. Mosty* **2018**, *17*, 181–191.

44. Manko, Z.; Beben, D. Full-scale field tests of soil-steel bridge structure in two stages of its construction. *Arch. Civ. Mech. Eng.* **2006**, *6*, 57–76. [CrossRef]

45. Machelski, C. Effectiveness of surveying measurements during the construction of soil-steel structures. *Mosty* **2016**, *5*, 51–54.

46. Manko, Z.; Beben, D. Static load tests of a road bridge with a flexible structure made from super cor type steel corrugated plates. *J. Bridg. Eng.* **2005**, *10*, 604–621. [CrossRef]

47. Beben, D. Experimental study on the dynamic impacts of service train loads on a corrugated steel plate culvert. *J. Bridg. Eng.* **2013**, *18*, 339–346. [CrossRef]

48. Sobótka, M. Numerical simulation of hysteretic live load effect in a soil-steel bridge. *Stud. Geotech. Mech.* **2014**, *36*, 104–110. [CrossRef]

49. Bobkowka, K.; Nykiel, G.; Tysiąc, P. DMI measurements impact on a position estimation with lack of GNSS signals during Mobile Mapping. *J. Physics Conf. Ser.* **2017**, *870*, 12010. [CrossRef]

50. Burdziakowski, P.; Tysiąc, P. Combined close range photogrammetry and terrestrial laser scanning for ship hull modelling. *Geosciences* **2019**, *9*, 242. [CrossRef]

51. Ossowski, R.; Przyborski, M.; Tysiąc, P. Stability Assessment of coastal cliffs incorporating laser scanning technology and a numerical analysis. *Remote Sens.* **2019**, *11*, 1951. [CrossRef]

52. Ziółkowski, P.; Szulwic, J.; Miśkiewicz, M. Deformation analysis of a composite bridge during proof loading using point cloud processing. *Sensors* **2018**, *18*, 4332. [CrossRef]

53. Miśkiewicz, M.; Pyrzowski, L. Load test of new European record holder in span length among extradosed type bridges. *E3S Web Conf.* **2018**, *63*, 00006. [CrossRef]

54. Bureick, J.; Alkhatib, H.; Neumann, I. Robust spatial approximation of laser scanner point clouds by means of free-form curve approaches in deformation analysis. *J. Appl. Geod.* **2016**, *10*, 10. [CrossRef]

55. Hou, W.; Chan, T.; Ding, M. Denoising point cloud. *Inverse Probl. Sci. Eng.* **2012**, *20*, 287–298. [CrossRef]

56. Lague, D.; Brodu, N.; Leroux, J. Accurate 3D comparison of complex topography with terrestrial laser scanner: Application to the Rangitikei canyon (N-Z). *ISPRS J. Photogramm. Remote Sens.* **2013**, *82*, 10–26. [CrossRef]

57. Zienkiewicz, O.; Taylor, R.; Zhu, J. *The Finite Element Method: Its Basis and Fundamentals*; Elsevier: Oxford, UK, 2013.

58. Dassault Systèmes Simulia Corp. *Abaqus 6.14 Documentation*; Dassault Systèmes: Providence, RI, USA, 2014.

59. Chróścielewski, J.; Witkowski, W. Four-node semi-EAS element in six-field nonlinear theory of shells. *Int. J. Numer. Methods Eng.* **2006**, *68*, 1137–1179. [CrossRef]

60. Iwicki, P.; Tejchman, J.; Chróścielewski, J. Dynamic FE simulations of buckling process in thin-walled cylindrical metal silos. *Thin Walled Struct.* **2014**, *84*, 344–359. [CrossRef]

61. Chróścielewski, J.; Burzyński, S.; Sabik, A.; Sobczyk, B.; Witkowski, W. Nonlinear FEM analysis of irregular shells composed of fiber metal laminates. In *Shell Structures: Theory and Applications Volume 4*; Pietraszkiewicz, W., Ed.; CRC Press: London, UK, 2018; pp. 363–366, ISBN 978-1-138-05045-7.

62. Maleska, T.; Beben, D. Behaviour of corrugated steel plate bridge with high soil cover under seismic excitation. *MATEC Web Conf.* **2018**, *174*, 04003. [CrossRef]

63. National Standards Body in Poland. *PN-81/B-03020 Building Soils. Foundation Bases. Static Calculation and Design*; Polish Committee for Standardization, Measures and Quality Control: Warsaw, Poland, 1981.

64. Zani, G.; Martinelli, P.; Galli, A.; Di Prisco, M. Three-dimensional modelling of a multi-span masonry arch bridge: Influence of soil compressibility on the structural response under vertical static loads. *Eng. Struct.* **2020**, *221*, 110998. [CrossRef]

65. Loria, A.F.R.; Coulibaly, J.B. Thermally induced deformation of soils: A critical overview of phenomena, challenges and opportunities. *Geomech. Energy Environ.* **2020**, 100193. [CrossRef]

 *materials*

Article

Application of Elastic Waves and Neural Networks for the Prediction of Forces in Bolts of Flange Connections Subjected to Static Tension Tests

Piotr Nazarko *,† and Leonard Ziemiański †

Department of Structural Mechanics, Faculty of Civil and Environmental Engineering and Architecture, Rzeszow University of Technology, Powstancow Warszawy 12, 35-959 Rzeszow, Poland; ziele@prz.edu.pl
* Correspondence: pnazarko@prz.edu.pl; Tel.: +48-178-651-621
† These authors contributed equally to this work.

Received: 30 June 2020; Accepted: 11 August 2020; Published: 14 August 2020

Abstract: There is a group of measurement techniques that can be used in the task of force identification in steel bolts. In this paper, the potential of elastic wave propagation signals was studied for possible application in force monitoring systems. A series of laboratory tests was carried out on flange connections subjected to static tensile tests. Each one contained six screws of the same diameter. Four bolts were equipped with washer load cells. Alternatively, selected bolts were equipped with piezoelectric transducers (actuator and sensor) in order to measure the elastic wave signals. Principal components analysis, time of arrival, and neural network compression were used for dimensionality reduction of the measured signals. Examples of the obtained results with respect to the studied connections show that the tension forces in bolts can be estimated with relatively good accuracy.

Keywords: elastic waves; signal processing; neural networks; force prediction, flange connection; static test

---

1. Introduction

Measurements of the physical quantities that describe structural elements' state are significant in many engineering and industrial applications, including civil engineering. They are often conducted pending trial loads of new structures and prototype solutions in laboratory tests. This allows a better understanding of the behavior of the entire structure and its individual elements. Non-destructive tests (NDT) and structural health monitoring systems (SHM) are very useful in this task. Their constant use also increases the safety and reliability of the structure. There are also several tasks that can be carried out with their help [1–6]:

- Anomaly or damage detection;
- Damage type classification;
- Load or internal force prediction;
- Material parameter identification.

There is a group of structure connections where the pretension force level has an influence on the strength of the slip resistance connection. Changing pretension forces, i.e., due to the structure's usage over time, may become a highly relevant matter for the structural integrity. It is especially important in cyclically loaded constructions like telecommunication towers, bridges, and wind turbines.

Besides the necessity to monitor the state of strategic structures, the precision of the tightening method implemented to accomplish an accurate preload level is equally important. One of the hazards related to the occurrence of failure or structural damage may be the insufficient stress ratio of screw connections. Among the methods used to control their tightening level, two main features can be distinguished: accuracy and practical applicability. Determining the prestressing force in everyday applications is most often based on the tightening torque measured with a dynamometer wrench. The accuracy of this approach is defined as ±25%. The best accuracy here is accomplished by using ultrasonic sensing (±1%) or strain gauges. There are already a few commercial solutions that can be used to estimate the axial forces in bolts. One of them concerns ultrasonic load-monitoring devices, which most often require access to bolts in a joint from both sides. The second solution is related to piezoelectric load cells or washer-type strain. The measurement uncertainty in this case amounts to ±10%. Fric et al. compared some of these methods [7]. However, their costs cannot be ignored, and therefore, in long-term research, strain gauges are used, either fitted inside the bolt shank or glued onto the bolt. For this reason, the article presents the results of laboratory research aimed at the use of commonly available and relatively inexpensive piezoelectric transducers (PZT) that can be used to excite and receive elastic waves.

In this work, the elastic wave propagation phenomenon was implemented due to the possibility of adjusting their parameters to the applied task, their sensitivity, and their ease of implementation in SHM systems. To measure and introduce the time signals of elastic waves, piezoelectric transducers were used. Nazarko and Ziemiański [8] used this idea also in the field of non-destructive damage detection in a variety of elements and materials. Chaki and Bsource [9] examined guided ultrasonic waves to monitor and observe the stress levels in steel strands. In these approaches, including ultrasonic testing, the most frequently used parameters are changes in longitudinal and transverse wave propagation velocities or time of flight/arrival (ToF/ToA). In the discussed problem of identifying axial forces in bolts of flange connections, this turned out to be insufficient. Therefore, in order to obtain reliable signal parameters, their deep compression was applied. Additionally, as the inference tool, artificial neural networks (ANNs) were used [10].

The results presented in the article concern a series of laboratory tests carried out on flange connections under tension in a static test machine. It appeared that force changes have an impact on the signals measured by the sensors [11]. This was reflected also in the calculated principal components, which are often used to compress signals (see Nazarko [12] or Chen et al. [13], where other methods concerning signal compression were compared). This article presents new results where ANNs were used both for signal compression and as an inference tool. The value of these results is increased by the fact that they refer to experimental studies, where there are a number of factors (measurement noise, apparatus limitations) that are quite difficult to examine in numerical simulations and analytical calculations. In this case, the research concerned the possibility of predicting axial forces in tension bolts of flange connections. The elastic waves were regularly excited and recorded (by PZT sensors) during the connection tensile test. After extending the set of patterns (compressed signal parameters) with information about the elongation of one of the bolts, it became possible to estimate (using ANNs) the magnitude of axial forces not only within a single connection, but also to generalize the results within several analyzed connections. Moreover, these preliminary results show that ANNs are able to predict the axial forces in bolts with reasonably good accuracy. This also indicates the significant potential of the approach being developed for real-life NDT inspections.

2. Materials and Methods

2.1. The Idea of Force Monitoring in Bolts

The idea of bolt force monitoring discussed in this paper is based on measurements of elastic waves. It was confirmed by Kim and Hong [14] and Ding et al. [15] that even a relatively small change in a bolt force will affect the signals measured (its time of flight, amplitude, frequency, etc.). Changes in the tension force have an influence on the shortening/elongation of the screw. This results in variations occurring in the distance traveled by waves and the time related to it. Thus, the propagation velocity and ToF are the parameters frequently used to determine and detect the stresses or forces of the monitored elements [16].

In this paper, two single PZT were implemented to excite and to measure the response signals in the bolts of the studied flange connections (Figure 1). These transducers were placed in pitch–catch configurations. At the beginning, a single screw was equipped with additional transducers (one on the end of the shank and two on the screw's head) in order to get a pulse–echo configuration. Nevertheless, due to the fact that both of them provided similar accuracy considering force identification, it was decided to use pitch–catch in subsequent studies because this enables one to obtain a higher signal-to-noise ratio (SNR).

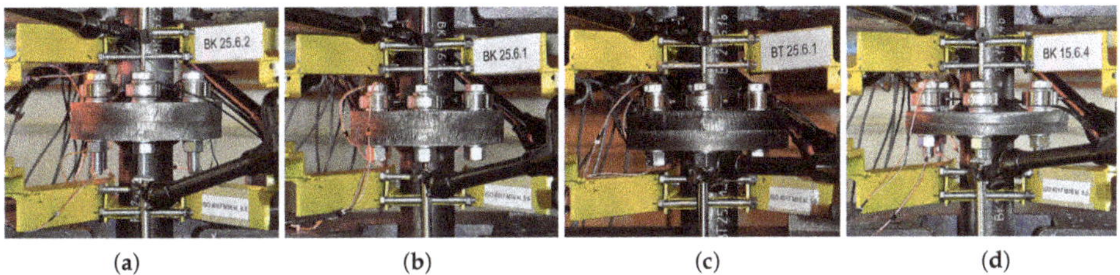

Figure 1. Example of connections subjected to static tests and the elastic wave propagation measurements in bolts: (**a**) P1, (**b**) P2, (**c**) P3, (**d**) P4.

Taking into account the simplicity of conducting on-site inspections, the most convenient placement of measurement units would be probably an impulse–echo configuration, but this was not considered in this study. In the case of SHM systems, whether or not the structure is equipped with instrumentation during its construction can be seen as irrelevant (actuators and sensors can even be mounted inside the bolts).

A typical washer force sensor consists of three or four standard piezoelectric measuring elements or strain gauges inside the unit. This is to average the results over the entire sensor surface. This approach should be considered in future work related to the approach under consideration because of the possibility of the simultaneous tensioning and bending of bolts (Figure 2).

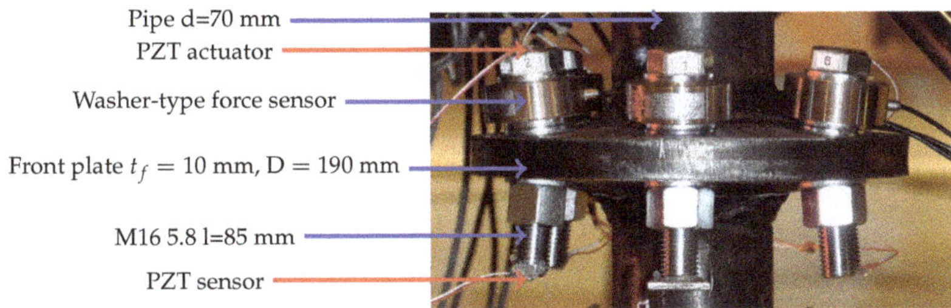

Figure 2. An example of visible bolts deformation at the end of the static test (P6 connection).

Parameters describing the force changes (e.g., physically measured or determined based on the measured signals) can be used for training an efficient diagnosis system that is based on ANNs. The advantage of this over existing solutions is that the application of the proposed approach may allow one to estimate not only changes in forces over time, but also detect yielding and fatigue damages and monitor the structural integrity.

2.2. Laboratory Setup

The laboratory setup consisted of a signal generator (40 MHz DDS Function/Arbitrary Generator, TG4001, Thurlby Thandar Instruments Ltd., United Kingdom), in which an excitation was defined in the form of a 2.5 sine wave modulated by a Hanning window. The operational frequency was set to 52 kHz. Then, the signal was amplified and split into actuators and synchronization channels. Two digital oscilloscopes (WaveSurfer 424 and WaveRunner 104MXi, LeCroy, NY, USA) were used to store the signals received from all the sensors. Piezoelectric transducers (CMAP6, plate $5 \times 5 \times 2$ mm, Noliac) were mounted on the bolt heads (excitation) and at the ends of their shanks (response). Sensor wax was used to attach them (this enabled trouble-free recovery of all the sensors), while their cables were fixed to a single spot with a weak adhesive to hold them during the test, allowing for non-invasive removal.

As usual, the measuring system used (piezoelectric transducer, amplifier, signal generator) has its limitations. This applies to both the excitation frequency and the signal amplification. This is one of the most important factors that affects the obtained accuracy of identification. In this case, the frequency and gain were selected in such a way that the response signal had the highest SNR at the highest possible frequency.

A set of static tests was carried out using the testing machine Instron J1D 1200 kN (Figure 3). To evenly load the oscilloscopes that store the measurement data (saving data at the same time), it was assumed that each one would register the pattern of the excitation signal (in order to synchronize the recorded signals in time) and the response from the two screws.

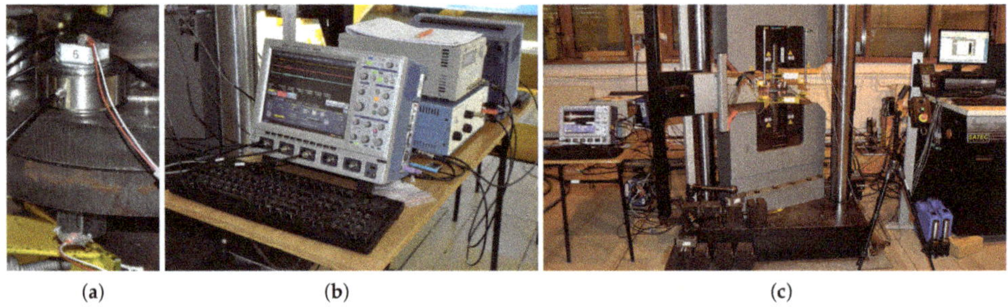

(a)   (b)   (c)

Figure 3. Laboratory setup: (**a**) Piezoelectric transducers mounted at the bolt ends (a washer-type force sensor is also visible in this case). (**b**) Digital oscilloscopes, function generator, and amplifier. (**c**) Static test machine Instron J1D 1200 kN.

2.3. Flange Connections under Static Tests

The investigated set of flange connections consisted of six M16 bolts (Figure 4). In each connection, only four of them were equipped with washer strain load cells (Bolt Nos. 1, 2, 4, and 5; see Figure 4). The other two had spacers that imitated the force sensors in order to ensure the similar nature of bolt work. The dimensions of the considered connections and the types of screws are listed in the Table 1.

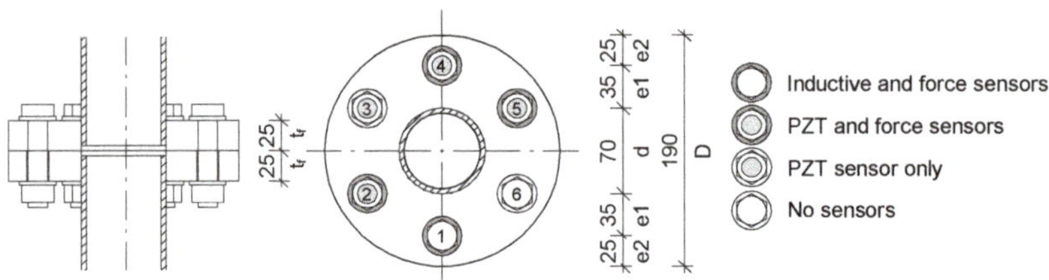

Figure 4. A scheme of the flange connection (P1/P2), screw numbers, and sensor locations (all dimensions are given in mm; washer-type sensors are not included).

Unfortunately, as a result of technical complications (a lack of free hard disk space during the measurements) and the resulting differences in the number of recorded signals between the two oscilloscopes, measurement data from the P5* connection were not used in this article.

Table 1. Dimensions of the investigated flange connections.

Connection	Symbol	Plate Geometry ($t_f/e_1/e_2/D$) (mm)	Screws	
			Class	Length (mm)
P1	BK25.6.2	25/35/25/190	6M16 8.8	125
P2	BK 25.6.1	25/35/25/190	6M16 5.8	85
P3	BT25.6.1	25/35/25/190	6M16 5.8	85
P4	BK 15.6.4	15/45/25/210	6M16 8.8	85
P5 *	BK 15.6.3	15/45/25/210	6M16 5.8	85
P6	BT 10.6.1	10/35/25/190	6M16 5.8	85

* data related to this connection were not used – details are given in the text.

Two inductive sensors were mounted on screw S1 (from above and below; see Figure 1) to measure its displacement. Thanks to this, it was possible to determine the elongation of this screw during the tensile test. The other two inductive sensors were attached on the right and left side of the pipe rod. They were used to monitor possible rotational deformation, which was considered insignificantly small. On this basis, it was assumed that the investigated flange connections generally deform symmetrically, and the elongation of a single bolt can be used as an additional parameter describing their current condition. The conducted tests did not take into account the occurrence of a complex stress state, in which this assumption could be false.

The forces obtained during the static tests and the elongation in one representative screw were used to define an output vector for ANN training. The obtained relationship between these values in the case of single connections is shown in Figure 5. It can be seen there that the forces measured in the bolts were not exactly the same. This may be related to the geometric imperfections of the connection (non-symmetric holes) or measurement uncertainty (this type of applied force sensor is sensitive to the washer hardness or a screw touching the edge of the hole).

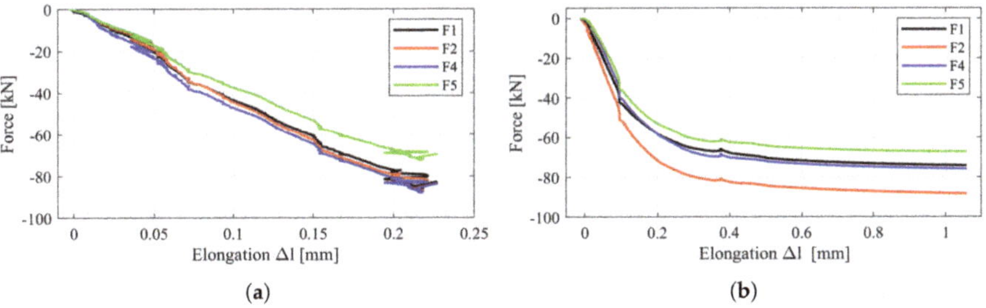

Figure 5. Elongation and forces measured in the bolts: (**a**) P1 connection; (**b**) P2 connection (see Table 1).

In addition, four bolts were equipped with piezoelectric transducers, but their numbers were different with respect to the force sensor locations, i.e., 2, 3, 4, and 5 (Figure 4). It was not possible to mount them at the S1 bolt because the inductive sensors were placed there. As a consequence, the signals measured in the S3 bolts (in all connections), which were not equipped with the washer load cell, were used to define the database with unknown force magnitudes. After the process of ANN training, these data were used to predict the forces in those bolts.

It is worth mentioning here that the completed test plan did not assume cyclical loading/unloading of the connections. In this particular case, each individual test was intended to cause damage in one of the following forms: bolts yielding or breaking, plastic deformation of pipes or head plates. The failure model depended on the connection dimensions and the class of bolts used. For example, in the P1 connection, there was plastic deformation of the pipe observed (Figure 5a), while in the case of the P2 connection (see Figure 5b), the deformation of the pipes was accompanied by yielding of the bolts. This explains the differences that can be seen in these charts. Despite this fact, an attempt was made to estimate the bolt forces regardless of the connection operation phase.

In the experimental studies carried out, it was not possible to obtain measurement data from the unloading phase of a damaged joint in the case of all considered models. When the load capacity of the connection was being approached, some of the sensors were disassembled due to the risk of their damage. Therefore, it was decided not to include these data in the ANN training process. However, the issue of cyclic loading is very important, and further research in this area is needed. Examples of experimental studies carried out so far and the preliminary results of the axial forces' identification can be found in the literature [11,17].

2.4. Signal Analysis

The elastic wave signal comparison in the initial stage showed that they were not exactly the same (Figure 6). There may be several reasons for this: the piezoelectric transducers may not have been positioned precisely in the same place (the central axis of the bolt); the thickness of the wax layer was not the same; the signals were affected by the closest components of the connection (washers, brackets on the pipe, weld thickness, etc.). Despite these differences, an attempt was made to train the ANNs for the purpose of force prediction. It was even decided to perform signal normalization within the range ±0.9 (with respect to the initial stage), but this did not significantly affect the results of the predicted force values.

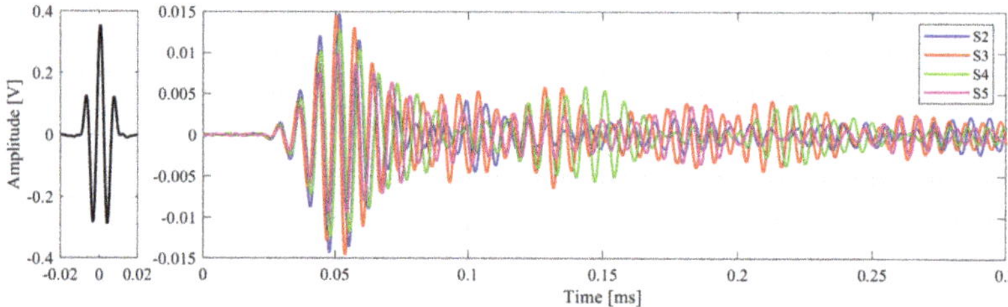

Figure 6. Excitation signal and the received responses measured at the S2 to S5 bolts in the P1 connection.

As a result of the length of each signal, it was necessary to determine the parameters sensitive to load changes in the connections. Thus, the measured signals were transformed into the domain of the principal components [8,12]. This allows one to compute the linear transformation:

$$\mathbf{y} = \mathbf{W} \cdot \mathbf{s} \tag{1}$$

which maps data from a high-dimensional space $\mathbf{s} \subset R_N$ to a lower dimensional space $\mathbf{y} \subset R_K$ of the principal components, without much loss of information. In this way, the elastic wave signals, each containing 10,002 points, after decimation at the rate $r = 2$, were reduced to only 12 principal components.

On the basis of the preliminary test results [11], it turns out, however, that the principal components alone may not be sufficient for proper training of the diagnosis system. The point is for the ANNs to acquire generalization capabilities that enable the prediction of force values in bolts in which no control measurements were carried out. In other words, the obtained values of the estimated forces in the screw, which did not participate in the training of ANNs, differed from the nature of the work of the other screws, for which these forces were physically measured using washer sensors. Therefore, an attempt was made to determine other parameters from the signals that could improve the generalization abilities of ANNs. The pattern database was extended to include data related to the time and amplitude of the occurrence of subsequent maxima in the response signals (Figure 7a). These arrival times were expressed as the distance between the maximum excitation amplitude and the local extremes of the response amplitudes. Their values were determined for all signals and analyzed connections. To illustrate the nature of the changes in arrival times, the value t_0 obtained without a load was subtracted from each of them ($\Delta t = t_i - t_0$, where $i = \{1, 2, \ldots, 7\}$). An example of the obtained time variations in the case of the P1 connection is shown in Figure 7b. Examples in the literature indicate that the relationship between ToF changes and load is most often almost linear [18,19]. In this particular case, with the type of sensors used and the established excitation signal parameters, these relationships are non-linear.

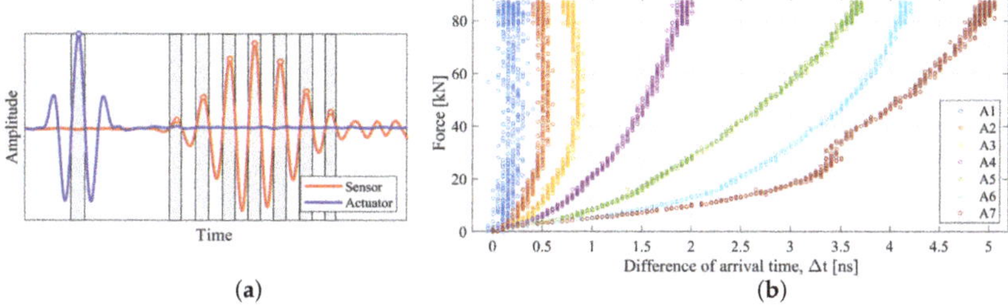

Figure 7. Arrival time and amplitudes: (**a**) an example of how they were determined; (**b**) differences of arrival time in the P1 connection.

Moreover, although this type of graph does not indicate this, along with the increase in the axial force, the determined arrival times of the individual peaks in the signal were generally constantly reduced. It was expected that the tensioning of the connection would cause screw elongation and a longer wave path would increase the related propagation time. Although there are examples in the literature where the impact of stress on the propagation speed was analyzed [16], a wider discussion of this issue, which is very interesting, is not the main subject of this study.

2.5. Artificial Neural Networks

Tasks related to the prediction or identification of parameters are related to the category of so-called inverse problems [12]. Most often, they consist of determining unknown parameters on the basis of values measured experimentally or obtained from numerical simulations. In a situation where knowledge about the observed phenomenon or the degree of its complexity does not allow building an expert system, the most frequently used tools are ANNs.

ANNs are widely used in many areas and tasks. The assumption is that ANNs are able to learn an unknown relationship between input and output data. This typically requires large amounts of data acquired in computational or experimental investigations. In recent years, much attention has been paid to so-called deep learning (DL) [20], and one of the major benefits of using deep models is successful NN learning from fewer training data. In addition, deep learning approaches have proven to be suitable for big data analysis, and hierarchical learning systems show superior performance in several engineering applications [21]. In the approach described in this article, two applications are demonstrated: DL multi-layer perceptron (MLP) was used for signal compression, and DL regression ANNs were trained to predict the axial forces in the screws of the investigated connections.

The learning process consisted of minimizing the computed error value between the target and the network outputs obtained for successive iterations. Testing and validation were carried out based on data that the network had never seen before. The ability to produce such a prediction for the training set is called network generalization [10]. In this article, the mean squared error (MSE) and standard deviation were used as measures of the error obtained at the network output.

The force identification provides information about the predicted value of that force with respect to parameters that are sensitive enough to its changes. The correct selection of these parameters is the most important issue in any identification task. Then, the accuracy of the neural predictor may be obtained by tuning the architecture or different training strategies. For the aforementioned task, feed forward ANNs are commonly used [10]. They consist of an input (first) layer, (usually) a few hidden layers, and an output layer. The number of elements in the input and output layers is determined by the size of the training datasets.

As an alternative to the designated wave parameters (PCA, time and amplitude (TA)), an MLP was trained to reproduce the inputs in the output layer. This kind of network is called an autoencoder. The network is trained so that the signal feed at the input is reproduced at the output, so the input layer and the output layer have the same dimensions. In the middle of the network, there is a hidden layer—called the bottleneck layer—that has fewer neurons than the input layer. This auto-associative neural network is used to compress signal data. After the learning process, the output and hidden layers after the bottleneck are neglected, and only the input, hidden, and bottleneck layers are used. This part of the network is called the encoder, which is used to reduce the dimension of the measured signals. It can be assumed that the parameters obtained from the encoder represent all the information contained in the signals. Moreover, this approach allowed us to perform compression on the input data and to reduce the data dimensionality. In other words, the encoder extracts the most important components of the signal and ignores the less important parts. The encoder, with only one hidden layer (the bottleneck), can be compared by computing the PCA of the input data, while the additional hidden layers introduce some non-linear transformation of the signal [22], and the encoder performs the PCA on a non-linear version of the signals. In this work, various encoder architectures were trained: $inl - 850 - 250 - bl$, where $inl = \{8192, 4096, 3072\}$ is related to the number of data taken from the measured signal (starting from its beginning) and $bl = \{24, 12, 6, 4\}$. It was found that the most suitable architecture for the DNN for the compression task was $4096 - 850 - 250 - 12 - 250 - 850 - 4096$, and the DNN for the regression task was $13 - 50 - 50 - 50 - 1$.

3. Results and Discussion

3.1. Force Prediction in Single Connections

The first task was to identify the forces in the bolts in a single connection. Each contained a screw, which was not equipped with a force sensor, but the elastic wave signals were recorded there during a static tensile test. With reference to this particular screw, the axial forces were predicted.

In previous works [11,17], it was assumed that data from all bolts (2, 4, 5) were to be used for the purpose of ANN training, assuming a constant distribution of patterns for testing and validation. The new approach proposed herein involves the learning, testing, and validation patterns being separated into individual bolts from a given connection. Among the possible combinations, the following division of patterns was adopted (short names are provided in brackets; they are also used in the data descriptions in the charts and tables):

- S4 for learning (learn);
- S2 for testing (test);
- S5 for validation (valid);
- S3 for prediction (predict).

This is to achieve the repeatability of force identification for various bolts in the connection and then possibly extend the database to include bolts from the other connections.

After the first series of simulations, it turned out that none of the signal parameter sets (PCA, TA, encoder) allowed us to obtain a satisfactory level of identification accuracy for the axial forces in the screws. This was due to the fact that there were relatively significant differences between the signals measured in the individual bolts. Of course, this can be influenced not only by the excitation pattern and its parameters (especially the frequency), but also by the likely differences in the attachment of the screws (some of them may touch the edge of the hole, while others may not). However, these are limitations arising both in the measurement capabilities and factors that occur in real constructions. For this reason, the obtained pattern database derived from laboratory experiments of connection models with real dimensions is a good test of the proper operation of the developed diagnostic system.

At the current stage of research, it was decided to extend the database by the elongation measured on one of the screws. This measurement was made using inductive sensors applied to both ends of the S1 screw. In this way, the input data gained an additional physical sense, which seems to be the key issue in this case.

The input vectors consisted of the elongation of S1 and:

- Twelve principal components (PCA);
- Six amplitudes of the response signals and theirs six arrival times (TA);
- Twelve parameters obtained from the encoder.

The obtained values of the identification errors for two exemplary connections (P1) with respect to the input data used are shown in Table 2. It can be seen that the smallest learning errors were obtained for the PCA, but this also led to the largest validation errors. Time and amplitude (TA) had the lowest validation error, but reasonably good accuracy was obtained in the case of the encoder. A graphic comparison of the results for the encoder and PCA is shown in Figure 8. Therein, the F2 and F5 forces measured in the S2 and S5 bolts, as well as the respective force values obtained from the ANN (test and valid) are shown. It can be seen there that they are in good agreement, especially the input vector taken from the encoder. In addition, the predicted values of the axial forces in bolt S3 (predict) show the nature of the changes to be very similar to the other screws.

Table 2. Comparison of the learning, testing, and validation errors: P1.

Method	Learn		Test		Valid	
	MSE $\cdot 10^5$	$\sigma \cdot 10^3$	**MSE $\cdot 10^5$**	$\sigma \cdot 10^3$	**MSE $\cdot 10^5$**	$\sigma \cdot 10^3$
PCA	**14.6**	**12.1**	41.1	20.3	134.9	36.7
TA	139.4	37.3	164.3	35.2	**13.2**	**11.5**
Encoder	27.5	16.6	**32.7**	**18.1**	16.6	12.9

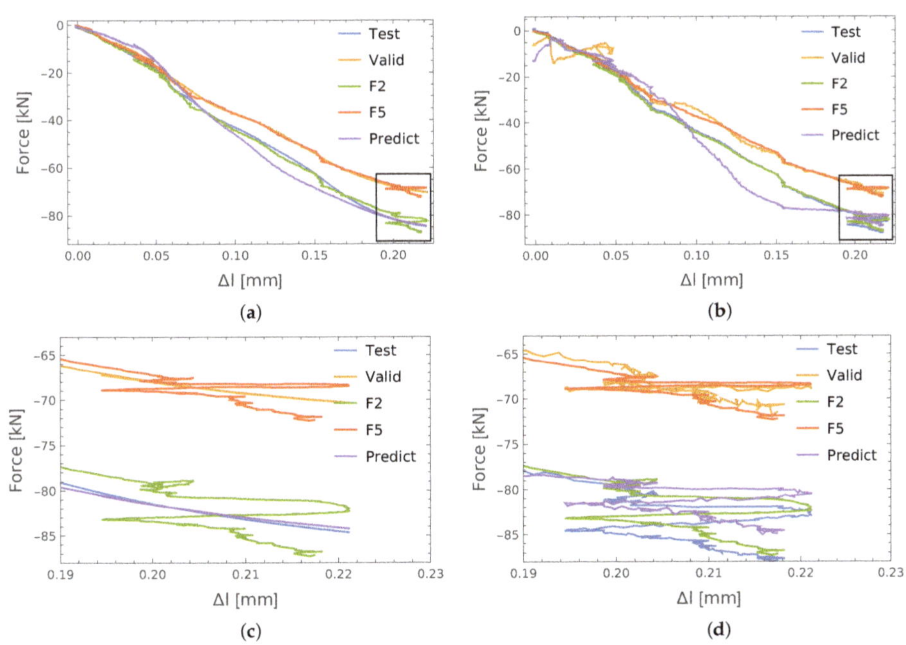

Figure 8. The results of identification forces: P1. (**a**) Encoder; (**b**) PCA; (**c**) details (**a**); (**d**) details of (**b**).

If we look at the fragment of results that are enlarged (Figure 8c,d), we see that training the ANN using the input data taken from the encoder led to the averaging of the identified quantities (Figure 8c), which is very interesting. However, in the case of the principal components (Figure 8d), the results obtained followed the measured values, which in the case of the measurement errors may lead to them becoming stronger.

The same approach was also repeated for the other connections. For each case, it was possible to identify the axial forces in the bolts at a similar level of accuracy. As an example, the list of errors obtained for the P3 connection can be analyzed (Table 3). This time, the best validation results were obtained for the encoder data. Although they were very similar to the TA results, the errors for learning and testing were at a lower level. Therefore, in other cases presented in the paper, encoder data were used as the input to the ANN.

Table 3. Comparison of the learning, testing, and validation errors: P3.

Method	Learn		Test		Valid	
	MSE $\cdot 10^5$	$\sigma \cdot 10^3$	**MSE $\cdot 10^5$**	$\sigma \cdot 10^3$	**MSE $\cdot 10^5$**	$\sigma \cdot 10^3$
PCA	**2.75**	**5.24**	4.24	7.23	36.2	19.1
TA	34.1	18.5	39.5	19.8	5.98	7.73
Encoder	3.83	6.20	**4.03**	**3.55**	**3.78**	**6.11**

3.2. Force Prediction in Sets of Connections

In the next task, an attempt was made to combine the data from two and more connections. The idea was to check whether it was possible to train the diagnostic system on one connection and use it to assess the forces in the bolts of the other connections. Therefore, connections with the same bolt lengths were grouped, and the results of the simulations carried out are presented in the following sections.

3.2.1. P2P3

The first set of connections consisted of two connections (P2P3). They corresponded in terms of the dimensions of the end plates and the class of screws. In this case, two sets of input vectors were also analyzed:

1. The ANN used data from the P2 = {S2, S4, S5} connection to learn, while data from P3 = {S4, S5} were used for testing and P3 = {S2} for validation;
2. The ANN used data from the P3 = {S2, S4, S5} connection to learn, while data from P2 = {S4, S5} were used for testing and P2 = {S2} for validation.

In both cases, the prediction was made with respect to the S3 screws from both connections (P2, P3). The obtained results of testing, validation, and prediction are shown in Figure 9. There was a good agreement between the measured axial forces (F2, F5) and the values obtained from the ANNs.

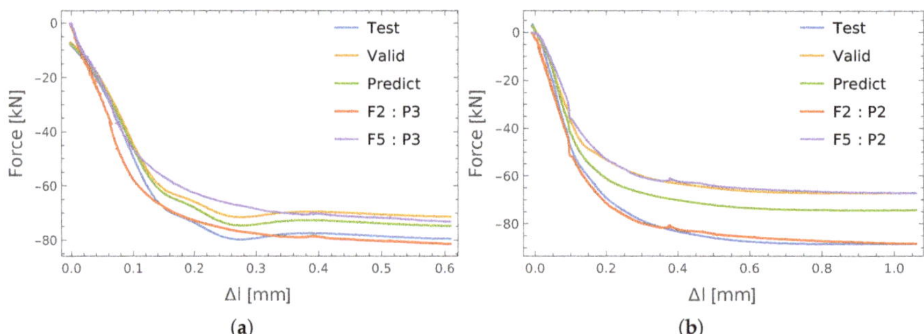

Figure 9. The results of force identification: (**a**) P2 learning, P3 testing; (**b**) P3 learning, P2 testing.

It can be seen in Table 4 that lower statistical parameters were observed when data from P3 were used for learning and P2 for training and validation.

Table 4. Comparison of the learning, testing, and validation errors: P2P3.

Method	Learn		Test		Valid	
	MSE $\cdot 10^5$	$\sigma \cdot 10^3$	MSE $\cdot 10^5$	$\sigma \cdot 10^3$	MSE $\cdot 10^5$	$\sigma \cdot 10^3$
P2-learn, P3-test and valid	41.6	24.2	287.4	53.6	290.1	53.8
P3-learn, P2-test and valid	39.8	19.9	54.3	23.3	30.0	17.3

3.2.2. Set of P2P3P4 Connections

The promising results obtained for the first set prompted us to extend it with data related to the P4 connection, which differed not only in the dimensions of the front plate, but also in the class of screws. Training patterns in this case were separated as follows: learning was performed on patterns from P2 and P3 connections (i.e., P2 = {S2, S4, S5}, P3 = {S4, S5}), while testing and validation involved a set of patterns related to P3 and P4 connections (i.e., [P3 = {S2}, P4 = {S2}], and P4 = {S4, S5}, respectively). An example of the results obtained from testing and prediction is shown in Figure 10. As before, there is good agreement between the measured axial forces F2 and the values estimated by the ANN (Figure 10a). The predicted values of the forces in the S3 screws (which were not equipped with force sensors) for all three connections are shown in Figure 10b. The nature of their changes is similar to the other bolts in the analyzed connections.

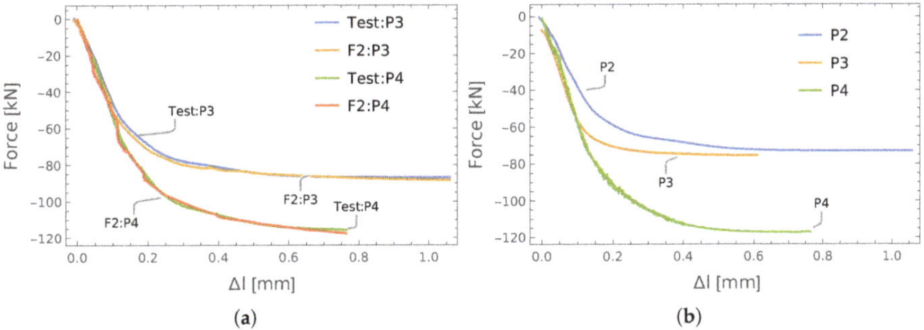

Figure 10. The results of force identification: (**a**) testing; (**b**) prediction.

3.2.3. Set of P2P3P4P6 Connections

The dataset used to train the ANN was expanded to include results from the next connection (P6). The patterns in this process were divided according to the following scheme: the learning was carried out on data from bolts P2 = {S2, S4, S5}, P3 = {S4, S5}, P4 = {S4, S5}; the testing involved data related to [P3 = {S2}, P4 = {S2}, P6 = {S2}]; while the validation was performed using data from P6 = {S4, S5}. The obtained results from the testing and prediction of the axial forces in the S2 bolts in the analyzed connections are shown in Figure 11.

The characteristic pattern in this case was the noticeable smoothing of the test results. The determined elongation values indicated that in a certain phase of the tensile test, the length of the S1 screw was temporarily shortened. The real reason, however, may be the fact that the end plates of the P6 connection were deformed and the screws were bent (see Figure 2). Thus, the changes registered by the sensors were the result of the bolt end rotating and not its actual shortening.

Summarizing the results obtained so far, it can be stated that adding information related to the elongation of one of the bolts allows for good estimation of the identified values of the axial forces in the bolts of flange connections. However, taking into account that the application of this approach in practice is very difficult or even impossible, an attempt was also made to replace the bolt elongation with its increments. This idea and preliminary results are briefly described in the next section.

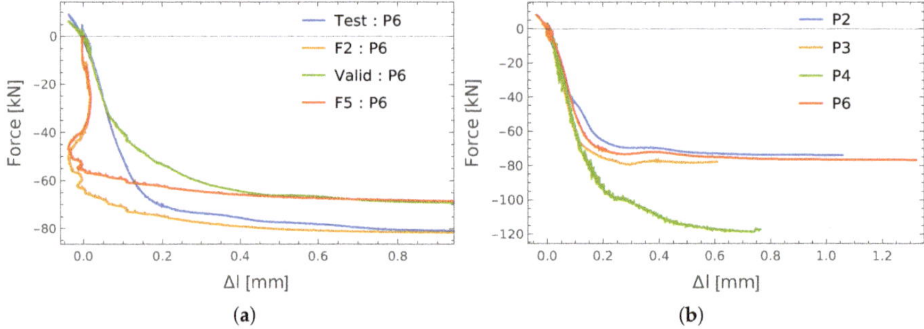

Figure 11. The results of force identification: (**a**) testing; (**b**) prediction.

3.3. Force Prediction Using Load/Elongation Increments

The concept of identifying forces in bolts using load or elongation increments consists of the fact that the load change in the tested connection is introduced. This is accompanied by the simultaneous measurement of elastic wave propagation and the elongation in one of the screws. Then, the obtained elongation increment is added to the other parameters provided in the input vector to the ANN.

To show the potential of the proposed approach, preliminary simulations were performed on the data related to the P3 connection. For this purpose, the database was divided into relatively small increments. It turned out that better identification results were obtained when these increments were not uniform, but variable with a random Gaussian distribution. Thus, the database used for ANN training consisted of elongation increments of the S1 bolt and the parameters of signals (taken from the encoder) measured at the following bolts: S4 for learning; S2 for testing; S5 for validation; and S3 for prediction.

The results of the preliminary tests are presented in Figure 12a together with the results corresponding to the ANN trained directly using the elongation of the S1 screw (Figure 12b). It can be seen that there were some differences in the accuracy of testing, validation, and prediction of the axial forces. The error values calculated on this basis are collected in Table 5. Despite the clearly superior results in the case of training the ANN with the direct elongation of the S1 screw, the results of both approaches remained in good agreement with the results of the experimental measurements of the F2 and F5 axial forces in the S2 and S5 screws. Therefore, in future studies, this approach will also be extended to other connections.

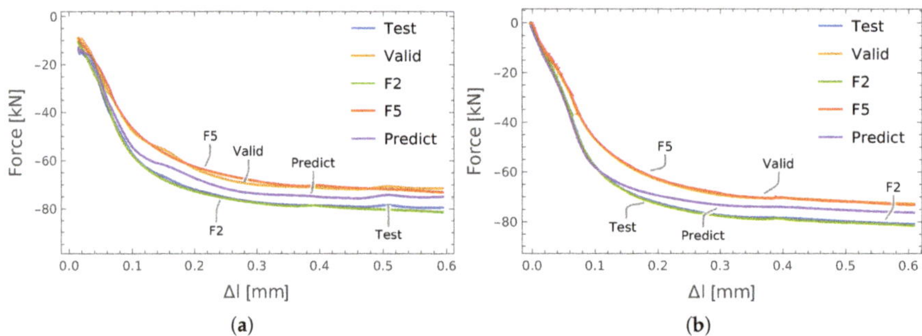

Figure 12. The results of force identification in the P3 connection based on: (**a**) load/elongation increments; (**b**) direct elongation of the bolt S1.

Table 5. Comparison of the learning, testing, and validation errors: P3 and P3 increments.

Method	Learn		Test		Valid	
	MSE $\cdot 10^5$	$\sigma \cdot 10^3$	MSE $\cdot 10^5$	$\sigma \cdot 10^3$	MSE $\cdot 10^5$	$\sigma \cdot 10^3$
P3	3.83	6.20	4.03	3.55	3.78	6.11
P3 increments	18.8	13.7	26.8	16.4	29.7	17.2

As a summary of the approaches analyzed so far, the results of axial force prediction in the S3 bolt of the P3 connection (which was not equipped with a force sensor) are summarized in Figure 13a. There, we can see the results obtained for the ANNs trained on the basis of data from a single connection (P3 with elongation, P3in with elongation increments), two connections (P2P3), and three connections (P2P3P4). The results related to P3 and P2P3P4 are similar to each other (which is clearly seen in Figure 13b), while P3in and P2P3 slightly differ from them.

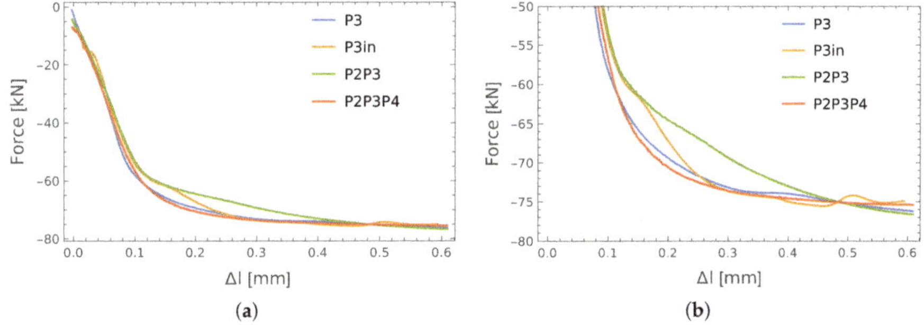

Figure 13. The result of the prediction forces with respect to the P3 connection: (**a**) comparison of the studied input vector scenarios; (**b**) details of (**a**).

4. Conclusions

One of the great advantages of the research presented herein is that it is based on experimental measurements carried out on a set of laboratory models of real flange connections. The main assumption of the proposed approach was that elastic waves carry information about the axial force in the bolt in which they propagate. Due to the significant differences in the measured signals, their compression using the MLP encoder was of great importance. By adding the variable of the elongation of one screw to the input vector, it became possible not only to achieve satisfactory identification results, but to achieve generalization capabilities. Moreover, this was not only possible with respect to one single connection, but after merging the signal parameters, it was possible for several investigated connections. Therefore, it can be concluded that the use of the MLP encoder (signal compression) and the addition of a bolt elongation (a physical variable) made it possible to find the relationship (ANN trained) between the propagating elastic wave and the axial force in the bolt subjected to tension.

On the basis of the trained ANN, axial force prediction was carried out for bolts that were not equipped with force sensors. The values obtained were consistent with those of other bolts, which was not possible in previous studies [11,17]. In this work, the dimensionality reduction of measured wave signals (neural network encoder) gave meaningful improvement to the results of the identification of the force in the bolts. Moreover, this new approach enabled: (1) the prediction of forces in bolts where they were not directly measured; (2) simultaneous analysis of data related to several connections. Obtaining the expected values of the prediction of axial forces was completely impossible in the previously considered stage of the research due to a serious problem with achieving the generalizing ability of ANNs [11,17]. The approach presented in this paper has eliminated this problem.

The article also presents the preliminary results of the identification of the axial forces, in which instead of the elongation of a single screw, elongation increments were used. Thanks to this, the proposed approach also has the potential for practical applications in real non-destructive test (NDT) measurements. Therefore, in future studies, this idea will also be extended to other connections. It will also be a good opportunity to include, in addition to processing elastic wave signals, the filtering of data related to the measurement of axial forces in bolts, which also influence the prediction accuracy.

Furthermore, it is worth mentioning that during the static tensile test, the force applied by the testing machine was also recorded. Therefore, it is also possible to use this in the pattern database information regarding load increments, which, in practice, may be more useful than measurements of elongation changes.

Author Contributions: Conceptualization, P.N. and L.Z.; methodology, P.N. and L.Z.; software, L.Z.; validation, P.N. and L.Z.; investigation, P.N.; resources, P.N.; writing, original draft preparation, P.N.; writing, review and editing, L.Z.; visualization, P.N. and L.Z.; supervision, L.Z.; project administration, P.N.; funding acquisition, L.Z. All authors read and agreed to the published version of the manuscript.

Funding: This research was supported by the Polish Ministry of Science and Higher Education grant to maintain research potential.

Conflicts of Interest: The authors declare no conflict of interest.

Abbreviations

The following abbreviations are used in this manuscript:

ANN	Artificial neural networks
Pj	Connection number, where $j = \{1, 2, \ldots, 6\}$
DL	Deep learning
Fi	Force magnitude in the Si screw, where $i = \{1, 2, \ldots, 6\}$
DNN	Deep neural network
MLP	Multi-layer perceptron
MSE	Mean squared error
NDT	Non-destructive test
PCA	Principal components analysis
PZT	Piezoelectric transducer
Si	Screw number, where $i = \{1, 2, \ldots, 6\}$
SHM	Structural health monitoring
SNR	Signal-to-noise ratio
TA	Arrival time and amplitude
ToF	Time of flight
bl	A bottleneck layer of the neural network
inl	An input layer of the neural network
σ	Standard deviation

References

1. Rucka, M. Monitoring steel bolted joints during a monotonic tensile test using linear and nonlinear Lamb wave methods. A feasibility study. *Metals* **2018**, *8*, 683. [CrossRef]
2. Sale, M.; Rizzo, P.; Marzani, A. Semi-analytical formulation for the guided waves-based reconstruction of elastic moduli. *Mech. Syst. Signal Proc.* **2008**, *25*, 2241–2256. [CrossRef]
3. Słoński, M. Bayesian Machine Learning in Analysis of Selected Identification Problems in Mechanics of Materials and Structures. In *Monografia–Politechnika Krakowska im. Tadeusza Kościuszki: Inżynieria Lądowa*; Wydawnictwo PK: Krakow, Poland, 2014.
4. Yu, L.; Giurgiutiu, V.; Marzan, A. Advanced signal processing for enhanced damage detection with piezoelectric wafer active sensors. *Smart Struc. Syst.* **2005**, *1*, 185–215. [CrossRef]
5. Zima, B.; Rucka, M. Guided waves for monitoring of plate structures with linear cracks of variable length. *Arch. Civ. Mech. Eng.* **2016**, *16*, 387–396. [CrossRef]
6. Żak, A.; Radzieński, M.; Krawczuk, M.; Ostachowicz, W. Damage detection strategies based on propagation of guided elastic waves. *Smart Mater. Struct.* **2012**, *21*, 035024. [CrossRef]

7. Fric, N.; Budjevac, D.; Miskovic, Z.; Veljkovic, M.; Markovic, Z.; Dobric, J. Calibration of the high strenght bolts for measuring of the pretension force. In Proceedings of the Eighth International Conference on Advances in Steel Structures, Lisbon, Portugal, 22–24 July 2015; pp. 1–10.

8. Nazarko, P.; Ziemiański, L. Damage detection in aluminum and composite elements using neural networks for Lamb waves signal processing. *Eng. Fail. Anal.* **2016**, *69*, 97–107. [CrossRef]

9. Chaki, S.; Bourse, G. Guided ultrasonic waves for non-destructive monitoring of the stress levels in prestressed steel strands. *Ultrasonics* **2009**, *49*, 162–171. [CrossRef] [PubMed]

10. Waszczyszyn, Z.; Ziemiański, L. Neural Networks in the Identification Analysis of Structural Mechanics Problems. In *Parameter Identification of Materials and Structures*; CISM Courses and Lectures, 469; Mróz, Z., Stavroulakis, G.E., Eds.; Springer: Vienna, Austria, 2005; pp. 265–340.

11. Nazarko, P.; Ziemiański, L. Force identification in bolts of flange connections for structural health monitoring and failure prevention. *Procedia Struct. Integr.* **2017**, *5*, 460–467. [CrossRef]

12. Nazarko, P. Soft computing methods in the analysis of elastic wave signals and damage identification. *Inverse Probl. Sci. Eng.* **2013**, *21*, 945–956. [CrossRef]

13. Chen, R.; Chen, S.; Yang, L.; Wang, J.; Xu, X.; Luo, T. Looseness diagnosis method for connecting bolt of fan foundation based on sensitive mixed-domain features of excitation-response and manifold learning. *Neurocomputing* **2017**, *219*, 376–388. [CrossRef]

14. Kim, N.; Hong, M. Measurement of axial stress using mode-converted ultrasound. *NDT&E Intern.* **2009**, *42*, 164–169.

15. Ding, X.; Wu, X.; Wang, Y. Bolt axial stress measurement based on a mode-converted ultrasound method using an electromagnetic acoustic transducer. *Ultrasonics* **2014**, *54*, 914–920. [CrossRef] [PubMed]

16. Miao, R.; Shen, R.; Zhang, S.; Xue, S. A Review of Bolt Tightening Force Measurement and Loosening Detection. *Sensors* **2020**, *20*, 3165. [CrossRef] [PubMed]

17. Nazarko, P. Axial force prediction based on signals of the elastic wave propagation and artificial neural networks. *MATEC Web Conf.* **2019**, *262*, 10009. [CrossRef]

18. Sun, Q.; Bo, Y.; Mu, X.; Sun, W. Bolt preload measurement based on the acoustoelastic effect using smart piezoelectric bolt. *Smart Mater. Struct.* **2019**, *28*, 055005. [CrossRef]

19. Yasui, H.; Kawashima, K. Acoustoelastic Measurement of Bolt Axial Load with Velocity Ratio Method. In Proceedings of the Proceeding of the 15th World Conference on Nondestructive Testing, Roma, Italy, 15–21 October 2000.

20. LeCun, Y.; Bengio, G. Hinton, Deep learning. *Nature* **2018**, *521*, 436–444. [CrossRef] [PubMed]

21. Miller, B.; Ziemiański, L. Optimization of dynamic behavior of thin-walled laminated cylindrical shells by genetic algorithms and deep neural networks supported by modal shape identification. *Adv. Eng. Softw.* **2020**, *147*, 102830. [CrossRef]

22. Marsland, S. *Machine Learning: An Algorithmic Perspective*; Chapman & Hall/CRC: Boca Raton, FL, USA, 2009.

Article

2D Digital Image Correlation and Region-Based Convolutional Neural Network in Monitoring and Evaluation of Surface Cracks in Concrete Structural Elements

Marek Słoński * and Marcin Tekieli

Faculty of Civil Engineering, Cracow University of Technology, ul. Warszawska 24, 31-155 Kraków, Poland; Marcin.Tekieli@pk.edu.pl
* Correspondence: Marek.Slonski@pk.edu.pl; Tel.: +48-12-628-2562

Received: 30 June 2020; Accepted: 5 August 2020; Published: 10 August 2020

Abstract: This paper shows how 2D digital image correlation (2D DIC) and region-based convolutional neural network (R-CNN) can be combined for image-based automated monitoring and assessment of surface crack development of concrete structural elements during laboratory quasi-static tests. In the presented approach, the 2D DIC-based monitoring enables estimation of deformation fields on the surface of the concrete element and measurements of crack width. Moreover, the R-CNN model provides unmanned simultaneous detection and localization of multiple cracks in the images. The results show that the automatic monitoring and evaluation of crack development in concrete structural elements is possible with high accuracy and reliability.

Keywords: digital image correlation; region-based convolutional neural network; machine learning; crack monitoring; crack detection and localization

---

1. Introduction

Much of the important concrete structures that are in use today were erected several years ago and now they are close to their design life [1]. As a result, the structures require regular condition assessment for understanding of the current state of their structural components. Condition assessment of a concrete structural element involves monitoring of displacements and evaluation of crack development visible on the surface of the element during laboratory mechanical tests. Monitoring is often done using wired contact sensors such as linear-variable-differential transformers (LVDTs). However, these sensors are often difficult to install and maintain.

Crack evaluation typically involves visual inspection by trained staff and simple measuring tools such as a Brinell magnifier. However, such inspection methods can be expensive, dangerous and time-consuming. On the other hand, computer vision methods allow for fully automated extraction of important information from digital images [2,3]. As a consequence, various methods leveraging image-processing techniques and computer vision methods such as digital image correlation and convolutional neural networks have been developed and applied in past decades [4–9].

Monitoring and measurement of displacements using computer vision methods are often performed by applying optical flow-based algorithms such as digital image correlation [10]. A detailed review of DIC applications can be found in [7,11]. DIC have been applied for measuring displacements and strains of specimens made of various materials. Mróz et al. [12] presented a feasibility study of DIC in determining strains in concrete exposed to fire.

DIC methods have also been used for assessment of concrete crack development. Helm in [13] shown how to use DIC for assessment of specimens with multiple growing cracks. Similarly, Rui et al.

in [14] presented DIC-based measurement system of crack generation and evolution during static testing of concrete sleepers. In [15] Gehri et al. shown a study on an automated crack detection and measurement based on DIC. Finally, DIC techniques allow for measurements and calculations of strains localization and the width of the fracture process zones on the surface of notched concrete beams [16,17].

Crack assessment using DIC methods is very precise but also require huge computational resources and is very time-consuming. As a result, it is mainly used off-line for assessment after the tests. It is also possible to apply other non-destructive monitoring techniques such as Acoustic Emission (AE) [18] or microwave sensors [19]. On the other hand, in recent years, convolutional neural networks (CNN) have been developed and applied for online automatic detection of concrete cracks and structural damage. See for example, recent state-of-the-art reviews [5,6]. In [20] Cha et al. described an autonomous system for structural visual inspection using region-based deep learning for detecting multiple damage types. In [21] a system for real-time crack assessment with wall-climbing unmanned aerial system is presented. Roberts et al. in [22] shown a system for low-cost pavement condition health monitoring and analysis. In [23] Deng et al. presented a region-based CNN with deformable modules for visually classifying concrete crack. Finally, an application of CNN for detection of flaws in concrete using ultrasonic tomography is described in [24].

These two approaches to monitoring and evaluation of surface cracks in concrete structural elements can be combined for better description and assessment of concrete elements. To the best knowledge of the authors of the paper, there is no such a study on combining DIC and CNN algorithms in this context. As a result, this paper is organized as follows. In this first section, the motivation and main goals of the research study undertaken are given. In Section 2, a detailed description of the proposed new method is presented. In Section 3, CNN model development and deployment is outlined. In Section 4, the experimental procedure is described. In Section 5, the results and discussion are given. Finally, in Section 6, the most relevant conclusions are drawn.

2. Methodology

This section provides a description of the proposed methodology. The main goal of the proposed approach is the automatic assessment of the development of concrete cracks by combining two computer vision methods: 2D digital image correlation and region-based convolutional neural network. Figure 1 shows the flowchart of the proposed approach. During the 1st and 2nd step, the tested element is prepared and the vision system is set. In the 3rd step, during the experiment, the digital images are taken and stored using DSLR cameras. In the 4th step, the stored images are pre-processed and the Faster R-CNN model is developed using the stored images. During the 5th step, the developed Faster R-CNN is used for detection and localization of surface cracks on the tested element. In the 6th and 7th step, DIC method is used to compute the displacements, strains and the crack width, respectively. Finally, the automatic assessment of the cracks during the experiments done with Faster R-CNN and DIC can be saved for the later analysis.

2.1. Region-Based Convolutional Neural Network

In this paper, to detect and localize multiple cracks on the surface of a tested concrete element a region-based convolutional neural network or regions with CNN features (R-CNN) architecture is applied. The R-CNN model works by performing computations in four steps [25]:

1. a selective search on the input image to select proposed regions containing objects,
2. a pre-trained CNN transforms each proposed region and computes the features extracted from the proposed regions,
3. the extracted features and labeled category of each proposed region are combined to train support vector machine (SVM)-based classifiers for object classification,
4. the extracted features and labeled bounding box of each proposed region are combined to train a linear regression model for bounding box prediction.

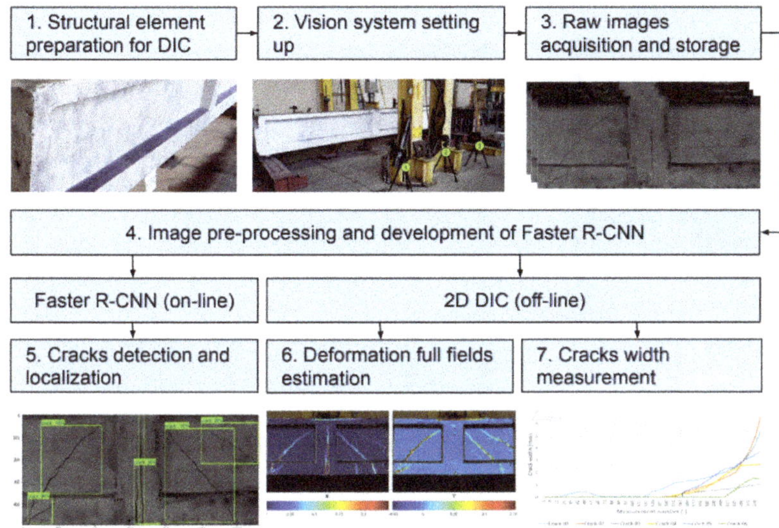

Figure 1. Flowchart for assessment of concrete surface cracks using Faster R-CNN and 2D DIC.

This approach was proposed by Girshick et al. in 2013 [26]. The schematic architecture of this model is presented in Figure 2.

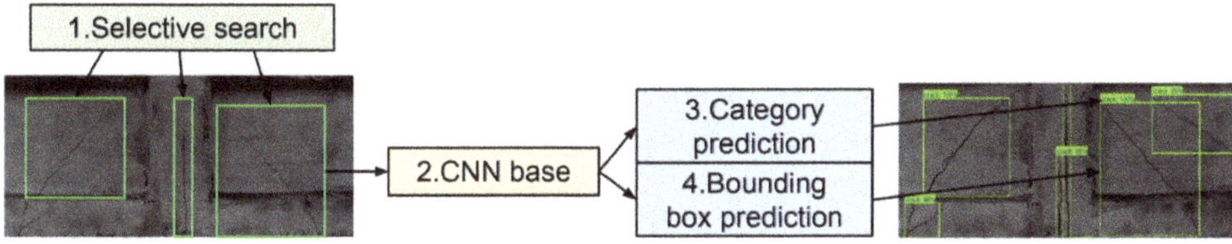

Figure 2. Diagram of the R-CNN model architecture.

The main bottleneck of the R-CNN model is the need to extract features for each proposed region. The R-CNN model was improved by performing CNN forward computation on the whole image. This improved model is known as a Fast R-CNN model [27]. For obtaining precise object detection, Fast R-CNN requires, in general, many proposed regions in selective search. The Fast R-CNN model was improved by the Faster R-CNN model [28] which replaces selective search with a region proposal network (RPN) and reduces the number of generated proposed regions [25]. The schematic architecture of this model is presented in Figure 3.

The main part of these two architectures is a convolutional neural network (CNN). CNN is a special class of a layered feed-forward artificial neural network. CNN was designed for processing images and audio signals [29]. The typical CNN has an input layer, several hidden layers with nonlinear units, and an output layer with linear units (for regression) or nonlinear units (for classification). Each unit computes a weighted sum of its inputs (activation of the unit). The activation is sent to a transfer function, an S-shaped function such as sigmoid function or rectified linear unit (ReLU) function [29].

The training process use the backpropagation algorithm for efficiently computing the gradient of the loss function and stochastic gradient descent (SGD) algorithm for learning the weights of the CNN model. During training of the convolutional neural network with several millions of parameters, the main issue is the overfitting of the neural model to the dataset. Fortunately, there are several techniques to cope with the overfitting. For example, on can apply transfer learning to a pre-trained model [29]. Transfer learning is a technique used for adaptation of the pre-trained model to another dataset. It is performed by additional training of selected convolutional layers of CNN while the rest of the layers are preserved [29].

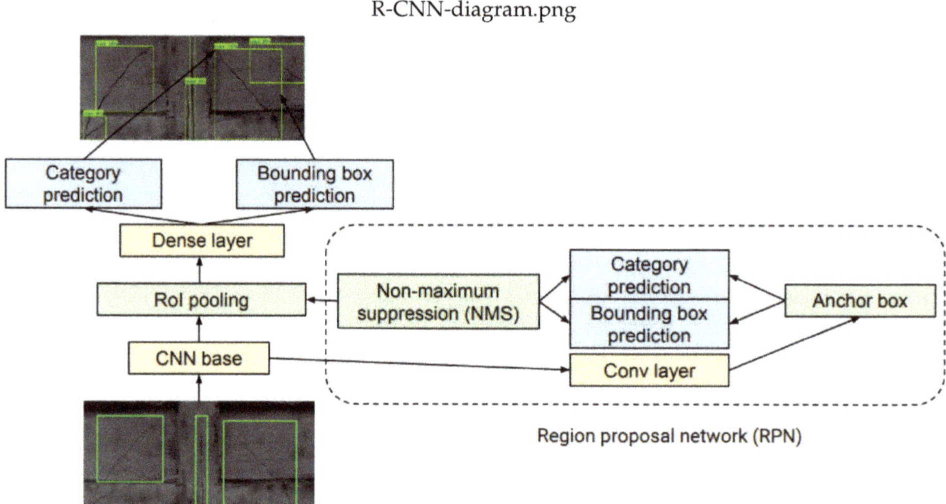

Figure 3. Diagram of the Faster-R-CNN model architecture.

In this paper, as a CNN base the Inception V2 architecture is applied. Inception V2 is a convolutional neural network proposed by Szegedy et al. from Google Research in 2016 [30]. These architectures are available through the TensorFlow object detection API [31] which is an open-source library developed by Google Research and built upon TensorFlow [32]. It allows easy development and deployment of CNN-based models for object detection and other computer vision problems.

2.2. Digital Image Correlation

Digital Image Correlation is a well-developed and popular tool for evaluation of surface deformations [7,8,10]. It can also be used as a non-destructive method for full-field measurements of displacements of a tested specimen surface. DIC was developed at the University of South Carolina in the early 80s [33,34].

DIC works by processing images taken during the deformation of an object. Then it tries to establish a mapping between the image coordinates of the reference (undeformed) object image and the image coordinates of the deformed object image by searching for the mapping which gives the highest correlation between the reference image and the current image. The mapping is then used for calculating full-field strains [35].

The images are stored as a 2D matrix of pixels and each image is correlated with the reference (undeformed) image. The points of the grid based on the specified image subsets are matched and identified as that associated with the highest value of the correlation coefficient. This coefficient is calculated between the reference subset "f" and the target subset "g", whose dimensions are equal and are $M \times N$ pixels using the zero-mean normalized cross-correlation criterion defined in Equation (1). Illustration of the basic principle of digital image correlation is shown in Figure 4.

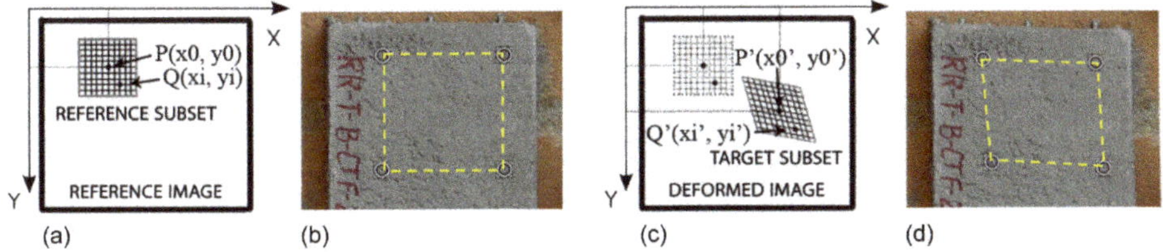

Figure 4. Illustration of the basic principle of the digital image correlation method: sketch of the subset (**a**,**c**) and measurement points on the surface of a specimen (**b**,**d**) before (**a**,**b**) and after (**c**,**d**) deformation [36].

$$CC^{ZMN} = \frac{\sum\limits_{i=1}^{M}\sum\limits_{j=1}^{N}\left[\left(f(i,j)-u_f\right)\times\left(g(i,j)-u_g\right)\right]}{\sqrt{\sum\limits_{i=1}^{M}\sum\limits_{j=1}^{N}\left(f(i,j)-u_f\right)^2\times\sum\limits_{i=1}^{M}\sum\limits_{j=1}^{N}\left(g(i,j)-u_g\right)^2}},\tag{1}$$

where u_f is the intensity of the reference subset and u_g is the intensity of the target subset form.

3. Model-Development Workflow for Crack Detection Based on Faster R-CNN

In this work we have developed a predictive model for crack detection based on Faster R-CNN architecture described in the previous section. The development of the model consisted of several stages. In the first stage we collected and annotated several images of concrete elements with surface cracks. For model development all annotated images were divided randomly into three sets. The first set was used for training, the second for validation images and third for testing the Faster R-CNN model. Finally, the developed model was deployed for crack detection during experiments described in the next section.

3.1. Dataset Collection and Annotation

Collection and annotation of images for building a concrete crack detector is very important stage because the accuracy of the trained model depends to large extent on the quality of the prepared dataset. For this research, 1058 images containing cracked concrete elements were collected from different laboratory experiments performed at Cracow University of Technology (CUT). Figure 5 shows selected images of concrete element containing cracks.

Figure 5. Example of images of concrete elements containing cracks.

The images were then manually annotated using a graphical image annotation tool LabelImg [37]. It works by defining a bounding box coordinates and the corresponding label. Figure 6 shows an example of image of concrete element containing several thin cracks.

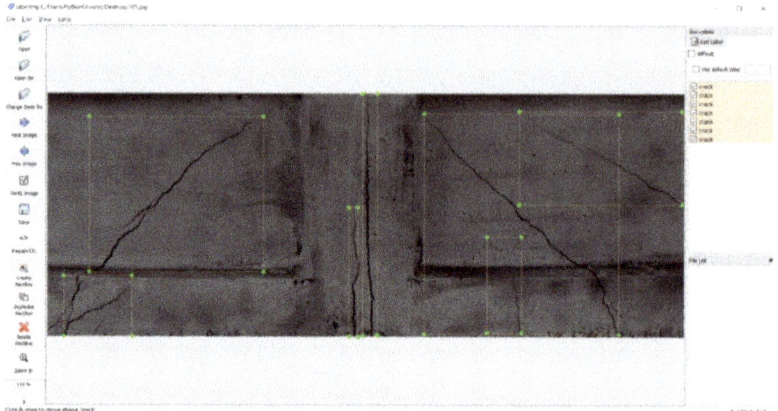

Figure 6. Example of annotation of an image of concrete element containing several cracks.

3.2. Model Development

Model development from scratch requires a large number of annotated images. For an object detection task where only small number of training data is available, a common solution is to perform fine-tuning on a CNN which is pre-trained with related source data. In this work, we adopted transfer learning and used a pre-trained convolutional neural network called Inception V2 described in the previous section.

The annotated images were converted to the record format to be used within TensorFlow and the dataset was randomly split in the ratio of 75% for training the model and 25% for testing. This ensures that the model do not overfit to the dataset and would therefore be able to perform well on unseen data.

The training and validation process of the Faster R-CNN model was performed on a laptop computer with GPU working under the 64-bit Windows 10 operating system using TensorFlow environment and monitored by using TensorBoard system. Figure 7 shows changes of two examples of the monitored losses with and without smoothing (classification loss and localization loss) during the training process. This process was manually stopped after 11,000 epochs which took about 7 h. From the plots, it can be observed that after about 30,000 epochs the smoothed losses started to decrease with constant rates.

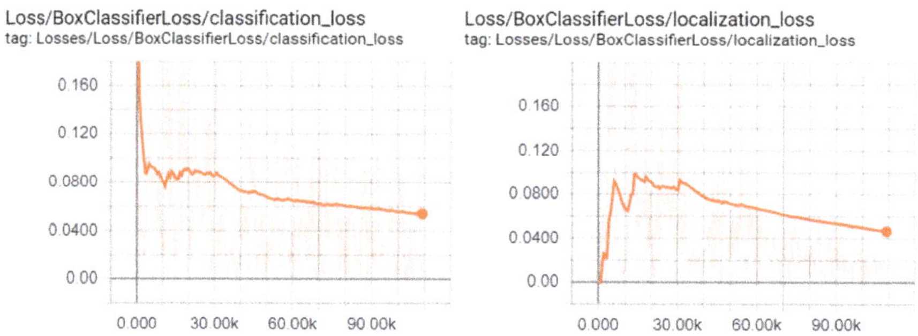

Figure 7. Changes of classification loss and localization loss (with and without smoothing) during the training process (stopped after 11,000 epochs).

After training, the model was tested by detecting cracks in new images. The model produces a bounding boxes on the images along with a percentage of how accurate this bounding box is based on the trained model. This value provides users with an assessment of how good the detection of cracks is. Figure 8 shows tested images containing cracks. It can be noted that while the crack on the left image was properly detected and localized as one crack with certainty 99% (inside the green bounding box) and also not properly divided into two cracks: inside the blue bounding box (with certainty 96%) and inside the red bounding box (with certainty 81%).

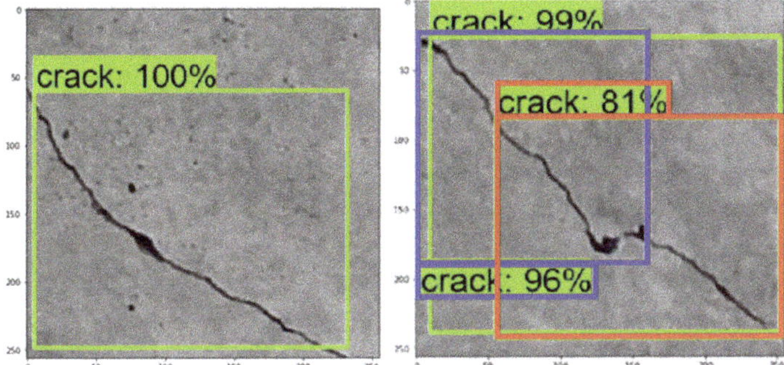

Figure 8. Example of two tested images: the crack on the left image properly detected and localized as one crack (with certainty 100%), the crack on the right image simultaneously properly detected as one crack (with certainty 99%) and not properly detected as two cracks (with certainty 96% and 81%, respectively).

After the model was developed it was deployed for automatic surface crack detection during mechanical tests in laboratory.

4. Experiments

In this section, we present the experiments for testing the proposed approach to computer vision-based assessment of surface cracks in concrete structural elements during laboratory experiments. The first part of the assessment consists of monitoring deformation of the element and analyzing development of crack width formed on the surface of the beam to determine the moment in which each of them was created. This part was based on optical measurements and DIC method. In the second part, crack detection and localization using Faster R-CNN model was carried out.

In this work, the proposed methodology is verified by applying it to the laboratory assessment of post-tensioned, precast crane runway beams after more than 50 years of exploitation. They were produced between 1962 and 1963 and were disassembled from the structure at the industrial hall [1]. The three-point bending tests were conducted on girders made of two precast segments, with the total length of the span L = 2 × 290 cm + 20 cm = 600 cm. The girder has an I-section with height H = 80 cm. The segments were connected due to the action of the force and welding of steel sheets to steel marks. The joint was filled by applying fine-grained concrete.

The static three-point bending test yields the maximum bending moment at the joint of the segments. The loading cycle consisted of two stages. In the first stage a relatively small load value was applied and then the beam was unloaded. In the second stage the loading was carried out to the complete collapse of the beam.

The experiments were conducted in the Research Laboratory for Building Materials and Structures at the Cracow University of Technology. The deformation of the beam was monitored by applying optical measurements carried out using 3 synchronized DSLR cameras. The cameras were located at the center of the beam. Before experiments, the side surface of the beam was prepared for vision-based measurements by adding random distribution of black dots with spray paint on whitewashed surface of the beam. Figure 9 shows the test stand prepared for this research with the numbering of acquisition devices.

The CivEng Vision system, developed at Cracow University of Technology (CUT), was used for acquisition, storing and processing the images [38,39]. The images were then processed using DIC method for computing deformation fields and crack width visible on the surface of the beam.

Finally, the images were processed by trained R-CNN model to automatically detect and localize the cracks on the surface. Figure 10 shows four images of the side surface of the beam, taken using the CivEng Vision system, during the three-point bending test.

Figure 9. Test stand prepared for monitoring post-tensioned, precast crane runway beam during the three-point bending test using the CivEng Vision system.

Figure 10. Images of the side surface of the beam, taken using the CivEng Vision system, during the three-point bending test.

Figure 11 shows grid of subsets for monitoring deformation of the surface of the tested crane runway beam using DIC.

Figure 11. Grid of subsets placed on the surface of the tested crane runway beam.

5. Results and Discussion

In this section, we present the results of application of digital image correlation and region-based convolutional neural networks in evaluation of surface cracks in the crane runway beam during experiments described in the previous section.

5.1. Monitoring of Beam Side Surface Deformation Fields

The deformation fields of the beam side surface was monitored using DIC method. Figure 12 shows the changes of deformation fields in X and Y direction during the loading of the beam. As can be seen in the first row and the first column, the first vertical crack formed exactly in the middle of the beam span (opening of the joint) in the first phase of the loading and was observed under the load P = 287 kN. It is worth mentioning that no cracks were observed, during the initial phase of the loading, by visual inspection of the surface.

The next two rows present the development of vertical and diagonal cracks during the experiments. The first diagonal crack was observed using DIC after applying loading P = 454.8 kN in the second phase. The last row shows the deformation fields just before the failure. The element was damaged at the load of P = 837.7 kN. Failure of the element was signalized through the opening of the joint and by occurring diagonal cracks. In the final phase of beams operation, perpendicular cracks have occurred connecting with diagonal cracks and a large increase of deflection was observed. The girder was finally damaged at the joint as a result of significant strains in prestressing cables.

It is worth adding that deflection measured by the DIC was very close to the same as the deflection indicated by LVDT device. The difference was below 1 mm.

5.2. Assessment of Crack Width

Optical measurements and DIC were also used to analyze the development of cracks visible on the surface of the beam and to measure their width. It is also possible to determine the moment the crack starts to develop. In Figure 13 shows the numbering of the analyzed cracks.

Figure 12. Deformation fields on the surface of the beam in X and Y direction during loading phases measured by DIC.

Figure 13. Numbering of the analyzed cracks [40].

As can be seen in Figure 14, the main vertical crack (Crack 01) formed exactly in the middle of the beam span and appeared in the first phase of the loading. Then the load was reduced to zero and the crack was almost completely closed. It appeared again during the second phase of the experiment.

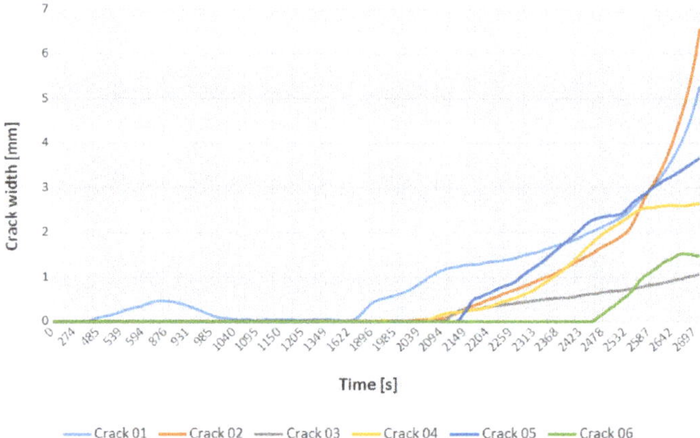

Figure 14. The crack width development during the loading phases.

5.3. Cracks Detection and Localization

To provide an assessment of the trained Faster R-CNN for detecting and localizing concrete cracks, new images of cracked concrete elements taken during laboratory experiments on prestressed crane runway beam in the three-point bending test were provided to the trained model.

Figure 15 shows detected and localized cracks in concrete beam with the Faster R-CNN. As can be seen, the network correctly detected and localized all the cracks (diagonal and vertical) visible in the image (green rectangles) with high certainty (more than 98%).

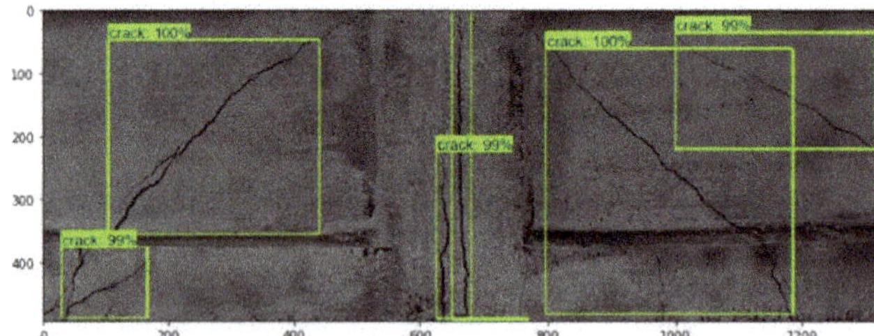

Figure 15. Detected and localized surface cracks (green rectangles) in the testing image of concrete beam using trained Faster R-CNN model.

6. Summary and Final Conclusions

In this paper, we have developed a computer vision system for automatic assessment of crack development visible on the surface of a concrete structural element during laboratory quasi-static tests. This approach combines 2D digital image correlation for monitoring the development of crack width and trained region-based convolutional neural network for automated detection and localization of multiple cracks. The intention of this work was to provide researchers and engineers with a description of easy-to-use computer vision-based system for quick assessment not only the crack width that are visible on the surface of the tested concrete element but also an automatic approach to crack detection and localization for monitoring purposes.

The computer vision system was evaluated during static tests performed in the CUT laboratory, to set up a system capable of carrying out this task. The images were captured using three DSLR cameras connected by one automatic trigger. The system showed high accuracy in assessment of the surface cracks. The system has the ability to automatic identify the cracks number and their localization. This process can be repeated for static tests of other concrete elements. Once the system was created it can be deployed for use by researchers and engineers for the concrete crack development analysis.

Author Contributions: Conceptualization, M.S.; methodology, M.S. and M.T.; software, M.T.; validation, M.T.; formal analysis, M.S. and M.T.; investigation, M.T. and M.S.; resources, M.T.; data curation, M.T.; writing—original draft preparation, M.S.; writing—review and editing, M.S. and M.T.; visualization, M.S. and M.T.; supervision, M.S. All authors have read and agreed to the published version of the manuscript.

Funding: This research received no external funding.

Acknowledgments: We acknowledge support given by P. Bełda by providing results of experiments from his Master Thesis on the application of R-CNN to crack detection and localization.

Conflicts of Interest: The authors declare no conflict of interest.

References

1. Derkowski, W.; Walczak, R. Problem of condition assessment of precast, post-tensioned concrete crane beams in an extended period of use. In *Concrete—Innovations in Materials, Design and Structures, Proceedings of the FIB Symposium 2019, Kraków, Poland 27–29 May 2019*; Derkowski, W., Ed.; International Federation for Structural Concrete: Lausanne, Switzerland , 2019; pp. 1507–1514.
2. Solem, J.E. *Programming Computer Vision with Python: Tools and Algorithms for Analyzing Images*; O'Reilly Media, Inc.: Newton, MA, USA, 2012.
3. Szeliski, R. *Computer Vision: Algorithms and Applications*; Springer Science & Business Media: Berlin, Germany 2010.
4. Valença, J.; Dias-da Costa, D.; Júlio, E. Characterisation of concrete cracking during laboratorial tests using image processing. *Construct. Build. Mater.* **2012**, *28*, 607–615.
5. Azimi, M.; Eslamlou, A.D.; Pekcan, G. Data-Driven Structural Health Monitoring and Damage Detection through Deep Learning: State-of-the-Art Review. *Sensors* **2020**, *20*, 2778. [CrossRef]
6. Spencer, B.F.J.; Hoskere, V.; Narazaki, Y. Advances in computer vision-based civil infrastructure inspection and monitoring. *Engineering* **2019**, *5*, 199–222. [CrossRef] [PubMed]
7. Pan, B. Digital image correlation for surface deformation measurement: historical developments, recent advances and future goals. *Meas. Sci. Technol.* **2018**, *29*, 082001. [CrossRef]
8. Zhao, J.; Sang, Y.; Duan, F. The state of the art of two-dimensional digital image correlation computational method. *Eng. Rep.* **2019**, *1*, e12038. [CrossRef]
9. Dorafshan, S.; Thomas, R.; Maguire, M. Comparison of deep convolutional neural networks and edge detectors for image-based crack detection in concrete. *Construct. Build. Mater.* **2018**, *186*, 1031–1045. [CrossRef]
10. Sutton, M.; Orteu, J.J.; Shreier, H. *Image Correlation for Shape, Motion and Deformation Measurements*; Springer: Berlin, Germany, 2009. [CrossRef]
11. Khoo, S.W.; Karuppanan, S.; Tan, C.S. A review of surface deformation and strain measurement using two-dimensional digital image correlation. *Metrol. Meas. Syst.* **2016**, *23*, 461–480.

12. Mróz, K.; Tekieli, M.; Hager, I. Feasibility Study of Digital Image Correlation in Determining Strains in Concrete Exposed to Fire. *Materials* **2020**, *13*, 2516. [CrossRef]

13. Helm, J.D. Digital image correlation for specimens with multiple growing cracks. *Exp. Mech.* **2008**, *48*, 753–762. [CrossRef]

14. Rui, L.; Zappa, E.; Collina, A. Vision-based measurement of crack generation and evolution during static testing of concrete sleepers. *Eng. Fract. Mech.* **2020**, *224*, 106715. [CrossRef]

15. Gehri, N.; Mata-Falcón, J.; Kaufmann, W. Automated crack detection and measurement based on digital image correlation. *Construct. Build. Mater.* **2020**, *256*, 119383. [CrossRef]

16. Kozicki, J.; Tejchman, J. Experimental investigations of strain localization in concrete using Digital Image Correlation (DIC) technique. *Arch. Hydro Eng. Environ. Mech.* **2007**, *54*, 3–24. [CrossRef]

17. Skarżyński, Ł.; Syroka, E.; Tejchman, J. Measurements and calculations of the width of the fracture process zones on the surface of notched concrete beams. *Strain* **2011**, *47*, 319–332.

18. Lacidogna, G.; Piana, G.; Accornero, F.; Carpinteri, A. Multi-technique damage monitoring of concrete beams: Acoustic Emission, Digital Image Correlation, Dynamic Identification. *Construct. Build. Mater.* **2020**, *242*, 118114.

19. Gkantou, M.; Muradov, M.; Kamaris, G.S.; Hashim, K.; Atherton, W.; Kot, P. Novel Electromagnetic Sensors Embedded in Reinforced Concrete Beams for Crack Detection. *Sensors* **2019**, *19*, 5175. [CrossRef]

20. Cha, Y.J.; Choi, W.; Suh, G.; Mahmoudkhani, S.; Büyüköztürk, O. Autonomous structural visual inspection using region-based deep learning for detecting multiple damage types. *Comput. Aided Civ. Infrastruc. Eng.* **2018**, *33*, 731–747. [CrossRef] [PubMed]

21. Jiang, S.; Zhang, J. Real-time crack assessment using deep neural networks with wall-climbing unmanned aerial system. *Comput. Aided Civ. Infrastruc. Eng.* **2020**, *35*, 549–564. [CrossRef]

22. Roberts, R.; Giancontieri, G.; Inzerillo, L.; Di Mino, G. Towards Low-Cost Pavement Condition Health Monitoring and Analysis Using Deep Learning. *Appl. Sci.* **2020**, *10*, 319. [CrossRef]

23. Deng, L.; Hong-Hu, C.; Shi, P.; Wang, W.; Kong, X. Region-Based CNN Method with Deformable Modules for Visually Classifying Concrete Cracks. *Appl. Sci.* **2020**, *10*, 2528. [CrossRef]

24. Słoński, M.; Schabowicz, K.; Krawczyk, E. Detection of Flaws in Concrete Using Ultrasonic Tomography and Convolutional Neural Networks. *Materials* **2020**, *13*, 1557. [CrossRef]

25. Zhang, A.; Lipton, Z.C.; Li, M.; Smola, A.J. Dive into Deep Learning. 2020. Available online: https://d2l.ai (accessed on 3 August 2020). [CrossRef]

26. Girshick, R.; Donahue, J.; Darrell, T.; Malik, J. Rich feature hierarchies for accurate object detection and semantic segmentation. *arXiv* **2013**, arXiv:1311.2524.

27. Girshick, R. Fast R-CNN. *arXiv* **2015**, arXiv:1504.08083.

28. Ren, S.; He, K.; Girshick, R.; Sun, J. Faster R-CNN: Towards Real-Time Object Detection with Region Proposal Networks. *arXiv* **2015**, arXiv:1506.01497.

29. Goodfellow, I.; Bengio, Y.; Courville, A. *Deep Learning*; MIT Press: Cambridge, MA, USA, 2016.

30. Szegedy, C.; Vanhoucke, V.; Ioffe, S.; Shlens, J.; Wojna, Z. Rethinking the inception architecture for computer vision. In *Proceedings of the IEEE Conference on Computer Vision and Pattern Recognition*; IEEE: Piscataway, NJ, USA, 2016; pp. 2818–2826.

31. Huang, J.; Rathod, V.; Sun, C.; Zhu, M.; Korattikara, A.; Fathi, A.; Fischer, I.; Wojna, Z.; Song, Y.; Guadarrama, S.; Murphy, K. Speed/accuracy trade-offs for modern convolutional object detectors. In *Proceedings of the IEEE Conference on Computer Vision and Pattern Recognition*; IEEE: Piscataway, NJ, USA, 2017; pp. 7310–7311.

32. Abadi, M.; Agarwal, A.; Barham, P.; Brevdo, E.; Chen, Z.; Citro, C.; Corrado, G.S.; Davis, A.; Dean, J.; Devin, M.; et al. TensorFlow: Large-Scale Machine Learning on Heterogeneous Systems. 2015. Available online: tensorflow.org (accessed on 12 June 2020).

33. Peters, W.; Ranson, W. Digital imaging techniques in experimental stress analysis. *Opt. Eng.* **1982**, *21*, 427–431.

34. Sutton, M.; Wolters, W.; Peters, W.; Ranson, W.; McNeill, S. Determination of displacements using an improved digital correlation method. *Image Vis. Comput.* **1983**, *1*, 133–139. [CrossRef]

35. Olufsen, S.N.; Andersen, M.E.; Fagerholt, E. μDIC: An open-source toolkit for digital image correlation. *SoftwareX* **2020**, *11*, 100391. [CrossRef]

36. Tekieli, M. A Vision-Based Measurement System for the Analysis of Structural Element Deformation Fields. Ph.D. Thesis, Faculty of Civil Engineering, Cracow University of Technology, Cracow, Poland, 2019. [CrossRef]

37. Tzu Ta, L. LabelImg. *Free Software: MIT License*; 2015 Available online: https://github.com/tzutalin/labelImg (accessed on 4 May 2020).
38. Tekieli, M.; Słoński, M. Particle filtering for computer vision-based identification of frame model parameters. *Comput. Assis. Methods Eng. Sci.* **2014**, *21*, 39–48.
39. Tekieli, M.; Słoński, M., Digital image correlation and Bayesian filtering in inverse analysis of structures. In *Recent Advances in Civil Engineering : Computational Methods*; Cecot, W., Ed.; Publishing House of Cracow University of Technology: Kraków, Poland, 2015; pp. 111–124.
40. Bełda, P. Reinforced Concrete Element Crack Detection by Application of Convolutional Neural Network Using the TensorFlow Library. Master's Thesis, Cracow University of Technology, Cracow, Poland, 2019. (In Polish)

 *materials*

Article

Use of Time-Frequency Representation of Magnetic Barkhausen Noise for Evaluation of Easy Magnetization Axis of Grain-Oriented Steel

Michal Maciusowicz * and Grzegorz Psuj *

Department of Electrical and Computer Engineering, Faculty of Electrical Engineering,
West Pomeranian University of Technology, ul. Sikorskiego 37, 70-313 Szczecin, Poland
* Correspondence: michal.maciusowicz@zut.edu.pl (M.M.); gpsuj@zut.edu.pl (G.P.);
  Tel.: +4891-449-4967 (M.M.); +4891-449-4727 (G.P.)

Received: 10 June 2020; Accepted: 28 July 2020; Published: 31 July 2020

Abstract: The paper presents a new approach to non-destructive evaluation of easy/hard magnetization axis in grain-oriented SiFe electrical steels based on the Barkhausen phenomenon and its time-frequency (*TF*) characteristics. Anisotropy in steels is influenced by a number of factors that formulate the global relationship and affect the Barkhausen effect. Due to the observed high variability in the dynamics of magnetic Barkhausen noise (MBN) over time, obtained for various directions in grain-oriented steel, it becomes justified to conduct MBN signal analyses in the time-frequency domain. This representation allows not only global information from MBN signal over entire period to be expressed, but also detailed relationships between properties in time and in frequency to be observed as well. This creates the opportunity to supplement the information obtained. The main aspect considered in the work is to present a procedure that allows an assessment of the resultant angular characteristics in steel. For this purpose, a sample of a conventional grain-oriented SiFe sheet was used. Measurements were made for several angular settings towards the rolling and transverse directions. A data transformation procedure based on short-time Fourier transform (STFT) as well as quantitative analysis and synthesis of information contained in the *TF* space was presented. Angular characteristics of selected *TF* parameters were shown and discussed. In addition, an analysis of the repeatability of information obtained using the proposed procedure under various measurement conditions was carried out. The relationship between the selection of calculation parameters used during transformation and the repeatability of the obtained *TF* distributions were demonstrated. Then the selection of the final values of the calculation parameters was commented upon. Finally, the conclusions of the work carried out were discussed.

Keywords: non-destructive testing; magnetic Barkhausen noise; magnetic anisotropy; grain oriented steel; time-frequency representation; signal processing; data mining methods

---

1. Introduction

Magnetic Barkhausen noise (MBN) is a phenomenon that occurs when a ferromagnetic material is being magnetized. During that process, when the magnetic field strength increases, the magnetic domain structure is gradually reorganized. This process is associated with the movement of domain walls (DW), followed by the rotation of magnetization vectors inside domains, resulting in discontinuous changes in the magnetic flux inside a material [1–3]. Those changes in the magnetic flux can be observed by a coil located at the surface of the material, in the form of voltage noise induced in its turn as a result of rapid changes in the degree of magnetization [4]. This method is used for various industrial applications for non-destructive condition testing and quality assessment of ferromagnetic materials, e.g., for evaluation of residual stress, active tensile and compressive stress, microstructure and micro hardening [5–19]. In addition to the possible applications mentioned above, the MBN is also used to evaluate the directional properties of ferromagnetic materials. Within this group of applications, an important aspect is the assessment of magnetic anisotropy of materials and the detection of easy magnetization axes. Recently, the use of MBN to evaluate magnetic anisotropy in steels, which is the result of a number of different factors, has become an equally important issue. The key in this aspect is the possibility of non-destructive and quick assessment of directional magnetic characteristics (introducing the axis of easy/hard magnetization) of steel formulated specially by means of surface engineering methods.

Determining the axis of easy magnetization is particularly important in the process of designing electrical machines, in essence the transformer cores or electric motors. Anisotropic properties allow optimizing their constructions and in consequence increasing effectiveness and reduce the costs of transformation and distribution of energy [20,21]. The resultant easy magnetization axis is influenced by a number of factors, including: residual stress, grain shape and size, etc. [22,23]. Depending on the influence of each factor, the value of anisotropy will change, which changes the magnetization ability in different directions. Within the currently conducted research on the application of the MBN to assess the magnetic anisotropy of materials, two approaches can be distinguished. The first one is related to the analysis of the impact of individual factors and its rate on the scale of anisotropy [24–26]. The second one refers to possible application for the need of rapid assessment of the resultant axis of easy magnetization [27,28]. This paper considers the latter aspect.

In the literature, different approaches to assessing the easy magnetization axis can be distinguished by omnidirectional observation of MBN activity. The following description presents several literature examples of the use of MBN to detect the easy magnetization axis. The first, classic approach is to observe the MBN for various settings of directional magnetic field acting on the examined material. In paper [28], the MBN burst expressed in the time domain was analyzed to obtain information about anisotropic properties. A classical grain-oriented electrical steel sheet was used for the study. The authors observed a higher amplitude value of MBN signal envelope for the rolling direction (RD), which was characterized by the occurrence of two distinct peaks in comparison to the signals obtained for other directions. They explained this phenomenon with much higher MBN activity for RD due to the large number of 180° domain walls (DWs), and associated the maximum values with the process of nucleation and annihilation of DWs. Confirmation of these observations was a distribution of the root-mean-square (RMS) value of the measured MBN signal versus the measuring angle, showing the highest value along the RD direction. An attempt to overcome the need for mechanical rotation of the measuring head is to introduce an appropriate modification of the MBN transducer's excitation system. Examples of applications of such original solutions were presented in papers [27,29]. In [29], the authors used two C-shaped cores with coils placed perpendicularly to each other to control the excitation characteristics, a method named rotational Barkhausen noise (RBN). In this way, it is possible to generate a two-dimensional magnetizing field imitating the characteristics occurring during the classical MBN angular measurement, without the need for mechanical rotation of the magnetizing system. A disadvantage of this approach is the problem to obtain full homogeneity of the magnetic field. In paper [27], a rotatable transducer was used, in which the rotating permanent magnets generated a magnetic field, and the centrally located coil measured the MBN signal. The transducer collected data of the generated in the material magnetic induction disturbances, and then the measured MBN signal was analyzed in the time domain using features extracted from the MBN signal's envelope. This method, referred to as continuous rotational Barkhausen noise (CRBN), allows straight observation of directions for which there is a clear increase in magnetic emission. The directional characteristics were related to the anisotropy of the material. The performance of the proposed system was verified based on the comparison with the classical approach. Despite the high correlation between the results obtained and those of the classic angular measurement, the key issue of the proposed CRBN method is the selection of the speed of the transducer motion and the rotation of the magnets. Both factors have a large impact on the sensitivity of the measurement and, as a result, the discrimination between the directional magnetic properties of the material (directions of magnetization). Mindful of the need to increase the effectiveness of assessing the anisotropic properties of magnetic materials, the idea of sequential analysis of the MBN burst has recently appeared. Its purpose is to divide a single burst into bands (sections) defined in relation to the magnetization period or magnetic field level. An example of such an approach was presented in [25,30,31]. In those works, the area of greatest MBN activity (referred to main MBN peak) was associated by authors with the motion of 180° DWs referring to the effect of roll magnetic anisotropy (RMA). The two other sub-periods occurring before and after the major one were associated, respectively, with the process of nucleation of reverse domains influenced by magneto-crystalline anisotropy (MCA) energy and motion of 90° DWs. In that work, from each section the energy value of MBN was calculated, and then the angular characteristics of the energy for individual sub-periods were achieved. The results obtained confirmed the possibility of a more accurate analysis of the phenomenon of anisotropy. However, the main problem of the method is the appropriate determination of signal divisions into sub-periods. Additionally subsequent interpretation requires consideration of many factors. Regardless of the approach finally applied, the named factors that define anisotropy are not the only ones that affect the course of angular characteristics of the MBN. The measurement conditions also have a significant impact on the effectiveness of anisotropy detection and measurement sensitivity, as presented and discussed in [31]. Similar observations were made in [23], where the need to maintain a constant value of magnetic flux density was underlined. It was also stated that during the tests it is necessary to observe not only the height of the MBN envelope but also its shape. All of the above factors result in a requirement for a broad analysis of the dynamics of the observed Barkhausen phenomenon, including both time and frequency characteristics. This entails the need for sequential monitoring of time and frequency characteristics in successive short time windows. The synchronous approach creates the opportunity to supplement the acquired information. Therefore, it becomes justified to carry out time-frequency (*TF*) analysis of measured MBN signals. This way of presentation allows more detailed information about the analyzed phenomenon to be obtained, because in addition to the extraction of parameters integrating global information (such as rms, energy or number of pulses) about the signal within the entire MBN period, it is possible to observe detailed relationships between properties in time and in frequency for subsequent time moments. As a result, it is possible to perform additional analysis and provide more detailed data on the nature of the phenomenon, which may be important for carrying out the final assessment. Recently, the authors presented the results of works on the application of the *TF* transformation to quantify the MBN signal information [24,32]. The [32] presents a broad description of the proposed *TF* analysis, *TF* features extraction and their interpretation along with a comparison of the information they carry with respect to the classic features of the MBN signal determined in the time or frequency domain. The influence of measuring conditions on the discrimination level was also discussed. The second paper of the authors [24] proposes a procedure based on *TF* characteristics enabling a clear division of the MBN signal into the aforementioned three sub-bands and determining the corresponding distributions of directional characteristics. However the described procedure did not allow the global anisotropy characteristic to be defined, being an assembly of all factors. Therefore, this paper proposes a procedure for determining the resultant easy magnetization axis based on the *TF* representation as well as a detailed analysis of the factors affecting the results. The successive sections present a setup of the measuring system, a proposed procedure for analyzing and evaluating the resultant easy magnetization axis. Ultimately, the results will be discussed and conclusions will be presented.

2. Samples and Measuring System Setup

In the experiment, the conventional cold-rolled 3%SiFe electrical steel sheet (in reference to EN 10107 and IEC 60404-8-7 Standards) of 0.27 mm thickness, with silicon contents of 3% by mass and iron being almost a complement (over 96%) was used. Additionally, a trace content including carbon or other impurities (not exceeding tenths of a percent in overall) is also expected in this steel. The specific density of steel was 7650 kg/m^3. Measurements were made for 16 angles at the location on the sample depicted with the letter A. The Figure 1 shows the view of the sample as well as the visualization of the first 5 measurement steps and rolling direction RD. The axis of first measurement is along to the transverse direction (TD) which corresponds to hard magnetization axis. The transducer was rotated by 22.5° after each measurement until reaching a full turn of 360°. In order to perform the measurement, the MBN system controlled by a personal computer (PC) equipped with a data acquisition board DAQ (NI USB-6251) was used. The system has been described in detail in paper [32]. The scheme of the measuring system setup is shown in Figure 2.

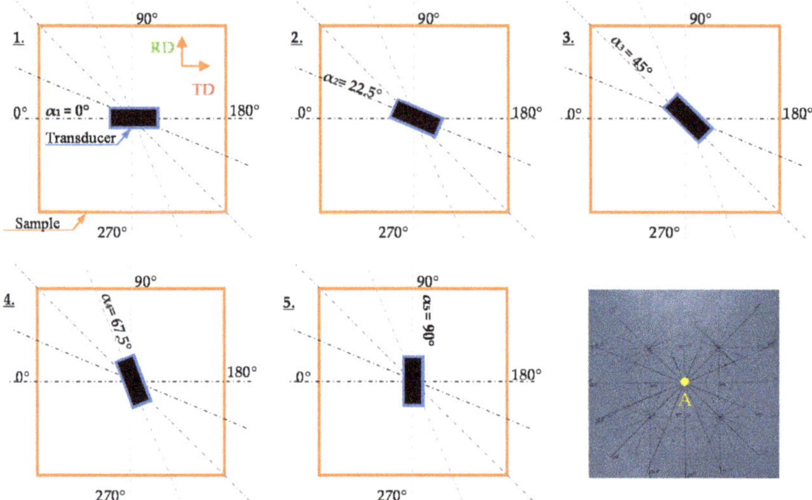

Figure 1. Measurement steps and view of conventional electrical steel sheet with marked measurement location.

During the measurements, an MBN transducer having two sections was used (Figure 2c): the magnetizing and measuring ones. The magnetizing section consists of a coil wound on a C-shaped ferrite core. For the need of MBN signal sensing, a setup of coils wound on a rod core placed between the pole pieces of the electromagnetic yoke was utilized. In addition, a coil wound on one of the columns of the electromagnet and a Hall sensor located between its columns were used to control the magnetizing field conditions (magnetic induction and tangential component of magnetic field strength). For the needs of excitation, the sinusoidal waveform was generated using a digital-to-analog D/A converter of a DAQ board and then amplified using Power Amplifier (PA–OPA549) before driving the magnetizing coil. The excitation current frequency was set to 10 Hz, and the amplitude of the current was kept constant during the measurements. The estimated value of the magnetizing filed strength was 1.8 kA/m. The detailed discussion of the measuring conditions and analysis of the influence of the excitation field parameters on the MBN *TF* characteristic, as well as the choice of utilized settings (applied also in this paper) were presented and discussed elsewhere [24]. The MBN signal (U_{BN}) induced in measuring coils was fed to the input of the analog processing system (APS) unit [24,32]. First, it was bandpass filtered in 0.6–97.5 kHz range and then amplified. The measurement was made in the angular steps, each time acquiring signals for the successive 10 periods of the magnetization process. The measurement was repeated 10 times, which made it possible to average the obtained results and minimalize the influence of interfering factors. Signal acquisition was carried out using the analog-to-digital A/D converter of DAQ board at sampling frequency of 250 kHz. Then, all measured signals were further processed using digital signal processing procedures.

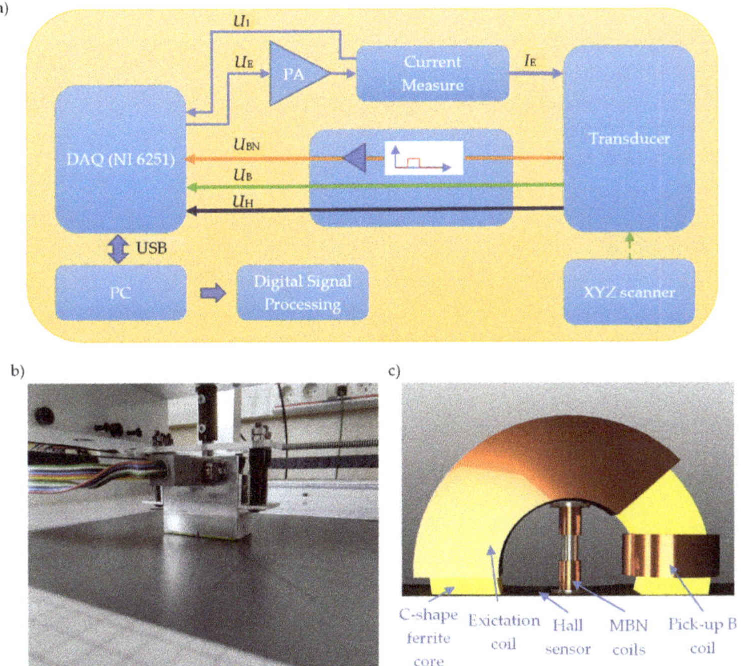

Figure 2. The measurement system and XYZ scanner with mounted transducer: (**a**) schematic view of the configuration, (**b**) photo and (**c**) 3D view of the transducer.

3. The Time-Frequency Based Procedure for Evaluation of Easy Magnetization Axis

Figure 3 presents the successive stages of the procedure for evaluation of magnetic anisotropy based on the *TF* representation of the measured MBN signals along with exemplary MBN signals received for three different transducer's orientations. The early step is to perform the initial signal processing and determine the *TF* representation.

First, in order to minimize low-frequency instrumentation disturbances, the digital signal filtration was performed by the Butterworth high pass filter (with a cutoff frequency f_C of 2 kHz). Then the measurements obtained were divided into half periods of magnetization, each corresponding to the individual MBN burst. In the next step, for each individual burst, transformation of the U_{BN} (time representation) to the *TF* domain was performed. For this purpose, the STFT transformation was used, which enables a homogeneous division of the computational grid in the *TF* space to be obtained and, in consequence, the dynamics of the Barkhausen phenomenon to be observed in detail [24]. The Kaiser type of the computational window having a length of 512 samples along with the overlapping technique with a rate of 0.75 was used during the STFT transformation. The given above parameters of the window resulted in a computational time step ΔT of 512 μs and frequency step ΔF of 488 Hz, what allowed to precisely observe the changes in dynamics of the MBN activity for various measurement angles. A more detailed analysis of the impact of the window width and the value of the filter's cutoff frequency f_C on the quality of the acquired angular characteristics and the possibility of assessing the examined dependence is presented in the following sections of this paper. Finally, as a result, the complex *TF* representations $S_{BN}(t, f)$ of U_{BN} voltage signals were obtained. In next step, the spectrograms expressed as $|S_{BN}(t, f)|^2$ were calculated.

Multiple measurements for a single angular orientation enabled implementation of a smoothing procedure of the results obtained. Its purpose was to average all the spectrograms achieved for a single angular orientation of the transducer. Selected averaged spectrograms BN_{STFT_S} calculated for the measurements acquired within the half range of the transducer rotation (with respect to the TD from 0° to 180°) are shown in Figure 4. There is a noticeable difference between spectrograms for subsequent angles in the activity of the measured MBN. The highest activity was obtained for orientation of the transducer in accordance with the rolling direction of steel and along its easy magnetization axis (Figure 4, $\alpha = 90°$ i.e., RD). In this case, the highest energy level is visible practically throughout the entire MBN period, excluding the time span in which the magnetizing field changes its direction (around a time of 25 ms, see Figure 3b). At the same time, the frequency band occupied by the highest activity level of MBN is also the widest. Moreover, it can be seen that along with the gradual rotation of the transducer into the transverse direction to the RD (TD direction: $\alpha = 0°$ and $\alpha = 180°$), a gradual delay of the beginning of the MBN activity area (arises more slowly) as well as the general decrease in its area can be seen. The bandwidth of the MBN activity is also noticeably reduced. On all the distributions presented, regardless of the angle of measurement, one can notice two distinct areas of MBN activity, one at the beginning of the period around 15 ms and the other at its end around 40 ms. In addition, a third area of activity is strongly visible at the RD angle. This follows the observations presented by other researchers. Considering this, those activity regions could be associated with the aforementioned (three) processes of nucleation of reverse domains, 180° and 90° DWs motion and the growth of the MBN activity for RD could be explained in reference to the large quantity of 180° DWs [2,25,28,30,31,33]. Despite clearly visible variation of the MBN activity in spectrograms, for proper assessment it is necessary to carry out a detailed quantitative analysis enabling the quantification of observed relationships.

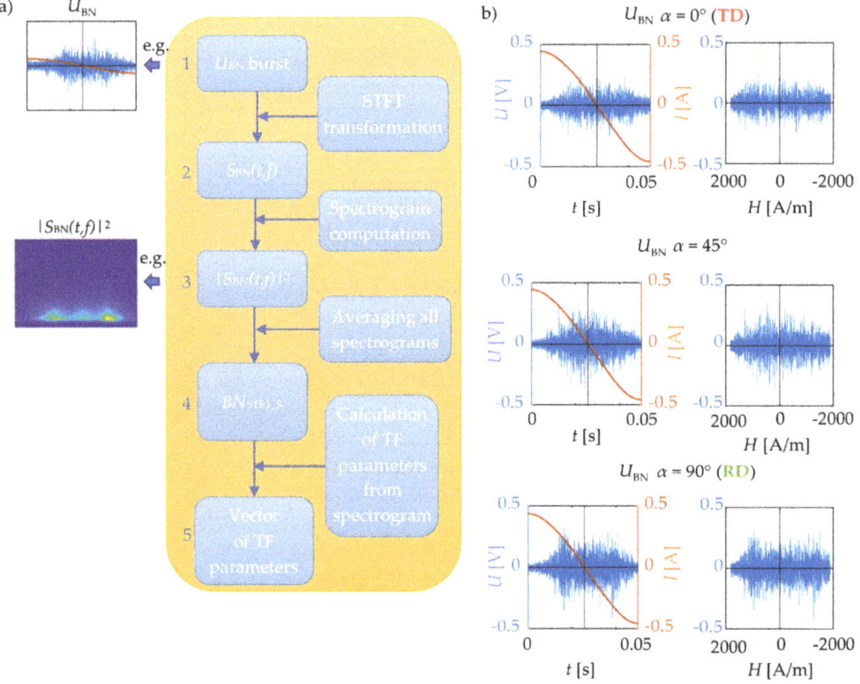

Figure 3. Time-frequency (TF) calculation procedure: (**a**) diagram of the procedure, (**b**) exemplary U_{BN} signals for orientation $\alpha = 0°$ (transverse direction, TD), $\alpha = 45°$, and $\alpha = 90°$ (rolling direction, RD) vs. time (left column) and magnetic field (right column—presented in accordance to descending half of magnetizing period).

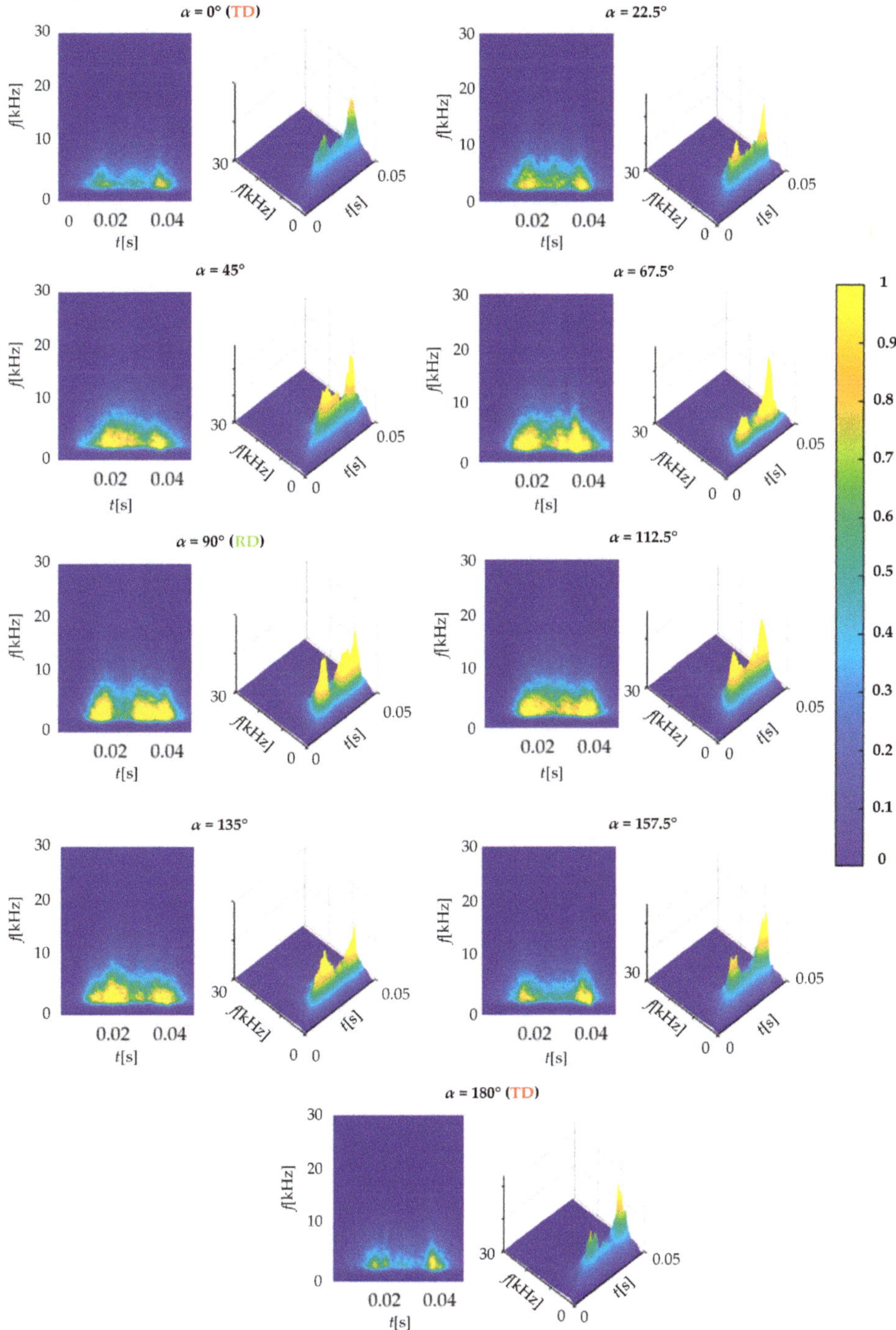

Figure 4. View of spectrogram activity in angle range between 0–180 degrees.

The magnetic anisotropy of the material ends up with differences in the magnetic properties occurring for different test angles. Based on the spectrograms, it can be noticed that these changes in properties are then reflected in the course and the intensity of the observed Barkhausen phenomenon. In consequence, the dynamics of the energy distribution, concentration, centroid shift in time and frequency, and the degree of order/disorder or scope of changes of the MBN *TF* representation may be affected. Therefore, in order to quantify information expressing the angular variations of magnetic characteristics, a multi-parameter extraction was used to define vectors of *TF* features for each test angle. As a result, the set of several parameters carried by the *TF* representation were calculated from the $BN_{\text{STFT\_S}}$. The first group refers to some statistical properties, including various forms of mean values (i.e., arithmetic, geometric, etc.), centroid, variance or standard deviation, skewness or kurtosis. The next subset are the features describing of the shape of the *TF* spectrogram and its energy distribution or entropy, thus allowing the dynamics of variance, uniformity of distribution or the degree of disorder of the MBN spectral content to be assessed. The last, but no less important part of the *TF* features vector is parameters that indicate different characteristics values of the $BN_{\text{TF\_S}}$, such as: symmetry, center shift in the *t* or *f* axis, flatness, homogeneity or monotonicity, etc. The definition and properties of all proposed *TF* features and the used calculation procedure have been introduced earlier and discussed in detail in [32]. The results of the angular distributions of selected features will be presented in the following chapter.

4. Angular Distribution Results of Time-Frequency (TF) Features

Figure 5 presents results of four selected parameters obtained from the *TF* representation: spectral flatness $BN_{\text{TF\_SF}}$, concentration measures $BN_{\text{TF\_CM}}$, spectral entropy $BN_{\text{TF\_SE}}$ and mean value $BN_{\text{TF\_MEAN}}$ of the spectrogram. The chosen parameters refer to all three groups described above, allowing generalization of the *TF* characteristic. The definitions of the presented parameters were shown in Table 1. Before presentation all features were normalized, to eliminate the isotropic part of information (presented in 0–1 scale).

Table 1. Definition of selected features based on spectrogram $BN_{\text{TF\_S}}$ of *TF* representation $S_{\text{BN}}(t, f)$.

Feature	Formula
Spectral Flatness	$BN_{\text{TF\_SF}} = M \cdot N \dfrac{\left(\prod_{i=1}^{N} \prod_{j=1}^{M} \left\| BN_{\text{TF\_S}_{i,j}} \right\| \right)^{\frac{1}{N \cdot M}}}{\sum_{i=1}^{N} \sum_{j=1}^{M} BN_{\text{TF\_S}_{i,j}}}$
Concentration Measure	$BN_{\text{TF\_CM}} = \left(\sum_{i=1}^{N} \sum_{j=1}^{M} \left\| BN_{\text{TF\_S}_{i,j}} \right\|^{\frac{1}{2}} \right)^{2}$
Spectral Entropy	$BN_{\text{TF\_SE}} = \dfrac{1}{1-3} \log_2 \sum_{i=1}^{N} \sum_{j=1}^{M} \left(\dfrac{BN_{\text{TF\_S}_{i,j}}}{\sum_{i=1}^{N} \sum_{j=1}^{M} BN_{\text{TF\_S}_{i,j}}} \right)^{3}$
Mean	$BN_{\text{TF\_MEAN}} = \dfrac{1}{N \cdot M} \sum_{i=1}^{N} \sum_{j=1}^{M} BN_{\text{TF\_S}_{i,j}}$

All feature distributions were approximated using a piecewise fitted curve to facilitate the analysis and interpretation of the obtained angular characteristic. It should be emphasized that obviously interpreting the distributions obtained is not the same in each case and the minimum values of some presented features do not have to refer to the minimum values of others, and the same for maximums. First, two of the selected parameters relate to the assessment of the energy concentration of the analyzed spectrogram. The $BN_{\text{TF\_SF}}$ parameter specifies the ratio of the geometric to the arithmetic mean and obtains higher values when the distribution is homogeneous (e.g., random). In reference to the presented *TF* spectrograms, this parameter takes the highest values for the hard magnetization axis. As the transducer orientation approaches to 90 degree angle (easy magnetization axis), value of the parameter begins to decrease, which indicates a growth of MBN activity (energy) area. Confirmation of this observation is found in the distribution of the second parameter, the BN_{TF_CM}, which takes higher values in case of the evenly distributed energy over the entire *TF* plane. However, at the same time, the parameter is not sensitive to small quantities. Thus when the level of MBN activity is growing in all spectral bands in general, the parameter value is reaching higher values as well. According to the received angular distribution, the BN_{TF_CM} value increases as the angle increases within the 0–90 degrees range. This parameter achieves the largest value in the direction of the easy magnetization axis, which is consistent with the observation for the angular distribution of the BN_{TF_SF} parameter. The spectral entropy allows the rate of disorder of the spectrogram to be assessed. It can be noticed that BN_{TF_SE} assumes the highest value for relatively wide orientation range around TD ($\alpha = 0°$ and $\alpha = 180°$), at the same time showing a sudden decrease of value for the angle close to RD ($\alpha = 90°$ and $\alpha = 270°$). This can be understood as leading the spectral distribution BN_{TF_S} to a higher degree of order and to accommodate the existing energy states for easy magnetization direction. The last presented parameter, BN_{TF_MEAN}, relates to the statistical quantity. The angular characteristic of the feature is well correlated with the course of the BN_{TF_CM}. This confirms the earlier observations regarding the increase in value levels practically throughout the whole spectrogram space for angles consistent with the RD direction or close to it. All obtained distributions allow to draw similar conclusions. One can see a general indication of the directions of high and low activity of MBN, which are also consistent with the direction of, respectively, easy and hard magnetization. In reference to Figure 4, for the RD angle, one can observe a clear increase in MBN activity, expressed by a global increase in its level (reflecting among others, in BN_{TF_MEAN} and BN_{TF_SF} values). At the same time, the greatest energy values accumulate within the three mentioned sub-periods (affecting among others, the BN_{TF_CM}), with a growing difference between areas with low and high MBN activity (accommodation of existing energy states depicted by among others, BN_{TF_SE}). Moreover, this increase in activity is most visible on the presented spectrograms near the central part of the entire MBN signal period. From extensive research reported in the literature, RMA has a particular impact on activity in this area [25,30,31]. This would indicate the greatest impact of RMA on the distribution of resultant anisotropy and the occurrence of the easy magnetization axis. The information contained in the proposed *TF* features can be validated by classical MBN features. Figure 6 presents the distribution of two frequently utilized MBN parameters, i.e., number of events BN_N and energy BN_{EN}. The distributions obtained confirm the directional properties of the tested steel, showing a clear increase in energy for the direction consistent with RD. An increase in activity for the RD direction was also observed in other publications [26,28], where authors explained the higher MBN activity (determined by the higher amplitude of the MBN envelope and RMS or energy values) by a much larger number of 180° domain walls in this direction. Similar conclusions were also presented in a number of other works [25,27,30,31], under consideration of only a middle part of the MBN signal associated with the movement of the 180° DWs and related to the influence of RMA. Based on the obtained angular characteristics (Figures 5 and 6) one can notice good agreement with the presented *TF*-based results. Recently the authors presented detailed comparison of the various features obtained from the time, frequency and time-frequency domain [32]. A good correlation between the spectral flatness and number of events, and between concentration measure and energy as well was reported. The presented results in Figures 5 and 6 confirmed the previous observations. The observed increase for RD in activity also translates into energy carried by MBN, while the decreasing number of events can be explained by the increase in the MBN phenomenon intensity and the superposition of smaller impulses into a larger cluster. This confirms the observations made for the parameters of the *TF* characteristics. Higher pulse values obtained for the RD direction compared to the TD direction cause a significant increase in energy value, but also a noticeable increase in the bandwidth occupied by the areas of highest activity, which (considering the scale of both factors) finally affects an increase in the value of the BN_{TF_CM} parameter. Furthermore, the increase in the difference between the energy states of the highest activity areas and the rest of the spectrogram determines the decrease in the BN_{TF_SF} parameter value. In addition, the overlapping of MBN events and accumulation of energy lead to a decrease in the parameter $BN_{\text{TF\_SE}}$. Considering all aspects, the results obtained indicated the possibility of using this proposed method based on the *TF* representation of full-period for the analysis of the resultant anisotropy in SiFe steel.

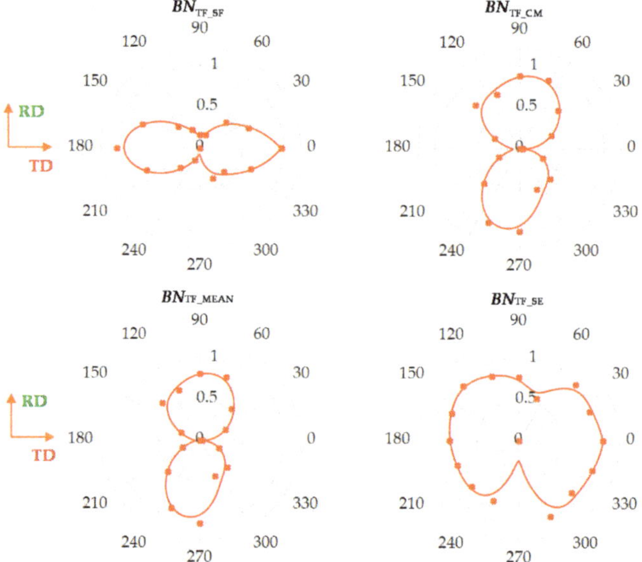

Figure 5. View of calculated parameters for 2 kHz and 512 window size: Markers represents the features values obtained for selected test angles; solid line refers the piecewise fitting result; all results are normalized with respect to maximum value.

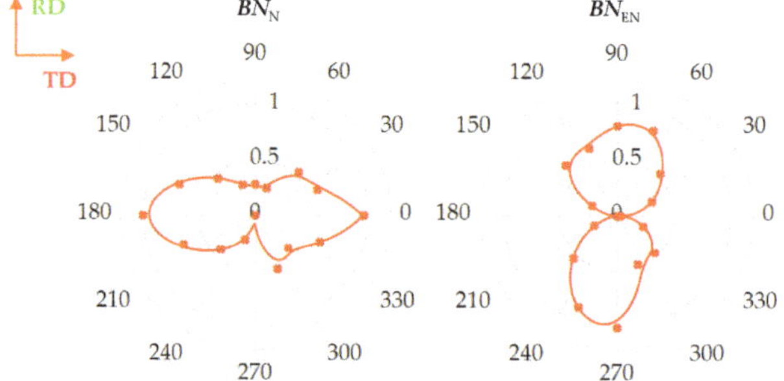

Figure 6. Results of the angular distributions of classical magnetic Barkhausen noise (MBN) parameters, that is the number of events BN_N and the energy BN_{EN} derived from time domain representation: markers represents the features values obtained for selected test angles; solid line refers the piecewise fitting result; all results are normalized with respect to maximum value.

Finally, despite the sparse measurements (22.5° angular step) and the occurrence in some cases of non-compliance of measuring points with angular characteristics, which also affects the achieved approximations, the RD and TD axes can in almost all cases be uniquely identified by the location of maximum or minimum values of individual parameters.

However, due to the observed discrepancies, it becomes justified to verify the repeatability of the results obtained. As a significant number of MBN bursts were registered each time (10 measurements, 10 magnetization periods each) the assessment of the divergence of individual *TF* features obtained for subsequent measurement angles can be performed. Therefore, in the next stage of the work, the impact of possible causes affecting the range of the dispersion of parameter values obtained for a single angular orientation of the transducer was analyzed.

5. Analysis of the *TF* Features Dispersion

Several factors may influence the dispersion of the parameters' distributions described above and in the previous section. They concern both the stage of measurement and hardware configuration of the system as well as the subsequent stage of data processing and the selection of computational procedure parameters. It should be emphasized that one of the key factors affecting the smoothness of the characteristic transitions between successive measurement angles is the number of adopted angular steps. The step applied in the paper is relatively large, however, it still makes it possible to determine general trends of the characteristic. On the other hand, it can be noted that errors for individual measurement points have a significant impact on the local course of the characteristic. Another key factor affecting errors is the relatively large size of the transducer used, which means that the measured values should not be treated as spot, but measurements in a certain vicinity of the given point. Furthermore, the minimization of possible influence of random factors on the recorded signals was obtained by repeating the measurements sequence several times under the same conditions (excitation parameters, transducer orientation angle in relation to the sample axis) and then by implementing the procedure of data averaging. At the same time, the influence of alternating interfering components having a frequency constant over time (if appears) would result in constant energy level over the entire spectrogram period for a given frequency band, and thus could also be clearly identified on the time-frequency distributions. Some systematic error may also be the result of a small angular shift between the sample edges and the actual magnetization direction, but it is replicated for all orientations in very similar scale and has no major impact on values dispersion or on the achieved shape of the distributions, but possibly only on the angular shift of the characteristics. Another possible source of measurement errors may also be a small misalignment of the transducer's symmetry axis in relation to its rotation axis. This misalignment error would result in an asymmetrical angular relationship of the calculated *TF* parameters. However, it would not affect the assessment of the direction of the easy or hard magnetization axis, but would end in obtaining a different scale of values on both sides (assuming high repeatability of the angular magnetic properties of the steel the difference would be negligible) of the symmetry of *TF* features' characteristics. Nevertheless, it would also not results in dispersion variation between orientations of the *TF* parameters. Therefore, bearing in mind the above considerations, the assessment of the impact of the computational procedure configuration on the repeatability and dispersion of the results still remains. There are two key elements within this aspect. The first concerns the choice of window width used during the transformation of data into the *TF* domain. The second is related to the pre-transformation data preparation stage and refers to the adjustment of the digital high-pass filter (HPF) band f_C. These parameters are crucial in the process of obtaining information.

The window size affects the resolution of time and frequency steps of the computed *TF* representation during the STFT transformation. For a narrow window, a high resolution in the time domain and a lower one in the frequency domain is achieved. In the opposite case, the wider the window is, the lower the resolution in time (larger time steps) and the higher in frequency is achieved. Therefore, the adjustment of the size is crucial to sensitivity for variations in analyzed spectrograms. As discussed in the introduction, anisotropy can be caused by a number of factors affecting the various stages of the reorganization of the domain structure during a single period of magnetization, and consequently appearing with different activity of the Barkhausen phenomenon in the time sequences. Thus, when analyzing individual factors, there is a need to obtain a high resolution over time, enabling unambiguous distinction between subperiods of increased Barkhausen phenomenon activity. Recently, the authors have considered the possibility of using *TF* analysis to distinguish time spans of MBN activity corresponding to various factors affecting anisotropy [24]. In that work, the key aspect was to maintain high resolution over time. Therefore, a relatively narrow window (guaranteeing steps $\Delta T = 128$ μs and $\Delta F = 1.952$ kHz) was used during STFT transformation. However, the purpose of this work is to explore the possibility of conducting a quick assessment of the resultant anisotropy. The entire period of MBN activity is then analyzed, not its subperiods. Therefore, in this situation, a high resolution over time is not so crucial, and higher resolution in frequency may have a greater impact on the effectiveness of the proposed analysis.

The second factor that may affect the amount of dispersion of features in a given measurement group is the value of the lower band f_C of the MBN signal frequency range. As can be noticed in Figure 4, the highest MBN activity is obtained within the lower frequency range of presented spectrograms BN_{TF_S}. This means that the selection of the lower band value can be decisive for the *TF* parameter distributions obtained during the analysis. In this range, interference from measuring instrumentation can affect the signal. In addition, distortions of a frequency close to the lower range of the hardware filter may also occur. Therefore, it is crucial to properly filter the signal prior further analysis. The cutoff frequency f_C of the digital high-pass filter should be high enough to effectively cut off the effect of low-pass interference, but at the same time low enough not to lose information about the magnetic anisotropy of the material.

Finally, during the analysis of the impact of both factors on the spread of *TF* parameters, a series of calculations was carried out for four cases determined by two window sizes: 128 and 512 samples and two digital filter border frequencies: 0.5 kHz and 2 kHz. The choice of window width was made taking into account the resolution in time and frequency that they guarantee. In the first case, it is possible to obtain relatively small steps of $\Delta T = 128\mu s$ and large of $\Delta F = 1.952$ kHz, while in the second one inversely, relatively large steps of $\Delta T = 512\mu s$ and small steps of $\Delta F = 488$ Hz. In this way, it was possible to estimate the impact of the individual resolutions of the computational grid on the repeatability of the obtained distributions. When choosing the f_C value, the cutoff frequency of the hardware filter (0.6 kHz) was taken into account, as well as the values of the frequency steps for both computational grids. Therefore, finally the analysis was carried out for f_C less than 0.6 kHz and greater than 1.952 kHz, which allowed evaluation of the spread of *TF* parameters for two extreme settings.

In the first stage of the analysis the averaged spectrograms BN_{TF_S} were calculated for each out of 10 measurements made for a given angular orientation. Next, for each averaged spectrogram the *TF* parameters were calculated. Then, spreads of *TF* features values were determined in reference to their average values achieved from all measurements. Finally, the values obtained for individual orientation angles (individual subsets) were further averaged and presented in the form of a common bar representation. Figure 7 presents a bar graph illustrating the dispersion range expressed in percentage scale (*%CumRange* indicator) of selected *TF* parameters accumulating the results achieved for all transducer orientations. Each bar specifies a dispersion range received under different data processing conditions. It can be noted that in most cases the smallest range value is obtained for the f_C of 2 kHz (first two columns of Figure 7). Considering the dispersion values of BN_{TF_MEAN}, one can notice that when the f_C is 2 kHz the window size is not so significant, as the *%CumRange* values are comparable for both widths considered. However, the situation changes when the f_C is equal to 0.5 kHz. In that case, a smaller *%CumRange* is achieved for wider window. This confirms the high importance of the low-frequency range of the spectrogram on the *TF* features distributions. For a window width of 128 samples, the frequency step ΔF is 1.952 kHz. As a result, the use of a cutoff frequency of 0.5 kHz causes that any undesirable components of the transformed U_{BN} signal can be reflected in the lower range of the spectrogram band. Therefore, it can be concluded that the improvement of the robustness can be obtained when the f_C is set to greater value than the first frequency step of the calculation grid, i.e., for $f_C = 2$ kHz. On the other hand, in the case of window size of 512 sample the ΔF is equal to 488 Hz. Thus both considered cutoff frequencies are higher than the first frequency step of the spectrogram grid. Therefore, mostly similar results of the *%CumRange* are obtained for both f_C values, and the distributions of BN_{TF_MEAN} are more convergent in those cases.

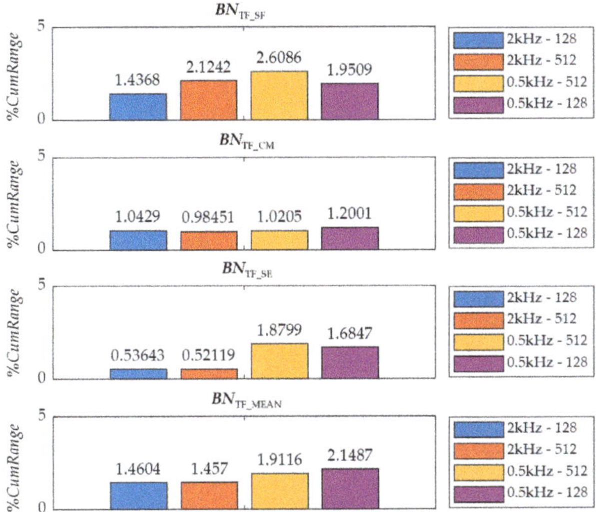

Figure 7. Bar graph of %CumRange indicator values of used features.

In the case of the spectral flatness parameter, smaller spreads were obtained using a smaller size of window. However, as in the case of the mean parameter, also in this case the use of a higher cut-off frequency significantly reduces the dispersion range. In the case of the BN_{TF_CM} parameter, convergent results were obtained to those obtained for the BN_{TF_MEAN} parameter. The smallest spread value was obtained for a window of 512 samples and f_C of 2 kHz. However, it must be noted that the BN_{TF_CM} presents the highest stability over various computation parameters. The values of *%CumRange* obtained for feature BN_{TF_SE} are the smallest for window width 512 samples and $f_C = 2$ KHz. Thus, the BN_{TF_CM} feature proves high repeatability and resistance to interfering factors.

In order to further examine the influence of the cutoff frequency on the information contained in the lower frequency ranges of spectrograms, one more analysis was performed. Its purpose was to determine the angular distributions (characteristics) of *TF* parameters as a function of successive frequency ranges of the spectrogram's computational grid. Graphic visualization of the calculation procedure is shown in Figure 8. For a given spectrogram, the entire time vector (single row) is considered, representing a single frequency range. Then the *TF* parameter value is calculated for this vector and the received value represents then a single cell of the resulting distribution presented in the form of the heat map.

Figure 9 presents two heat maps obtained for the BN_{TF_MEAN} parameter under two considered cutoff frequencies. In order to compare both conditions, the input spectrograms were computed using a window of 512 samples. Then the distributions were normalized to the (0–1) range and plotted in individual scales. If the lower frequency band value is used, the resulting distribution (top row) in the lower frequency ranges undergoes quite rapid changes of a rather random nature and not dependent on the angle of testing. Only for 90 and 270 degrees angles (along RD), in the higher frequency ranges, the characteristics present the angular relationship more visibly. In the second case, after applying a high-pass filter with a cutoff frequency of 2 kHz, a clear angular relationship in all compartments characterized by smooth and repeatable transitions between individual angles is obtained. Moreover, for angles that determine the axis of easy magnetization, large parameter values that at the same time cover a much wider frequency band are clearly visible. Attention should be paid to normalized ranges of values of both distributions as well. The unwanted components of low-frequency signals, not containing anisotropic information, reach much higher values. Therefore, they affect the sensitivity of the *TF* parameters and distort the crucial information.

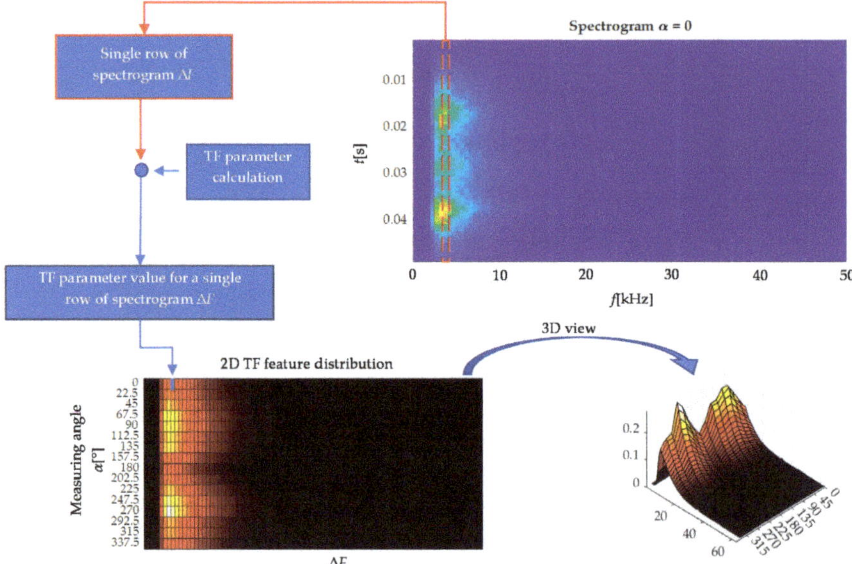

Figure 8. Diagram of the computational procedure of TF feature angular distribution in successive frequency ranges.

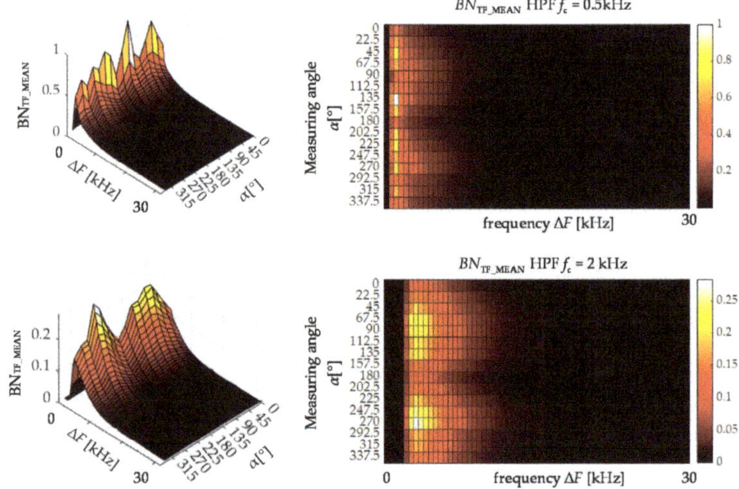

Figure 9. Visualization of the BN_{TF_MEAN} angular distribution in successive frequency ranges of the spectrogram; left column is a perspective presentation of the results in the right column.

Summarizing, the results of the analysis confirmed the impact of the computational parameters selection on the achieved dispersion of *TF* features values in successive subgroups, which in turn affects the repeatability of the entire method. In addition, it should be emphasized that the choice of the digital filter cutoff frequency f_C value is closely related to the window width value used during the STFT transformation. Therefore, in order to obtain the highest convergence of results and minimize the impact of external factors on parameter distribution, during the final analysis all results were (presented in previous chapter) achieved using the width of the window equal to 512 samples. This allowed the frequency resolution of spectrograms to be obtained enabling observation of even small variations of the MBN band, and at the same time to minimize the influence of the f_C value on the results. Nevertheless, aiming at maximum reduction of the influence of external factors on the computed *TF* distributions, additionally during the signal-processing procedure a cutoff frequency of 2 kHz was used. With these values of calculation parameters, the percentage dispersion of *TF* features values practically do not exceed 2%.

6. Conclusions

The ability to carry out rapid validation of the angular distribution of the magnetic properties of electrical steel is an important practical issue. Therefore, the purpose of this article was to investigate the possibility of using the Barkhausen effect and its time-frequency representation as a non-destructive tool to develop a procedure that allows retrieval of information about the resultant magnetic anisotropy in a classic grain-oriented electrical steel sheet. By assumption, this method is to be an alternative to the classic MBN approach, and ought to enable broad assessment of changes in the dynamics of the MBN phenomenon synchronously considering both time and frequency characteristics. For this purpose, tests were performed for a sample of 3% SiFe grain-oriented electrical steel. The measurements were made for 16 angles of transducer orientation, equally spaced within the range corresponding to full rotation. The paper presents the procedure of signal transformation into the *TF* domain, presentation of *TF* spectrograms and quantification of information contained therein. The obtained *TF* parameters expressing the change of the spectrogram allow the angle corresponding to the easy and hard magnetization axis to be determined. The possibility of extensive analysis of properties and observation of changes in the MBN activity makes this method possible to detect even small variations in the dynamics of the phenomenon over time resulting from the changes in magnetic properties. This creates a chance to obtain complementary information about the properties of the material, which can allow the formation of more complete knowledge.

The proposed method, based on the analysis of time-frequency characteristics in the full period of the MBN signal, allows detailed observation of the relationship between MBN properties expressed in time and in frequency for subsequent time moments. In consequence, it enabled the three characteristic MBN activity areas to be noticed. The occurrence of these areas in the MBN period has already been widely analyzed in many other works, and the angular characteristics of *TF* parameters obtained in this work are consistent with the other results reported. According to observations, these areas were associated with nucleation of reverse domains and, furthermore, the movement of the 180° and 90° DWs. The course of the first two was associated respectively with the magnetocrystalline anisotropy MCA and roll magnetic anisotropy RMA. Considering this, the angular distributions obtained by the authors underline the key importance of RMA for the alignment of resultant easy and hard magnetization axes in the tested steel. The angular characteristics of Barkhausen noise energy and numbers of events can be treated as confirmation of the effectiveness of the proposed *TF* method. These parameters, being used many times by other authors in the classic analysis, show convergent results with the *TF* ones. However, the results of both methods show some asymmetries and disturbances in the characteristics. The relatively small number of utilized measurement steps may have a decisive impact on the smoothness of the transitions of the characteristics obtained between successive measurement angles. The step applied in the paper makes it possible to determine general trends of the characteristic. On the other hand, errors for individual angles have a significant impact on the local course of the characteristics. Nevertheless, due to the sequencing of calculations procedures and the use of a sliding window over the analyzed signal, the STFT transformation leads to generalization of the phenomenon characteristics. In results, it can affect the increase of the robustness of the method for interfering external factors. This could be an explanation for obtaining slightly smoother courses of approximated *TF* characteristics.

In addition, the paper presents a detailed analysis of measurement errors and assesses the repeatability of the method in the context of the dispersion of *TF* parameters values within individual subsets of measurement signals. The choice of calculation parameters of the transformation procedure enables optimization of resolution in time and frequency, in reference to the nature of the observed changes. Based on the analysis conducted, it is crucial to emphasize the importance of the concentration measure parameter (carrying convergent information to MBN energy), which, regardless of the calculation parameters used, was characterized by high repeatability. The study also showed that the frequency value of the high-pass filter used for signal conditioning before *TF* analysis should not be lower than the value of the first step of the applied *TF* computational grid, which in turn allows for a significant increase in *TF* parameter robustness.

The presented method of time-frequency analysis is largely derived from the development of commonly applied methods of time or frequency analysis and is intended to present a possible path of development. The parameters defining the time-frequency characteristics are for the most part an extension of the classically used statistical coefficients. In principle, classical methods of analysis, in a single time or frequency domain, can be expressed as a generalization of the trends of changes presented in the time-frequency characteristics. Of course, at the same time it is not said that obtaining more detailed information will significantly improve the effectiveness of the MBN method in the future. However, the possibility of sequential data analysis itself (due to the potential to supplement the knowledge about the tested material) affects the validity of *TF* analysis usage and should be further examined. In the context of the presented arguments and already published papers, it can be seen that the use of time-frequency methods certainly does not affect the loss of information. On the contrary it can provide significant arguments to the discussion, as it gives a much wider perspective and brings a broader analysis of complex relationships of time and frequency characteristics.

In addition to the content of the work, one more aspect should be noted related to the development of future measurement systems. The classic method of analysis, despite many years of use, is still not standardized, and often the calculation procedures differ between researchers. The problematic element is, among others, the way of determining the MBN signal's background level or procedure of counting the number of events. Therefore, the time-frequency calculation procedures are not showing greater complexity in terms of the practical implementation of the MBN method, especially currently when the advancement and miniaturization of computing units is common. It should be also considered that today's diagnostic systems are increasingly based on monitoring many parameters and formulating multi-variate rules and correlation dependencies. Under these circumstances the results presented here show good potential. However the investigation with a large number of angular steps for steels having various textures and under influence of various anisotropy factors should be reaped. The issue, which also requires further research, is the synthesis method of the knowledge contained in *TF* representations, enabling greater efficiency to generalize and detailed analysis of information. The authors will present work on multiple samples and on multi-threaded analysis and synthesis procedures in the future.

Author Contributions: Conceptualization, G.P. and M.M.; methodology, G.P. and M.M.; software, G.P. and M.M.; validation, G.P. and M.M., formal analysis, G.P. and M.M.; investigation, G.P. and M.M.; resources, G.P.; data curation, M.M.; writing—original draft preparation, M.M.; writing—review and editing, G.P. and M.M.; visualization, G.P. and M.M.; supervision, G.P. All authors have read and agreed to the published version of the manuscript.

Funding: The APC was funded by Research Funds of Faculty of Electrical Engineering of West Pomeranian University of Technology, Szczecin, Poland.

Conflicts of Interest: The authors declare no conflict of interest.

References

1.   Jiles, D. *Introduction to Magnetism and Magnetic Materials*, 3rd ed.; CRC Press, Taylor & Francis Group: Boca Raton, FL, USA, 2016; ISBN 978-1-4822-3887-7.

2.   Jiles, D.C. Dynamics of domain magnetization and the Barkhausen effect. *Czech J. Phys.* **2000**, *50*, 893–924. [CrossRef]

3. Hubert, A.; Schäfer, R. *Magnetic Domains: The Analysis of Magnetic Microstructures*; Springer: Berlin, Germany; New York, NY, USA, 1998; ISBN 978-3-540-64108-7.
4. Cullity, B.D.; Graham, C.D. *Introduction to Magnetic Materials*, 2nd ed.; IEEE/Wiley: Hoboken, NJ, USA, 2009; ISBN 978-0-471-47741-9.
5. Nahak, B. Material Characterization using Barkhausen Noise Analysis Technique—A Review. *Indian J. Sci. Technol.* **2017**, *10*, 1–10. [CrossRef]
6. Deng, Y.; Li, Z.; Chen, J.; Qi, X. The effects of the structure characteristics on Magnetic Barkhausen noise in commercial steels. *J. Magn. Magn. Mater.* **2018**, *451*, 276–282. [CrossRef]
7. Makowska, K.; Kowalewski, Z.L.; Augustyniak, B.; Piotrowski, L. Determination of mechanical properties of P91 steel by means of magnetic Barkhausen emission. *J. Theor. Appl. Mech.* **2014**, *52*, 181–188.
8. Yamazaki, T.; Furuya, Y.; Nakao, W. Experimental evaluation of domain wall dynamics by Barkhausen noise analysis in Fe30Co70 magnetostrictive alloy wire. *J. Magn. Magn. Mater.* **2019**, *475*, 240–248. [CrossRef]
9. Gatelier-Rothea, C.; Chicois, J.; Fougeres, R.; Fleischmann, P. Characterization of pure iron and (130 p.p.m.) carbon–iron binary alloy by Barkhausen noise measurements: Study of the influence of stress and microstructure. *Acta Mater.* **1998**, *46*, 4873–4882. [CrossRef]
10. Bartošová, I.; Veterníková, J.; Slugeň, V. Study of candidate materials for new reactor systems using positron annihilation spectroscopy and Barkhausen noise. *Nucl. Eng. Des.* **2014**, *273*, 376–380. [CrossRef]
11. Drehmer, A.; Gerhardt, G.J.L.; Missell, F.P. Case depth in SAE 1020 steel using barkhausen noise. *Mater. Res.* **2013**, *16*, 1015–1019. [CrossRef]
12. Sorsa, A.; Santa-aho, S.; Vippola, M.; Lepistö, T.; Leiviskä, K. Utilization of frequency-domain information of Barkhausen noise signal in quantitative prediction of material properties. *Aip Conf. Proc.* **2014**, *1581*, 1256–1263.
13. Maass, P.; Teschke, G.; Willmann, W.; Wollmann, G. Detection and classification of material attributes-a practical application of wavelet analysis. *IEEE Trans. Signal Process.* **2000**, *48*, 2432–2438. [CrossRef]
14. Tomkowski, R.; Jonsson, S.; Lundin, P.; Nerman, P. *Penetration Depth Investigation of Barkhausen Noise Signal for Case-Hardened Components*; ICBM: Dresden, Germany, 2017.
15. Ding, S.; Tian, G.; Dobmann, G.; Wang, P. Analysis of domain wall dynamics based on skewness of magnetic Barkhausen noise for applied stress determination. *J. Magn. Magn. Mater.* **2017**, *421*, 225–229. [CrossRef]
16. Roskosz, M.; Fryczowski, K.; Schabowicz, K. Evaluation of Ferromagnetic Steel Hardness Based on an Analysis of the Barkhausen Noise Number of Events. *Materials* **2020**, *13*, 2059. [CrossRef]
17. Chmielewski, M.; Piotrowski, L. Application of the Barkhausen effect probe with adjustable magnetic field direction for stress state determination in the P91 steel pipe. *J. Electr. Eng.* **2018**, *69*, 497–501. [CrossRef]
18. Qiu, F.; Ren, W.; Tian, G.Y.; Gao, B. Characterization of applied tensile stress using domain wall dynamic behavior of grain-oriented electrical steel. *J. Magn. Magn. Mater.* **2017**, *432*, 250–259. [CrossRef]
19. Łopato, P.; Psuj, G.; Herbko, M.; Maciusowicz, M. Evaluation of stress in steel structures using electromagnetic methods based on utilization of microstrip antenna sensor and monitoring of AC magnetization process. *Inf. Cntrl. Meas. Econ. Environ. Prot.* **2016**, *4*, 32–36. [CrossRef]
20. Elleuch, M.; Poloujadoff, M. Anisotropy in three-phase transformer circuit model. *IEEE Trans. Magn.* **1997**, *33*, 4319–4326. [CrossRef]
21. Lee, J.-J.; Kwon, S.-O.; Hong, J.-P.; Ha, K.-H. Cogging Torque Analysis of the PMSM for High Performance Electrical Motor Considering Magnetic Anisotropy of Electrical Steel. *World Electr. Veh. J.* **2009**, *3*, 365–369. [CrossRef]
22. Clapham, L.; Heald, C.; Krause, T.; Atherton, D.L.; Clark, P. Origin of a magnetic easy axis in pipeline steel. *J. Appl. Phys.* **1999**, *86*, 1574. [CrossRef]
23. Pal'a, J.; Bydžovský, J.; Stoyka, V.; Kováč, F. Barkhausen noise study of microstructure in grain oriented FeSi steel. *J. Electr. Eng.* **2008**, *59*, 4.
24. Maciusowicz, M.; Psuj, G. Time-Frequency Analysis of Barkhausen Noise for the Needs of Anisotropy Evaluation of Grain-Oriented Steels. *Sensors* **2020**, *20*, 768. [CrossRef]
25. Pérez-Benítez, J.A.; Espina-Hernández, J.H.; Man, T.L.; Caleyo, F.; Hallen, J.M. Identification of different processes in magnetization dynamics of API steels using magnetic Barkhausen noise. *J. Phys. D Appl. Phys.* **2015**, *48*, 295002. [CrossRef]

26. Chávez-Gonzalez, A.F.; Martínez-Ortiz, P.; Pérez-Benítez, J.A.; Espina-Hernández, J.H.; Caleyo, F. Comparison of angular dependence of magnetic Barkhausen noise of hysteresis and initial magnetization curve in API5L steel. *J. Magn. Magn. Mater.* **2018**, *446*, 18–27. [CrossRef]
27. Caldas-Morgan, M.; Padovese, L.R. Fast detection of the magnetic easy axis on steel sheet using the continuous rotational Barkhausen method. *NDT E Int.* **2012**, *45*, 148–155. [CrossRef]
28. De Campos, M.F.; Campos, M.A.; Landgraf, F.J.G.; Padovese, L.R. Anisotropy study of grain oriented steels with Magnetic Barkhausen Noise. *J. Phys. Conf. Ser.* **2011**, *303*, 012020. [CrossRef]
29. Tsuchida, Y.; Oka, M.; Enokizono, M. Rotational Barkhausen signals under rotating magnetic flux. In Proceedings of the 35th Annual Review of Progress in Quantitative Nondestructive Evaluation, Chicago, IL, USA, 20–25 July 2008; AIP: New York, NY, USA, 2009; Volume 1096, pp. 459–466.
30. Martínez-Ortiz, P.; Pérez-Benítez, J.A.; Espina-Hernández, J.H.; Caleyo, F.; Hallen, J.M. On the estimation of the magnetic easy axis in pipeline steels using magnetic Barkhausen noise. *J. Magn. Magn. Mater.* **2015**, *374*, 67–74. [CrossRef]
31. Martínez-Ortiz, P.; Pérez-Benítez, J.A.; Espina-Hernández, J.H.; Caleyo, F.; Mehboob, N.; Grössinger, R.; Hallen, J.M. Influence of the maximum applied magnetic field on the angular dependence of Magnetic Barkhausen Noise in API5L steels. *J. Magn. Magn. Mater.* **2016**, *401*, 108–115. [CrossRef]
32. Maciusowicz, M.; Psuj, G. Use of Time-Dependent Multispectral Representation of Magnetic Barkhausen Noise Signals for the Needs of Non-Destructive Evaluation of Steel Materials. *Sensors* **2019**, *19*, 1443. [CrossRef]
33. Shilling, J.; Houze, G. Magnetic properties and domain structure in grain-oriented 3% Si-Fe. *IEEE Trans. Magn.* **1974**, *10*, 195–223. [CrossRef]

 *materials*

Article

Nondestructive Damage Testing of Beam Structure Based on Vibration Response Signal Analysis

Xiaohao Li [1,2,*], Deyu Shi [1] and Zihang Yu [1]

[1] School of Mechanical Engineering and Automation, Northeastern University, Shenyang 110819, China; qaz_ln@126.com (D.S.); yu201809neu@yeah.net (Z.Y.)

[2] Key Laboratory of Vibration and Control of Aero-Propulsion Systems Ministry of Education of China, Northeastern University, Shenyang 110819, China

* Correspondence: xhli@me.neu.edu.cn; Tel.: +86-13478358901

Received: 20 June 2020; Accepted: 16 July 2020; Published: 24 July 2020

Abstract: Nondestructive damage-testing technology based on vibration signal analysis makes full use of the response characteristics of wave and energy. With the advantages of wide bandwidths of response frequency and high sensitivity, the nondestructive testing technology based on vibration signal analysis has a superiority in the application for the detection and characterization of structural defects, and has become one of the important methods for the nondestructive testing of structural material defects and damage. This paper presents a novel method of detection localization and quantitative analysis for local damage in beam structures, based on the response analysis of vibration signals. A damage-detection and -identification algorithm based on a unscented Kalman filter (UKF) was designed, which greatly reduces the computational workload in the process of damage identification over that in conventional methods. The method presented in this paper has significances to widen the application scope of the nondestructive testing method, and increase the recognition efficiency and effectiveness of this kind of method in engineering.

Keywords: vibration signal analysis; nondestructive testing; Kalman filtering; parameter identification

---

1. Introduction

With the rapid development of modern industry, nondestructive testing (NDT) technology for machinery, architecture and other structures has been rapidly developed and widely applied. However, the emergence of new materials and technologies, as well as the harsh working conditions of high temperature, high pressure and high load, have put higher requirements on the development of nondestructive testing technologies [1,2]. The effective testing of materials and structural defects is crucial to ensure the safety of materials and the reliability of structures. In the early stage of defect, crack propagation is very likely to occur in the material components under cyclic loading, which may result in the local deformation and failure of materials, introducing a huge risk of the failure of the overall system structure [3–5]. Therefore, it is of great practical significance to use nondestructive testing technology to monitor the quality of engineering structures in the early stage without invasive sampling.

Currently, traditional nondestructive testing technologies such as ultrasonic methods, thermal wave-imaging methods, ray methods and acoustic emission methods are widely used in mechanical engineering and the construction industry [6–8]. However, there are some limitations in the quantitative detection and defect expression in these methods. With nondestructive testing using vibration signal wave, its vibration response can effectively excite the vibration waves of various modes in the components being tested, such as the longitudinal wave, transverse wave and surface wave [9,10]. The reference [11] took a three-layer rack structure with six layers of storage as an example, carried out the dynamic analysis on it under horizontal and vertical loads, and completed the verification of force distribution and sag moment under test. By analyzing the interaction mechanism between the vibration wave and the structural material defects, the location and quantitative expression of the structural defects can be achieved. Compared to the traditional nondestructive testing technology, the nondestructive testing technology based on the vibration signal makes full use of the characteristics of wave and energy response. It has strong advantages in the detection and characterization of structural defects. In addition, due to its response in a wide frequency bandwidth, high sensitivity, and high spatial resolution, it has become one of the most important tools in the field of the structural defects of the material and nondestructive testing of damage [12–14].

In recent decades, researchers have done a lot of research on nondestructive testing technologies based on vibration signal analysis and obtained many meaningful research results. Cawley and Adams [15] found that the ratio of any two-order natural frequencies would change after the damage, which was only related to the damage location and had no relevance to the degree of damage when there was only single damage or multiple damage in the same degree in the structure, based on the assumption that the damage only caused changes in stiffness. Salawu [16] pointed out that different forms of damage might cause the same change of response frequency, especially in a situation of the damage of a symmetrical position in a symmetrical structure where the frequency index creates difficulty in distinguishing the damage. Yang et al. [17] proposed a method to add mass to solve the aforementioned issue. However, this method required the addition of known mass to the structure, which essentially changes the symmetric structure into asymmetric structure. Guan et al. [18] combined strain mode with wavelet transform and applied it to the damage identification of frame structures. Gu et al. [19] proposed a working strain mode identification technology under environmental excitation, and the corresponding damage-identification method. Pandey and Biawas [20] proposed a damage-identification method based on a modal flexibility matrix and studied the impact of damage on the flexibility matrix through numerical examples. The results showed that the damage location could be accurately determined by only using the first two order modes. However, the structure of the flexibility matrix was complex, which needs further improvement for engineering applications.

This paper presents a method for the detection of the local damage position in a structure and quantitative analysis based on vibration signal analysis. A novel algorithm of the nondestructive damage identification based on the unscented Kalman filter (UKF) was designed. This paper also expands a discussion about the disadvantages of the conventional application of the extended Kalman filter (EKF) in this field, which needs to spend high computational cost to calculate the Jacobi matrix. Compared to the EKF, the algorithm presented in this paper was able to substantially reduce the computing workload in the process of the structure damage identification, and significantly broaden the scope of application of nondestructive testing methods. The proposed method can also significantly improve the recognition efficiency and effectiveness in the engineering. At the same time, the method studied in this paper can effectively avoid: (1) the problem of low accuracy caused by the limitation of modal parameters in traditional damage-identification methods; and (2) the traditional method has a large amount of calculation work, which is not conducive to online loss identification. The effectiveness of the method was demonstrated in a computational application. The results show that the damage location identification, quantitative characterization and the anti-noise performance of UKF meets the requirements for practical engineering applications.

2. Design of the Nondestructive Testing Algorithms Based on Vibration Signal Analysis

2.1. Assumption of Nondestructive Testing Based on Vibration Signal Analysis

After damage to the structure of an equipment, the mass, stiffness and other characteristics of the equipment may all be changed. If all of these changes are considered, it is bound to increase the complexity of the research and analysis. Therefore, it is necessary to reasonably and concisely characterize the impact of damage on the structure. In engineering structures, common damages, such as cracks, have a great impact on local stiffness, while the change of local mass is usually minimal.

Therefore, it can be assumed that the damage only leads to change in the local stiffness, and the change of local mass can be ignored. This assumption can be expressed as in Equation (1);

$$\begin{cases} \mathbf{K}_D \neq \mathbf{K}_U \\ \mathbf{M}_D = \mathbf{M}_U \end{cases} \tag{1}$$

where, $\mathbf{K}_D$ and $\mathbf{M}_D$ are the local stiffness matrix and mass matrix after damage; $\mathbf{K}_U$ and $\mathbf{M}_U$ are the local stiffness matrix and the mass matrix before damage. The assumptions described as Equation (1) are basically in line with current engineering practice [21–24], and this can greatly simplify the difficulty of the research. According to Equation (1), a factor of local damage in a structure can be further defined with Equation (2):

$$\mathbf{K}_D = (1 - \mathbf{D})\mathbf{K}_U \tag{2}$$

where, $\mathbf{D}$ is the factor matrix of local damage, and it ranges $0 \sim 1$, where 0 means no damage and 1 means complete destruction. Formula (2) describes the reduction of local stiffness caused by the damage, which is the basis for defining and simulating damage.

Based on the above assumptions, this paper has designed a nondestructive testing algorithm as follows.

2.2. Design of Algorithms of the Nondestructive Testing Based on Vibration Signal Analysis

With known excitation, the motion equations of a linear structure with multiple degrees of freedom can be written as

$$\mathbf{M}\ddot{\mathbf{X}}(t) + \mathbf{C}\dot{\mathbf{X}}(t) + \mathbf{K}\mathbf{X}(t) = \mathbf{B}\mathbf{f}(t) \tag{3}$$

where, $\mathbf{X}(t)$, $\dot{\mathbf{X}}(t)$ and $\ddot{\mathbf{X}}(t)$ are the displacement, velocity and acceleration responses of the structure, respectively; $\mathbf{M}$, $\mathbf{C}$ and $\mathbf{K}$ are the mass matrix, damping matrix and stiffness matrix of the structure, respectively; $\mathbf{f}(t)$ is an external excitation vector; $\mathbf{B}$ is the influence matrix of external excitation (position matrix). An augmented structural state variable is introduced as

$$\mathbf{X} = \left[\mathbf{X}_1^T, \mathbf{X}_2^T, \mathbf{X}_3^T, \mathbf{X}_4^T\right]^T \tag{4}$$

where, $\mathbf{X}_1 = x$, $\mathbf{X}_2 = \dot{x}$ and $\mathbf{X}_3 = [k_1, k_2, \cdots, k_m]^T$ which is a vector consisting of all non-zero elements in the stiffness matrix; $\mathbf{X}_4 = [c_1, c_2, \cdots, c_m]^T$ is a vector consisting of non-zero elements in the damping matrix. Note that the effect of local damage on mass is ignored. Assuming that the damping is Rayleigh damping, it obtains:

$$\mathbf{C} = \alpha\mathbf{M} + \beta\mathbf{K} \tag{5}$$

where, α and β are the mass and stiffness damping coefficients, respectively, and $\mathbf{X}_4 = [\alpha, \beta]^T$. Equation (3) is rewritten as Equation (6) which is represented by augmented state variables:

$$\begin{Bmatrix} \mathbf{X}_1 \\ \mathbf{X}_2 \\ \mathbf{X}_3 \\ \mathbf{X}_4 \end{Bmatrix} = \begin{Bmatrix} \mathbf{X}_2 \\ \mathbf{M}^{-1}\left[\mathbf{B}\mathbf{f}(t) - \left[(\mathbf{C})_{\mathbf{X}_1}\mathbf{X}_3 + (\mathbf{K})_{\mathbf{X}_2}\mathbf{X}_4\right]\left[(\mathbf{C})_{\mathbf{X}_4}\mathbf{X}_2 - (\mathbf{K})_{\mathbf{X}_3}\mathbf{X}_1\right]^T\right] \\ \mathbf{K}^T \\ \mathbf{C}^T \end{Bmatrix} \tag{6}$$

where, $(\mathbf{C})_{\mathbf{X}_1}$ and $(\mathbf{C})_{\mathbf{X}_4}$ is the damping matrix based on $\mathbf{X}_1$ and $\mathbf{X}_4$; $(\mathbf{K})_{\mathbf{X}_2}$ and $(\mathbf{K})_{\mathbf{X}_3}$ is the stiffness matrix based on $\mathbf{X}_2$ and $\mathbf{X}_3$. Since $(\mathbf{C})_{\mathbf{X}_1}$, $(\mathbf{C})_{\mathbf{X}_4}$, $(\mathbf{K})_{\mathbf{X}_2}$ and $(\mathbf{K})_{\mathbf{X}_3}$ contain state variables, Equation (6) is a nonlinear equation for augmented state variables, which can be shortened as

$$\dot{\mathbf{X}} = \mathbf{g}(\mathbf{X}, \mathbf{f}, t) \tag{7}$$

The equation of the state is obtained by integrating Equation (7) with time as

$$\mathbf{X}_k = \mathbf{X}_{k-1} + \int_{t_{k-1}}^{t_k} \mathbf{g}(\mathbf{X}, \mathbf{f}, t)\mathrm{d}t + \mathbf{w}_k \tag{8}$$

where, t_k denotes the time at which the k state is; $\mathbf{w}_k$ is the process noise that here is assumed as Gauss white noise, thus the covariance matrix is constant matrix $\mathbf{Q}$. Equation (8) can be realized by a dynamic direct integration method in the process of designing a nondestructive testing and analysis program. Considering the placement of acceleration sensors on the structure [25,26], the observation equation is expressed as

$$\mathbf{Z}_k = \mathbf{D}\dot{\mathbf{X}} = \mathbf{D}\mathbf{M}^{-1}\left\{ \begin{array}{c} \mathbf{M}(\mathbf{X}_2)_k \\ \mathbf{B}\mathbf{f}_k - \left[(\mathbf{C})_{\mathbf{X}_1}\mathbf{X}_3 + (\mathbf{K})_{\mathbf{X}_2}\mathbf{X}_4\right]_k\left[(\mathbf{C})_{\mathbf{X}_4}\mathbf{X}_2 - (\mathbf{K})_{\mathbf{X}_3}\mathbf{X}_1\right]_k^{\mathrm{T}} \\ 0 \\ 0 \end{array} \right\} + \mathbf{v}_k \tag{9}$$

where, $\mathbf{f}_k$ represents the excitation at the time of k state; $\mathbf{v}_k$ is the measurement noise that here is assumed as Gauss white noise, thus the covariance matrix is a constant matrix $\mathbf{R}$. Equation (9) is also a nonlinear equation, which can be shortened as

$$\mathbf{Z}_k = \mathbf{h}(\mathbf{X}_k, \mathbf{f}_k) + \mathbf{v}_k \tag{10}$$

The aforementioned deduction shows that after introducing the augmented state variables into the linear structure, the corresponding state equations and measurement equations become nonlinear, thus a nonlinear filtering technology must be adopted. If the extended Kalman filter (EKF) is used, two Jacobi matrices as shown in Equations (11) and (12) need to be computed:

$$\mathbf{A} = \left.\frac{\partial \mathbf{g}(\mathbf{X}, \mathbf{f}, t)}{\partial \mathbf{X}}\right|_{\mathbf{X}=\mathbf{X}_{k-1|k-1}} \tag{11}$$

$$\mathbf{H} = \left.\frac{\partial \mathbf{h}(\mathbf{X}, \mathbf{f}, t)}{\partial \mathbf{X}}\right|_{\mathbf{X}=\mathbf{X}_{k|k-1}} \tag{12}$$

Based on Equations (6) and (9), the calculation of the Jacobi matrix involves differentiating the stiffness matrix and damping matrix. The process is closely related to the position of non-zero elements in the matrices. Once the structure changes, the position of the non-zero elements in the matrices will also change, thus Equations (11) and (12) must be deducted again. Therefore, this method of deduction lacks universality. When the expression of the stiffness matrix and damping matrix is very complex, the derivation process becomes even more complicated and prone to making large errors. If the unscented Kalman filter (UKF) is used, it just needs simple matrix operations on Equations (8) and (10) in the processing. Jacobi Matrix $\mathbf{A}$ and Matrix $\mathbf{H}$ are not required to be deduced again. The workload is significantly reduced, and the program becomes more universal, indicating that UKF has obvious advantages in damage identification. Based on this, a damage-identification algorithm of the vibration signal analysis, based on the UKF, is designed as follows:

(1) Establish augmented state variables according to Equation (4); construct the state equations according to Equations (6)–(8); and construct the observation equation according to Equation (9).
(2) Estimate the initial mean $\mathbf{X}_{0|0}$ and covariance matrix $\mathbf{P}_{\mathbf{X},0|0}$ of the state variables, and estimate the covariance matrix $\mathbf{Q}$ and $\mathbf{R}$ of process noise and measurement noise.
(3) Identify the parameters of structural stiffness and damping by the UKF filtering algorithm:

The main processing of the UKF filtering is to use the Unscented Transformation (UT) transform to deal with the nonlinear transfer of means and covariance in a Kalman filtering process. For the nonlinear equation of state and observation described as Equations (8) and (9), the UKF is processed as follows

(I) Set the initial parameters:

$$\mathbf{X}_{0|0} = \mathbf{E}\left[\mathbf{X}_{0|0}\right] \tag{13}$$

$$\mathbf{P}_{\mathbf{X},0|0} = \mathbf{E}\left[\left(\mathbf{X}_{0|0} - \overline{\mathbf{X}}_{0|0}\right)\left(\mathbf{X}_{0|0} - \overline{\mathbf{X}}_{0|0}\right)^{\mathrm{T}}\right] \tag{14}$$

where, $\overline{\mathbf{X}}_{0|0}$ and $\mathbf{P}_{\mathbf{X},0|0}$ represent the mean and covariance matrix of the initial estimated state variables, respectively.

(II) Construct the set of sigma points According to $\overline{\mathbf{X}}_{k-1|k-1}$ and $\mathbf{P}_{\mathbf{X},k-1|k-1}$ of state $k-1$, the sigma point set $\chi_{k-1|k-1}^{i}$ can be constructed according to Equation (15), where superscript i denotes the order of Point i in the set:

$$\chi_i = \begin{cases} \overline{x}, i = 0 \\ \overline{x} + \sqrt{(n-\kappa)(n+\kappa)}\left(\sqrt{\mathbf{P_x P_x^T}}\right)_i, i = 1, \cdots, n \\ \overline{x} - \sqrt{(n-\kappa)(n+\kappa)}\left(\sqrt{\mathbf{P_x P_x^T}}\right)_{i-n}, i = n+1, \cdots, 2n \end{cases} \tag{15}$$

(III) Predict By substituting the sigma point set $\chi_{k-1|k-1}^{i}$ into the Kalman filter [27,28], the nonlinear stochastic difference Equation (16) can be obtained as

$$\mathbf{X}_k = \mathbf{f}(\mathbf{X}_{k-1}, \mathbf{u}_{k-1}) + \mathbf{w}_{k-1} \tag{16}$$

Then, the transformed sigma point set $\chi_{k|k-1}^{i}$ can be obtained as

$$\chi_{k|k-1}^{i} = \mathbf{f}\left(\chi_{k|k-1}^{i}, \mathbf{u}_{k-1}\right) \tag{17}$$

According to $\chi_{k|k-1}^{i}$, the predicted mean of state variables is shown as Equation (18) and the covariance matrix is shown as Equation (19):

$$\overline{\mathbf{X}}_{k|k-1} = \sum_{i=0}^{2n} \mathbf{W}_i^{(m)} \chi_{k|k-1}^{i} \tag{18}$$

$$\mathbf{P}_{x,k|k-1} = \sum_{i=0}^{2n}\sum_{m=0}^{2n} \mathbf{W}_i^{(m)}\left[\chi_{k|k-1}^{i} - \mathbf{X}_{k|k-1}\right]\left[\chi_{k|k-1}^{i} + \overline{\mathbf{X}}_{k|k-1}\right]^{\mathrm{T}} + \sum_{i=0}^{2n}\left(\chi_{k|k-1}^{i} - \mathbf{I}\right)\left(\chi_{k|k-1}^{i} + \mathbf{I}\right)^{\mathrm{T}} 10^{-5}\mathbf{I} \tag{19}$$

where, $\mathbf{W}_i^{(m)}$ is the calculated mean weight coefficient and we set $\sum_{i=0}^{2n}\left(\chi_{k|k-1}^{i} - \mathbf{I}\right)\left(\chi_{k|k-1}^{i} + \mathbf{I}\right)^{\mathrm{T}} 10^{-5}\mathbf{I}$ as $\mathbf{Q}$.

By substituting $\chi_{k|k-1}^{i}$ into the nonlinear observation equation, Equation (9), we obtain:

$$\gamma_{k|k-1}^{i} = \mathbf{h}\left(\chi_{k|k-1}^{i}\right) \tag{20}$$

The mean value of the observation variables is calculated as

$$\mathbf{Z}_{k|k-1} = \sum_{i=0}^{2n} \mathbf{W}_i^{(m)} \gamma_{k|k-1}^{i} \tag{21}$$

(IV) Correct Compute the covariance matrix of the observation variables:

$$\mathbf{P}_{z,k|k-1} = \sum_{i=0}^{2n}\sum_{c=0}^{2n} \mathbf{W}_i^{(c)}\left[\gamma_{k|k-1}^i - \mathbf{Z}_{k|k-1}\right]\left[\gamma_{k|k-1}^i + \overline{\mathbf{Z}}_{k|k-1}\right]^{\mathrm{T}} + \sum_{i=0}^{2n}\left(\gamma_{k|k-1}^i - \mathbf{I}\right)\left(\gamma_{k|k-1}^i + \mathbf{I}\right)^{\mathrm{T}}\mathbf{I} \qquad (22)$$

where, $\mathbf{W}_i^{(c)}$ is the variance weight coefficient and we set $\sum_{i=0}^{2n}\left(\gamma_{k|k-1}^i - \mathbf{I}\right)\left(\gamma_{k|k-1}^i + \mathbf{I}\right)^{\mathrm{T}}\mathbf{I}$ as $\mathbf{R}$.

Compute the covariance matrix between the state variables and observation vectors;

$$\mathbf{P}_{k|k-1} = \sum_{i=0}^{2n}\sum_{c=0}^{2n} \mathbf{W}_i^{(c)}\left[\gamma_{k|k-1}^i + \mathbf{Z}_{k|k-1}\right]\left[\overline{\mathbf{X}}_{k|k-1} - \overline{\mathbf{Z}}_{k|k-1}\right]^{\mathrm{T}}\left[\mathbf{X}_{k|k-1}^i - \overline{\mathbf{X}}_{k|k-1}\right]^{\mathrm{T}} \qquad (23)$$

Then, the Kalman gain matrix can be obtain by

$$\mathbf{K}_k = \mathbf{P}_{z,k|k-1}\mathbf{P}_{z,k|k-1}^{-1} \qquad (24)$$

Update the mean of state variables:

$$\mathbf{X}_{k|k} = \mathbf{X}_{k|k-1} + \mathbf{K}_k\left(\mathbf{Z}_k - \mathbf{Z}_{k|k-1}\right) \qquad (25)$$

Update the covariance matrix of the state variables:

$$\mathbf{P}_{\mathbf{X},k|k} = \mathbf{P}_{\mathbf{X},k|k-1} - \mathbf{K}_k\mathbf{P}_{z,k|k-1}\mathbf{K}_k^{\mathrm{T}} \qquad (26)$$

From the steps mentioned above, it only needs to calculate the state equation and observation equation in the UKF filtering process, and does not need to calculate the Jacobi matrices. This method has obvious advantages in terms of the ease of use and flexibility when the state equation is complex and non-differentiable.

(V) Judge structural damage status based on the identification results.

The flow chart of the UKF algorithm is shown in Figure 1.

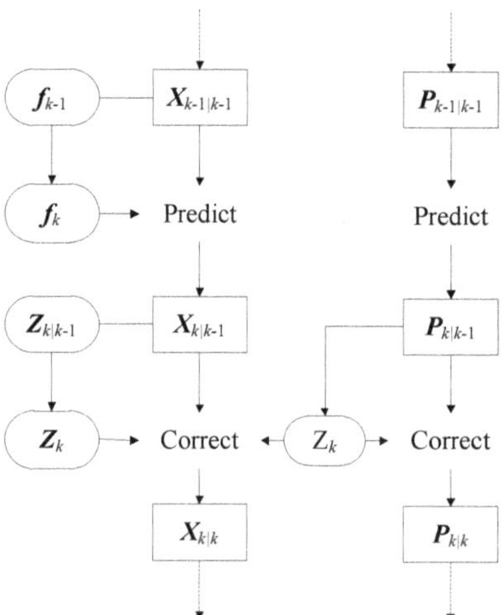

Figure 1. Flow chart of the unscented Kalman filter (UKF) algorithm under known excitation.

3. Application Example Analysis

3.1. Experimental Setting

As shown in Figure 2, the four-layer frame model is composed of plexiglass plates and aluminum columns, which are connected by bolts (due to the universality of the application of a frame beam in practical production, the frame structure is taken as an example. The non-destructive testing algorithm is used to evaluate and monitor the strength, stiffness, damping and other technical parameters of the frame beam, which is of great practical significance to eliminate potential safety hazards and ensure the safety of the frame structure [29]). The structure sits on orbits and is only allowed to move in the x direction. Each layer of the structure consists of four aluminum columns ($25 \times 25 \times 1$ mm) connected with two up-and-down plexiglass plates ($900 \times 450 \times 2.5$ mm), respectively, forming a four-degree-of-freedom system.

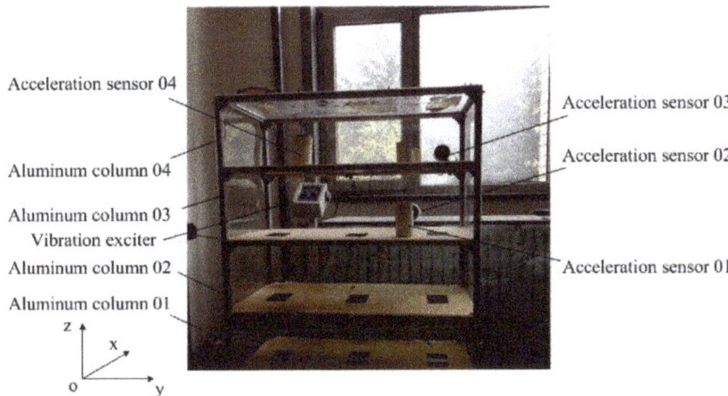

Figure 2. Frame structure and the sensor arrangement.

In Figure 2, an electromagnetic exciter (SA-JZ-50, Wuxi Shiao Technology Co., Ltd, Wuxi, China) is used to apply the lateral excitation to the bottom plate along the center line of the structure. The structure and the exciters are installed together on the base plate. A force sensor (BK-2Y, China Academy of Aerospace Aerodynamics, Beijing, China) (sensitivity 2.23 mV/N) is installed between the rod of the exciter and the structure to measure the excitation force. Four acceleration sensors (AD100T, Qinhuangdao Xinhua Technology Co., Ltd, Qinhuangdao, China) (sensitivity 100 mV/g) are installed in the positions shown in Figure 2 to measure the acceleration response of each layer. The excitation signal is the random excitation with a bandwidth of 5~50 Hz and the excitation level is 2.6 V as shown in Figure 3.

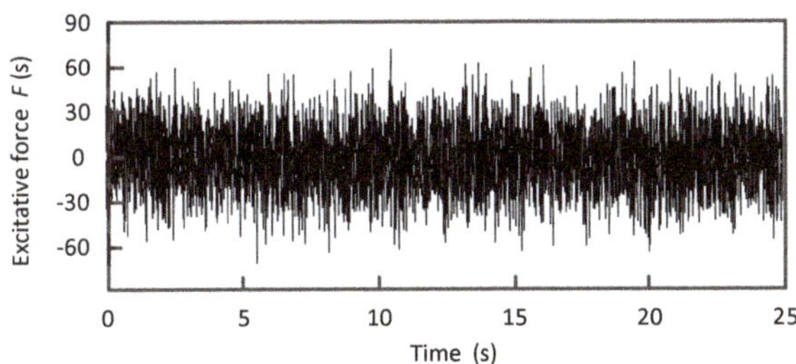

Figure 3. Measured random excitation signal.

The sampling frequency and sampling time of the experimental test system were 160 Hz and 25.6 s respectively, and 4096 points were sampled. The experiments were tested on 18 different working conditions, including increasing the mass of a certain layer, reducing the stiffness of a certain column, and introducing nonlinear damage into a buffer gap. This paper only analyzed the identification effectiveness of the algorithm on the stiffness reduction of Column 2 in Figure 2. We only selected four representative working conditions for research in this paper, as shown in Table 1.

Table 1. Physical experiment condition of the frame structure.

Working Condition Serial Number	Damage Condition
1	Nondestructive
2	The stiffness of No. 2 aluminum column decreases by 64.9%
3	The stiffness of No. 3 aluminum column decreases by 64.9%
4	The stiffness of No. 4 aluminum column decreases by 64.9%
...	...

3.2. Nondestructive Testing Based on Vibration Signal

In order to validate the damage identification of the algorithm proposed in this paper, a numerical model of the frame structure in Figure 2 was established as shown in Figure 4, in which the friction between the bottom plate and the track is neglected.

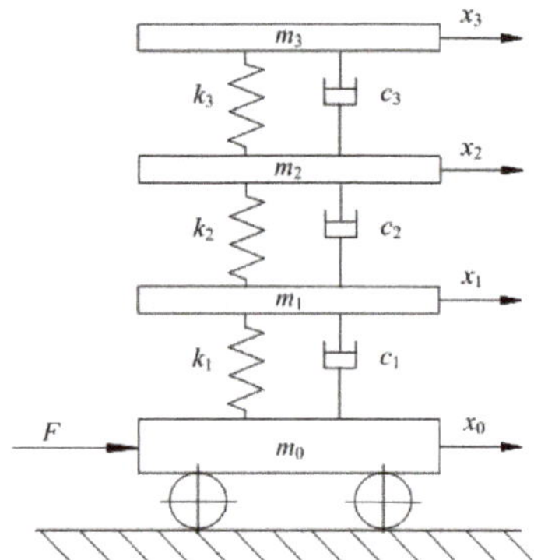

Figure 4. Mathematic dynamical model of the frame structure.

The motion equation of the numerical model in Figure 4 is established as follows:

$$\mathbf{M}\ddot{\mathbf{x}} + \mathbf{C}\dot{\mathbf{x}} + \mathbf{K}\mathbf{x} = \mathbf{f}(t) \tag{27}$$

where, $\mathbf{M}$, $\mathbf{C}$ and $\mathbf{K}$ are the mass, damping and stiffness matrices, respectively; $\mathbf{f}(t)$ is the input excitation vector; $\mathbf{x} = [x_0, x_1, x_2, x_3]^{\mathrm{T}}$ is the displacement of each layer, and the 0th floor represents the bottom plate; $\dot{\mathbf{x}}$ and $\ddot{\mathbf{x}}$ denote the speed and acceleration, respectively. The expressions of $\mathbf{M}$ and $\mathbf{K}$ are as follows:

$$\mathbf{M} = \begin{bmatrix} m_0 & 0 & 0 & 0 \\ 0 & m_1 & 0 & 0 \\ 0 & 0 & m_2 & 0 \\ 0 & 0 & 0 & m_3 \end{bmatrix}, \; \mathbf{K} = \begin{bmatrix} k_1 & -k_1 & & \\ -k_1 & k_1 + k_2 & -k_2 & \\ & -k_2 & k_1 + k_2 & -k_3 \\ & & -k_3 & k_3 \end{bmatrix} \tag{28}$$

where, $m_0 \sim m_3$ denotes the mass of each layer; $k_0 \sim k_3$ denotes the interlayer stiffness. Damping matrix $\mathbf{C}$ can be assumed to be the Rayleigh damping matrix, as shown in Equation (5).

The augmented state variables as shown in Equation (4) are established by selecting the relevant displacement, velocity, story stiffness and Rayleigh damping coefficient.

For the numerical model of Figure 4, the initial values of the parameters are set according to Figure 2 as follows: (1) it is considered that the mass matrix **M** remains unchanged and the density of aluminum is 2700 kg/m^3 in the processing; thus, $m_0 = m_1 = m_2 = m_3 = 6.7$ kg is calculated according to the structure size; (2) The modulus of the elasticity of aluminum is set 70 Gpa so that $k_0 = k_1 = k_2 = k_3 = 4.2 \times 10^5$ N/m; (3) By analyzing the excitation response test of the beam element shown in Figure 2, it can be found that the measured mode damping ratio usually has less influence than the inertia and stiffness of the structure. We can determine the Rayleigh damping coefficient by using the orthogonality experiment between the damping matrix and the vibration response mode shape of the structure. Rayleigh damping coefficients are estimated as $\alpha = 3.12 \times 10^{-3}$, $\beta = 1.77 \times 10^{-4}$; (4) $\mathbf{Q} = 1.2 \times 10^{-7}\mathbf{I}$, $\mathbf{R} = 1.15\mathbf{I}$, and these will be slightly and appropriately adjusted in the calculation. (5) Take the initial displacement and velocity as 0, so that $\mathbf{X}_{0|0} = \left[0,0,0,0,0,0,0,0,4.2 \times 10^5, 4.2 \times 10^5, 4.2 \times 10^5, 3.12 \times 10^{-3}, 1.77 \times 10^{-4}\right]^{\mathrm{T}}$. By applying the magnitude balance technology, we can set $\kappa_{\mathrm{amp}} = 4.2 \times 10^5$, $\alpha_{\mathrm{amp}} = 3.12 \times 10^{-3}$, $\beta_{\mathrm{amp}} = 1.77 \times 10^{-4}$ and adjust $\mathbf{X}_{0|0}$ as $[0,0,0,0,0,0,0,0,1,1,1,1,1]^{\mathrm{T}}$; (6) $\mathbf{P}_{\mathbf{X},0|0}$ is set as the identity matrix **I**.

By substituting the above parameters into the structural state equations and the observation equations that are expressed as Equations (6)~(10), the results of the NDT analysis can be obtained as described in the following section.

3.3. Analysis of the Nondestructive Testing Results

3.3.1. Working Condition without Damage

Under the conditions without damage, $k_1 \sim k_3$, α and β are obtained as shown in Figure 5 when the proposed recognition algorithm is used in the situation.

It can be seen from Figure 5 that structural damage can be identified based on the algorithm in the paper. The computation converges very quickly so that, $k_1 \sim k_3$ reaches the steady values within 14.5 s. Moreover, α and β converge to steady values within the 20 s. The differences between the recognition results and the initial values show that the initial estimation of the structure parameters is not accurate. However, by using the UKF it can give a more accurate structural parameter identification even under the condition with observation noises.

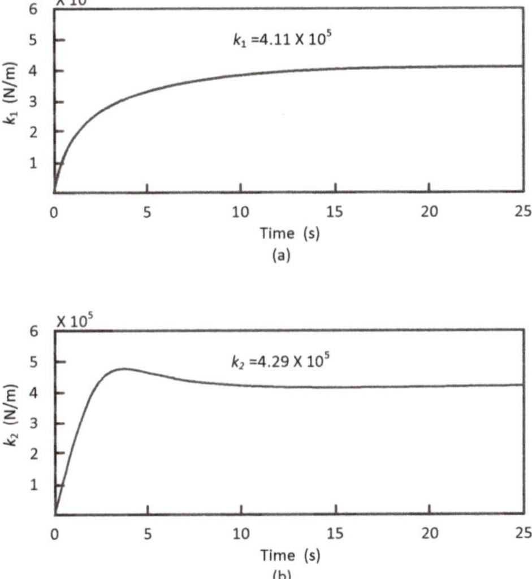

Figure 5. *Cont.*

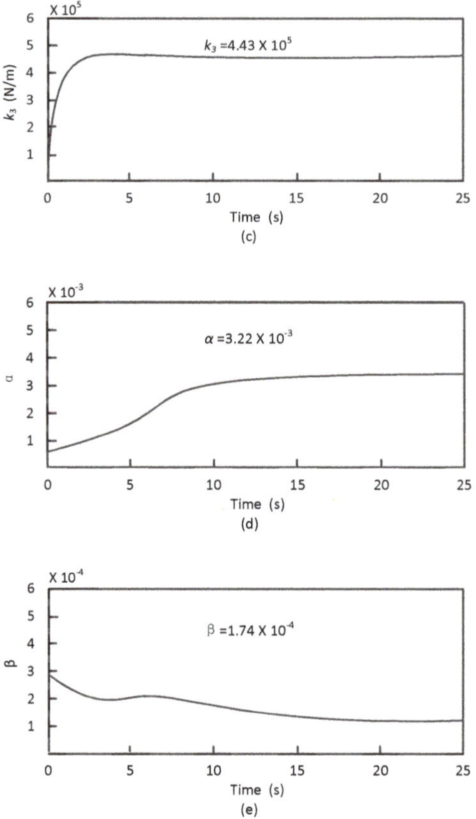

Figure 5. Identification results of the system parameters under the nondestructive working conditions: (**a**) recognition results of k_1, (**b**) recognition results of k_2, (**c**) recognition results of k_3, (**d**) recognition results of α and (**e**) recognition results of β.

3.3.2. Conditions of Damage on Column 2

When column 2 is damaged, the proposed algorithm is also used to identify $k_1 \sim k_3$, α and β. The recognition results are compared with that in Figure 5, as shown in Figure 6.

It can be seen from Figure 6 that: (1) among the stiffness parameters, only k_1 decreased considerably compared with that in the nondestructive condition. It was about 4.11×10^5 N/m without damage, while about 2.53×10^5 N/m after the damage, with a decrease rate of 38.4%. According to Table 1, when the stiffness of Column 2 decreases by 64.9% under Condition 2, the theoretical reduction of k_1 between the first layer and second layer should be 47%. There is a small difference between the identification by UKF and theoretical values, and this does not affect the location of damage. (2) The mass damping coefficient α does not have significant change, but the stiffness damping coefficient β changes significantly after the damage. This phenomenon was caused by the stiffness change of the frame structure which is reasonable. In conclusion, the damage-identification method based on UKF has achieved good performance, which can not only locate the damage position effectively, but also give an accurate damage degree estimation.

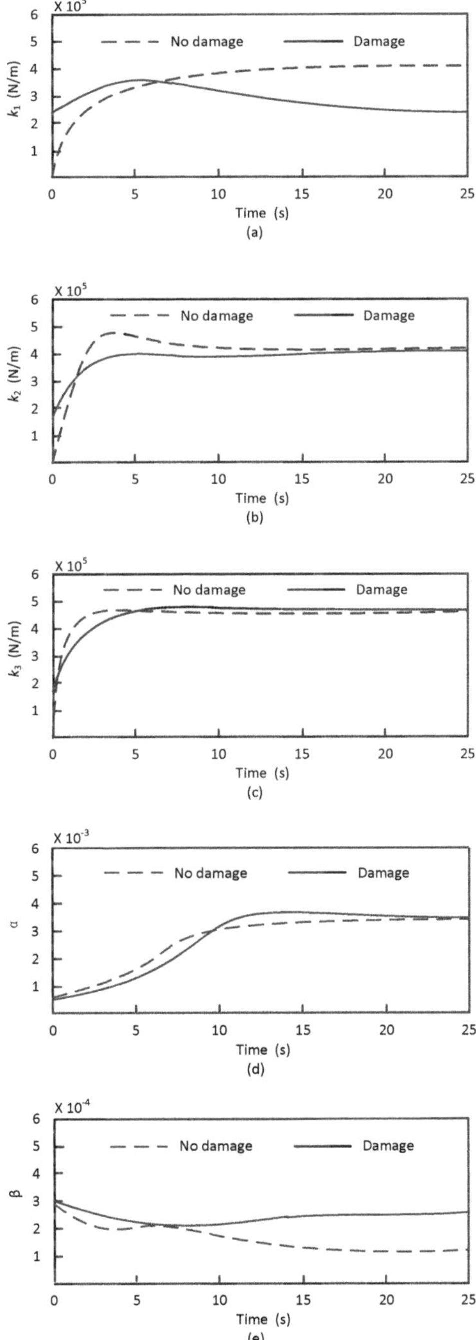

Figure 6. Identification of the frame's structural parameters in the case of damage on Column 2: (**a**) recognition results of k_1, (**b**) recognition results of k_2, (**c**) recognition results of k_3, (**d**) recognition results of α and (**e**) recognition results of β.

4. Conclusions

This paper presents a vibration signal analysis method based on UKF. A simulation was conducted to analyze the interaction process between the vibration and the frame structure defects. A nondestructive damage-detection algorithm based on UKF was designed in this paper. Taking the beam structure unit widely used in production as an example, the damage-identification effect of the UKF method under known excitation and unknown excitation was investigated, and the damage location and quantitative evaluation of the damage degree of the frame structure using the proposed algorithm were realized. The results show that the research method in this paper has obvious advantages in damage location and quantitative performance. Compared with the traditional Kalman filter method, the biggest advantage of the UKF method is that it does not need to calculate the Jacobi matrix, and the calculation amount is equivalent to that of EKF, but its accuracy is obviously higher than that of EKF. Thus, the calculation workload in the process of the damage identification of the frame structure is greatly reduced and the application is much simpler. In the field of structural material damage identification, the method shows a better universality. The proposed method also significantly increases the efficiency and effectiveness, as well as the damage location accuracy of damage identification in current engineering applications. The analysis of the response signal in the structure under additional noise shows that the nondestructive testing method presented in this paper can still give reliable damage-identification results under the condition of interference, showing a characteristic of strong anti-interference and high robustness. In the follow-up study, we will comprehensively analyze the material and beam structure characteristics, and use the finite element method to study the deformation, principal stress, tensile stress and shear stress of the beam element, so as to improve the practical application effect of the identification algorithm.

Author Contributions: Conceptualization, X.L.; data curation, D.S. and Z.Y.; investigation, X.L.; theoretical analysis, X.L.; formal analysis, X.L., D.S. and Z.Y.; writing—original draft, X.L.; writing—review and editing, X.L, D.S.; supervision, Z.Y.; project administration, X.L. All authors have read and agreed to the published version of the manuscript.

Funding: This research was supported by the National Natural Science Foundation of China (Grant Nos. 51105066), the Natural Science Foundation of Liaoning Province, China (Grant Nos. 2019-MS-119).

Conflicts of Interest: The authors declare no conflict of interest.

References

1. Yam, L.H.; Li, Y.Y.; Wong, W.O. Sensitivity studies of parameters for damage detection of plate-like structures using static and dynamic approaches. *Eng. Struct.* **2002**, *24*, 1465–1475. [CrossRef]
2. Giridhara, G.; Rathod, V.T.; Naik, S.; Roy Mahapatra, D.; Gopalakrishnan, S. Rapid localization of damage using a circular sensor array and Lamb wave based triangulation. *Mech. Syst. Signal Process.* **2010**, *24*, 2929–2946. [CrossRef]
3. Wang, L.; Yang, Z.C.; Tan, G.H. Improved Structural damage detection method based on natural frequency vector. *J. Mech. Strength.* **2008**, *25*, 709–713.
4. Nie, Z.H.; Hao, H.; Ma, H.W. Structural damage detection based on the reconstructed phase space for reinforced concrete slab: Experimental study. *J. Sound Vib.* **2013**, *332*, 1061–1078. [CrossRef]
5. Liu, W.G.; Yan, L.; Guo, L.Q. Damage classification of elastic thin plate based on modal strain energy method. *Noise Vib. Control* **2016**, *36*, 164–168.
6. Aydin, K.; Kisi, O. Damage detection in Timoshenko beam structures by multilayer perceptron and radial basis function networks. *Neural Comput. Appl.* **2014**, *24*, 583–597. [CrossRef]
7. Jiang, J.T.; Yu, H.L. Study on identification of offshore platform component damage based on strain modal difference. *J. Catastr.* **2010**, *25*, 67–69.
8. Zhang, C.; Cheng, L.; Xu, H.; Qiu, J.H. Structural damage detection based on virtual element boundary measurement. *J. Sound Vib.* **2016**, *372*, 133–146. [CrossRef]
9. Satpal, S.B.; Guha, A.; Banerjee, S. Damage identification in aluminum beams using support vector machine: Numerical and experimental studies. *Struct. Control Health Monit.* **2016**, *23*, 446–457. [CrossRef]

10. Zhao, Y.X.; Xu, Y.G.; Gao, L.X. For fault pattern recognition of rolling bearing acoustic emission technique based on harmonic wavelet packet and BP neural network. *J. Vib. Shock* **2010**, *29*, 162–165. [CrossRef]

11. Kostrzewski, M. Loads Analysing In Pallet Racks Storage Elevation. In Proceedings of the Carpathian Logistics Congress Proceedings (Reviewed Version), Cracow, Poland, 9–11 December 2013.

12. Friswell, M.; Mottershead, J.E. *Finite Element Model Updating in Structural Dynamics*; Springer Science and Business Media: Berlin, Germany, 2013; pp. 25–97.

13. Grip, N.; Sabourova, N.; Tu, Y.M. Sensitivity-based model updating for structural damage identification using total variation regularization. *Mech. Syst. Signal Process.* **2017**, *84*, 365–383. [CrossRef]

14. Yan, A.M.; Kerschen, G.; De Boe, P.; Golinval, J.C. Structural damage diagnosis under varying environmental conditions–part I: A linear analysis. *Mech. Syst. Signal Process.* **2005**, *19*, 847–864. [CrossRef]

15. Cawley, P.; Adams, R.D. The location of defects in structures from measurements of natural frequencies. *J. Strain Anal. Eng. Des.* **1979**, *14*, 49–57. [CrossRef]

16. Salawu, O.S. Detection of structural damage through changes in frequency: A review. *Eng. Struct.* **1997**, *19*, 718–723. [CrossRef]

17. Yang, Q.W.; Liu, J.K. Structural damage identification by adding given masses. *Eng. Mech.* **2009**, *26*, 159–163.

18. Guan, D.Q.; Huang, Y. Damage identification of frame structure by means of wavelet analysis of strain mode. *Chin. J. Comput. Mech.* **2010**, *27*, 325–329, 341.

19. Gu, P.Y.; Deng, C.; Tang, L. Experimental study on damage identification based on operational strain modal shape. *J. Vib. Shock* **2011**, *30*, 175–178.

20. Pandey, A.K.; Biswas, M. Experimental verification of flexibility difference method for locating damage in structure. *J. Sound Vib.* **1995**, *184*, 311–328. [CrossRef]

21. Zhang, Q.W. Statistical damage identification for bridges using ambient vibration data. *Comput. Struct.* **2007**, *85*, 476–485. [CrossRef]

22. Zong, Z.H.; Niu, J.; Wang, H. Research progress of structural Probabilistic Damage Identification Method Based on model validation. *China Civ. Eng. J.* **2012**, *45*, 121–130. [CrossRef]

23. Sajid, H.; Hossam, H.A. A robust method for coupling detection among process variables. *Int. J. Process. Syst. Eng.* **2012**, *2*, 93–110.

24. Widodo, A.; Yang, B.S. Support vector machine in machine condition monitoring and fault diagnosis. *Mech. Syst. Signal Process.* **2007**, *21*, 2560–2574. [CrossRef]

25. Wang, L.; Chan, T.H.T. Review of vibration-based damage detection and condition assessment of bridge structures using structural health monitoring. In Proceedings of the Second Infrastructure Theme Postgraduate Conference: Rethinking Sustainable Development: Planning, Engineering, Design and Managing Urban Infrastructure, Queensland University, Queensland, Australia, 26 March 2009.

26. Xu, H.; Cheng, L.; Su, Z.Q.; Guyader, J.L. Identification of structural damage based on locally perturbed dynamic equilibrium with an application to beam component. *J. Sound Vib.* **2011**, *330*, 5963–5981. [CrossRef]

27. Challamel, N. Higher-order shear beam theories and enriched continuum. *Mech. Res. Commun.* **2011**, *38*, 388–392. [CrossRef]

28. Yan, Y.J.; Cheng, L.; Wu, Z.Y.; Yam, L.H. Development in vibration-based structural damage detection technique. *Mech. Syst. Signal. Process.* **2007**, *21*, 2198–2211. [CrossRef]

29. Kostrzewski, M. Securing of safety by monitoring of technical parameters in warehouse racks, in high-bay warehouses and high storage warehouses–literature review of the problem. *LogForum,* **2017**, *13*, 125–134. [CrossRef]

 *materials*

Article

Testing the Influence of the Material Bonding System on the Bond Strength of Large-Format Tiles Installed on Concrete Substrate under Mechanical Loading

Libor Topolář [1], Dalibor Kocáb [1], Jiří Šlanhof [1,*], Pavel Schmid [1], Petr Daněk [1] and Jaroslav Nováček [2]

[1] Faculty of Civil Engineering, Brno University of Technology, Veveří 95, 602 00 Brno, Czech Republic; Libor.Topolar@vutbr.cz (L.T.); Dalibor.Kocab@vutbr.cz (D.K.); Pavel.Schmid@vutbr.cz (P.S.); Petr.Danek@vutbr.cz (P.D.)

[2] Profibaustoffe CZ, s.r.o, Vídeňská 140/113c, 619 00 Brno, Czech Republic; jaroslav.novacek@profibaustoffe.cz

* Correspondence: slanhof.j@fce.vutbr.cz; Tel.: +420-541-147-979

Received: 30 June 2020; Accepted: 14 July 2020; Published: 17 July 2020

Abstract: The paper describes an experiment focusing on the way the material system influences the bond strength of large-format tiles installed on concrete substrate during mechanical loading under conditions that correspond to real-life application. This involves a controllable mechanical load applied over an area of a test model while observing its condition using non-destructive methods (ultrasonic pulse velocity test, acoustic emission method, strain measurement, and acoustic tracing). The model consisted of a concrete slab onto which were mounted four different systems with large-format tiles with the dimensions of 3 m × 1 m. The combinations differed in the thickness of the tile, the adhesive, and whether or not a fabric membrane was included in the adhesive bed. The experiment showed that the loading caused no damage to the ceramic tile. All the detected failures took place in the adhesive layer or in the concrete slab.

Keywords: acoustic nondestructive methods; large format tiles; real-life application; mechanical load; failure

---

1. Introduction

The history of tiling started in the ancient states of the Mediterranean. The first tiles were used in 4000 BC by the Egyptians; from there they slowly spread to Italy, which can be considered a true treasure trove of beautiful tiles. Both public and ecclesiastical buildings of Medieval Italy were the first to receive decorated tiles, usually of small dimensions. Larger tiles began to be produced in the 12th century in Florence, Sienna, Orvieto, and Faenza. Since the beginning of the 12th century, the development of ceramic tiles had sought to make production more efficient. This meant emphasis on planning, manufacturing, as well as logistics. Thanks to this, the first manufacturing companies began to emerge, some of which still exist to this day [1–3].

While at the end of the 20th and beginning of the 21st century, the largest tile was roughly 0.5 m × 0.5 m, nowadays there are tiles with the dimensions of up to 1.6 m × 3.2 m. This marks a significant increase in tile size. The arrival of large-format tiles is one of the main trends of today and goes hand in hand with the modernization of manufacturing [4]. In recent years, the technical properties of the tiles have improved as well; properties such as water absorption, freeze–thaw resistance, wear resistance, or anti-slip safety [5,6]. Large-format tiles are therefore almost exclusively manufactured as the so-called rectified tiles. Their large size is useful, especially in the tiling of large areas, thanks to the minimal number of grout lines in between. At the same time, the thickness of some of the tiles is also being reduced. The most modern technology makes it possible to produce tiles of only 3 mm in thickness. These ultra-thin tiles are usually reinforced with a fibreglass mesh and possess the same technical properties as tiles of the standard thickness of 10 mm, while remaining lighter and thus easier to handle [7,8].

A downside of large-format tiles is the fact that they are very sensitive to the quality and levelness of the substrate, the quality of the substrate priming, adhesives, grouts, and tiling methods.

The goal of this experiment was to determine how the material system between the substrate concrete and the large-format tiles affects the bond strength during mechanical loading. The substrate consisted of a steel-reinforced concrete slab of 6 m × 4 m (nominal dimensions), onto which five large-format tiles of 3 m × 1 m were installed. The slab with the tiles was subjected to area loading using a vacuum chamber. Strain gauges were used to continuously measure the relative strain of both the concrete and the tiles. At the same time, acoustic emission was observed, mainly to record the development of mechanical damage in the concrete and the tiles. After the loading, the integrity of the bond was tested by acoustic tracing. The severity of damage to the tiles themselves was tested using the ultrasonic pulse velocity test. The experiment operated with two types of tiles—one with the thickness of 6 mm and the other 3 mm. The former was chosen because it is one of the most widespread types and the latter because it has been known to crack in real-life application. At the same time, two adhesives of different composition were used to attach the tiles onto the substrate. The first is a lightweight, highly deformable cementitious adhesive modified by a powder polymer binder (12 wt.% of the dry mix) and the other is a deformable cementitious adhesive with extended open time without any lightweight filler and modified by a powder polymer binder (3.5 wt.%).

2. Test Methods

2.1. Ultrasonic Pulse Velocity Test

This is one of the basic non-destructive methods used to test building materials and structures. It involves the repeated sending of ultrasonic pulses into the material and measuring their velocity. This velocity is influenced by the properties of the material as well as its quality; in materials of poorer quality, the ultrasonic pulse travels at a lower velocity. The ultrasonic pulse velocity test (UPV) can be used to determine e.g., the homogeneity of the material, material properties (modulus of elasticity, compressive strength, etc.), changes in these properties over time (for instance due to degradation), or the presence of cracks or air voids in the material [9–11]. The main advantages of UPV include its purely non-destructive character, the possibility to repeat the measurement in the same place at different times (different days, months, or years), or the simple and easy application in a laboratory and on-site [12].

There are two basic applications of the UPV test in civil engineering—the transmission and the pulse echo method [13]. When it comes to testing concrete, ceramics, and similar materials, the more common is the transmission method, which uses two transducers—a transmitter and a receiver. The device then measures the time it takes for the pulse to travel from one transducer to the other. The ultrasonic wave has the highest energy in a direction perpendicular to the position of the transmitter, but in cases where both sides of the tested member are not easily accessible, the transit time of the ultrasonic pulses can be measured in different directions as well. Depending on the position of the transducer, there can be several testing techniques:

- Direct—the transducers are facing each other directly; this is the best arrangement,
- semi-direct—the transducers are not placed on opposite sides of the member, or they are not facing each other directly,
- indirect—the receiver is parallel to the transmitter on the same side of the member, and the measurement is repeated several times with the receiver being shifted by a set distance every time [14].

Indirect testing, which was used in the experiment described herein, is highly sensitive to external interference. This is why it is used when only one side of the structure is accessible (as was the case of this experiment), or when it has been well established that the measured surface is perfectly level and defect-free [15] (as is also true here).

2.2. Acoustic Emission Method

Acoustic emission (AE) belongs among the most modern non-destructive methods used in material engineering and fatigue testing. The source of AE may originate from many phenomena depending on the type of material. Most sources of acoustic emission are damage-related [16]. The detection and observation of AE is commonly used for predicting material failure [17]. Its benefit is in the fact that it is a global rather than local method, which means that it observes the overall structure of the material rather than a small area. Moreover, the monitoring can be performed over a short time and is not very labour-intensive. However, a downside is its dependence on the way the material is being loaded [18]. This means that certain discontinuities may not generate a detectable AE at certain types or levels of loading. Unlike most other non-destructive testing methods, AE observes only active defects developing inside the material structure. These defects can only occur when the structure is under load. Passive defects or the structure's shape have no major influence on localizing AE [19]. The source of AE is the release of energy resulting from stimulation by internal or external stress. An AE event is emitted by irreversible dislocations and degradations created in the material's micro- and macrostructure. The energy thus released transforms into a mechanical stress pulse propagating through the material as an elastic wave. When the wave reaches the surface, part of it rebounds back and part of it transforms into one or more wave modes. In slabs, waves travel mostly in the form of Rayleigh (or surface) waves. Besides surface waves, there are also e.g., Lamb (or plate) waves. Both of these waves travel at different speeds. The signal that is detected by the AE receiver and converted to electrical current is known as an AE signal [20].

2.3. Strain Measurement

Relative strain that develops at a chosen point in the structure or element is measured by a number of devices and sensors which are typically called strain gauges. Strain gauges, the basic types of which are mechanical, wire, and resistance strain gauges, are used mainly for stress analysis and the measurement of important material properties. These measurements always involve relative size change in a given part of the member; i.e., the detection of changes in the measured length (whether an increase or decrease) caused by a change in loading or by the external environment. The length of strain gauges is chosen depending on their design and the particular constraints of the element being measured. This experiment uses resistance strain gauges, which is why the information below applies only to this type.

Measurements of homogeneous materials (such as metals) are typically performed with gauges of 3–10 mm; whereas non-homogeneous materials (e.g., concrete or wood) require longer strain gauges, usually 50–200 mm. How accurately the relative strain is measured (and by extension stress) depends on the quality of the contact of the strain gauge and the material, on the compensation or correction of errors caused by parasitic waves influencing the reading (mainly temperature), and on the necessary calibration and verification of the technique [21–23].

2.4. Acoustic Tracing

The method studies sub-surface air voids, separated layers beneath the surface, and surface treatment. Its principle involves introducing acoustic energy onto the surface by percussion or by dragging a hard object on the exposed face of the structure. The researcher then observes the acoustic response of the material. The acoustic response can be a ringing, hollow, or crunching sound. This method is used as an auxiliary verification of the bonding strength of the tiling. A ringing response indicates adequate adhesive strength of the bond (a strong enough bonding layer holding the materials together) and satisfactory cohesive strength of the individual materials (the individual cohesion of each separate material). A hollow response is a nearly certain indicator of failures and defects caused by unsatisfactory bond strength (bond failure) [24].

3. Experiment

The goal of the experiment was the non-destructive assessment of the quality of the bonding system between large-format tiles and a concrete substrate. The test model consisted of a steel-reinforced concrete slab of 6 m × 4 m × 0.2 m, which held five large-format tiles with the dimensions of 3 m × 1 m attached by different means. Each system was different—they consisted of two different adhesives (from one manufacturer), two different tiles without a reinforcing fibreglass mesh (thickness of 6 mm and 3 mm), and, in one case, the use of a fabric membrane.

3.1. Preparation

The Faculty of Civil Engineering of the University of Technology in Brno houses a research centre equipped with a vacuum chamber with the dimensions of 6.18 m × 4.18 m × 0.44 m. It is connected to an Edwards GXS250/2600 dry screw pump (Edwards Group Ltd., Stockholm, Sweden) with a Roots booster mechanism. The pump has enough suction performance to create a vacuum of 700 mbar in the chamber. An advantage of applying a load using a vacuum is that the load is uniformly applied on the entire surface of the specimen. Depending on the performance of the pump, the system can deliver different levels of loading up to 100 kN/m^2 [25]. The vacuum was created underneath the slab, causing it to deflect downwards and thus introduce strain to bonding systems and tiles.

The floor area of the vacuum chamber was a factor that limited the size of the test specimen. In order to preserve the airtight seal of the chamber (the walls and floor were connected and sealed together), it was not possible to safely remove the formwork of the specimen if it were to be cast inside the chamber. It was thus necessary to design a monolithic slab with lost formwork consisting of 5 reinforced pre-slabs of 1.2 m × 4.05 m × 0.05 m with additional reinforcement and additional concrete layer of 0.15 m on top. The inner surface of the pre-slabs was fitted with 3 strain gauges (Figure 1)—each slab bore 1 gauge (the leftmost and rightmost slabs were not fitted). The additional top layer was made with C 20/25 XC1 concrete with B500B reinforcement, with a concrete cover of 30 mm.

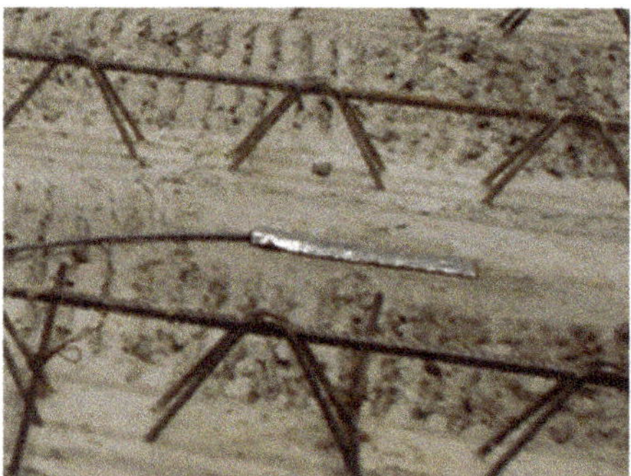

Figure 1. Fitting a strain gauge onto the surface of the lost formwork (reinforced concrete pre-slabs).

Figure 2 shows the structure of the reinforcement. Until they became fully integral to the slab, the pre-slabs were temporarily supported for additional stability. The supports consisted of two massive frames from which descended suspending rods. The rods were cut once the concrete had hardened. The result was a concrete member that behaved like a monolithic slab. The test chamber has a floor area 60–80 mm smaller than the dimensions of the slab. The gaps were sealed with extruded polystyrene.

Figure 2. Structure and distribution of the reinforcement.

After the concrete slab had hardened (i.e., when it reached the age of 28 days), its top surface was ground down by 3–5 mm to create a smooth and level surface (the maximum deviation per 1 m was 1 mm). After that, the surface was cleaned using a powerful industrial-grade vacuum cleaner. Besides these, the surface received no other treatment before the installation of the large-format tiles. The test slab, which had the nominal dimensions of 6 m × 4 m (6 m × 4.05 m to be precise), was divided into five symmetrical test fields. Each field consisted of a bed of adhesive and a large-format tile of 3 m × 1 m with cutouts (see Figure 3). One of the tiles was used in another experiment, which is not part of this paper. This is why this article only discusses measurements performed on four of the tiles designated A through D. The first adhesive, identified as type I, is a highly deformable cement adhesive with an extended open time (class C2ES2 according to EN 12004-1 [26]), with standard setting and a large portion of lightweight filler. This adhesive is heavily modified by powdered polymer binder at an amount of 12 wt.% of the dry mix. The second adhesive, identified as type II, is a deformable cementitious adhesive with extended open time and reduced slip (class C2TES1 according to [26]). It is a normally-setting adhesive without any lightweight filler, modified by a powdered polymer binder at 3.5 wt.% of the dry mix. In the case of tile A, the composite included an additional layer of adhesive and a separating membrane from a synthetic non-woven fabric. Table 1 shows the composition of the material system in each test field. The tiles were installed when the slab reached 35 days of age.

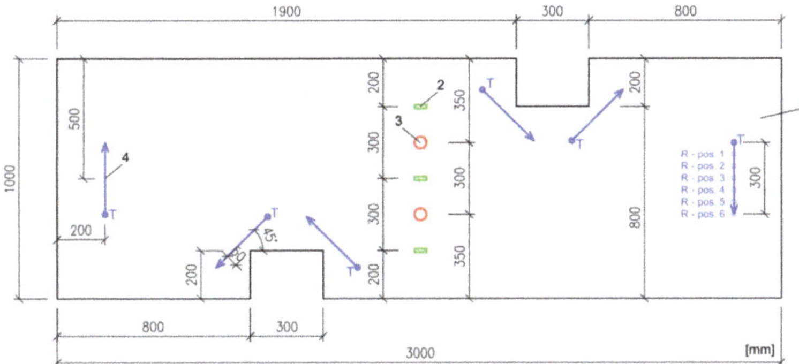

Figure 3. Tile dimensions (1) including the positions of the cutouts, strain gauges (2), acoustic emission (AE) sensors (3) and measurement lines for the ultrasonic pulse velocity (UPV) test (4) with the position of the transmitter (T) and, for illustration, receiver positions indicated at one of the lines (R).

Table 1. An overview and composition of the test fields.

Field ID	A	B	C	D
Ceramic element and	Iris-Iron Corten	Iris-Iron Corten	Iris-Iron Corten	Levantina Techlam
its thickness	6 mm	6 mm	6 mm	3 mm
Adhesive	Type I	Type I	Type II	Type I
Fabric membrane	Yes	No	No	No
Adhesive	Type I	-	-	-

The diagram (Figure 3) shows the precise size and position of the cutouts. The cutouts serve to concentrate stress; i.e., to act as a possible source of failure. Figure 4 shows a diagram of the position of the tiles on the slab as well as the mounting of the slab on the vacuum chamber.

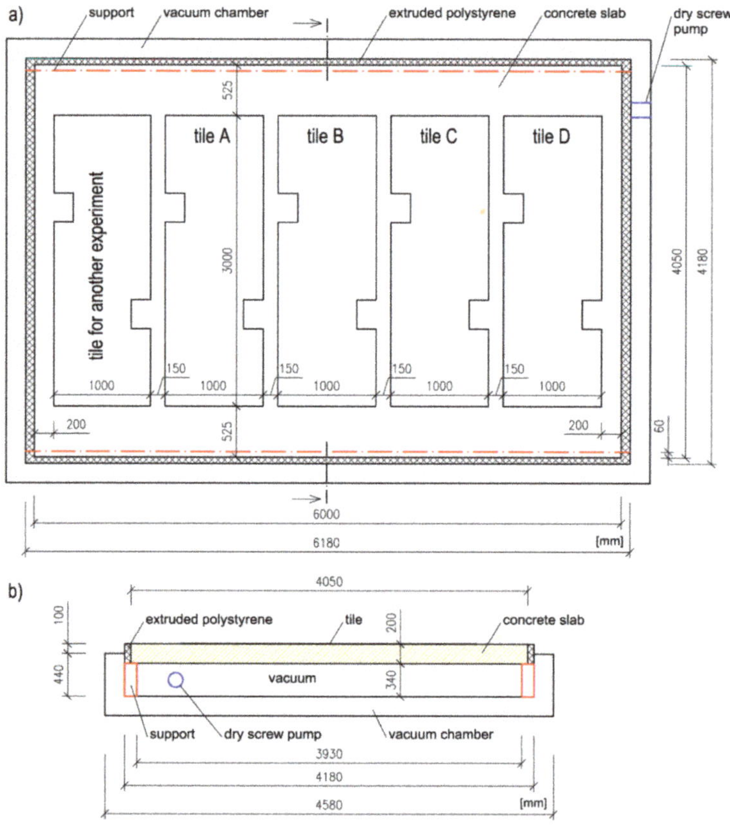

Figure 4. Diagram showing the mounting of the substrate slab with the tiles on the vacuum chamber (**a**) top view, (**b**) cross-section.

3.2. Loading

While mounted on top of the vacuum chamber, the concrete slab was supported by its longer 6-m sides. This meant that when vacuum was introduced to its bottom side, it would deflect at a span of 4 m (3.93 m to be exact), assuming the theoretical shape of a cylinder. The slab was subjected to gradual cyclic loading, where the load was lifted at the end of every cycle. During the first cycle, the load was 10 kN/m² lasting for 10 min, followed by subsequent cycles that were originally planned to be applied in 5 kN/m^2 increments. However, it was decided prior to commencing the test that the loading cycles would be adjusted depending on how damage and deformations develop. The maximum load was indeed adjusted after the third cycle (both the load and the holding time)—in the end, the structural model was loaded by eight cycles; see Table 2. The loading and unloading rate remained constant throughout the whole test—5 kN/m^2/min.

Table 2. List of loading cycles.

Cycle No.	1	2	3	4	5	6	7	8
Maximum load (kN/m^2)	10.0	15.0	20.0	22.5	25.0	27.5	30.0	35.0
Load-holding time (min)	10	10	10	5	5	5	5	10

The cyclic loading was performed when the adhesive had reached the age of 42 days (the slab was 77 days old by this time). Before the loading, each tile was tested by UPV. Six uniformly spaced measurement lines were drawn on each tile. The lines consisted of 7 points, the first of which marked the position of the transmitter. The remaining 6 indicated the six positions of the receiver during the measurement. This made up a total of 6 measurement points along one line, which thus had the total length of 300 mm. The distance between the point was always 50 mm. Each line was measured three times, which produced a total of 18 pulse velocities per tile. At the same time, the surface of each tile was fitted with two types of sensors—three surface-mounted strain gauges and two AE sensors (AE measuring parameters: frequency range 80–400 kHz; pre-amplifier 35 dB; threshold 400 mV). Two more AE sensors were also placed onto the surface of the concrete slab. The number and spacing of the UPV measurement lines, strain gauges, and AE sensors was identical for all the tiles and is pictured in Figure 3.

4. Results and Discussion

4.1. Results Obtained during Cyclic Loading

Figure 5a shows the real progress of loading. The graph shows that the original load increment of 5 kN/m^2 was reduced by half. The reason for this was the formation of the first cracks in the substrate slab. During the following four cycles, the step-increase in strain did not occur (i.e., it corresponded to load), which is why the last loading during cycle No. 8 was increased by the original 5 kN/m^2 to a total of 35 kN/m^2. Figure 5b shows a graph depicting the dependence of AE counts on time (blue dots) and the dependence of the relative strain of the slab on time (red curve). The strain gauges were placed beneath the additional concrete layer, which is why the values represent tensile strain. This is an average value made up from data from three strain gauges. However, the record of the relative strain is not complete but ends at a point where one of the gauges stopped measuring; i.e., just before reaching the maximum load during the eighth cycle (the second strain gauge stopped measuring during the holding of the maximum load of the eighth cycle, the third worked until the end of the loading). The highest number of AE overshoots was recorded during the 2nd and 3rd loading cycle, during which the relative strain increased dramatically as well. It therefore appears that the concrete slab suffered the most substantial cracking during these cycles. During the other loading cycles, nothing significant in terms of acoustic emission had occurred.

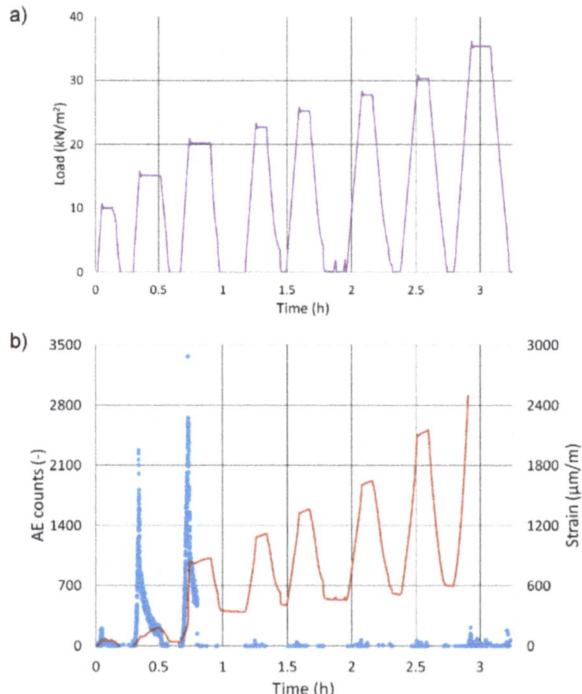

Figure 5. Real progress of loading (**a**); AE counts and relative strain of the concrete slab (**b**).

The values of relative strain in the tiles were measured using three strain gauges glued onto their surface (see Figure 3); the values therefore represent compressive strain. The final values are an average of all three strain gauges. For the sake of clarity, graph axes in Figure 6 show the same range of values. System A, as opposed to the other systems, shows very small relative strain and very low acoustic emissions throughout the loading. A major increase in relative strain occurred only during the last cycle.

Systems B and C show a greater increase in relative strain during the third cycle. It is probable that the slab suffered cracks in these areas and that a part of the tile became detached from the concrete substrate. This is further confirmed by the increased AE counts, which most likely indicate a failure within the adhesive or a failure of the substrate/adhesive or tile/adhesive interface. Another more significant increase in relative strain only occurs during the eighth cycle.

System D showed no response to crack development in the concrete in terms of relative strain or AE counts. The reason may be its lower stiffness (i.e., high flexibility) that comes with its thickness of 3 mm—the tile followed the curvature of the slab's deflection and, unlike the 6-mm tiles, had a lower tendency to become detached. System D showed a major change during the sixth loading cycle where its AE counts saw a marked increase, and since the seventh cycle it also showed substantial strain.

The analysis of relative strain shows that the stress in system A (containing the membrane) had been markedly reduced since the third cycle. The difference between the system with the membrane and the ones without is visible since the third cycle and very pronounced since the seventh. The membrane reduced the stress transfer from the concrete substrate onto the tile down to a mere 40% of the relative strain of the other tiles (i.e., strain created in tiles that were glued directly onto the substrate). Even when the slab deflected to an extreme 66 mm, tile A suffered no damage. However, during the third cycle, increased AE counts were observed—a clear phenomenon occurred involving the membrane, possibly its elongation or partial damage to its fibres.

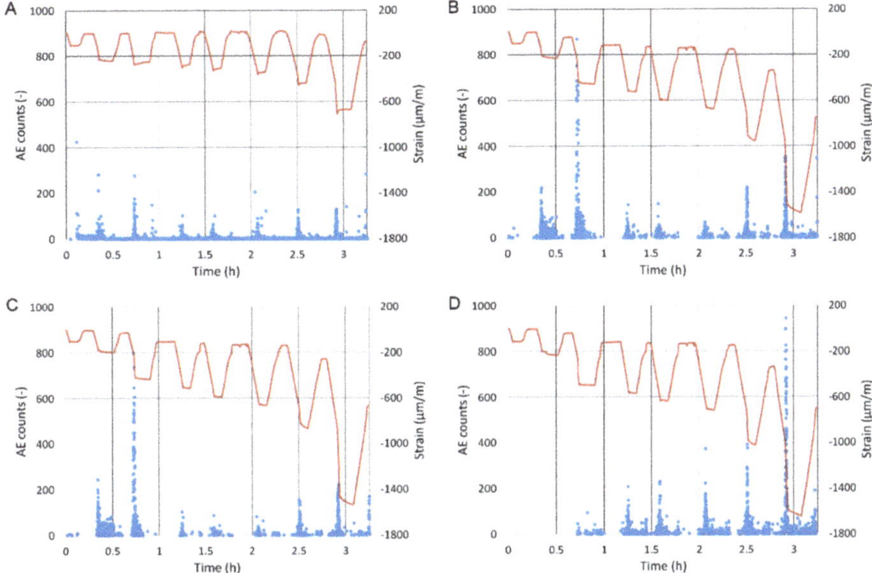

Figure 6. AE counts and development of relative strain in each tile (**A**) material system A, (**B**) material system B, (**C**) material system C and (**D**) material system D.

Figures 7 and 8 show the dependence of cumulative AE counts on permanent strain per every loading cycle. The loading was divided into three groups of cycles:

- Cycle 1–3 (solid line)—until the slab had suffered first cracks; the substrate concrete was undergoing the greatest changes and thus the tiles were subjected to the greatest stress;
- Cycle 3–6 (dashed line)—the loading increment was reduced (a step of 5 kN/m^2 was changed to 2.5 kN/m^2); the slab did not suffer significant cracking; smaller changes in relative strain and lesser AE counts were detected;
- Cycle 6–8 (dotted line)—these cycles brought the substrate concrete close to overall destruction and it was clear that some system suffered a bond failure.

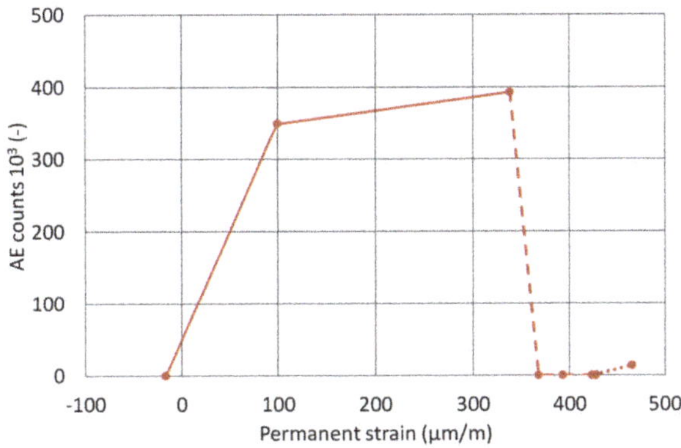

Figure 7. Dependence of AE counts on the development of permanent strain in the concrete slab.

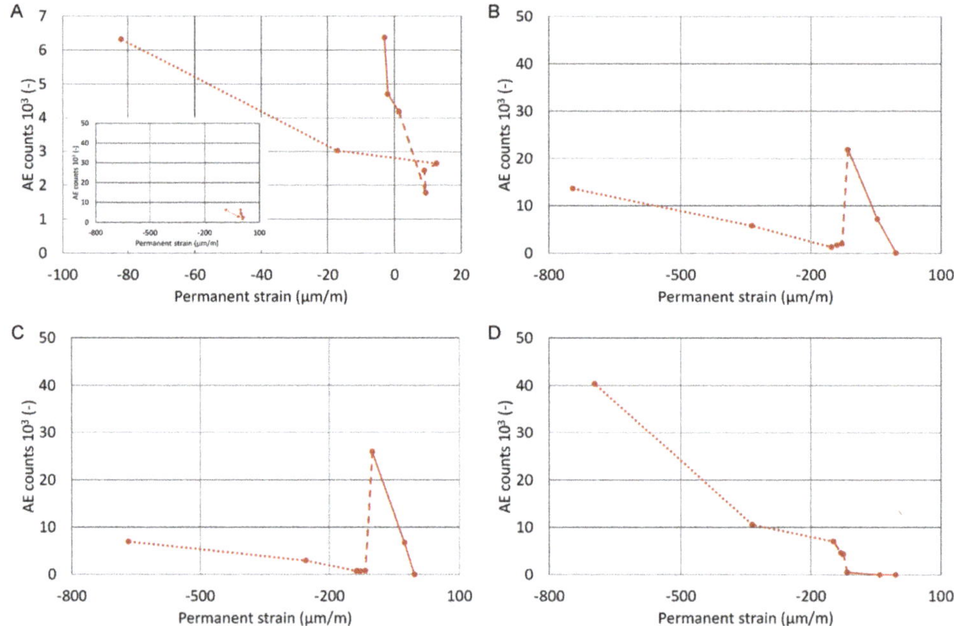

Figure 8. Dependence of AE counts on the permanent strain in (**A**) material system A, (**B**) material system B, (**C**) material system C and (**D**) material system D.

The figures obtained from measuring the slab (Figure 7) clearly show that the first three cycles caused significant damage, accompanied by an extreme increase in cumulative AE counts. There is also a substantial decrease in AE counts, but also in permanent strain during later cycles after the reduction in the loading cycle. The final loading cycles again show an increase not only in AE counts, but also in permanent strain.

All graphs in Figure 8 have the same scale, except for system A where the scale was changed due to the low number of AE counts. System A, which contains the membrane, shows only a small increase in permanent strain and AE counts compared to the other systems. The membrane seems to have a dampening effect on the strain transfer from the substrate. A small inserted graph shows what the results of system A would have looked like if they were plotted to scale. The results in Figure 8 also show a striking similarity between systems B and C. Thanks to its elasticity, system D with the 3-mm tile stresses the adhesive to a smaller degree than systems with the 6 mm tile. The first six loading cycles did not damage the adhesive enough to affect the bond

The curves in Figure 8 were used in a calculation that determined, using linear regression, the slopes for each loading section, which were later compiled into a correlation matrix; see Table 3. It shows a clear correlation between the behaviour of the concrete slab and system D with the 3-mm slab; i.e., the fact, the system copied the slab's deflections during loading. It also shows an indirect correlation between the slab and systems B and C. This demonstrates the minimal difference between adhesive I and II. The influence of the textile membrane in system A is also visible, as it causes a certain degree of independence on the shape changes in the substrate.

Table 3. A correlation matrix of the groups of loading steps from Figure 8.

	Concrete Slab	A	B	C	D
Concrete Slab	1.000	-	-	-	-
A	−0.296	1.000	-	-	-
B	−0.960	0.550	1.000	-	-
C	−0.991	0.418	0.989	1.000	-
D	0.985	−0.456	−0.994	−0.999	1.000

4.2. Results Obtained after the Loading

After all the cycles were completed, the surface of all the tiles was examined using acoustic tracing, where a percussive force was delivered by an impact hammer. The surface of the large-format tiles showed no cracks or defects that would be visible by the naked eye. Figure 9 shows a map of damage that was detected by acoustic tracing. System A, with the membrane, showed no damage to the tile or loss of bond strength. The other tiles that were attached directly showed a separation from the substrate around the cutouts. Systems B through D differed in the size of the unbonded areas. Figure 9 shows the percentages of the damaged areas compared to the total area of the tiles.

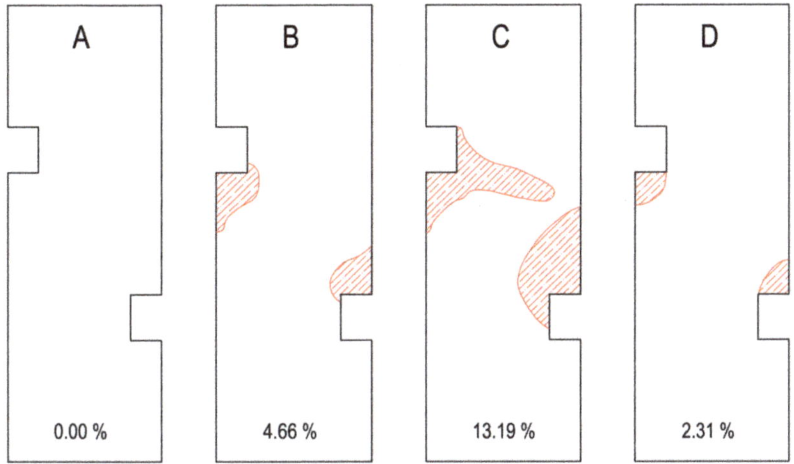

Figure 9. Areas and their percentages where acoustic tracing revealed a loss of bond strength after the eight loading cycles for system A, system B, system C and system D.

After the loading, the tiles were again tested using UPV in the same way as before the loading. Figure 10 shows a boxplot of the pulse velocities measured in each tile before and after loading. The statistical analysis of all the systems showed that the ultrasonic pulse velocity before and after the loading did not differ in a statistically significant way, which means that the internal structure of the tiles was not damaged. All the damage thus occurred in the bond or the slab.

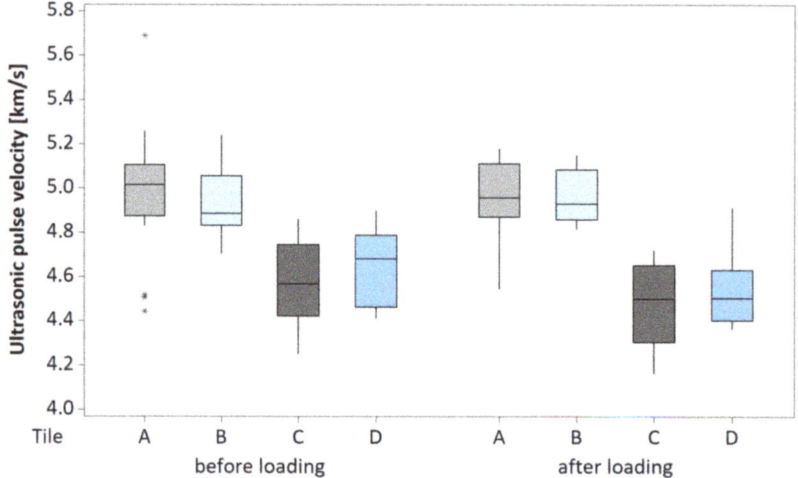

Figure 10. Boxplots of UPV determined by surface measurement on each tile before and after the cyclic loading.

5. Conclusions

The experiment examined the influence of the bonding system between a concrete substrate and large-format tiles, and observed the bond strength during mechanical loading. The ultrasonic pulse velocity test showed that the loading did not cause damage to the internal structure of the tiles. All the damage that was discovered by other methods thus occurred in the bonding system between the concrete substrate and the tiles, or in the substrate itself.

The results of AE measurements taken during the loading show that the method is useful for observing structural changes in "sandwich" structures. This method was able to indicate the approaching damage in the bonding system in time. The results corresponded with both the outcomes of strain measurement (greater deflection due to cracks in the concrete slab and the increase in permanent strain) and the findings of acoustic tracing, which was performed after the loading was finished.

Brief conclusions for every bonding system:

- System A, with the separating membrane from unwoven fabric and adhesive I, showed the greatest resistance to the effect of strain in the substrate and an ability of this adhesive bed to protect a large-format tile of 6 mm in thickness even during extreme deflection of the slab. The conclusion is confirmed both by the results of strain measurement performed on the surface of the tile and the lowest recorded AE counts during loading. Acoustic tracing revealed no damage to the tile anywhere throughout its entire area.
- System B, which consisted of a 6-mm tile glued directly onto the substrate with adhesive I, was the second best option for installing a large-format tile of this thickness.
- System C, which consisted of a 6-mm tile glued directly onto the substrate with adhesive II, proved to be more susceptible to damage due to substrate deformation than system B, even though fewer AE counts were recorded. During acoustic tracing, this system showed the highest degree of damage from all the systems examined herein, which would explain the lower AE counts (since the detached areas suffer less strain). The type of adhesive used in this case is not suitable for large-format tiles of 3 m length.
- System D, which consisted of a 3-mm tile glued onto the substrate with adhesive I, showed, to an extent (in this case it was the sixth loading cycle), minimal differences in strain compared to the concrete slab, and AE counts were also lower than in the case of systems B and C. Since the sixth

cycle, the stress caused by the deformation of the substrate concrete slab locally exceeded the bond strength of the system and a part of the tile had become detached.

This experiment shows that the choice of adhesive is critical in the application of large-format tiles in high-risk floor structures (i.e., disregarding substrate expansion and contraction, flexible substrate, underfloor heating, shrinking substrate due to insufficiently mature concrete, performing cutouts, etc.), but more importantly, the fabric membrane should be included in the adhesive bed. Furthermore, the condition of large structures should be regularly monitored, mainly to eliminate high financial costs associated with damage. At the same time, it is recommended to use several methods simultaneously to arrive at a clearer idea of what is happening in the material instead of relying on just one method, since its results may be interpreted wrongly.

Author Contributions: P.S. and J.N. designed the experiments; P.D., D.K. and L.T. performed the experiments and analyzed the data; L.T. and D.K. wrote the paper; P.S., J.N. and J.Š. revised the manuscript; J.Š. was responsible for project administration. All authors have read and agreed to the published version of the manuscript.

Funding: This research was funded by the Ministry of Industry and Trade (Grant Trio FV20606).

Conflicts of Interest: The authors declare no conflict of interest.

References

1. Van Lemmen, H. *5000 Years of Tiles*; Smithsonian Inst Press: Washington, DC, USA, 2013; ISBN 978-1-58-834398-7.
2. Casasola, R.; Rincón, M.J.; Romero, M. Glass–ceramic glazes for ceramic tiles: A review. *J. Mater. Sci.* **2012**, *47*, 553–582. [CrossRef]
3. Confindustria Ceramica. Ceramic Tiles. In *Statistical Survey on Italian Industry*; Confindustria Ceramica: Modena, Italy, 2010.
4. Albalat, M.P. *Current and Future Challenges of the Ceramic Tile Firms*; UJI: Castellón de la Plana, Spain, 2018.
5. Bondioli, F.; Taurino, R.; Ferrari, A.M. Functionalization of ceramic tile surface by sol–gel technique. *J. Colloid Interface Sci.* **2009**, *334*, 195–201. [CrossRef] [PubMed]
6. Schabbach, L.M.; Bondioli, F.; Ferrari, A.M.; Manfredini, T.; Petter, C.O.; Fredel, M.C. Influence of firing temperature on the color developed by a (Zr, V) SiO4 pigmented opaque ceramic glaze. *J. Eur. Ceram. Soc.* **2007**, *27*, 179–184. [CrossRef]
7. Jolliet, O.; Margni, M.; Charles, R.; Humbert, S.; Payet, J.; Rebitzer, G.; Rosenbaum, R. IMPACT 2002+: A new life cycle impact assessment methodology. *Int. J. Life Cycle Assess.* **2003**, *8*, 324. [CrossRef]
8. Pini, M.; Ferrari, A.M.; Gamberini, R.; Neri, P.; Rimini, B. Life cycle assessment of a large, thin ceramic tile with advantageous technological properties. *Int. J. Life Cycle Assess.* **2014**, *19*, 1567–1580. [CrossRef]
9. Bungey, J.H.; Grantham, M.G. *Testing of Concrete in Structures*; CRC Press: Boca Raton, FL, USA, 2006; ISBN 978-0-20-396514-6.
10. Helal, J.; Sofi, M.; Mendis, P. Non-destructive testing of concrete: A review of methods. *Electron. J. Struct. Eng.* **2015**, *14*, 97–105.
11. Faseeva, G.R.; Nafikov, R.M.; Lapuk, S.E.; Zakharov, Y.A.; Novik, A.A.; Vjuginova, A.A.; Kabirov, R.R.; Garipov, L.N. Ultrasound-assisted extrusion of construction ceramic samples. *Ceram. Inter.* **2017**, *43*, 7202–7210. [CrossRef]
12. Balayssac, J.P. *Non-Destructive Testing and Evaluation of Civil Engineering Structures*; Garnier, V., Ed.; Elsevier: Amsterdam, The Netherlands, 2017; ISBN 978-1-78-548229-8.
13. Carino, N.J. *Handbook on Nondestructive Testing of Concrete*; Malhotra, V.M., Ed.; CRC Press: Boca Raton, FL, USA, 2004; ISBN 978-0-84-931485-8.
14. European Committee for Standardization. *EN 12504-4 Testing Concrete—Part 4: Determination of Ultrasonic Pulse Velocity*; European Committee for Standardization: Brussels, Belgium, 2004.
15. Stawiski, B.; Kania, T. Testing quality of ceramic tiles in order to evaluate condition of the manufacturing process. *Procedia Eng.* **2016**, *161*, 937–943. [CrossRef]
16. Nakamura, H. *Practical Acoustic Emission Testing*; Springer: Tokyo, Japan, 2016; ISBN 978-4-431-55072-3.
17. Cunha, R.; Maciel, R.; Nandi, G.S.; Daros, M.R.; Cardoso, J.P.; Francis, L.T.; Ramos, V.F.C.; Marcelino, R.; Fröhlich, A.A.; De Araujo, G.M. Applying Non-destructive Testing and Machine Learning to Ceramic

Tile Quality Control. In Proceedings of the IEEE 2018 VIII Brazilian Symposium on Computing Systems Engineering (SBESC), Salvador, Brazil, 5–8 November 2018; pp. 54–61. [CrossRef]

18. Ohtsu, M.; Enoki, M.; Mizutani, Y.; Shigeishi, M. Principles of the acoustic emission (AE) method and signal processing. In *Practical Acoustic Emission Testing*; Springer: Tokyo, Japan, 2016; pp. 5–34. ISBN 978-4-431-55072-3.

19. Shamsudin, M.F.B. Structural Health Monitoring of Fatigue Cracks Using Acoustic Emission Technique. Ph.D. Thesis, Brunel University, London, UK, 2019.

20. Grosse, C.U.; Ohtsu, M. (Eds.) *Acoustic Emission Testing*; Springer Science & Business Media: Berlin, Germany, 2008; ISBN 978-3-540-69972-9.

21. Constantinescu, D.M.; Sandu, M.; Volceanov, E.; Găvan, M.; Sorohan, S. Experimental and numerical analysis of the behaviour of ceramic tiles under impact. *Key Eng. Mater.* **2009**, *399*, 161–168. [CrossRef]

22. Walley, S.M. Historical review of high strain rate and shock properties of ceramics relevant to their application in armour. *Adv. Appl. Ceram.* **2010**, *109*, 446–466. [CrossRef]

23. Gillespie, J.W., Jr.; Carlsson, L.A.; Gawandi, A.A.; Bogetti, T.A. Fatigue crack growth at the face sheet-core interface in a discontinuous ceramic-tile cored sandwich structure. *Compos. Struct.* **2012**, *94*, 3186–3193. [CrossRef]

24. Schabowicz, K. Non-Destructive Testing of Materials in Civil Engineering. *Materials* **2019**, *12*, 3237. [CrossRef] [PubMed]

25. Melcher, J.; Karmazínová, M. *Experimental Verification of Process of Deformation and Load-Carrying Capacity of Structural Components and Members Placed in Metal Chamber in the Sloping or Vertical Position, Respectively, with the Use of the Vacuum Test Method*; Z/B - Verified technology, Faculty of Civil Engineering, Brno University of Technology: Brno, Czech Republic, 2009; RIV/00216305:26110/09:PR24352.9.

26. Czech Office for Standards, Metrology and Testing. *ČSN EN 12004-1. Adhesives for Ceramic Tiles—Part 1: Requirements, Assessment and Verification of Constancy of Performance, Classification and Marking*; UNMZ: Prague, Czech Republic, 2018. (In Czech)

Article

The Recognition of the Micro-Events in Cement Composites and the Identification of the Destruction Process Using Acoustic Emission and Sound Spectrum

Dominik Logoń * and Krzysztof Schabowicz

Faculty of Civil Engineering, Wrocław University of Science and Technology, Wybrzeże Wyspiańskiego 27, 50-370 Wrocław, Poland; krzysztof.schabowicz@pwr.edu.pl
* Correspondence: dominik.logon@pwr.edu.pl

Received: 8 June 2020; Accepted: 1 July 2020; Published: 4 July 2020

Abstract: This paper presents the recognition of micro-events and their concentration in quasi-brittle cement composites and the identification of the destruction process based on acoustic emission and sound spectrum. The tests were conducted on a quasi-brittle composite of a cement paste reinforced with a high volume of dispersed polypropylene fibers. The possibility of identifying the destruction process based on acoustic emission and sound spectrum was confirmed. This paper focused on the identification of micro-events using the 3D spectrum. It was shown that the identification of the concentration of micro-events precedes the occurrence of critical crack f_{cr}, ending the Hooke's law range. The ability to recognize this phenomenon with the use of the 3D spectrum makes it possible to predict the structure destruction process and subsequently to assess the structure destruction (micro and macro-cracks) and the reinforcement destruction (pull-off, breaking). It was confirmed that the three-dimensional spectrum provided additional information, enabling a better recognition of micro and macro-changes in the structure of the samples based on the analysis of sound intensity, amplitudes, and frequencies.

Keywords: quasi-brittle cement composites; acoustic emission; acoustic spectrum; micro events

---

1. Introduction

Acoustic emission (AE) is a method that has been used for a very long time in concrete engineering [1]. Acoustic emission measurements can be applied to recognize the early hydration of cement [2,3]. Most papers have focused on determining the destruction process (cracks, maximum load, and failure of reinforcement in cement composites) [4–40]. It was noticed that AE is an effective method in determining the critical stress f_{cr} of cement composites corresponding to the first crack [4–6]. It is used for the accurate definition of the elastic range corresponding to Hooke's law [7,8]. The continuous AE evaluation in composites and acoustic emission AE events sum (sum AE) has been applied to determine the first crack [4,9,10], micro and macro-cracks, and their propagation in the fracture process in cement composites with [11,12] and without reinforcement [13,14].

The test results indicate that AE is a good method for crack formation monitoring in mechanically loaded traditional [15,16] and high strength [17,18] cement composites. It was demonstrated that this method is effective during compression [19,20] and bending tests [21,22].

Acoustic emission measurements also focus on the possibility of identifying crack orientation [23–25] q, thus enabling the recognition of cracks occurring as a result of compression, tensile, and shear stress.

AE is used to identify the destruction process of different materials (e.g., reinforced geopolymer mortars) [26]. The effectiveness of acoustic emission measurements in structures control was confirmed [27–29]. This method is used for example to monitor and control bridges condition [30,31]. The AE is still improved for the purpose of the identification of failure processes in different materials and structures [32–34].

Our previous own works have focused on the correlation between AE and the individual failure processes of each of the different composite components based on the sound spectrum [7,35,36]. The conclusions indicate that for the accurate recognition of composite failure processes, the AE recording should be expanded to include the analysis of each sound separately (also a single signal in a very small range of frequencies) and the analysis of the range of sounds corresponding to a given mechanical effect with the use of acoustic spectrum. It was noticed that the acoustic spectrum 2D and 3D should be correlated with the load-deflection curve and with other acoustic effects, which enables the identification of the failure process.

The presented paper confirms that there is a possibility of correlation between AE and the failure process in quasi-brittle cement composites. This correlation enables a determination of the stage of damage in cement composites, increasing the safety of using the composite and a decision whether or not the damaged composite can be repaired.

The main innovation of this research is the possibility of identifying AE micro events in the area preceding the occurrence of critical crack initiating the destruction process in cement composites. The main idea in this paper is the possibility of micro event recognition in the Hooke's law range, which enables the prediction of the destruction process.

The paper presents the results of our own studies on a selected paste sample in a four-point bending tensile test conducted at IPPT PAN in Warsaw, taking into consideration the current recommendations [37–41].

The results of tests carried out on a number of cement composites (cement paste, mortar, concrete, with and without dispersed reinforcement [40]) confirmed the general conclusions presented in the paper, indicating the possibility of their generalization.

2. Testing

2.1. Materials Used for Tests

High-strength cement composites w/(c + Sf) = 0.31, Portland Cement CEM I 42.5R (c), silica fume S_f = 10%c, siliceous fly ash 20%c, superplasticizer Sp, tap water (w).

Synthetic-structural polypropylene fibers (ASTM C-1116): density 0.91 kg/dm^3, f_t = 620–758 MPa, E = 4.9 GPa, l = 54 mm, d = 0.48 mm, l/d = 113.

The specimens were reinforced with a polypropylene fiber volume of V_f = 5%. The samples with dimensions 40 × 40 × 160 mm were cut out from pre-formed slabs. Each beam was turned by 90° and cut to the depth of 7 mm.

2.2. Preparation of Specimens for Tests

Four-point bending tensile tests were carried out at IPPT PAN in Warsaw [40]. The measurements were conducted on sample beams with the dimensions of 40 × 40 × 160 mm (Figure 1). In the middle of the span, in the lower part, a cut (7 mm deep) was made in the specimens, in accordance with American Society for Testing and Materials ASTM 1018 [37]. This paper presents the results for quasi-brittle cement composite of paste with high volume of polypropylene fiber reinforcement.

2.3. Description of the Test Stand

The tests were conducted on three specimens. The results obtained for each of the three specimens separately confirm the conclusions contained in the paper.

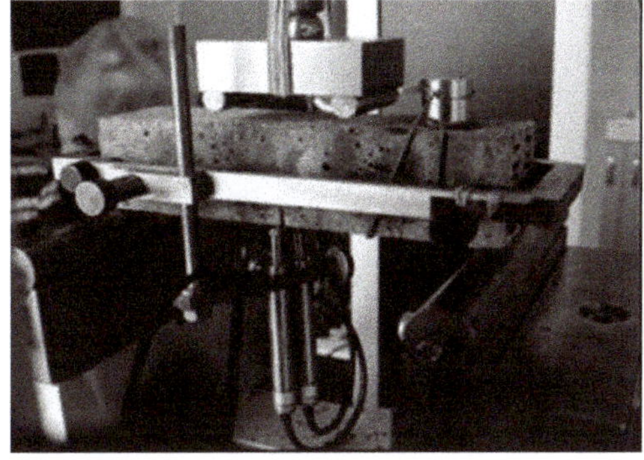

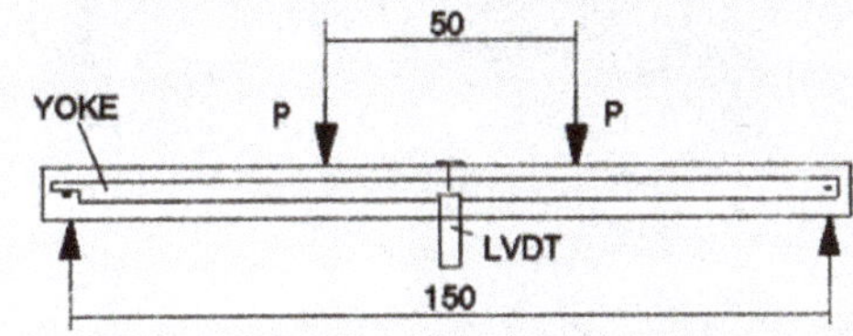

Figure 1. Four-point bending test [40].

The loading process was carried out with a controlled, constant displacement speed equaling 0.05 mm/min. Deflection was recorded by means of two LVDT sensors located under the beam, using a "Yoke" clamp (Figure 2), [40]. During the test, the bending load and deflection of the specimen were measured. The testing procedure corresponded to the requirements of the ASTM C 1018 standard.

At the end of the test, a 5 mm deflection of the specimens was recognized, which was determined in relation to the neutral axis. The acoustic emission sensor was fixed to the top surface of the beams with the use of an elastic band and the surface was coated with coupling graphite grease (Figure 1).

The broadband AE sensor manufactured by Physical Acoustic Corp. enabled the recording of the AE signal within the frequency range of 10–1000 kHz.

The AE signal was recorded with the use of the ADLINK 9112 card with the sampling rate of 88.2 kHz, the 12-bit resolution, and the function of a continuous recording on a computer disk. Source files were saved in the format (.wav), which makes it possible to listen to the recorded signal with the use of the computer sound card speakers.

Thanks to the recorded data, it is possible to plot a load-deflection curve, record AE events, and aggregate them (total AE).

The acoustic emission effects were presented as a 2D and 3D acoustic spectrum (amplitude of the frequency depending on sound intensity). The 2D sound spectrum was achieved with the use of the Audacity program and the 3D spectrum using SpectraPLUS-SC (Pioneer Hill Software LLC, Poulsbo, WA, USA).

The quasi-brittle cement composites (ESD—Eng. elastic range, strengthening control, deflection control) were characterized by higher load and absorbed energy in the elastic range compared to the sample without reinforcement (Figure 2) [35]. The reinforcement effects may be presented by characteristic points f_x (F_x-load, ε_x-deflection, W_x-work) and areas A_X under the load-deflection curve.

Figure 2 presents the mechanical effects of the quasi-brittle cement composites with the corresponding acoustic effects and compiled acoustic spectra with various amplitudes corresponding to different mechanical effects (reinforcement breaking, pull-out, macrocracks, microcracking, and micro-events).

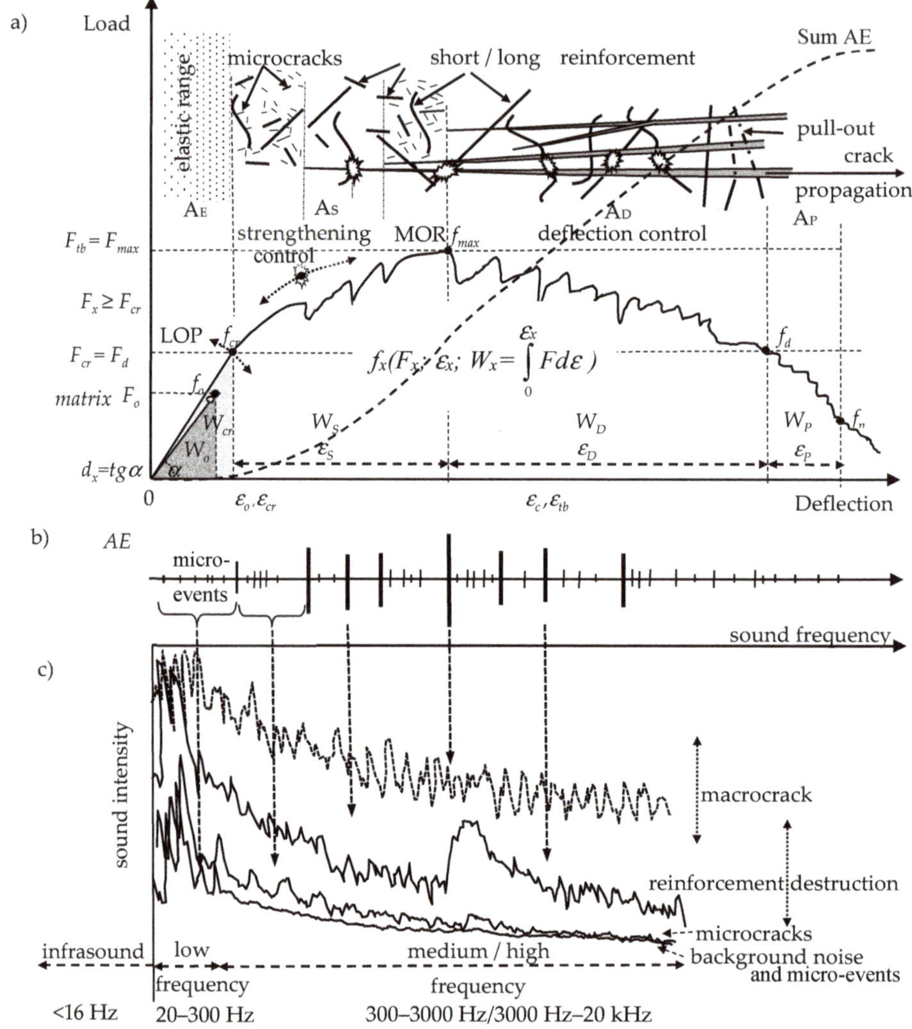

Figure 2. The quasi-brittle composite: (**a**) load-deflection curve, (**b**) AE, acoustic emission effects, (**c**) 2D acoustic spectrum (frequency amplitude depending on sound intensity) based on [36].

3. Test Results

The results are presented using the example of a quasi-brittle cement composite of slurry type with dispersed reinforcement in the form of structural polypropylene fibers (Figures 3 and 4).

In the top part of Figure 3, the recording of AE effects is presented. The figure shows charts from four-point bending tensile tests, force-deflection, and force–time; in addition, it shows the recorded events as a function of time (0–160 s) and the AE events total.

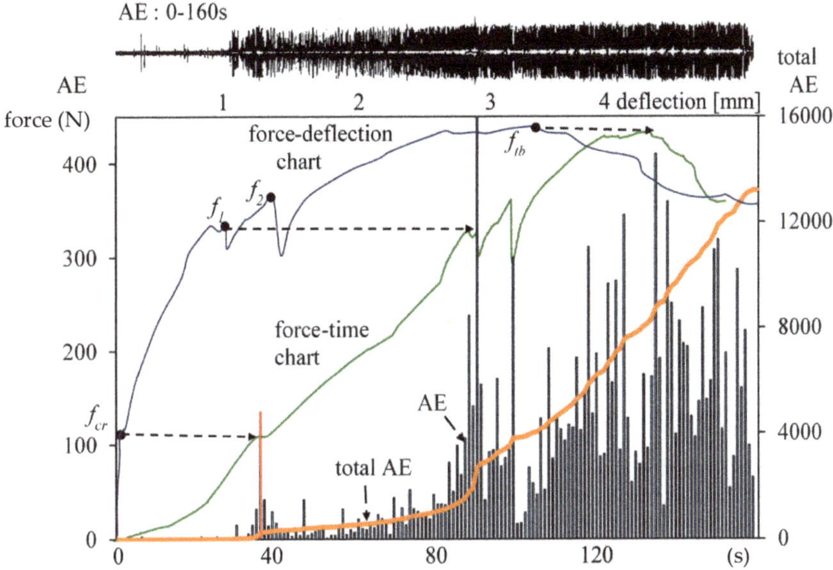

Figure 3. Results of tests on the composite in the range of 0–160 s: force-deflection, force-time curve, total acoustic emission AE, events, and AE recording.

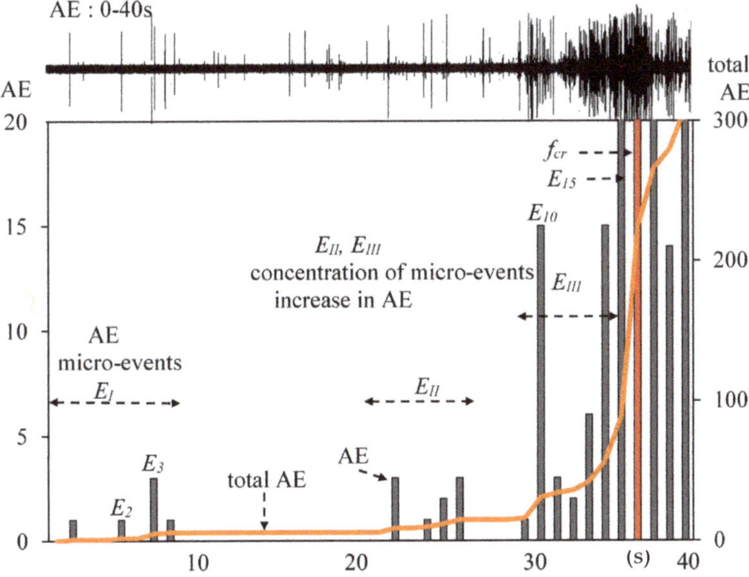

Figure 4. Results of tests on the composite Figure 1 in the range of 0–40 s: total AE, events, and AE recording.

The characteristic points and ranges of AE events are highlighted in the presented figure. In addition, the chart of total AE is shown. The red color marks the event f_{cr} (the end of Hooke's law), corresponding to the critical point f_{cr} on the force-deflection curve. Point f_1 corresponds to the sudden drop in stress in the strengthening area, and point f_{tb} corresponds to maximum stress.

In order to analyze the destruction process in the proportionality area shown in Figure 3, the effects obtained in that range of deflections (in the range of 0–40 s) were enlarged and presented in Figure 4.

The spectrogram of the AE signal of the tested composite is presented in Figure 5. Events are visible as vertical lines with marked characteristic points f_{cr}, f_1, f_{tb}, and image of events occurring in quasi-brittle composites, with highlighted areas of proportionality A_E, strengthening control A_S and deflection control A_D. The area with a dominating impact of event noise as well as the areas with the concentration of events E_{II} and E_{III} are marked. The spectrum is presented in the range of low, medium, and high frequency from 0–4000 Hz. The red intermittent line in the spectrogram marks an area that is linked to the possible occurrence of dispersed reinforcement breaking. Figure 5 enables the observation of the concentration of micro-events preceding f_{cr} (E_{II} and E_{III}).

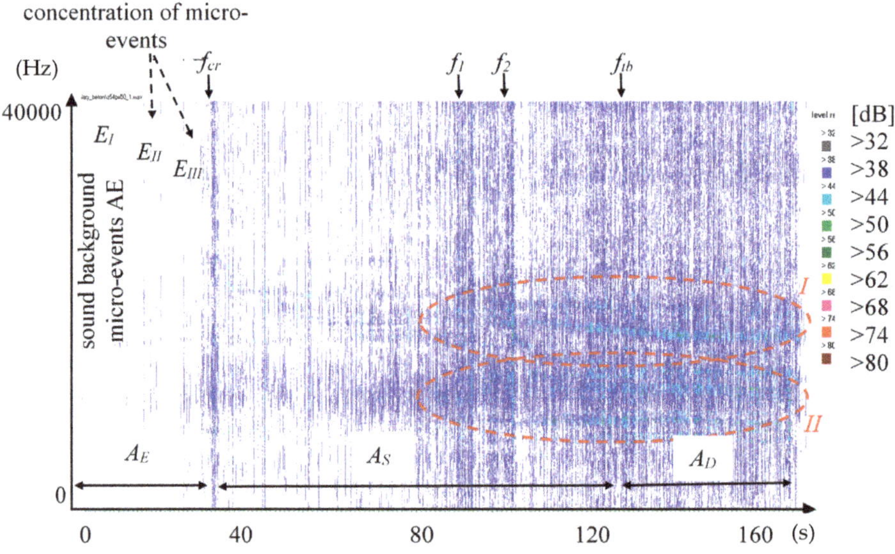

Figure 5. Spectogram of the AE signal.

Figure 6 presents the 2D sound spectra for different AE effects obtained at characteristic points or ranges of events by means of the Audacity program. Event E_0 corresponds to the background spectrum. Three characteristic areas were identified: E_I, E_{II}, and E_{III}. Area E_I groups single events from E_1 to E_4, area E_{II} includes events from E_5 to E_8, and E_{III} refers to events from E_9 to E_{15}. The spectra are presented in the wide range of low, medium, and high frequency from 0–45,000 Hz.

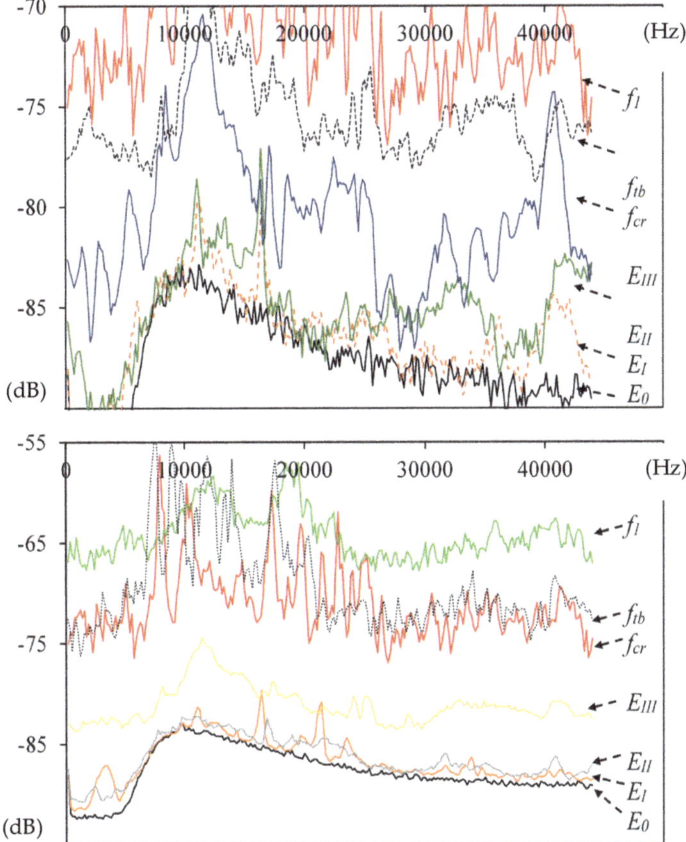

Figure 6. Sound spectra for characteristic events: sound background E_0 and ranges of events: E_I, E_{II}, E_{III}.

Figure 7a,b presents the 3D sound spectra from the Hooke's law range. Figure 7a shows the sound background spectrum and spectra of single AE events (groups of sounds E_{II}), with spectra in the time range of 5.4–33 s. Figure 7b presents the background spectra, and events spectra E_{II} and E_{III} in the time range 12.5–39.7 s. Figure 7c displays the background spectra and spectra of multicracking between points f_{cr} and f_1 in the time range 44.9–70.3 s. Relative amplitudes (of sound intensity components) were analyzed between 50–110 dB. The range of sound intensity components for background spectra and spectra of events E_{II} was from −110 to −85 dB, while the group of events E_{III} and multicracking corresponded to the level of −80 dB.

Figure 8a is an image of the sound spectra of the composite: its background, events E_{II}, E_{III}, and the critical point f_{cr}. Figure 8b presents the spectra of the area of sudden drops in stress in points f_1 and f_2 in the time range 74.8–102.3 s. Figure 8c shows the spectra of event in point f_{tb} and spectra of the deflection control area. The image is presented in the range 106.3–133.8 s. Relative amplitudes (sound intensity components) were analyzed in the range between 60–120 dB. The range of sound intensity components for background spectra and spectra of events E_{II} was from −110 to −85 dB, while the group of events E_{III} and multicracking corresponded to the level of −80 dB.

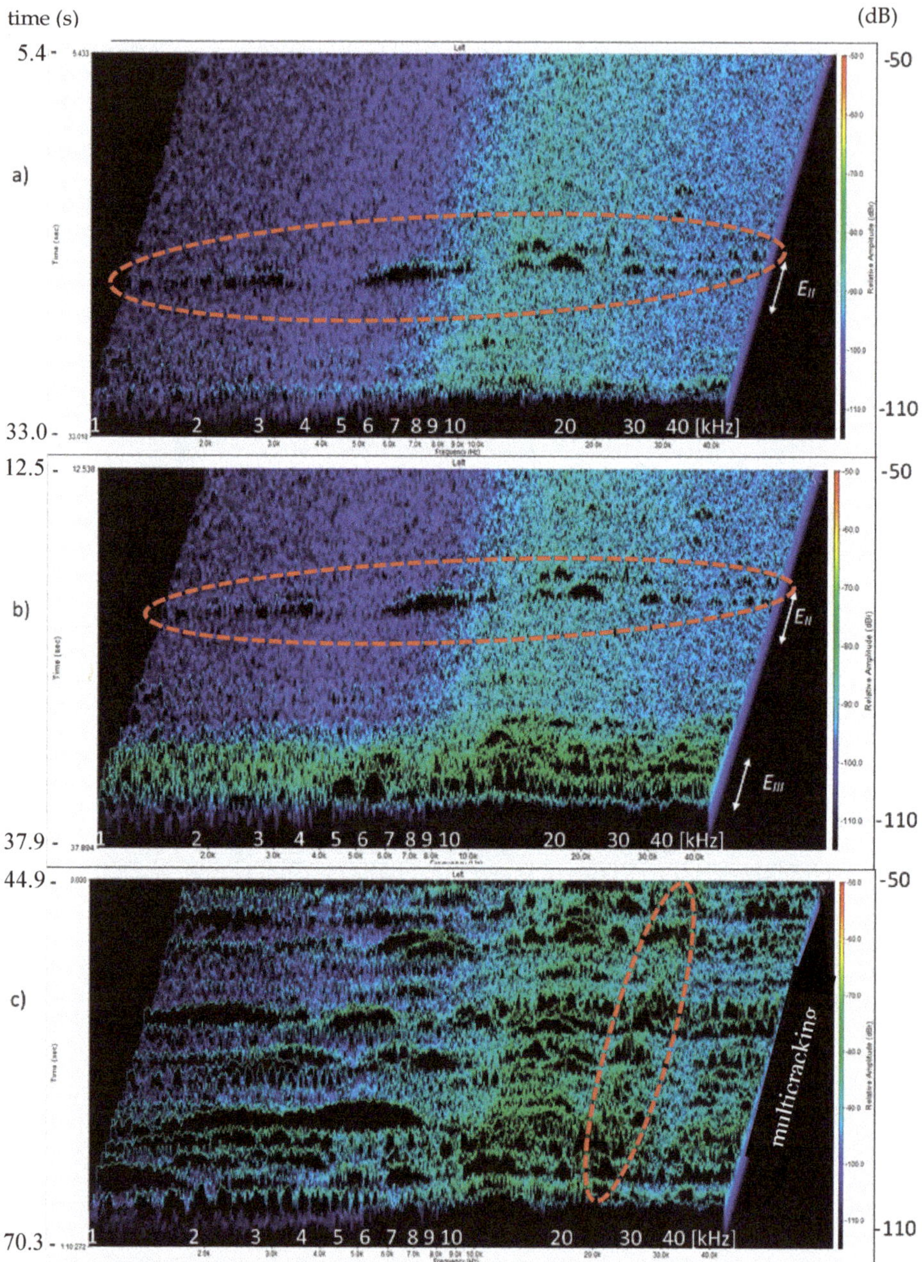

Figure 7. Composite sound spectra: (**a**) background, micro-events E_{II}, (**b**) micro-events E_{II} and E_{III}, (**c**) multicracking between points f_1 and f_2.

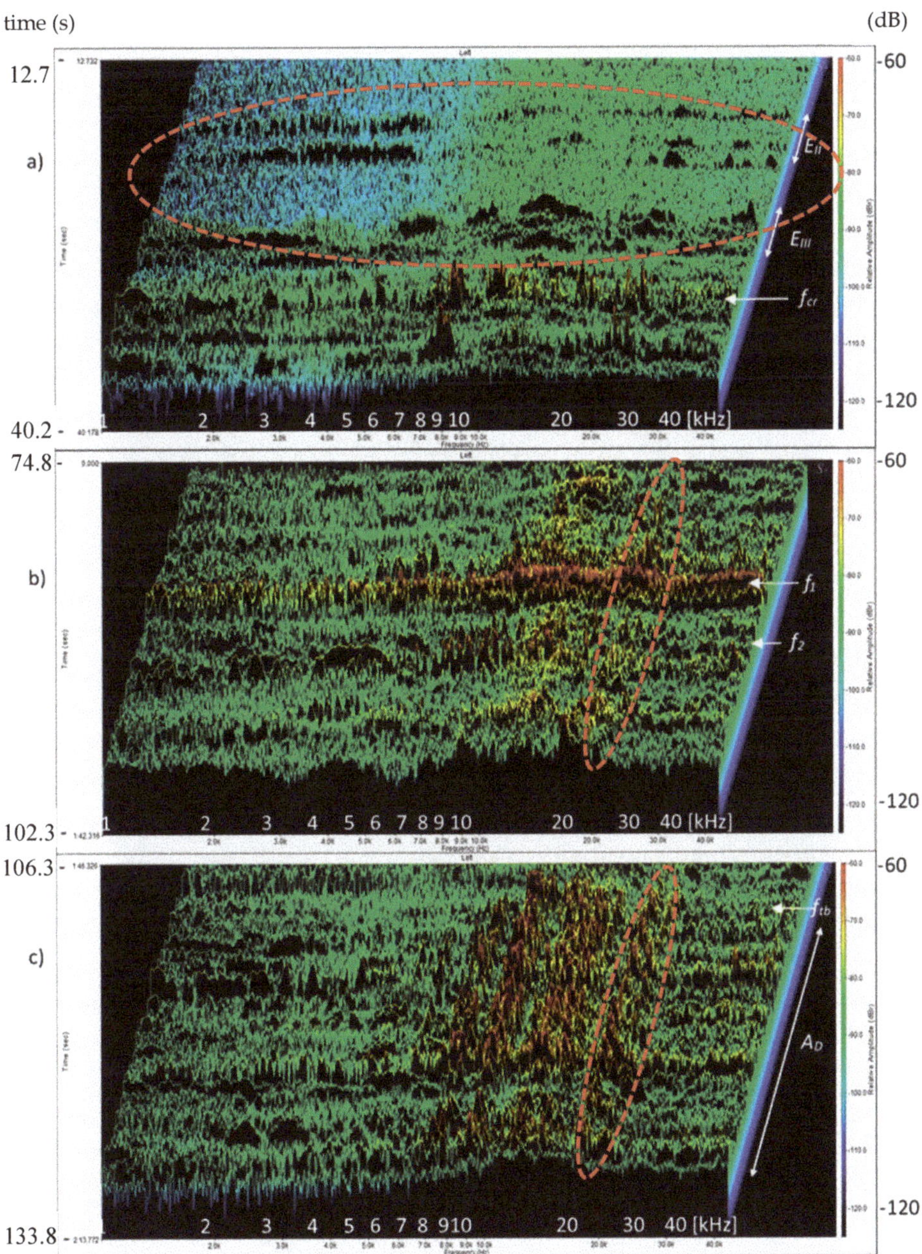

Figure 8. Composite sound spectra: (**a**) background, micro-events E_{II}, E_{III}, and f_c, (**b**) event f_1, (**c**) event f_{tb} and deflection control area A_D.

4. Discussion of the Results

Figure 3 presents force-deflection and force-time correlation obtained during bending–tension tests of a quasi-brittle cement composite with dispersed reinforcement. Presentation of these two charts in one figure enables a better recognition of the destruction processes. The force-deflection curve enables the identification of the proportionality, strengthening, deflection control, and crack propagation areas. The sudden decreases in the ability to carry stress that are recorded on the force-deflection curve indicate the appearance of macro-cracks/fractures including fiber breaking.

The force–time curve allows one to indicate the effects occurring at the same time as the recorded AE effects and the AE totals, which should be referred to the force-deflection curve. In order to precisely assess the composite destruction process, it is necessary to correlate both force-deflection and force–time curves with the measured AE effects. The recorded acoustic effects in the proportionality area is presented in 4, Figures 7b and 8a. The obtained data indicate the recording of signals and AE micro events in the proportionality area AE. The measured AE effects have been linked mainly to the sound background signal occurring during the tests and to the occurring events. In the figures, AE events can be observed that have been linked to the individual microcracks occurring within the area of the Hooke's law and micro-regrouping in the structure (micro-relaxation), which do not affect the stress-deflection linear correlation. The occurrence of microcracks before the occurrence of f_{cr}, has been confirmed in other publications [20,26,36].

A micro-event precedes the occurrence of critical point f_{cr}. As shown by Figures 4–8, in the initial period of the operation of Hooke's law, in addition to the background noise, single micro-events/signals appear. The spectra that correspond to sound backgrounds are characterized by the smallest relative amplitude (Figures 6 and 7). Slightly larger intensities of the sound component correspond to micro-events, which do not affect the stress-deflection linear relationship. If the correlating spectra do not occur in groups and there is no increase in sound intensity components, then they may be considered as insignificant (Figure 7a).

The ability to identify them makes it possible to avoid the catastrophic destruction process in traditional cement composites and in quasi-brittle composites to avoid exceeding f_{cr}. As is shown by the presented data, the sound spectrum corresponding to f_{cr} is characterized by the largest intensity of the sound spectrum component, which ends the concentration of events (Figures 6b and 8a).

Recognition of the destruction process by means of the 3D spectrum has already been presented earlier [36] with respect to various cement composites and has also been confirmed in this paper.

What has been observed in this study is a clear division of relative sound amplitudes at the level of circa 7 kHz. Medium- and low-frequency sounds are characterized by a lower intensity of sound spectrum component compared to high-frequency sounds.

The analysis of the low-frequency spectra was not clear (it results from the measurement range of the head recording the sound), which is why it was disregarded in the presentation of the 3D image (Figures 7 and 8). The analysis of sound spectra was conducted in the range of 1–40 kHz.

After exceeding f_{cr}, the multicracking effect was observed in the tested composite, resulting in the appearance of the intensities of sound components stronger than the background spectra and single events in the proportionality area, but significantly smaller than the spectrum of f_{cr} (Figure 7c). The appearance of a macrocrack at point f_{tb} and the deflection control process A_D resulted in spectra with the sound intensity similar to f_{cr} with larger amplitudes in the range of 8–20 kHz (Figures 6–8).

Figure 5 shows a 2D spectrogram that presents the possibility of the identification of destruction processes. It is a method of data imaging used previously by researchers. The obtained image does not show the velocity of increase in the relative amplitudes and contains too large ranges of amplitudes, which may contain various signals, especially those with small amplitudes, which makes it difficult to recognize them (especially with respect to single events and the phenomenon of their concentration). The spectrogram enables the recognition of AE effects with large differences in amplitudes. The use of frequency ranges makes it difficult to identify which area is responsible for the reinforcement breaking, and which is responsible for the pull-out of the reinforcement from the matrix. The 3D sound spectrum contains such information. The spectra responsible for fiber breaking and the occurrence of microcracks are characterized by a course of sharp, high amplitudes in the entire range of the spectrum 1–40 kHz (Figure 8b), whereas the pull-out of fibers and their partial destruction caused an increase in the relative amplitudes, mainly in the area of high frequencies Figure 8c.

The conducted tests indicate that it is possible to accurately identify the destruction processes in cement composites. It should be emphasised, however, that their correct recognition requires an appropriate correlation of the spectra with the individual micro and macro destruction effects.

5. Conclusions

It has been indicated that there is a possibility of predicting the occurrence of f_{cr} based on the analysis of sound spectra 3D, the occurring groups of micro-events that precede the end of the load-deflection proportionality area.

The control of micro destruction process before the first crack f_{cr} with the use of acoustic emission (AE) by means of recorded micro-events (increase sumAE) and spectrograms can be used. These analyses should be expanded by adding an interpretation of individual sound spectra (and not the groups of them) and correlated with the load-deflection curve.

The 3D spectra provide a better image of the whole destruction process (particularly with respect to micro-events). The identification of individual events should be correlated with the corresponding individual spectra including the analysis of spectrum in various frequency ranges.

It has been confirmed that the use of spectra with respect to grouped events (destruction processes), in particular in the case of 2D spectra, results in incorrect conclusions if it refers to different events.

Author Contributions: D.L. analyzed the test results and performed editing, prepared the specimens, and analyzed the test results, K.S. conducted the correction of results and applications. All authors have read and agreed to the published version of the manuscript.

Funding: This research received no external funding.

Conflicts of Interest: The authors declare no conflicts of interest.

References

1. Ohtsu, M. The history and development of acoustic emission in concrete engineering. *Mag. Concr. Res.* **1996**, *48*, 321–330. [CrossRef]
2. Dzayea, E.D.; Schutterb, G.D.; Aggelisa, D.G. Monitoring early-age acoustic emission of cement paste and fly ash paste. *Cem. Concr. Res.* **2020**, *129*, 105964. [CrossRef]
3. Assi, L.; Soltangharaei, V.; Anay, R.; Ziehl, P.; Matta, F. Unsupervised and supervised pattern recognition of acoustic emission signals during early hydration of Portland cement paste. *Cem. Concr. Res.* **2018**, *103*, 216–225. [CrossRef]
4. Brandt, A.M. Fibre reinforced cement-based (FRC) composites after over 40 years of development in building and civil engineering. *Compos. Struct.* **2008**, *86*, 3–9. [CrossRef]
5. Kucharska, L.; Brandt, A.M. Pitch-based carbon fibre reinforced cement composites. In *Materials Engineering Conference ASCE. Materials for the New Millenium*; Chong, K.P., Ed.; ASCE: Washington, DC, USA, 1996; Volume 1, pp. 1271–1280.
6. Logoń, D. Monitoring of microcracking effect and crack propagation in cement composites (HPFRC) using the acoustic emission (AE). In Proceedings of the The 7th Youth Symposium on Experimental Solid Mechanics, YSESM '08, Wojcieszyce, Poland, 14–17 May 2008.
7. Logoń, D. The application of acoustic emission to diagnose the destruction process in FSD cement composites. In Proceedings of the International Symposium on Brittle Matrix Composites BMC-11, Warsaw, Poland, 28–30 September 2015; Andrzej, M., Ed.; Institute of Fundamental Technological Research: Warsaw, Poland, 2015; pp. 299–308.
8. Paul, S.C.; Pirskawetz, S.; Zijl, G.P.A.G.; Schmidt, W. Acoustic emission for characterising the crack propagation in strain-hardening cement-based composites (SHCC). *Cem. Concr. Res.* **2015**, *69*, 19–24. [CrossRef]
9. Ranachowski, Z.; Jozwiak-Niedzwiedzka, D.; Brandt, A.M.; Debowski, T. Application of Acoustic Emission Method to Determine Critical Stress in Fibre Reinforced Mortar Beams. *Arch. Acoust.* **2012**, *37*, 261–268. [CrossRef]

10. Yuyama, S.; Ohtsu, M. Acoustic Emission evaluation in concrete. In *Acoustic Emission-Beyond the Millennium*; Kishi, T., Ohtsu, M., Yuyama, S., Eds.; Elsevier: Amsterdam, The Netherlands, 2000; pp. 187–213.

11. Kim, B.; Weiss, W.J. Using acoustic emission to quantify damage in restrained fiber-reinforced cement mortars. *Cem. Concr. Res.* **2003**, *33*, 207–214. [CrossRef]

12. Reinhardt, H.W.; Weiler, B.; Grosse, C. Nondestructive testing of steel fibre reinforced concrete. *Brittle Matrix Compos.* **2000**, *6*, 17–32.

13. Landis, E.; Ballion, L. Experiments to relate acoustic energy to fracture energy of concrete. *J. Eng. Mech.* **2002**, *128*, 698–702. [CrossRef]

14. Ouyang, C.S.; Landis, E.; Shh, S.P. Damage assessment in concrete using quantitative acoustic emission. *J. Eng. Mech.* **1991**, *11*, 2681–2698. [CrossRef]

15. Chen, B.; Liu, J. Experimental study on AE characteristics of free-point-bending concrete beams. *Cem. Concr. Res.* **2004**, *34*, 391–397. [CrossRef]

16. Aggelis, D.G.; Mpalaskas, A.C.; Matikas, T.E. Investigation of different modes in cement-based materials by acoustic emission. *Cem. Concr. Res.* **2013**, *48*, 1–8. [CrossRef]

17. Schabowicz, K.; Gorzelańczyk, T.; Szymków, M. Identification of the degree of fibre-cement boards degradation under the influence of high temperature. *Autom. Constr.* **2019**, *101*, 190–198. [CrossRef]

18. Granger, S.; Pijaudier, G.; Loukili, A.; Marlot, D.; Lenain, J.C. Monitoring of cracking and healing in an ultra high performance cementitious material using the time reversal technique. *Cem. Concr. Res.* **2009**, *39*, 296–302. [CrossRef]

19. Watanab, K.; Niwa, J.; Iwanami, M.; Yokota, H. Localized failure of concrete in compression identified by AE method. *Constr. Build. Mater.* **2004**, *18*, 189–196. [CrossRef]

20. Elaqra, H.; Godin, N.; Peix, G.; R'Mili, M.; Fantozzi, G. Damage evolution analysis in mortar, during compressive loading using acoustic emission and X-ray tomography: Effects of the sand/cement ratio. *Cem. Concr. Res.* **2007**, *37*, 703–713. [CrossRef]

21. Shahidan, S.; Rhys Pulin, R.; Bunnori, N.M.; Holford, K.M. Damage classification in reinforced concrete beam by acoustic emission signal analysis. *Constr. Build. Mater.* **2013**, *45*, 78–86. [CrossRef]

22. Soulioti, D.; Barkoula, N.M.; Paipetis, A.; Matikas, T.E.; Shiotani, T.; Aggelis, D.G. Acoustic emission behavior of steel fibre reinforced concrete under bending. *Constr. Build. Mater.* **2009**, *23*, 3532–3536. [CrossRef]

23. Ohtsu, M. Determination of crack orientation by acoustic emission. *Mater. Eval.* **1987**, *45*, 1070–1075.

24. Ohno, K.; Ohtsu, M. Crack classification in concrete based on acoustic emission. *Constr. Build. Mater.* **2010**, *24*, 2339–2346. [CrossRef]

25. Ohtsu, M. Elastic wave methods for NDE in concrete based on generalized theory of acoustic emission. *Constr. Build. Mater.* **2016**, *122*, 845–855. [CrossRef]

26. Šimonová, H.; Topolář, L.; Schmid, P.; Keršner, Z.; Rovnaník, P. Effect of carbon nanotubes in metakaolin-based geopolymer mortars on fracture toughness parameters and acoustic emission signals. In Proceedings of the BMC 11 International Symposium on Brittle Matrix Composites, Warsaw, Poland, 28–30 September 2015; pp. 261–288.

27. Schabowicz, K.; Gorzelańczyk, T.; Szymków, M. Identification of the degree of degradation of fibre-cement boards exposed to fire by means of the acoustic emission method and artificial neural networks. *Materials* **2019**, *12*, 656. [CrossRef]

28. Ranachowski, Z.; Schabowicz, K. The contribution of fibre reinforcement system to the overall toughness of cellulose fibre concrete panels. *Constr. Build. Mater.* **2017**, *156*, 1028–1034. [CrossRef]

29. Shiotani, T.; Li, Z.; Yuyama, S.; Ohtsu, M. Application of the AE Improved b-Value to Quantitative Evaluation of Fracture Process. *Concr. Mater. J. Acoust. Emiss.* **2004**, *19*, 118–133.

30. Parmar, D. *Non-Destructive Bridge Testing and Monitoring with Acoustic Emission (AE) Sensor Technology*; Final Report; Hampton University: Hampton, VA, USA, 2011.

31. Ono, K.; Gołaski, L.; Gębski, P. Diagnostic of reinforced concrete bridges by acoustic emission. *J. Acoust. Emiss.* **2002**, *20*, 83–98.

32. Van Steen, C.; Verstrynge, E.; Wevers, M.; Vandewalle, L. Assessing the bond behaviour of corroded smooth and ribbed rebars with acoustic emission monitoring. *Cem. Concr. Res.* **2019**, *120*, 176–186. [CrossRef]

33. Tsangouri, E.; Michels, L.; El Kadi, M.; Tysmans, T.; Aggelis, D.G. A fundamental investigation of textile reinforced cementitious composites tensile response by Acoustic Emission. *Cem. Concr. Res.* **2019**, *123*, 105776. [CrossRef]

34. Das, A.K.; Suthar, D.; Leung, C.K. Machine learning based crack mode classification from unlabeled acoustic emission waveform features. *Cem. Concr. Res.* **2019**, *121*, 42–57. [CrossRef]

35. Logoń, D. FSD cement composites as a substitute for continuous reinforcement. In Proceedings of the Eleventh International Symposium on Brittle Matrix Composites BMC-11, Warsaw, Poland, 28–30 September 2015; Andrzej, M.B., Ed.; Institute of Fundamental Technological Research: Warsaw, Poland, 2015; pp. 251–260.

36. Logoń, D. Identification of the Destruction Process in Quasi Brittle Concrete with Dispersed Fibers Based on Acoustic Emission and Sound Spectrum. *Materials* **2019**, *12*, 2266. [CrossRef]

37. ASTM 1018. *Standard Test Method for Flexural Toughness and First Crack Strength of Fiber-Reinforced Concrete*; ASTM: West Conshohocken, PA, USA, 1992; Volume 4.

38. EN 14651. Test Method for Metallic Fibre Concrete. Measuring the Flexural Tensile Strength (Limit of Proportionality (LOP), Residual). 11 July 2005. Available online: https://shop.bsigroup.com/ProductDetail/?pid=000000000030160504 (accessed on 6 January 2020).

39. JCI, Japan Concrete Institute Standard. *Method of Test for Bending Moment-Curvature Curve of Fiber-Reinforced Cementitious Composites, S-003-2007, 2003. Method of Test for Load-Displacement Curve of Fiber Reinforced Concrete by Use of Notched Beam, 2003. Method of Test for Fracture Energy of Concrete by Use of Notched Beam*; Japan Concrete Institute: Tokyo, Japan, 2003.

40. Brandt, A.M.; Ranchowski, Z.; Zieliński, M.; Dąbrowski, M.; Sobczak, M. *Report from Tests of Cracking Resistance of Bent Cement Composite Samples*; Polish Academy of Science Institute of Fundamental Technology Problems: Warsaw, Poland, 2010.

41. Logoń, D.; Schabowicz, K.; Wróblewski, K. Assessment of the mechanical properties of ESD pseudoplastic resins for joints in working elements of concrete structures. *Materials* **2020**, *13*, 2426. [CrossRef]

 *materials*

Article

Identification of Stress States in Compressed Masonry Walls Using a Non-Destructive Technique (NDT)

Radosław Jasiński

Department of Building Structures and Laboratory of Civil Engineering Faculty, Silesian University of Technology, ul. Akademicka 5, 44-100 Gliwice, Poland; radoslaw.jasinski@polsl.pl; Tel.: +48-32-237-1127

Received: 21 May 2020; Accepted: 22 June 2020; Published: 25 June 2020

Abstract: The structure safety can be assessed, but only indirectly, by identifying material properties, geometry of structures, and values of loads. The complete and comprehensive assessment can be done only after determining internal forces acting inside structures. Ultrasonic extensometry using an acoustoelastic effect (AE) is among the most common non-destructive techniques (NDT) of determining true stresses in structures. Theoretical bases of the method were described in the mid 20th century. They were founded on the correlation between ultrasonic waves and the value and direction of stresses. This method is commonly used to determine stresses mainly in homogeneous materials without any inherent internal defects. This method is rarely applied to porous or composite materials, such as concrete or rock due to a high dispersion of results. Autoclaved aerated concrete (AAC), characterized by high homogeneity and porosity, is the popular material in the construction sector, used to produce masonry units. The discussed tests involved the acoustoelastic effect to determine stresses in the masonry wall made of AAC. This paper presents a widely theoretical background for the AE method, and then describes the author's own research on AAC divided into two stages. At first, the empirical relationships between compressive stress and velocity of longitudinal ultrasonic wave, including humidity, were determined. In stage II, nine masonry walls were tested in axial compression. Mean compressive stresses in the masonry wall determined with the proposed method were found to produce a satisfactory confidence level up to ca. 50% of failure stresses. Results were significantly understated for stresses of the order of 75% of failure stresses.

Keywords: masonry structures; autoclaved aerated concrete masonry units (AAC); compressive strength; minor-destructive (MDT) techniques; non-destructive techniques (NDT); ultrasonic testing; acoustoelastic effect (AE)

---

1. Introduction

The ultrasonic technique is used for many purposes, but the most common purpose is diagnostic [1–4]. Ultrasounds are employed by many branches of the industry; they are a crucial tool for electronic engineering, telecommunications or material engineering. Generally, the application of ultrasounds is broad and covers active and passive uses. The passive use includes ultrasonic spectroscopy and defectoscopy, ultrasonic diagnostic for medical purposes, and hydrolocation. Ultrasonic waves are more and more often used to test kinetics of hardening of different types of substances. The active use includes ultrasonic coagulation and dispergation, ultrasound therapy, cavitation, development of sonoluminescence, or chemical reactions. Other active applications are: crushing and forming hard media, bonding, soldering, washing, extracting, and drying of substances. They are also used quite commonly to measure stresses in metal constructions. These methods have been elaborated to measure stresses caused by thermal treatment of rolled profiles or during welding. Ultrasonic stress measurements are based on the acoustoelastic effect (AE), that is, on the dependence of acoustic wave velocity on stress. Measurements of stress in bolts are the oldest application of the

AE method. Stress is determined on the basis of measured change in the times of flight of ultrasonic waves propagating along the stressed bolt. New tests and applications related to measurements of stresses in rails, train wheels and unit shafts, have been presented in [4].

Non-destructive techniques are used for other popular materials, such as concrete or ceramics, to determine time of set, and changes in the modulus of elasticity. Ultrasonic testing using minor-destructive techniques (MDT) can determine compressive strength [5,6]. Classic methods of damage detection have been intensively developed [7–10]. No attempts have been so far made to determine stress state in materials with porous structure, such as concrete or rock. These materials behave advantageously under compression, and the complex structure of ordinary concrete cause difficulties in interpreting the results. This aspect is completely different for autoclaved aerated concrete (AAC), whose compressive strength is relatively low and at the same time this material is more homogeneous than concrete despite its porous structure.

Autoclaved aerated concrete (AAC) contains cement, calcium, and lime as binding material, sand used as a filler and tiny quantities of aluminium powder (or paste) which is used as a blowing agent. Density of this type of concrete ranges from 300 to 1000 kg/m^3, and its compressive strength varies from 1.5 to 10 N/mm^2. AAC has been commonly used since the middle of the 1950s. This material (>40% of the construction segment in Europe) is used for masonry structures, precast wall or floor elements, and intels [11]. The open-pore structure explains why AAC is sensitive to direct exposure to moisture, which results in worse insulating and strength properties. The available articles, apart from general relationships specified in standards, do not contain detailed references expressed as empirical relations to determine strength properties of AAC using NDT and semi-NDT techniques. This work describes the practical application of selected issues on ultrasonic testing presented in the papers [12,13].

This paper is an attempt to evaluate changes in stress state of masonry units made of autoclaved aerated concrete built into small fragments of the masonry wall under axial compression. The masonry wall was also made of autoclaved aerated concrete with considerably more porous structure than ordinary concrete. The aim of the tests was to define empirical relationships concerning values of vertical stresses in the AAC masonry wall including the acoustoelastic (AE) effect well-known in practice [14]. This paper is divided into the theoretical part containing detailed bases of the AE method, and the research part consisting of stages I and II In stage I of the tests, experiments were performed on 24 small cube specimens (100 × 100 × 100 mm) of autoclaved aerated concrete with nominal densities of 400, 500, 600, and 700 kg/m^3. The elastooptic constant β_{111} was determined that showed the longitudinal wave c_{p0} depended on stress σ_{33}. In stage II, nine small models made of autoclaved aerated concrete with nominal density of 500 kg/m^3 were prepared and tested in the second phase of tests. They were used to measure velocity of the ultrasonic wave c_p. Relationships determined in the first phase were used to identify the stress state in the masonry wall and validate σ_{33}—c_p relationship.

2. Theoretical Basis

2.1. Propagation of Ultrasonic Waves in Linear-Elastic Material

Generally, an anisotropic body, e.g., crystal of defined symmetry, can be the solid medium. The propagation of waves in the anisotropic medium, particularly velocity, depends on the direction relative to the axis of coordinates usually related to the crystallographic arrangement that corresponds to the given symmetry. Hooke's law [15] describes elastic properties of the anisotropic arrangement in the linear relationship between the stress tensor σ_{ij} and the deformation tensor ε_{kl} in the following way:

$$\sigma_{ij} = c_{ijkl}\varepsilon_{kl} + c_{ijklmn}\varepsilon_{kl}\varepsilon_{mn} + \ldots \tag{1}$$

where: σ_i—components of stress state, ε_{kl}—components of deformation state.

Both quantities are symmetric tensors of second rank, which means they can have six independent components. Coefficients c_{ijkl} and c_{ijklmn} are constants of elasticity of second or third rank, respectively. They are symmetric tensors of fourth and sixth rank, respectively. The linear theory of elasticity assumes materials are elastic, and the relationship between stress and deformation is linear. All constants of elasticity of third order or higher are neglected. Even for such a simplification, the number of tensor components c_{ijkl} defining elastic properties is 36, but the number of independent components is 21. In the case of orthotrophic materials with three mutually perpendicular planes of symmetry, elastic properties are described by nine independent constants of elasticity in the following form:

$$c_{ij} = \begin{bmatrix} c_{11} & c_{12} & c_{13} & 0 & 0 & 0 \\ c_{12} & c_{22} & c_{23} & 0 & 0 & 0 \\ c_{13} & c_{23} & c_{33} & 0 & 0 & 0 \\ 0 & 0 & 0 & c_{44} & 0 & 0 \\ 0 & 0 & 0 & 0 & c_{55} & 0 \\ 0 & 0 & 0 & 0 & 0 & c_{66} \end{bmatrix} \tag{2}$$

Regarding isotropic materials with the infinite number of axes of symmetry planes, elastic properties can be comprehensively described by two independent constants of elasticity c_{12} and c_{44}. Other matrix coefficients (2) can be expressed as linear combinations using the Lamé coefficients:

$$c_{11} = c_{22} = c_{33} = \lambda + 2\mu, \; c_{12} = c_{23} = c_{23} = \lambda, \; c_{44} = c_{55} = c_{66} = \mu. \tag{3}$$

The force acting on any volume element in the solid medium, in which the disturbance is observed, can be expressed as the gradient of stress caused by the disturbance [1]. The Equation of the particle motion representing the equilibrium state between the restoring force and the inertial force is expressed by the following Equation:

$$\rho_0 \frac{\partial^2 \xi_i}{\partial t^2} = \frac{\partial T_{ij}}{\partial x_j} \rightarrow \rho_0 \frac{\partial^2 \xi_i}{\partial t^2} = c_{ijkl} \frac{\partial^2 \xi_k}{\partial x_j \partial x_l} \tag{4}$$

where: ρ_0—density of the body in the tensionless state. The expression (4) contains the equations of three components of the displacement, which describe components of the wave equation of vector quantity ξ described by three components. Assuming that coordinates of the plane harmonic wave are expressed by the relationship $\xi = \xi_0 e^{i(\omega t - kr)}$, Equation (4) can be expressed as:

$$-\omega^2 \rho_0 \xi_{0i} = -c_{ijkl} k_j k_l \xi_{0k} \rightarrow \left(c_{ijkl} k_j k_l - \delta_{ik} \omega^2 \rho_0 \right) \xi_{0k} = 0 \tag{5}$$

where: ω is the wave frequency, k_j, k_l,—wave vector (towards j, l), ξ_{0i}, ξ_{0k}—coordinates of the plane harmonic wave (towards i, k).

Expression (5) is the system of homogeneous algebraic equations, which due to unknown ξ_{0k} is described in the following form:

$$\begin{aligned} \left(c_{1j1l} k_i k_l - \omega^2 \rho_0 \right) \xi_{10} + c_{1j2l} k_i k_l \xi_{20} + c_{1j3l} k_i k_l \xi_{30} &= 0 \\ c_{2j1l} k_i k_l \xi_{10} + \left(c_{2j2l} k_i k_l - \omega^2 \rho_0 \right) \xi_{20} + c_{2j3l} k_i k_l \xi_{30} &= 0 \\ c_{3j1l} k_i k_l \xi_{10} + c_{3j2l} k_i k_l \xi_{20} + \left(c_{3j21} k_i k_l - \omega^2 \rho_0 \right) \xi_{30} &= 0 \end{aligned} \tag{6}$$

The system of equations is fulfilled when the determinant of the coefficients is equal to 0. The equation of third degree relevant to ω^2 is the solution for the determinant. The equation contains three roots that correspond to three different waves with mutually perpendicular displacements. When the simplest case of the isotropic body and waves travelling along one axis (x_3), the determinant of the Equation (6) takes the following form:

$$\begin{vmatrix} c_{44} k^2 - \omega^2 \rho_0 & 0 & 0 \\ 0 & c_{44} k^2 - \omega^2 \rho_0 & 0 \\ 0 & 0 & c_{11} k^2 - \omega^2 \rho_0 \end{vmatrix} = 0 \tag{7}$$

By solving the determinant, the following equation is obtained:

$$\left(c_{44}k^2 - \omega^2\rho_0\right)^2\left(c_{11}k^2 - \omega^2\rho_0\right) = 0 \tag{8}$$

It has two roots equal to $\omega_1^2 = \omega_2^2 = \frac{c_{44}k^2}{\rho_0}$, and the third one equal to $\omega_3^2 = \frac{c_{11}k^2}{\rho_0}$. Taking into account that $k = \omega/C$ (where C is wave velocity), the following roots are obtained:

$$C_1 = C_2 = \sqrt{\frac{c_{44}}{\rho_0}}, \; C_3 = \sqrt{\frac{c_{11}}{\rho_0}} \tag{9}$$

A solution to this issue indicates the propagation of three waves in the body. Two of them are characterized by mutually perpendicular oscillations and the same velocity $C_1 = C_2 = c_T$ is known as transverse waves as c_{44} is the shear. The third wave with the velocity c_p is the longitudinal wave because c_{11} is constant related to the component of the normal deformation. Taking into account relationships between material constants, the following expression is obtained:

$$c_T = \sqrt{\frac{\mu}{\rho_0}}, \; c_p = \sqrt{\frac{\lambda + 2\mu}{\rho_0}} \tag{10}$$

Constants λ and μ can be introduced into the system of Equations (4) by replacing coefficients c_{ijkl}. Then, the system of equations broken down into components is for the isotropic body as follows:

$$\begin{aligned}
\rho_0\frac{\partial^2\xi_1}{\partial t^2} &= (\lambda + 2\mu)\frac{\partial^2\xi_i}{\partial x_1\partial x_i} + \mu\frac{\partial^2\xi_1}{\partial x_1\partial x_i}\\
\rho_0\frac{\partial^2\xi_2}{\partial t^2} &= (\lambda + 2\mu)\frac{\partial^2\xi_i}{\partial x_2\partial x_i} + \mu\frac{\partial^2\xi_2}{\partial x_2\partial x_i}\\
\rho_0\frac{\partial^2\xi_3}{\partial t^2} &= (\lambda + 2\mu)\frac{\partial^2\xi_i}{\partial x_3\partial x_i} + \mu\frac{\partial^2\xi_2}{\partial x_2\partial x_i}
\end{aligned} \tag{11}$$

When the medium is incompressible (no changes in volume), the above equations give the wave equation for transverse waves in the following vector form:

$$\frac{\partial^2\xi}{\partial t^2} = \frac{\mu}{\rho_0}\nabla^2\xi \tag{12}$$

where ∇^2 is the Laplace operator of the second order in n- dimensional Cartesian coordinate system expressed as: $\nabla^2 = \Delta = \frac{\partial^2}{\partial x_i^2} + \frac{\partial^2}{\partial x_j^2} + \frac{\partial^2}{\partial x_k^2} + \ldots + \frac{\partial^2}{\partial x_n^2}$.

Assuming the irrotational medium, the wave equation for longitudinal waves is as follows:

$$\frac{\partial^2\xi}{\partial t^2} = \frac{\lambda + 2\mu}{\rho_0}\nabla^2\xi \tag{13}$$

2.2. Propagation of Ultrasonic Waves in Porous Material

Biot is regarded as the initiator of works on the theory and studies on ultrasonic waves in porous materials [16,17]. According to the theory, there are two compressional waves in the wet porous material—P1-wave (the fast wave) and P2-wave (the slow wave). Further works [18,19] have confirmed Biot's hypothesis. Other research works refer to other phenomena, including reflections and refractions, which are significant for testing and diagnosing materials. Currently, different aspects concerning wave propagation in the porous medium are examined. The issue of wave propagation and scattering in the inhomogeneous material is presented in, inter alia, the papers [20,21]. The works [22,23] present the mathematical model of propagation of low-frequency surface waves–the Stoney waves, in the porous material. Another paper [24] describes experiments on absorption and propagation of ultrasonic waves in materials with dual porosity, whereas the work [25] demonstrates test on the propagation of Rayleigh waves at liquid–solid interfaces.

Concrete, like rock media, is not ideally elastic. Therefore, the wave equation cannot be directly applied to this medium (12). The imperfect elasticity of concrete causes internal friction that transforms a part of energy into heat causing scattering and dispersion of velocity of elastic waves. The mathematical presentation of imperfection of the elastic medium is described in different ways. For example, the equation of the perfectly elastic medium can be replaced with the system of equations describing stresses and deformations. The equation of the plane longitudinal wave moving and scattered in the imperfectly elastic medium takes the following form:

$$\rho_0 \frac{\partial^2 \xi}{\partial t^2} = \frac{1}{\beta_{ad}} \frac{\partial^2 \xi}{\partial x^2} + \eta \frac{\partial^3 \xi}{\partial x^2 \partial t} \tag{14}$$

where β_{ad}—adiabatic compressibility coefficient, $\eta = \eta'' + \frac{4}{3}\eta'$—viscosity coefficient composed of coefficients (η'') of bulk and shear viscosity (η').

Generally, the solution to the wave Equation (14) is expressed as:

$$\xi(x,t) = A e^{-\alpha_\eta x} e^{i\omega(t-\frac{x}{c})} + B e^{\alpha_\eta x} e^{i\omega(t+\frac{x}{c})} \tag{15}$$

where: A, B—integration constants, α_η—integration constant depending on the value of viscosity coefficient, c—wave velocity, x—coordinate of wavefront, ω—wave frequency.

The velocity of longitudinal waves in viscoelastic medium can be described as:

$$c_p = \sqrt{\frac{K}{\rho_0}} \sqrt{\frac{2(1+\omega^2 r^2)\left(\sqrt{1+\omega^2 r^2}-1\right)}{\omega^2 r^2}} \tag{16}$$

where $r = \eta\beta_{ad}$, $K = 1/\beta_{ad}$.

The velocity of waves in the inhomogenuous granular medium, despite being the material constant, is related to its physical properties—density, elasticity defined by the Lamé coefficients also depends on wave scattered by the medium, wave frequencies, the medium structure, etc. Therefore, velocity not regarded as the constant value in contrast to the propagation of waves in perfectly elastic media. Granular media, such as rocks, concrete, or mortar, are characterized by:

- different dimensions and properties of components—matrix grains,
- different models (systems) of arrangement and connections of individual grains—they can have a direct contact or are connected with binder of other properties. In the case of chemically bonded materials, the binder changes its properties during the transformation from liquid to solid state.

Scattering of the elastic wave in granular media depends on many factors—mechanical and thermal processes caused by the propagating wave. There are three main reasons for energy loss during wave scattering:

- internal frictions in the medium—δ_r,
- thermal effects—δ_T,
- Rayleigh scattering δ_R.

The overall wave scattering is the sum of mentioned elements:

$$\delta = \delta_r + \delta_T + \delta_R, \tag{17}$$

The role of each of the three factors above in ultrasonic wave scattering in the homogeneous granular medium depends on the frequency and structure of that medium characterized by:

- dimensions of the matrix grains,

- thermal properties of components,
- elastic properties of components, and their density.

Wave velocity in granular media characterized by a large coefficient of wave scattering can be calculated from the following dependence:

$$C = C_0 \sqrt{1 - \left(\frac{\delta}{\omega}\right)^2} \tag{18}$$

where C_0—wave velocity in the linear-elastic medium, δ—total scattering coefficient, and ω—wave frequency.

The velocity of wave propagation in granular materials changes within a wide range and is subjected to fluctuations depending on the type of components and their distribution. It is caused by different values of elasticity constants E G v demonstrated by individual components of granular materials. Therefore, we obtain a certain mean velocity that results from the percentage contribution of velocity to individual components. Determining ultrasound velocity for different specimens cannot be neglected in that case. Greater scattering and more problems related to signal recording are expected in specimens with longer wave paths. Hence, the use in NDT methods requires the conversion of wave velocities.

3. Stress Measurements Using an Ultrasonic Technique

Material stress can affect velocity of the acoustic wave due to inhomogeneity and anisotropy of the material. That effect has been described for the first time by seismologist Biot [26] and experimentally verified by Hughes and Kelly [27] and Bergman and Shahbender [28]. It is demonstrated that the static stress can change velocity of the acoustic wave in the medium, and that effect is called the acoustoelastic (AE) effect [29,30].

The acoustoelastic effect is based on the relationship between the velocity of transverse wave propagation and stress in solid bodies found by Benson and Raelson in the 1970s [14]. Since then, this aspect has been widely developed [31–33]. The impact of stress on the velocity of transverse wave propagation is determined by the direction of wave propagation with reference to the stress direction and wave polarization. A change in the polarization plane depends on stress, similarly to a light wave in the elastooptic effect. Its mechanism was theoretically described on the basis of the non-linear theory of solid deformation [27]. According to that theory, constant elasticity of higher orders (than those observed in the theory of linear elasticity) was responsible for nonlinear effects. The propagation velocity in the stressed body can be expressed as the sum of velocities in the tensionless stress ($\sigma = 0$) and its change (increment) caused by stress. That change can be defined as dependent on stress including constant characteristics of elasticity of second or third order.

In accordance with the infinite deformation of elastic materials by Murnaghan [34], the stress-deformation relationship should be described by the function of free energy W_s defined as [27,35]:

$$W_s = \frac{1}{2}(\lambda + 2\mu)I_1^2 - 2\mu I_2 + \frac{1}{3}(l + 2m)I_1^3 - 2m I_1 I_2 + n I_3 \tag{19}$$

where: λ, μ—Lamé constants, l m n—elasticity constants of second and third order by Murnaghan, I_1, I_2, I_3—deformation invariants.

Taking into account the principle of energy conservation, Hooke's law can be expressed as:

$$\rho \delta W_s = \sigma_{ij} \frac{\partial \delta u_i}{\partial u_j}, \tag{20}$$

where δW and δu_i mean finite increments in the function of free energy and displacement area, ρ is density after deformation. The combination of Equations (19) and (20) produces the acoustoelastic equation, which binds the static load with velocity of the elastic wave under hydrostatic pressure P:

$$\begin{aligned}
\rho_0 c_p^2 &= \lambda + 2\mu - \frac{P}{3\lambda + 2\mu}(6l + 4m + 7\lambda + 10\mu)\,, \\
\rho_0 c_T^2 &= \mu - \frac{P}{3\lambda + 2\mu}(3m + 0,5n + 3\lambda + 6\mu)\,,
\end{aligned}$$
(21)

where: c_p and c_T are velocity of longitudinal and transverse waves respectively, a ρ_0—body density in the tensionless state.

Thus, the hydrostatic level of stress can be defined from Equation (20) [36] by measuring velocity of the longitudinal and transverse waves—Figure 1a. In the case of uniaxial stress, wave velocity depends on the direction of the stress and the square of velocity on Figure 1b-1f is as follows:

$$V_{111}^2 = \frac{\lambda + 2\mu}{\rho_0} - \frac{\sigma_1}{3K_0\rho_0}\left[\frac{\lambda + \mu}{\mu}(4\lambda + 10\mu + 4m) + \lambda + 2l\right],$$
(22)

$$V_{113}^2 = \frac{\lambda + 2\mu}{\rho_0} + \frac{\sigma_3}{3K_0\rho_0}\left[\frac{2\lambda}{\mu}(\lambda + 20\mu + m) - 2l\right],$$
(23)

$$V_{131}^2 = \mu - \frac{\sigma_1}{3K_0\rho_0}\left[4\lambda + 4\mu + m + \frac{\lambda n}{4\mu}\right],$$
(24)

$$V_{133}^2 = \mu - \frac{\sigma_3}{3K_0\rho_0}\left[\lambda + 2\mu + m + \frac{\lambda n}{4\mu}\right],$$
(25)

$$V_{132}^2 = \mu + \frac{\sigma_2}{3K_0\rho_0}\left[2\lambda - m + \frac{n}{2}\frac{\lambda}{2}\frac{n}{\mu}\right],$$
(26)

where: $K_0 = \frac{E}{3(1-2\nu)} = \frac{2\mu + 3\lambda}{3}$.

Knowing velocity of the ultrasonic wave in the loaded material and elasticity constants of the first ($\lambda\ \mu$), second and third order ($m\ n\ l$) normal stresses can be determined. Measurements of wave velocity do not cause any problems except for small specimens (due to high sensitivity of the recording equipment). However, determining material constants m, n, and l is difficult.

Using the equation [27], the precise method of determining material constants was presented in the papers [37,38]. Velocities of longitudinal and transverse waves under the uniaxial stress are presented in the following form:

$$\rho_0 V_{11}^2 = \lambda + 2\mu + \frac{\sigma_1}{E}[5\lambda + 10\mu + 2l + 4m - 2\nu(\lambda + 2l)] \rightarrow V_{11}^2 = V_0^2\left(1 + 2\alpha_{11}\frac{\sigma_1}{E}\right)$$
(27)

$$\rho_0 V_{12}^2 = \mu + \frac{\sigma_1}{E}\left[\lambda + 4\mu + m - \nu\left(2\lambda + 2\mu + 2m - \frac{n}{2}\right)\right] \rightarrow V_{11}^2 = \left(\frac{\mu}{\rho_0}\right)^2\left(1 + 2\alpha_{12}\frac{\sigma_1}{E}\right)$$
(28)

$$\rho_0 V_{21}^2 = \mu + \frac{\sigma_1}{E}\left[\lambda + \mu + m - \nu\left(2\lambda + 4\mu + 2m - \frac{n}{2}\right)\right] \rightarrow V_{21}^2 = \left(\frac{\mu}{\rho_0}\right)^2\left(1 + 2\alpha_{21}\frac{\sigma_1}{E}\right)$$
(29)

$$\rho_0 V_{22}^2 = \lambda + 2\mu + \frac{\sigma_1}{E}[\lambda + 2l - \nu(6\lambda + 10\mu + 4l + 4m)] \rightarrow V_{22}^2 = V_0^2\left(1 + 2\alpha_{22}\frac{\sigma_1}{E}\right)$$
(30)

$$\rho_0 V_{23}^2 = \mu + \frac{\sigma_1}{E}\left[\lambda + m - \frac{n}{2} - \nu(2\lambda + 6\mu + 2m)\right] \rightarrow V_{21}^2 = \left(\frac{\mu}{\rho_0}\right)^2\left(1 + 2\alpha_{23}\frac{\sigma_1}{E}\right)$$
(31)

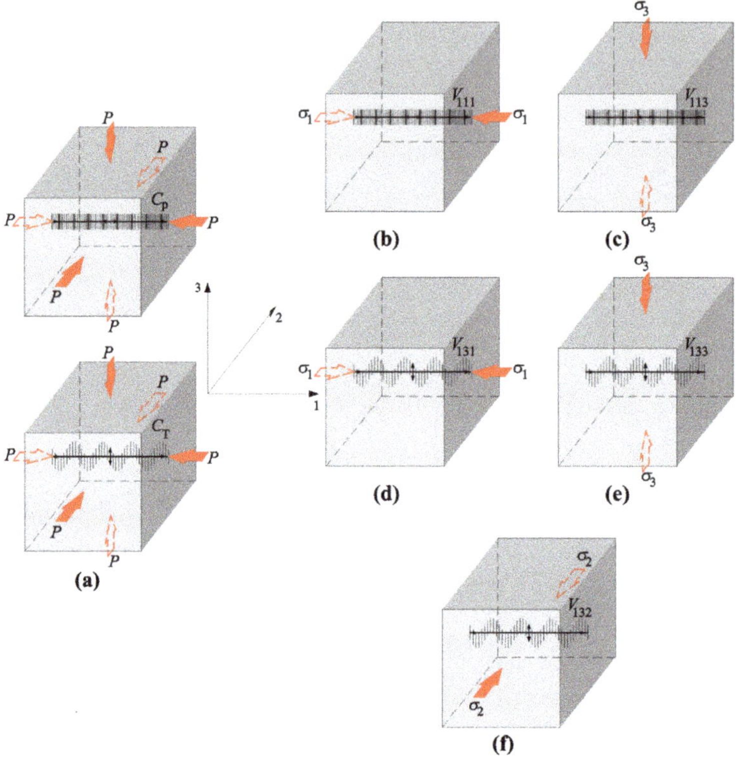

Figure 1. Identification of velocity of ultrasonic waves in isotropic material: (**a**) hydrostatic compression by pressure P, (**b**) longitudinal wave under stress σ_1, (**c**) transverse wave under stress σ_3, polarized in planes 1–3 (**d**) transverse plane under stress σ_1, polarized in planes 1–3, (**e**) transverse plane under stress σ_3, polarized in planes 1–3, (**f**) longitudinal wave under stress σ_2, polarized in planes 1–3.

Tests on the specimens of two lengths were the base to formulate relationships for determining constants α_{11}, α_{22}, α_{12}, α_{21}, and α_{23} from the following equations:

$$\alpha_{11} = 1 - \frac{E}{\sigma_1}\left[\frac{L_1}{L_2 - L_1}\left(\frac{\Delta t_1}{t_{01}}\right) - \frac{L_1}{L_2 - L_1}\left(\frac{\Delta t_2}{t_{02}}\right)\right] \tag{32}$$

where: L_1—length of specimens "1" and "2" used for calibration, $\Delta t_1 = t_1 - t_{01}$—difference in passing time of wave in specimen "1" after deformation (t_{01}) and before deformation (t_{01}), $\Delta t_2 = t_2 - t_{02}$—difference in passing time of wave in specimen "2" after deformation (t_{02}) and before deformation (t_{02}).

The equation for determining other material constants is as follows:

$$\alpha_{1j} = -\nu - \frac{E}{\sigma_1}\left(\frac{\Delta t}{t_0}\right) \tag{33}$$

where $\Delta t = t_1 - t_{01}$ is the difference in passing time of the wave in the specimen after and before its deformation.

My own research indicated the linear nature of changes in the ratio of passing time of the wave $\Delta t/t_0$ in relation to stress increase. Determining other constants consisted in solving the following system of equations:

$$l = \frac{(2\alpha_{11}-5)(\lambda+2\mu)}{2(1-2\nu)} - \frac{2m-\nu\lambda}{1-2\nu}, \quad m = \left[\frac{\alpha_{11}-\alpha_{22}}{2(1+\nu)} - 1\right](\lambda + 2\mu) - \frac{\mu}{2}$$
$$n_{12} = \frac{2}{\nu}[-(a + 4\nu) + 2\nu(a + \mu) + 2\mu\alpha_{12}], \quad n_{21} = \frac{2}{\nu}[-(a + 2\nu) + 2\nu(a + 2\mu) + 2\mu\alpha_{21}] \tag{34}$$
$$n_{23} = 2[a - 2\nu(a + 3\mu) - 2\mu\alpha_{23}]$$

where: $a = \lambda + m$.

Figure 2 shows changes in increment of propagation time of longitudinal and transverse waves described in the paper by Takahashi [38], who based this on his experience of developing and patenting the measuring apparatus to determine directly constants l, m, and n [39].

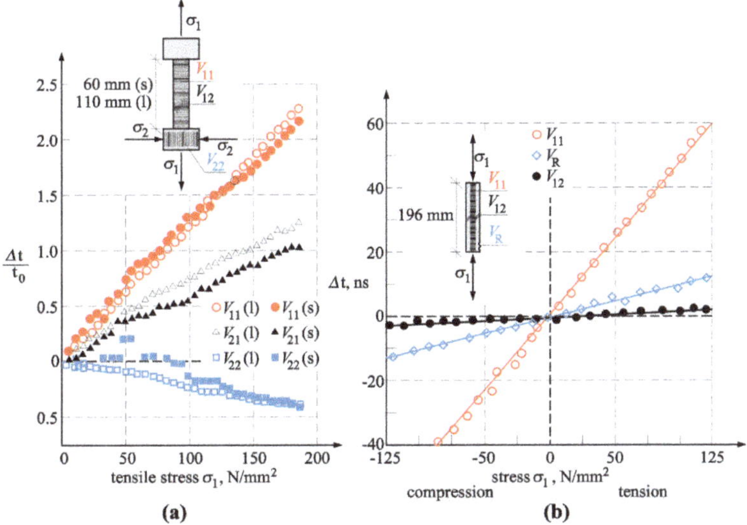

Figure 2. Selected test results for acoustoelastic effect: (**a**) ratio of changes in velocity of waves of different length obtained from tests (adapted from [38]), (**b**) changes in velocity of longitudinal, transverse, and Rayleigh waves obtained from tests (adapted from [40]).

As expected, the greatest increments in wave velocity were observed for longitudinal waves in the direction of stress. In addition, surface waves could be used to detect changes in stress states. The smallest gradients of velocity were obtained for transverse waves. As expected, the greatest increments in wave velocity were observed for longitudinal waves perpendicular to the stress direction. An increase in compressive stress caused an increase in wave velocity. Similar relationships were observed for waves propagating perpendicularly to the stress direction. Theoretical principles of the acoustoelastic effect are relatively well documented in the literature. There is also an apparatus to determine elasticity constants l, m, and n of the third order for metals and plastic in accordance with procedures described in, among others, papers [37,38,41]. Diagnosing stress states in structures using the NDT method requires the information on load direction and defined gradient of changes in velocity of longitudinal or transverse wave Knowing Muraghan coefficients is not essential.

4. Test Program and Results

The test program was divided into two stages. Stage I included the material tests on specimens made of autoclaved aerated concrete (AAC) to determine density ρ_0, elasticity modulus E and Poisson's ratio υ. Each cube specimen was subjected to axial compression until the failure and velocity of the longitudinal wave were determined at different normal stresses. The obtained results were used to determine linear correlations describing a σ–C_p relationship. In stage II, nine models of masonry walls were tested in axial compression. The velocity of the longitudinal wave was measured at different values of vertical loads. Then, vertical loads were determined on the basis of a correlation curve obtained during stage I. To interpret the results, they were compared with numerical calculations for 3D models of the masonry wall.

4.1. Stage I—Determination of Acoustoelastic Constant

The tests included four series of masonry units with thickness within the range of 180–240 mm and different classes of density: 400 kg/m^3, 500 kg/m^3, 600 kg/m^3, and 700 kg/m^3, each 20 masonry units were randomly selected. Six cores with a diameter of 59 mm and the height of 120 mm were taken from each type of the masonry unit using a drill. They were used to determine fundamental properties of tested autoclaved aerated concrete (AAC). All drilled cores were dried until constant weight at temperature of 105 ± 5 °C. Then, two vertical and horizontal electro-resistant tensometers were fixed to side surfaces of cylindrical specimens to measure deformations and determine modulus of elasticity E within the range of 0.1–0.33 σ_{max} and Poisson's ratio υ at the level of 0.33 σ_{max}. Tests were conducted using the testing machine, in which an increment in load was controlled manually, and the reading range of the dynamometer was 100 kN. Mean mechanical parameters obtained for all tested types of masonry units are shown in Table 1. The presented results from testing density were taken from the paper [12].

Table 1. Fundamental characteristics of masonry units.

No.	Nominal Class of Density kg/m^3	Density Range of AAC, kg/m^3	No. of Specimens (Cores φ59 × 120 mm)	Mean Density ρ_0, kg/m^3 (C.O.V) acc. to [12]	Mean Modulus of Elasticity, E, N/mm^2 (C.O.V)	Mean Poisson's Ratio ν, (C.O.V)
1	400	375–446	6	397 (6%)	1516 (9.6%)	0.19 (7.9%)
2	500	462–532	6	492 (3%)	2039 (8.9%)	0.21 (8.7%)
3	600	562–619	6	599 (2%)	2886 (10.5%)	0.20 (8.5%)
4	700	655–725	6	674 (3%)	4778 (10.1%)	0.19 (9.2%)

Besides the cores used to determine properties of AAC, four series of six cuboid specimens each (24 specimens in total) were drilled using a diamond saw 4. The specimens had dimensions of 100 × 100 × 100 mm, and were used as basic specimens for determining the strength f_B in accordance with Appendix B to the standard EN 771-4 [42]) harmonized with the European standard PN-EN 1996-1-1:2010 [43].

All specimens drilled from blocks to determine the correlation between vertical stresses and ultrasound velocity, were air-dried until constant weight at a temperature of 105 ± 5 °C (for at least 36 h). That way, the impact of moisture content on AAC was eliminated [13,44]. Generally, it tends to reduce significantly compressive strength and change velocity of the ultrasonic wave propagation [12].

The ultrasonic technique, commonly applied to test strength of concrete and masonry, was used to determine velocity of ultrasonic waves in AAC [45,46]. Ultrasonic testing was conducted on the block specimens 100 × 100 × 100 mm drilled from masonry units—Figure 3. The specimens in air-dry conditions and relative humidity w/wmax = 0% were used for testing. Each series of elements included at least six specimens, and 24 specimens in total were tested. PUNDIT LAB (Proceq SA, Schwerzenbach, Switzerland) instrument was used for tests. Exponential transducers with the waveguide length L = 50 mm, diameters ø1 = 4.2 mm and ø2 = 50 mm, and frequency 54 kHz were employed. The measurement accuracy of passing time of the ultrasonic wave was equal to ±0.1 μs. The used methodology of testing and equipment were typical for ultrasonic tomography for concrete and masonry [47,48]. Each specimen was placed on transducers of the testing machine (type FORM+TEST Prüffsysteme MEGA 3 with the range of 100 kN, class 1, reading accuracy ±1%)

through the vibration isolation washer and steel sheet of 3 mm thickness. The steel plate and the vibration isolation washer were placed on the top surface of the specimen. Vibration isolation was necessary for eliminating possible vibrations that could affect the results from measurements of ultrasonic waves. Then, the transducers were applied to opposite walls and the passing time of wave was measured using the transmission method. The transducers were in contact with the specimens at an angle of 90° within distance between the transducers measured every time with accuracy up to 1 mm. The tests were conducted for various loading of the specimens and the force was scaled every 2.5 kN.

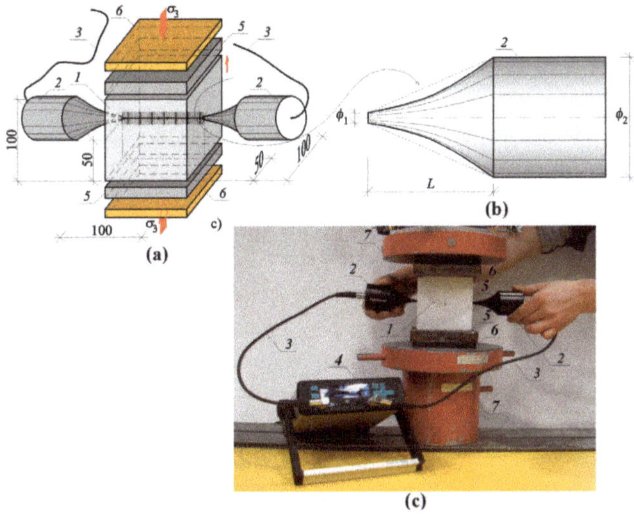

Figure 3. A test stand for measuring ultrasonic wave velocity in compressed specimens: (**a**) specimen geometry and elements of the stand, (**b**) geometry of exponential transducer, (**c**) a test stand; *1*—tested AAC specimen 100 × 100 × 100 mm, *2*—exponential transducers, *3*—cables connecting transducers with recording equipment, *4*—recording equipment, *5*—steel sheet, 15 mm thick, *6*—vibration isolation, *7*—heads of testing machine.

The selected results from measurements and maximum values of stress σ_{3max} are presented in Table 2. There are also empirical values of the longitudinal wave $^{obs}c_{p0}$ without the participation of compressive stress and ratios of normal stress σ_3/σ_{3max}, for which the measurements are presented in a tabular form. Figure 4a illustrates results from measured velocities of ultrasounds as the ratio $(c_p - {}^{obs}c_{p0})/{}^{obs}c_{p0}$ expressing the relative increment in ultrasound velocity as a function of stress σ_3. Relative increments in velocity of ultrasonic waves are presented in Figure 4b as a function of relative compressive stress σ_3/σ_{3max}.

As in previous tests [12], the specimens dried until constant weight demonstrated an increase in ultrasound velocity with increased density of AAC under stress $\sigma_3 = 0$. Velocity $^{obs}c_{p0}$ increased to 1875 m/s in concrete of a nominal class of 400 kg/m^3, and to 2225 m/s in concrete with density of 700 kg/m^3. Increased compressive stress in all specimens caused nearly proportional drop in ultrasound velocity. Under relatively low stress when $0 \leq \sigma_3 \leq 0.25\sigma_{3max}$, values of ultrasound velocity decreased by 2–4% when compared to $^{obs}c_{p0}$. When normal stress increased to the level of $0.25\sigma_{3max} \leq \sigma_3 \leq 0.50\sigma_{3max}$, the velocity of ultrasounds decreased by 5–7% when compared to the reference value of $0.25\sigma_{3max}$. Under greater values of relative stress $0.50\sigma_{3max} \leq \sigma_3 \leq 0.75\sigma_{3max}$, the greatest percentage drop in propagation of ultrasonic waves by 9–11% was found in concrete with nominal densities of 400 and 500 kg/m^3. The reduction in velocity of ultrasonic waves by 7–9% was observed in the specimens made of concrete with density of 600 and 700 kg/m^3. No clear reduction in wave velocity in concrete with densities of 600 and 700 kg/m^3 was observed for the stress level, at which slight noise was heard in the specimens and local crushing was apparent within the stress range of $0.75\sigma_{3max} \leq \sigma_3 \leq 0.95\sigma_{3max}$. The relative velocity of ultrasounds decreased by 11–12% in other specimens. In conclusion, a nearly linear drop in relative velocity of longitudinal ultrasonic wave was observed regardless of AAC density. The maximum reduction in relative velocity of ultrasounds was directly proportional to AAC density and changed within the range of 7–12%. As in tests conducted on metals [38,40], linear relationships were obtained, which defined the reduction in velocity of ultrasonic wave propagation as a function of applied normal stress. Considering the relationship (23), accurate physical relationships can be determined:

$$V_{113}^2 = V_0^2 + \frac{1}{3K_0\rho_0}\left[\frac{2\lambda}{\mu}(\lambda + 20\mu + m) - 2l\right]\sigma_3 \rightarrow c_p^2 = c_{p0}^2 + \frac{1}{3K_0\rho_0}\left[\frac{2\lambda}{\mu}(\lambda + 20\mu + m) - 2l\right]\sigma_3$$

$$c_p^2 - {}^{obs}c_{p0}^2 = \left(c_p - c_{p0}\right)\left(c_p + c_{p0}\right) \approx \left(c_p - c_{p0}\right)2c_{p0}, \tag{35}$$

$$\frac{\left(c_p - c_{p0}\right)2c_{p0}}{c_{p0}} = \frac{1}{6K_0\rho_0 c_{p0}}\left[\frac{2\lambda}{\mu}(\lambda + 20\mu + m) - 2l\right]\sigma_3 \rightarrow \frac{\left(c_p - c_{p0}\right)}{c_{p0}} = \frac{\frac{\lambda}{\mu}(\lambda + 20\mu + m) - 2l}{(\lambda + 2\mu)(3\lambda + 2\mu)}\sigma_3.$$

Table 2. Test results for ultrasound velocity in AAC at various compressive stresses.

No.	Mean Density ρ (Nominal Class of Density) kg/m^3	Mean Compressive Stress σ_3, N/mm^2	Mean Relative Compressive Stress σ_3/σ_{3max}	Mean Path Length L, mm	Mean Passing Time of Wave T, µs	Mean Ultrasound Velocity $c_p = L/t$, m/s	Standard Deviation, s, m/s	COV, %
1	2	3	4	5	6	7	8	9
1		0	0		53.5	${}^{obs}c_{p0} = 1875$	1.02	1.9%
2	397	0.75	0.27		55.7	1801	1.78	3.2%
3	(400)	1.33	0.48	100.2	57.3	1750	0.56	1.0%
4		2.08	0.75		60.4	1660	0.13	0.2%
5		2.58	0.93		60.9	1647	1.19	2.0%
6		0	0		53.0	${}^{obs}c_{p0} = 1893$	0.62	1.2%
7	492	0.83	0.24		54.3	1849	1.04	1.9%
8	(500)	1.66	0.48	100.3	57.1	1756	0.54	1.0%
9		2.59	0.75		58.2	1724	1.55	2.7%
10		3.33	0.96		59.3	1691	1.23	2.1%
11		0	0		49.5	${}^{obs}c_{p0} = 2031$	1.32	2.7%
12	599	1.25	0.24		51.7	1942	1.79	3.5%
13	(600)	2.58	0.50	100.4	52.9	1898	1.32	2.5%
14		3.92	0.75		54.6	1841	1.75	3.2%
15		5.00	0.96		53.9	1866	2.40	4.5%
16		0	0		45.1	${}^{obs}c_{p0} = 2225$	1.56	3.5%
17	674	2.00	0.24		46.5	2159	2.34	5.0%
18	(700)	4.17	0.50	100.2	47.5	2114	2.08	4.4%
19		6.17	0.74		48.3	2075	1.72	3.6%
20		8.17	0.98		48.6	2064	1.71	3.5%

The relationship after transformation can be expressed as:

$$\frac{\left(c_p - c_{p0}\right)}{c_{p0}} = \frac{\left(t_{p0} - t_p\right)}{t_p} = \frac{\frac{\lambda}{\mu}(\lambda + 20\mu + m) - 2l}{(\lambda + 2\mu)(3\lambda + 2\mu)}\sigma_3 = \beta_{113}\sigma_3, \tag{36}$$

where β_{113} is the acoustoelastic effect [40] related to the longitudinal wave perpendicular to the direction of the applied load.

If c_{p0} in the relationship (36) is replaced with the value determined in the tests, then the relationship illustrated in Figure 4a is obtained. By dividing both sides of the Equation (36) by the value of maximum stress σ_{3max}, the following relationship is developed:

$$\frac{\left(c_p - c_{p0}\right)}{c_{p0}} = \frac{\left(t_{p0} - t_p\right)}{t_p} = \gamma_{113}\frac{\sigma_3}{\sigma_{3max}}. \tag{37}$$

where $\gamma_{113} = \beta_{113}\,\sigma_{3max}$ can be called the relative acoustoelastic coefficient.

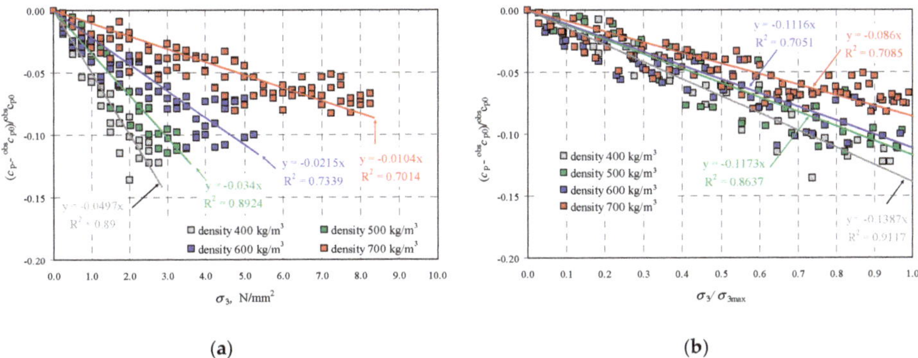

Figure 4. Results from measuring velocity of the longitudinal ultrasonic wave: (**a**) relative change in velocity of longitudinal wave as a function of compressive stress, (**b**) relative change in velocity of longitudinal wave as a function of relative compressive stresses.

The introduction of coefficient β_{113} considerably simplifies practical applications. By using relative values of passing time of the wave, the effect of wave scattering and other related effects described under point 2.2 could be neglected. If c_{p0} in the relationship (37) was replaced with the value determined in the tests, then the relationship illustrated in Figure 4b was obtained. It was adequate to know the coefficient γ_{113} to determine the maximum value of compressive stresses corresponding to normalized compressive strength of the masonry unit f_{Bw} in air-dry conditions. The obtained values of coefficients β_{113} and γ_{113} for straight lines determined from Equations (36) and (37) as a function of density are presented in Figure 5.

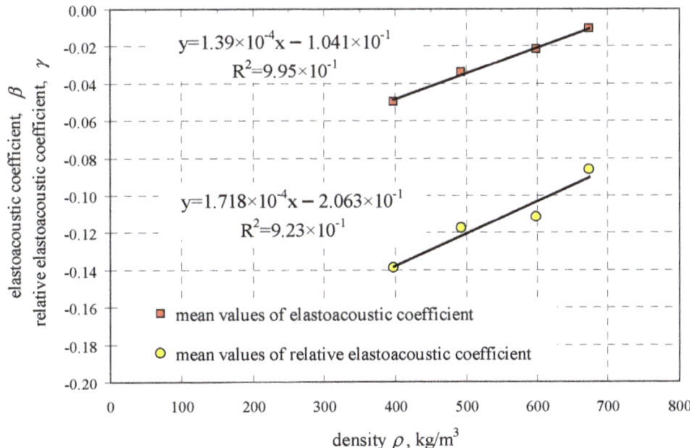

Figure 5. Values of coefficients β_{113} and γ_{113} as a function of AAC density.

Empirical relationships developed from obtained results were proposed to express values of coefficients β_{113} and γ_{113} as a function of AAC density at ($w = 0$)

$$\beta_{113} = 1.39 \times 10^{-4}\rho - 0.104,\ R^2 = 0.995,$$
$$\gamma_{113} = 1.72 \times 10^{-4}\rho - 0.206,\ R^2 = 0.923$$
$$\text{when } 397\,\tfrac{kg}{m^3} \le \rho \le 674\,\tfrac{kg}{m^3}$$

(38)

The practical applications required taking into account moisture content of AAC. The paper [12] demonstrated that the maximum moisture content in concrete depended on nominal density. At the density increase in the range from $\rho = 397$ kg/m^3 to 674 kg/m^3, the maximum moisture content was varying within $w_{\max} = 53.3$–89.9%, which made it possible to determine a straight line of the least square in the following form:

$$w_{\max} = -1.23 \times \frac{\rho}{1000} + 1.34, \text{ when } 397 \frac{\text{kg}}{\text{m}^3} \leq \rho \leq 674 \frac{\text{kg}}{\text{m}^3} \tag{39}$$

Moreover, relative changes in velocity of longitudinal ultrasonic waves were shown by the relationships illustrated in Figure 6.

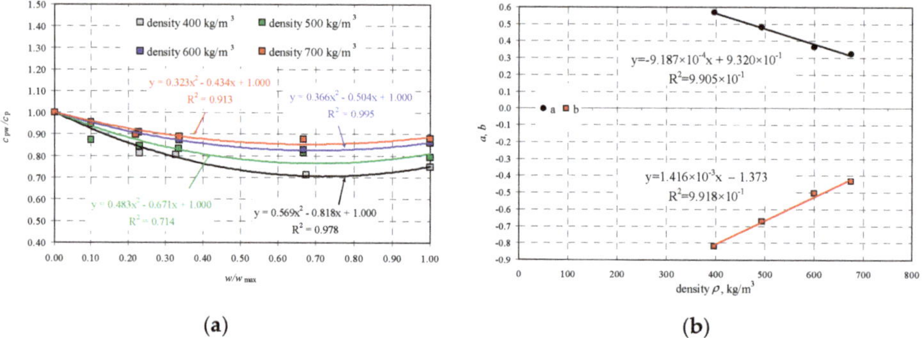

Figure 6. Relationship between velocity of ultrasonic wave propagation, moisture content and density: (a) relative changes in velocity of longitudinal wave as a function of relative moisture content w/w$_{\max}$, (b) values of coefficients as a function of AAC density.

The tests were used to develop the following relationships including velocity c_{pw} in wet AAC with reference to AAC in air-dry conditions c_p:

$$\begin{aligned}
\frac{c_{pw}}{c_p} &= 0.569 \frac{w}{w_{\max}} - 0.818 + 1, \text{ when } 397 \frac{\text{kg}}{\text{m}^3} \leq \rho \leq 446 \frac{\text{kg}}{\text{m}^3}, \\
\frac{c_{pw}}{c_p} &= 0.483 \frac{w}{w_{\max}} - 0.671 + 1, \text{ when } 462 \frac{\text{kg}}{\text{m}^3} \leq \rho \leq 532 \frac{\text{kg}}{\text{m}^3}, \\
\frac{c_{pw}}{c_p} &= 0.366 \frac{w}{w_{\max}} - 0.504 + 1, \text{ when } 562 \frac{\text{kg}}{\text{m}^3} \leq \rho \leq 619 \frac{\text{kg}}{\text{m}^3}, \\
\frac{c_{pw}}{c_p} &= 0.323 \frac{w}{w_{\max}} - 0.434 + 1, \text{ when } 655 \frac{\text{kg}}{\text{m}^3} \leq \rho \leq 725 \frac{\text{kg}}{\text{m}^3}.
\end{aligned} \tag{40}$$

After taking into account the obtained results, values of empirical coefficient defined the following linear relationships:

$$\begin{aligned}
a &= 9.187 \times 10^{-4} \rho + 0.932, \text{ when } 397 \frac{\text{kg}}{\text{m}^3} \leq \rho \leq 674 \frac{\text{kg}}{\text{m}^3}. \\
b &= 1.416 \times 10^{-3} \rho - 1.373, \text{ when } 397 \frac{\text{kg}}{\text{m}^3} \leq \rho \leq 674 \frac{\text{kg}}{\text{m}^3}.
\end{aligned} \tag{41}$$

4.2. Stage II—Test Results for Small Masonry Models

Stage II consisted of verifying empirical relationships developed in stage I. Small masonry walls made of AAC of nominal type of 600 kg/m^3, with thin joints laid in the ready-mixed mortar and with the strength f_m equal to 6.10 N/mm^2 [49] were used for that purpose. Nine test elements in total were prepared and divided into three series marked as I, II, and III. All elements had the same external dimensions: the length of 500 mm, the height of 724 mm, and the thickness of 180 mm. The presence of the head joint or its lack differentiated the models. This was required to highlight potential effects in changes of ultrasound wave velocity in the real wall near head joints. All models of series I were made from three masonry units without the head joint. The models of series II had the head joint in the central layer at the mid-length of the masonry units, and those of series III had the head joint at 1/4 of the masonry length. The view, shape, and dimensions of tests elements of series I, II, and III are shown in Figure 7.

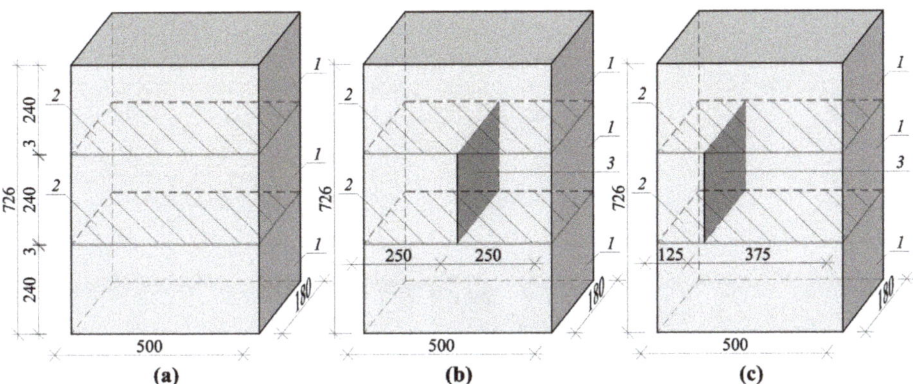

Figure 7. Geometry of models made of AAC tested in stage II (dimensions in mm): (**a**) models of series I without head joint, (**b**) models of series II with head joint at the mid-length of the element, (**c**) models of series III with head joint at 1/4 length; *1*—masonry units, *2*—bed joints, *3*—head joints.

Test models were placed in the strength testing machine with an operating range of 1000 kN (class 1). The applied load was perpendicular to the plane of bed joints and the machine piston displacement was monotonically increasing at a rate of 1 mm/min. The value of the applied load F was read from the dynamometer of the testing machine. Stress applied to top and bottom parts of the bed surface of the model was calculated from the equation $\sigma_3 = F/A$ (where A—area of bed surface of the element $A = 180 \times 500 = 90{,}000$ mm^2). During the tests, displacements and deformations were measured for two models of each series with the Digital Image Correlation (DIC) using the ARAMIS 6M system by GOM GmbH Braunschweig, Germany (the class of reading accuracy for displacements was 1%) [50–52]. To determine values of forces and stresses causing cracks (σ_{3cr}) and failure (σ_{3max}), some models of each series (I-3, II-3, III-3) were tested without measuring the velocity of ultrasonic wave propagation. Wave velocity c_p was measured in two other models at the following values: 0, $0.25\sigma_{3max}$, $0.50\sigma_{3max}$, $0.75\sigma_{3max}$. The transmission method was used to measure waves. Hence, the precise arrangement of ultrasonic transducers vis-à-vis each other was necessary. For that purpose, two plastic templates were used with holes having a diameter of 5 mm, made at the regular spacing adjusted to the model geometry—Figure 8a,b. Holes in the template (Figure 8b) were placed in horizontal and vertical configuration within a distance of ~30 mm. Before testing, apparent density ρ_0 in air-dry conditions, relative moisture content in the material used for preparing the models were calculated, and additionally the maximum moisture content w_{max} was calculated from the following relationship (39). Basic results for properties of the models and test results in the form of stresses causing cracks σ_{3cr}, and maximum stresses σ_{3max} are presented in Table 3, whereas relationships between compressive stress and deformation σ-ε are illustrated in Figure 9. All models were characterized by minor differences in obtained parameters. Density of models varied from 587 to 597 kg/m^3, and relative moisture content was within the range of 4.5–6.0%. At determined values of loading, the procedure of loading was stopped to measure passing time t_p of the ultrasonic wave, and then the propagation velocity was calculated from the relationship $c_p = L/t_p$ ($L = 180$ mm). The tests were performed only on one model of each series (highlighted rows in Table 3). No measurements were made when the measuring points overlapped with bed or head joints.

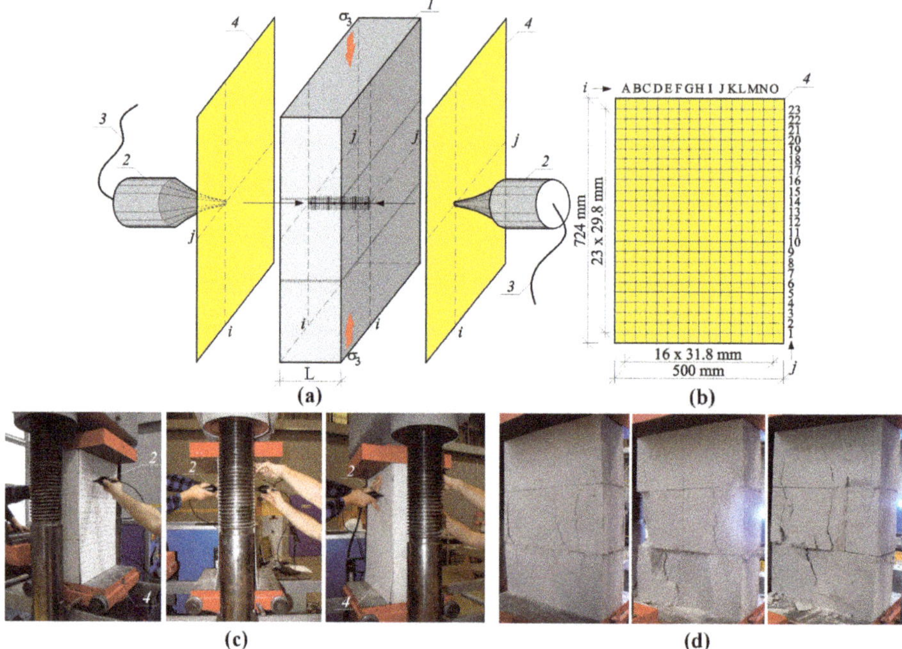

Figure 8. Testing methodology for models made of AAC used in stage II: (**a**) measurement of velocity of ultrasonic wave propagation at different stress values σ_3, (**b**) template geometry used for symmetric arrangement of ultrasonic transducers, (**c**) models during tests, (**d**) failure of selected models 1—masonry units, 2—ultrasonic transducers, 3—cables connecting transducers with recording equipment, 4—templates for symmetric location of ultrasonic transducers.

Table 3. Test results for all models.

No.	Series	Model	Mean Density ρ_0, kg/m³	Moisture Content w, %	Maximum Moisture Content (39) w, %	Compressive Stress Inducing Cracks σ_{3cr}, N/mm²		Maximum Compressive Stress σ_{3max}, N/mm²	
						of Model	Mean (COV)	of Model	Mean (COV)
1	2	3	4	5	6	7	8	9	10
1		I-1 *	594	6.0%	60.9%	2.93		2.97	
2	I	I-2	589	4.5%	61.6%	2.87	2.89 (1.1%)	3.04	3.01 (1.3%)
3		I-3	592	5.1%	61.2%	2.88		3.01	
4		II-1 *	588	6.0%	61.7%	3.00		3.00	
5	II	II-2	597	4.9%	60.6%	2.85	2.95 (2.8%)	2.87	2.96 (2.6%)
6		II-3	593	6.0%	61.1%	2.99		3.01	
7		III-1 *	594	5.1%	60.9%	2.97		2.99	
8	III	III-2	587	5.5%	61.8%	2.79	2.90 (3.3%)	2.90	2.97 (1.9%)
9		III-3	590	5.4%	61.4%	2.95		3.01	

*—models, for which the propagation of ultrasonic waves c_p was measured.

Nearly proportional increase in deformations was observed in all models exposed to increasing loading. Clear breaking of graphs illustrating stress–deformation relationships was only observed at the time preceding maximum stress that was reached under mean stress within the range of 2.96–3.01 N/mm². Cracks on external surfaces of masonry units were not observed until maximum stress that was reached in the weakening phase under mean stress within the range of 2.89–2.95 N/mm².

The transmission method was used to measure passing time of ultrasonic wave at stress levels (0, $0.25\sigma_{3max}$, $0.50\sigma_{3max}$, $0.75\sigma_{3max}$) shown in Figure 9. Results in the form of maps showing passing time of the wave t_p are illustrated in Figures 10–13.

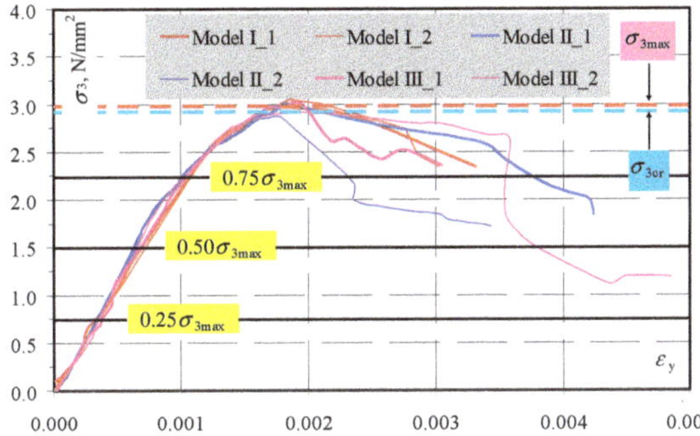

Figure 9. Relationships between stress and strain σ-ε for all tested models.

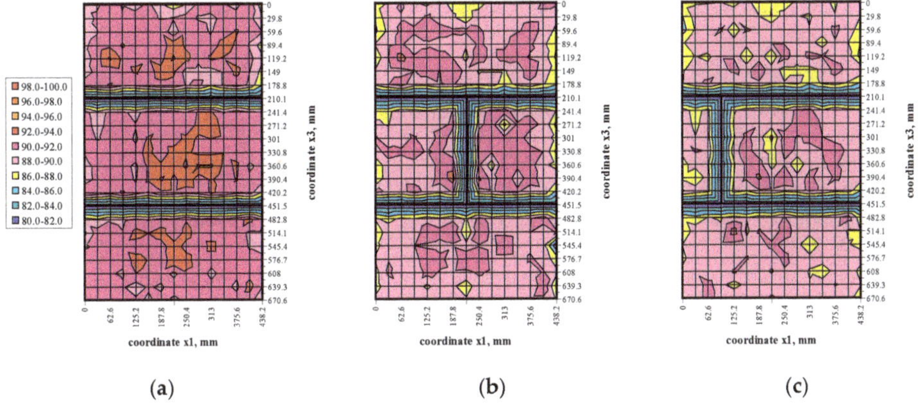

(a) (b) (c)

Figure 10. Results from measuring passing time of the ultrasonic wave under the load $\sigma_3 = 0$: (**a**) model I-1, (**b**) model II-1, (**c**) model III-1.

Basic results in the form of mean time of wave propagation for all points are compared in Table 4.

Table 4. Results from measuring propagation of ultrasonic waves.

Model	Number of Measuring Points in Each Step of Loading n	Passing Time of Ultrasonic Wave Under Various Levels of Loading t_p, µs (COV)											
		0			$0.25\sigma_{3max}$			$0.50\sigma_{3max}$			$0.75\sigma_{3max}$		
		t_{pmin}	t_{pmax}	t_{pmv}	t_{pmin}	t_{pmax}	t_{pmv}	t_{pmin}	t_{pmax}	t_{pmv}	t_{pmin}	t_{pmax}	t_{pmv}
1	2	3	4	5	6	7	8	9	10	11	12	13	14
I-1	315	86	94.2	90.8 (1.4%)	86.7	98.8	92.2 (1.3%)	90.5	99.2	93.9 (1.4%)	87.7	99.9	94.4 (1.4%)
II-1	308	82.2	92.9	89.2 (1.6%)	86.3	94.4	90.6 (1.2%)	89.0	97.4	92.2 (1.1%)	90.2	96.4	92.5 (1.1%)
III-1	308	85	92.9	88.8 (1.4%)	87.1	93.5	90.2 (1.2%)	88.8	95.1	91.6 (0.9%)	90.1	95.4	92.1 (0.9%)

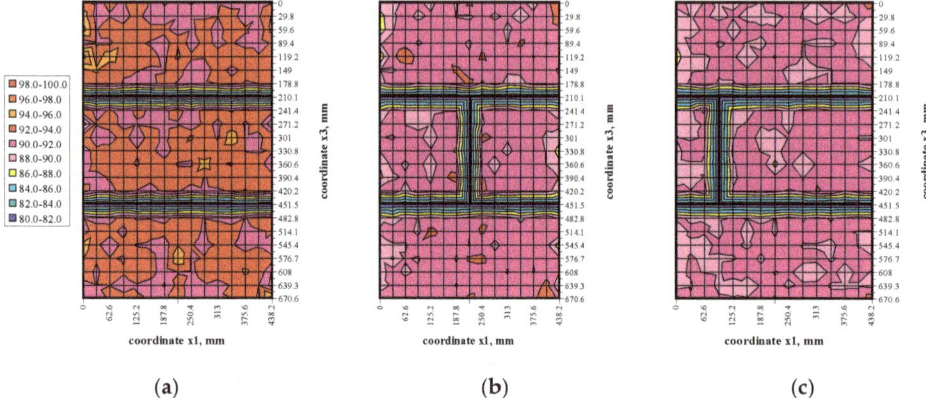

Figure 11. Results from measuring passing time of the ultrasonic wave under the load $\sigma_3 = 0.25\sigma_{3max}$: (**a**) model I-1, (**b**) model II-1, (**c**) model III-1.

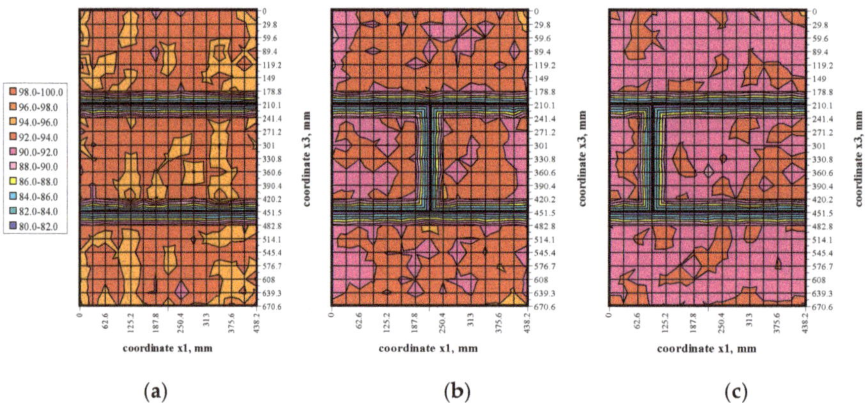

Figure 12. Results from measuring passing time of the ultrasonic wave under the load $\sigma_3 = 0.50\sigma_{3max}$: (**a**) model I-1, (**b**) model II-1, (**c**) model III-1.

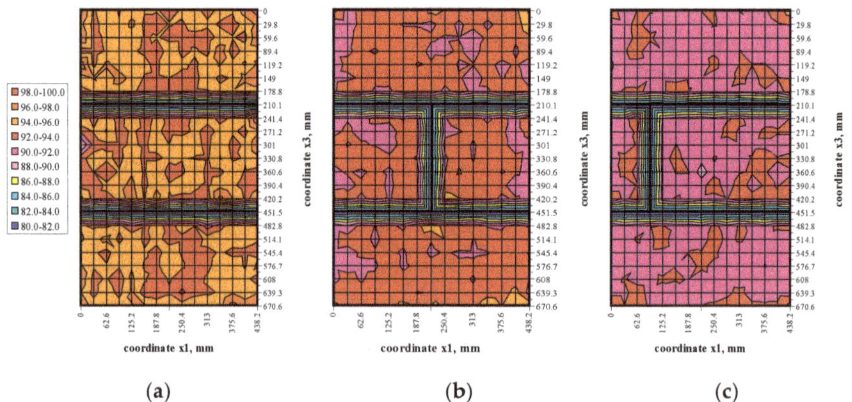

Figure 13. Results from measuring passing time of the ultrasonic wave under the load $\sigma_3 = 0.75\sigma_{3max}$: (**a**) model I-1, (**b**) model II-1, (**c**) model III-1.

The conducted tests indicated passing times of the ultrasonic wave in walls under zero loads were not constant, some fluctuations were observed—Figure 10. Usually, waves in central parts of the elements had the longest passing time. Clear disturbances at vertical edges and near bed joints were observed. However, the calculated coefficient of variation for all measurements, and from disturbed areas, was relatively low in the order of 1.4–1.6% due to a great number of performed measurements. An increase in loads to $0.25\sigma_{3max}$—Figure 11 caused an evident increase in passing time of the ultrasonic wave for all models. The effect of previous original disturbances was found on nearly whole surfaces of the units. The greatest difference in results was observed near edges of masonry units. As in the primary phase, the coefficient of variation was minor and ranged from 1.0–1.3%. An increase in loads to $0.50\sigma_{3max}$ and $0.25\sigma_{3max}$—Figures 12 and 13 produced a gradual increase in mean time of propagation, but did not cause apparent qualitative changes in maps presenting passing times. Similarly, coefficients of passing time of waves did not dramatically changes as the maximum value they reached was 1.4%.

5. Analysis of Test Results

On the basis of empirical relationships and those developed in the testing stage, an attempt was made to determine normal stresses in the tested models. The comprehensive approach based on all test results or the approach using a limited number of points was implemented for each model. In the first case, there were 315 (the model of series I) or 308 (the models of series II or III) measurement results for each step of loading. The calculations also included results for edges of the masonry units that demonstrated clear disturbances. The approach based on a limited number of points for determining stress involved only points located in the central area of the masonry units. That significantly limited the number of analyzed measuring points to 45 for model I, and 44 for models of series II and III. For successive levels of loading, the difference in passing time of the ultrasonic wave was calculated, and then acoustoelastic coefficient β_{113} was calculated from Equation (38). Finally, stress σ_3 from the transformed relationship (36) was calculated. The obtained values of stress are presented in Table 5.

Table 5. Results from calculating normal stress in the wall using all measuring points.

Model	Number of Measuring Points n	$0.25\sigma_{3max}$			$0.50\sigma_{3max}$			$0.75\sigma_{3max}$		
		$\frac{(t_p-t_{p0})}{t_{p0}}$	β_{113} mm^2/N (38)	$\sigma_3=\frac{(t_p-t_{p0})}{\beta_{113}\cdot t_{p0}}$ N/mm^2 (36)	$\frac{(t_p-t_{p0})}{t_{p0}}$	β_{113} mm^2/N (38)	$\sigma_3=\frac{(t_p-t_{p0})}{\beta_{113}\cdot t_{p0}}$ N/mm^2 (36)	$\frac{(t_p-t_{p0})}{t_{p0}}$	β_{113} mm^2/N (38)	$\sigma_3=\frac{(t_p-t_{p0})}{\beta_{113}\cdot t_{p0}}$ N/mm^2 (36)
1	2	3	4	5	6	7	8	9	10	11
I-1	315	−0.0154	−0.0215	0.715	−0.0310	−0.0224	1.437	−0.0388	−0.0215	1.800
II-1	308	−0.0149	−0.0215	0.664	−0.0322	−0.0224	1.441	−0.0359	−0.0215	1.603
III-1	308	−0.0150	−0.0215	0.696	−0.0306	−0.0224	1.419	−0.0360	−0.0215	1.672

The obtained coefficients depended on apparent density of AAC of the order -0.0215–-0.0224 mm^2/N. The values obtained for autoclaved aerated concrete aerated were many times greater than similarly determined acoustoelastic effect for metals [40] ($\beta_{113} = -0.99 \times 10^{-5}$–$-2.06 \times 10^{-5}$ mm^2/N—steel, $\beta_{113} = -7.75 \times 10^{-5}$ mm^2/N—aluminium, $\beta_{113} = -1.88 \times 10^{-5}$ mm^2/N—copper). The determined stress values were similar only at relatively low stress values equal to $0.25\sigma_{3max}$ and $0.50\sigma_{3max}$. Maximum differences in stress determined using the EA method did not exceed 11% (model II-1). For stress values of the order of $0.75\sigma_{3max}$, the estimated values of non -destructive stress were considerably lower than those determined from destructive testing. Stress values were underrated by no more than 28%.

In the second approach based on the limited number of results for central areas of all masonry units, the procedure was similar to the first one. The location of measuring points in the central part of the masonry units was determined by analysing the maps of passing times illustrated in Figures 10–13. Firstly, differentiation in passing time of ultrasonic waves was smaller in the central areas. Secondly, stress states in that area of masonry units was the most similar to stress states in the specimens

$100 \times 100 \times 100$ mm used to validate the AE method in stage I. In addition, the final aspect was purely practical because it was the easiest to determine centers of masonry units, apart from edge areas. For successive levels of loading, the difference in passing time of the ultrasonic wave was calculated. Then, acoustoelastic coefficient β_{113} was calculated from Equation (38), and finally stress values σ_3 were calculated from the relationship (36). The obtained values of stress are presented in Table 6.

Table 6. Results from calculating normal stress in the wall using a limited number of measuring points.

Model	Number of Measuring Points n	$0.25\sigma_{3max}$			$0.50\sigma_{3max}$			$0.75\sigma_{3max}$		
		$\frac{(t_p-t_{p0})}{t_{p0}}$	β_{113} mm^2/N (38)	$\sigma_3 = \frac{(t_p-t_{p0})}{\beta_{113}\cdot t_{p0}}$ N/mm^2 (36)	$\frac{(t_p-t_{p0})}{t_{p0}}$	β_{113} mm^2/N (38)	$\sigma_3 = \frac{(t_p-t_{p0})}{\beta_{113}\cdot t_{p0}}$ N/mm^2 (36)	$\frac{(t_p-t_{p0})}{t_{p0}}$	β_{113} mm^2/N (38)	$\sigma_3 = \frac{(t_p-t_{p0})}{\beta_{113}\cdot t_{p0}}$ N/mm^2 (36)
1	2	3	4	5	6	7	8	9	10	11
I-1	45	−0.0113	−0.0215	0.526	−0.0255	−0.0215	1.183	−0.0326	−0.0215	1.512
II-1	44	−0.0103	−0.0224	0.460	−0.0285	−0.0224	1.273	−0.0307	−0.0224	1.371
III-1	44	−0.0117	−0.0215	0.545	−0.0263	−0.0215	1.223	−0.0314	−0.0215	1.459

Using the approach of considerably decreased number of measuring points limited to central areas of the masonry units, much lower stress values were obtained. For the lowest level of stress of the order of 0.25 σ_{3max}, stress calculated for the model II-1 with the AE method was lower by 60% than in destructive tests. Stress underestimation for other models I-1 and III-1 was at the level of 36–43%. At the stress level of $0.50\sigma_{3max}$, the underestimation of stress was at the lowest level of 16–21%. As in the case of a greater number of points, stress values determined by the EA method at the stress level of $0.75\sigma_{3max}$ were the least accurate. Calculated compressive stress differed by 49–62% from experimentally obtained values. Compared results from destructive testing and calculated results are shown in Figures 14–16.

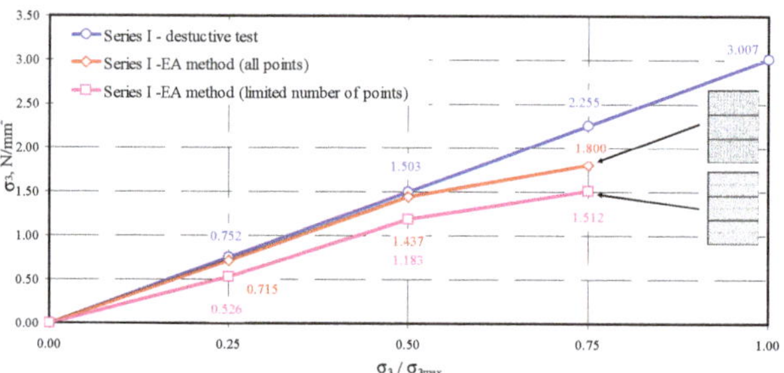

Figure 14. Measurement results of stress in the model I-1.

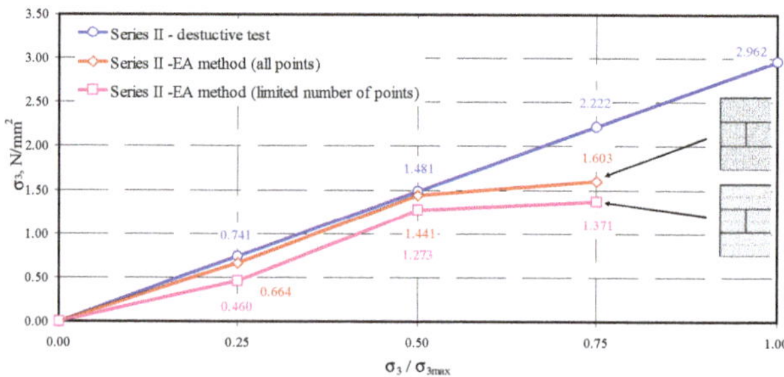

Figure 15. Measurement results of stress in the model II-1.

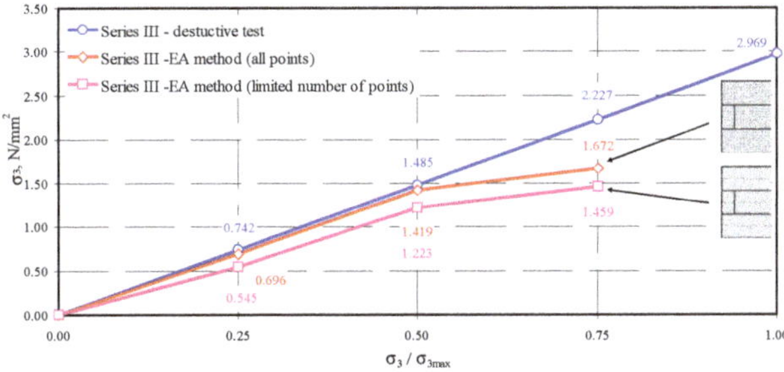

Figure 16. Measurement results of stress in the model III-1.

In conclusion, the most favorable results from measuring stress with the calibrated acoustoelastic method were obtained when all measuring points were used at stress levels within the range of 0–$0.5\sigma_{3max}$. The determined stress values were lower than those from destructive testing small wall models. Considering the approach based on the limited number of points, underestimation of compressive stress was considerably greater. The greatest differences in both methods were found at the stress level of $0.75\,\sigma_{3max}$, which resulted from an increase in effects of ultrasonic wave scattering, developing microcracks in AAC structure (invisible on the external surface of the models).

Statistical Estimation of Stress in Walls

The practical application of that method requires further tests mainly on location of measuring points and their minimal number. However, assuming only measuring points for central area of each masonry unit are used to determine stress in the masonry, then boundary values of strength could be determined with the probability that the obtained results were not lower than experimentally obtained results. Only values from the range of 0–$0.5\sigma_{3max}$ were used for the calculations. The selected range seems to be the most reasonable because at the operational stage force values in real walls can correspond to maximum stress of the order of 50% of the calculated compressive strength of the wall f_d. Thus, load-carrying capacity of the real wall [43] depends not on absolute values of compressive force generating stress σ_3, but on the stability expressed by the reduction factor for load-carrying capacity ($\Phi_{1,2}$ and Φ_{2m}). Boundary values in confidence intervals of the mean value [53] (at $n > 30$ and unknown variance σ) were determined form the general relationship at the statistical significance $\alpha = 0.1$:

$$P\left(t_p - u_{1-\alpha/2}\frac{S}{\sqrt{n}} < t_{pcal} < t_p + u_{1-\alpha/2}\frac{S}{\sqrt{n}}\right) = 1 - \alpha \tag{42}$$

where: t_p—mean time of wave propagation, S—standard deviation of propagation velocity for the specimen. $u_{1-\alpha/2}$—statistics with the random variable at the normal distribution N(0.1). When $n < 30$, the statistics $t_{1-\alpha/2}$ with the Student's t-distribution and n-1 degrees of freedom should be applied.

Only the upper value of confidence interval is suitable for practical applications, which in this case can be associated with the quantile of the order of 95%. In other words, the upper limit of the confidence interval for the mean value was assumed because it is commonly used in the construction sector. Stress values were determined with the AE method using calculated values of passing time of the wave. The obtained results were compared with true mean stress values of the masonry wall. Values for upper confidence intervals for passing time t_{pcal} and calculated stress values σ_{3cal} using the AE method are presented in Table 7 and compared with stress results obtained from testing the models σ_{3obs}. In that way, we obtain some estimation of the deviation between test and calculated results at the specified confidence level.

Table 7. Compared results from tests and upper values of confidence intervals.

Model	Number of Measuring Points n	$u_{1-\alpha/2}$	t_{pcal}	$\sigma_{3cal}=\frac{(t_{pcal}-t_{p0})}{\beta_{113}\cdot t_{p0}}$ N/mm^2	σ_{3obs} N/mm^2	$\frac{\sigma_{3obs}}{\sigma_{3cal}}$	t_{pcal}	$\sigma_{3cal}=\frac{(t_{pcal}-t_{p0})}{\beta_{113}\cdot t_{p0}}$ N/mm^2	σ_{3obs} N/mm^2	$\frac{\sigma_{3obs}}{\sigma_{3cal}}$
				$0.25\sigma_{3max}$				$0.50\sigma_{3max}$		
1	2	3	4	5	6	7	8	9	10	11
I-1	45		92.6	0.656	0.752	1.15	93.9	1.329	1.503	1.13
II-1	44	1.645	90.8	0.595	0.741	1.24	92.4	1.37	1.481	1.08
III-1	44		90.3	0.640	0.742	1.16	91.7	1.302	1.485	1.14
		on average:				1.18		on average:		1.12

Taking into account the statistical estimation of stress, it was underestimated but values were significantly reduced. It can suggest with the probability of not greater than 5% that determination of stress in the walls from central areas of the masonry units with the slightest disturbances will cause underestimation of the mean stress at $0.25\sigma_{3max}$ by 18%, and at $0.50\sigma_{3max}$ by ca. 12%. That underestimation can be acceptable for masonry structures.

6. Conclusions

This paper describes theoretical bases of the acoustoelastic method (AE) which is one of the methods of detecting stress in structures using NDT techniques. That method consists of the correlation between stress in the material and velocity of the wave propagation. It is commonly used in ultrasonic tensometry to determine own stresses usually in machine parts. Using that method for other materials has not been widely discussed in the literature so far. No results from tests and analyses in concrete, not mentioning masonry, are available. This lack of interest in using this method can only be explained by measuring difficulties (significant dispersion of measurement results) caused by inhomogeneity of that material. This work presents an attempt to use the AE method for autoclaved aerated concrete. It is a porous material with high homogeneity and repeatability of parameters due to the production of this material on an industrial scale. This work supplements comprehensive material tests for autoclaved aerated concrete [11]. The tests were divided into two stages: Stage I involved the suggestion of the procedure and the determination of acoustoelastic coefficient β_{113} linking the propagation of the longitudinal ultrasonic wave c_p with normal stress σ_3 acting towards the wave propagation. The standard cuboid specimens with the dimensions of $100 \times 100 \times 100$ mm were used for calibration. The effect of density ρ and relative humidity w was included on the basis of testing AAC of different density using correlations presented in [11]. Those considerations resulted in formulating the relationship β_{113} (ρ). The proposed procedure was verified in stage II, where destructive tests were conducted on small masonry walls made of autoclaved aerated concrete (AAC) with a nominal density of 600 kg/m^3. The models were divided into three series differing in the location of head joints in the masonry. Velocity of the ultrasonic wave propagation was measured for one model of each series at different values of compressive stress. The following stress levels were analyzed: $0.25\sigma_{3max}$, $0.50\sigma_{3max}$ and $0.70\sigma_{3max}$ because the range of the applied method was only limited to the elastic range. The performed measurements were used to determine values of acoustoelastic coefficients $\beta_{113} = -0.0215$–-0.0224, which were far lower than similarly determined acoustoelastic coefficients for metals. Mean stress values calculated with the proposed method using all measuring point for a given level ($n = 308$–315) were within the range of 93–96% of empirical values $0.25\sigma_{3max}$, $0.50\sigma_{3max}$. The highest underestimation of stress was found for the stress level of $0.75\sigma_{3max}$, for which the underestimation of mean stress values was equal to 24%. However, such a great number of measurements seem to be impractical for the applicable uses. Therefore, further analyses suggest determining stress values only on the basis of measurement results for central areas of each masonry unit. Then, the number of measuring points was significantly reduced to $n = 45$ and 44. As for all measuring points, the comparison indicated greater underestimation of the mean value of the order of 22–55%. It is not advantageous taking into account safety of the structure. Hence, it was decided to estimate the confidence interval of the mean value associated with the quantile of the order of 95%. Such a procedure caused the stress values were underestimated at the level of 12–18% within the stress range of 0–$0.50\sigma_{3max}$. In summary:

(a) the acoustoelastic method (AE) can be used to determine stress in autoclaved aerated concrete,

(b) correlations were obtained that bind the value of acoustoelastic coefficient β_{113} as a function of density and moisture content in AAC,

(c) the effect of scattering of the ultrasonic wave in medium can be neglected when the coefficient β_{113} is applied,

(d) rather precise values of mean stress in the wall were determined on the basis of measured velocity of ultrasonic wave propagation at a high number of measuring points,

(e) reduced number of measuring points resulted in a significant underestimation of mean stress,

(f) determination of the quantile equal to 95% for passing time of the ultrasonic wave was used to estimate stress in the wall with the underestimation of the order of 12–18%, which can be considered as satisfactory.

The formulation of explicit recommendations to diagnose in-situ structures requires additional tests on slender walls to evaluate the impact of stability and works on improving the selection of measuring points. The proposed procedure for selecting measuring points limited to central parts of masonry units can be inaccurate for slender walls. Tests are going to be performed on the acoustoelastic coefficient in the wall with a one-side access using transverse waves to determine the acoustoelastic coefficient β_{133}.

Funding: The research was financed from the personal funds of the Department of Building Structures and Laboratory of Civil Engineering Faculty.

Acknowledgments: The author would like to express particular thanks to Solbet Company for valuable suggestions and the delivery of masonry units and mortar, which were used to prepare test models and perform tests.

Conflicts of Interest: The author declares no conflict of interest.

References

1. Auld, B.A. *Acoustic Fields and Waves in Solid*; John Wiley and Sons: Hoboken, NJ, USA, 1973.
2. Bayer, R.T.; Letcher, S.V. *Phisical Ultarsonics*; Academic Press: Cambridge, MA, USA, 1969.
3. Malhotra, V.M.; Carino, N.J. *Handbook on Nondestructive Testing of Concrete*; CRC Press: Boca Raton, FL, USA, 2003.
4. Huan, H.; Liu, L.; Mandelis, A.; Peng, C.; Chen, X.; Zhan, J. Mechanical strength evaluation of elastic materials by multiphysical nondestructive methods: A review. *Appl. Sci.* **2020**, *10*, 1588. [CrossRef]
5. Hola, J.; Schabowicz, K. State-of-the-art non-destructive methods for diagnostic testing of building structures—Anticipated development trends. *Arch. Civ. Mech. Eng.* **2010**, *10*, 5–18. [CrossRef]
6. Breysse, D. Nondestructive evaluation of concrete strength: An historical review and a new perspective by combining NDT methods. *Constr. Build. Mater.* **2012**, *33*, 139–163. [CrossRef]
7. Forde, M.C. International practice using NDE for the inspection of concrete and masonry arch bridges. *Bridge Struct.* **2010**, *6*, 25–34. [CrossRef]
8. Cotic, P.; Jaglicic, Z.; Niederleithinger, E.; Effner, U.; Kruschwitz, S.; Trela, C.; Bosiljkov, V. Effect of moisture on the reliability of void detection in brickwork masonry using radar, ultrasonic and complex resistivity tomography. *Mater. Struct.* **2013**, *46*, 1723–1735. [CrossRef]
9. de Nicolo, B.; Piaga, C.; Popescu, V.; Concu, G. Non-invasive acoustic measurements for faults detecting in building materials and structures. *Appl. Meas. Syst.* **2012**, 259–292. [CrossRef]
10. Khan, F.; Rajaram, S.; Vanniamparambil, P.A.; Bolhassani, M.; Hamid, A.; Kontsos, A.; Bartoli, I. Multi-sensing NDT for damage assessment of concrete masonry walls. *Struct. Control. Health Monit.* **2015**, *22*, 449–462. [CrossRef]
11. Homann, M. *Porenbeton Handbuch. Planen und Bauen Mit System*, 6th ed.; Auflage: Hannover, Germany, 2008.

12. Jasiński, R.; Drobiec, Ł.; Mazur, W. Validation of selected non-destructive methods for determining the compressive strength of masonry units made of autoclaved aerated concrete. *Materials* **2019**, *12*, 389. [CrossRef]

13. Jasiński, R. Determination of AAC masonry compressive strength by semi destructive method. *Nondestruct Test Diagn.* **2018**, *3*, 81–85. [CrossRef]

14. Benson, R.W.; Raelson, V.J. Acoustoelasticity. *Prod. Eng.* **1959**, *20*, 565–569.

15. Lurie, A.I. *Nonlinear Theory of Elasticity*; Elsevier: Amsterdam, The Netherlads, 1991; ISBN 978-0444567710.

16. Biot, M.A. Theory of propagation of elastic waves in fluid-saturated porous solid. *J. Acoust. Soc. Am.* **1958**, *28*, 168–191. [CrossRef]

17. Biot, M.A. Mechanics of deformation and acoustic propagation in porous media. *J. Appl. Phys.* **1962**, *33*, 1482–1498. [CrossRef]

18. Plona, T.J. Observation of a second bulk compression wave in a porous medium at ultrasonic frequencies. *Appl. Phys. Lett.* **1980**, *36*, 259–261. [CrossRef]

19. Berryman, J.G. Confirmation of Biot's theory. *Appl. Phys. Lett.* **1980**, *37*, 382. [CrossRef]

20. Soczkiewicz, E.; Chivers, R.C. Propagation and scattering of acoustic waves in a turbulent medium. *J. Sound Vib.* **2001**, *241*, 197–205. [CrossRef]

21. Soczkiewicz, E. Aplication of quantum field theory methods in studies of ultrasonic wave propagation in random media. *Ultrason. Methods Eval. Inhomogeneous Mater.* **1978**, *126*, 163–173.

22. Edelman, I. On the existence of low-frequency surface waves in a porous medium. *Comptes Rendus Mécanique* **2004**, *332*, 43–49. [CrossRef]

23. Edelman, I. Bifurcation of the biot slow wave in a porous medium. *J. Acoust. Soc. Am.* **2002**, *114*, 90–97. [CrossRef]

24. Boutin, C.; Royer, P.; Aurialt, J.L. Acoustic absorbtion of porous surcacing with dual porosity. *Int. J. Solid Struct.* **1998**, *35*, 4709–4737. [CrossRef]

25. Mozhaev, V.G.; Weihnacht, M. Subsonic leaky rayleigh waves at liqud-solid interfaces. *Ultrasononics* **2002**, *40*, 927–933. [CrossRef]

26. Biot, M.A. The influence of initial stress on elastic waves. *J. Appl. Phys.* **1940**, *11*, 522–530. [CrossRef]

27. Huges, D.S.; Kelly, J.L. Second order elastic deformation of solids. *Phys. Rev.* **1953**, *92*, 1145–1149. [CrossRef]

28. Bergman, R.H.; Shahbender, R.A. Effect of statically applied stresses on the velocity of propagation of ultrasonic waves. *J. Appl. Phys.* **1958**, *29*, 1736–1738. [CrossRef]

29. Tokuoka, T.; Saito, M. Elastic wave propagations and acoustical birefringence in stressed crystals. *J. Acoust. Soc. Am.* **1968**, *45*, 1241–1246. [CrossRef]

30. Husson, D.; Kino, G.S. A perturbation theory for acoustoelastic effects. *J. Appl. Phys.* **1982**, *53*, 7250–7258. [CrossRef]

31. Nikitina, N.Y.; Ostrovsky, L.A. An ultrasonic method for measuring stresses in engineering materials. *Ultrasonics* **1998**, *35*, 605–610. [CrossRef]

32. Dorfi, H.R.; Busby, H.R.; Janssen, M. Ultrasonic stress measurements based on the generalized acoustic ratio technique. *Int. J. Solids Struct.* **1996**, *33*, 1157–1174. [CrossRef]

33. Man, C.; Lu, W.Y. Towards an acoustoelastic theory for measurement of residual stress. *J. Elast.* **1987**, *17*, 159–182. [CrossRef]

34. Murnagham, F.D. *Finite Deformation of an Elastic Solid*; Wiley Publishing: Hoboken, NJ, USA, 1951.

35. Murnaghan, F.D. Finite deformations of an elastic solid. *Am. J. Math.* **1937**, *59*, 235–260. [CrossRef]

36. Tylczyński, Z.; Mróz, B. The influence of uniaxial stress on ultrasonic wave propagation in ferroelastic $(NH_4)_4LiH_3(SO_4)_4$. *Solid State Commun.* **1997**, *101*, 653–656. [CrossRef]

37. Takahashi, S.; Motegi, R. Measurement of third-order elastic constants and applications to loaded structural materials. *Springer Plus* **2015**, *4*, 325. [CrossRef] [PubMed]

38. Takahashi, S. Measurement of third-order elastic constants and stress dependent coefficients for steels. *Mech. Adv. Mater. Mod. Process.* **2018**, *4*. [CrossRef]

39. Takahashi, S. Stress Measurement Method and its Apparatus. U.S. Patent 7299138, 10 December 2007.

40. Deputat, J. *Properties and Use of the Elastoacoustic Phenomenon to Measure Self-Stress*; Institute of Fundamental Technological Research Polish Academy of Sciences: Warsaw, Poland, 1987. (In Polish)

41. Egle, D.M.; Bray, D.E. Measurement of acoustoelastic and third-order elastic constants for rail steel. *J. Acoust. Soc. Am.* **1976**, *60*, 741–744. [CrossRef]

42. British Standards Institution. *EN 771-4:2011 Specification for Masonry Units—Part 4: Autoclaved Aerated Concrete Masonry Units*; British Standards Institution: London, UK, 2011.

43. CEN. *PN-EN 1996-1-1:2010+A1:2013-05P. Eurocode 6: Design of Masonry Structures. Part 1-1: General Rules for Reinforced and Unreinforced Masonry Structures*; Europeand Commettee for Standardization: Brusselles, Belgium, 2005. (In Polish)

44. Bartlett, F.M.; Macgregor, J.G. Effect of moisture condition on concrete core strengths. *ACI Mater. J.* **1993**, *91*, 227–236.

45. Suprenant, B.A.; Schuller, M.P. *Nondestructive Evaluation & Testing of Masonry Structures*; Hanley Wood Inc.: Washinton, DC, USA, 1994; ISBN 978-0924659577.

46. McCann, D.M.; Forde, M.C. Review of NDT methods in the assessment of concrete and masonry structures. *NDT E Int.* **2001**, *34*, 71–84. [CrossRef]

47. Haach, V.G.; Ramirez, F.C. Qualitative assessment of concrete by ultrasound tomography. *Constr. Build. Mater.* **2016**, *119*, 61–70. [CrossRef]

48. Zielińska, M.; Rucka, M. Non-destructive assessment of masonry pillars using ultrasonic tomography. *Materials* **2018**, *11*, 2543. [CrossRef]

49. EN 1015-11:2001/A1:2007. *Methods of Test for Mortar for Masonry. Part 11: Determination of Flexural and Compressive Strength of Hardened Mortar*; European Committee for Standardization: Brussels, Belgium, 1999.

50. Chu, T.C.; Ranson, W.F.; Sutton, M.A.; Peters, W.H. Application of digital-image-correlation techniques to experimental mechanics. *Exp. Mech.* **1985**, *25*, 232–244. [CrossRef]

51. Lord, J.D. *Digital Image Correlation (DIC), Modern Stress and Strain Analysis. A State of the Art Guide to Measurement Techniques, BSSM*; Evans, J.E., Dulie-Barton, J.M., Burguete, R.L., Eds.; Eureka Magazine: Cambridge, UK, 2009; pp. 14–15.

52. Frankovský, P.; Virgala, I.; Hudák, P.; Kostka, J. The use of the digital image correlation in a strain analysis. *Int. J. Appl. Mech. Eng.* **2013**, *4*, 1283–1292. [CrossRef]

53. Volk, W. *Applied Statistics for Engineers*; Literary Licensing LLC: Whitefish, MT, USA, 2013.

Article

Integrated Application of GPR and Ultrasonic Testing in the Diagnostics of a Historical Floor

Magdalena Rucka [1,*], **Erwin Wojtczak** [1] **and Monika Zielińska** [2]

[1] Department of Mechanics of Materials and Structures, Faculty of Civil and Environmental Engineering, Gdańsk University of Technology, Narutowicza 11/12, 80-233 Gdańsk, Poland; erwin.wojtczak@pg.edu.pl

[2] Department of Technical Fundamentals of Architectural Design, Faculty of Architecture, Gdańsk University of Technology, Narutowicza 11/12, 80-233 Gdańsk, Poland; monika.zielinska@pg.edu.pl

* Correspondence: magdalena.rucka@pg.edu.pl or mrucka@pg.edu.pl; Tel.: +48-58-347-2497

Received: 21 May 2020; Accepted: 1 June 2020; Published: 3 June 2020

Abstract: The paper presents the results of integrated ground penetrating radar (GPR) and ultrasonic testing (UT) measurements conducted on a historical floor in St. Nicholas' Church, Gdańsk, Poland. The described inspection was the first stage of the technical state assessment of the building. The aim of the study was the detection of underfloor air gaps, which were observed in a few trial pits. The condition of the ground under the floor was determined by localizing other inclusions such as rubble, human remains, brick walls and pipes. To identify the phenomenon of electromagnetic and ultrasonic wave propagation within the air gap, laboratory tests were conducted on physical models consisting of two concrete slabs stacked on top of each other and gradually moved apart to simulate a slot of varying thickness. The conducted research was supported by the numerical models of electromagnetic wave propagation. The obtained results showed that the integration of the GPR and UT methods provided an effective imaging of the floor and the area under it. Ultrasonic testing was proved to be a good technique for identifying air voids, while the GPR method allowed detecting concentrated anomalies and determining the degree of ground homogeneity under the floor.

Keywords: non-destructive testing; historical floor; integrated diagnostics; ground penetrating radar; ultrasonic testing; in situ surveys; finite-difference time-domain modeling

---

1. Introduction

Non-destructive testing (NDT) is commonly used for assessing the condition of components of engineering structures. It is a quick and efficient approach, the main advantage of which is the ability to examine a structure in a non-invasive way, without damaging or changing the composition or shape of the inspected object. NDT covers a variety of techniques based on a wide range of physical phenomena, including propagation of elastic waves, being the basis for ultrasonic testing (UT), and electromagnetic waves used in the ground penetrating radar (GPR) method. Non-destructive testing can be applied on a selected part of a structure or for a comprehensive inspection of large-size objects, like bridges [1–3], water gates [4], retaining walls [5] or archaeological sites [6–8].

Non-destructive testing is particularly suitable in the case of historical objects. Such an approach is more and more often applied in cultural heritage buildings due to the necessity to preserve such structures in an untouched condition for future generations. The exact structure of historical objects dated several centuries back is usually unknown, since the technical documentation is incomplete or entirely gone. An efficient method to collect information in such situations is in situ inspection, often supported by numerical analyses that provide design guidelines and recommendations for planned reconstruction, strengthening and restoration works [9,10]. Non-invasive testing conducted within the flooring area allows detecting crypts, tombs and hidden rooms, as well as evaluating the technical condition of the floors and ceilings. GPR was successfully used for the diagnostics and condition assessment of different historical objects [11–17]. Despite the advances in non-destructive testing technology, there is no one technique suitable for every situation. Researches are often carried out using several methods to ensure that the obtained results are correct. Drahor et al. [18] compared the results obtained using the GPR and electrical resistivity tomography (ERT) while searching for cracks and damages that could occur in the church floor. Perez-Gracia et al. [19] compiled two methods, GPR and the capacitively coupled resistivity method, to obtain 2D images of the shallow subsurface under and around the Cathedral of Mallorca. The ground penetrating radar technique and laser scanning technique were combined by Tapete et al. [20] in order to interpret the displacements influencing the condition of the archeological monuments. Moropoulou et al. [21] presented the implementation of the integrated non-destructive methods, such as digital image processing, infrared thermography, ground penetrating radar, ultrasonic testing and fibre-optic microscopy for the inspection of historical objects. They detected air voids, delamination, moisture, material wear and degradation in the evaluation of the effectiveness of interventions and the assessment of the compliance of the repair method used. Faella et al. [22] took comprehensive measurements of a church in Bethlehem and estimated the condition of the floors, walls and columns, employing GPR and ultrasonic methods as well as thermography.

Combining the GPR and UT methods is commonly used in imaging, monitoring and analyzing the condition of engineering structures. GPR is particularly useful to conduct surveys on large areas because it allows handling a large amount of data in a reasonable amount of time. In previous studies, the GPR method has shown high efficiency in the imaging of reinforcement bars [23,24], cracks [25], defects in the form of air voids, delamination and moisture [26–28] or systems applied for concrete strengthening [29,30]. On the other hand, the UT techniques were proved to work well in the detection of defects such as notches [31–33], micro- and macro-cracks [34], small air gaps [35] or minor scratches [36], as well as in the evaluation of plate-like structures [37,38] and adhesive materials connections [39,40]. What is more, ultrasounds can be applied for inspecting conductive materials, unlike electromagnetic waves used in the GPR method. If both methods are applied appropriately, they may be considered as complementary. The assessment of the examined element becomes more reliable by implementing these two measurement techniques. The literature shows that both methods ensure a unique way of imaging the studied surface, enabling thus to resolve multiple research problems. Guadagnuolo [41] juxtaposed the GPR and UT techniques while examining the walls and floors of a historical church. Binda et al. [42] implemented the UT and GPR research to verify the damages and possible preservation works of the walls and piers due to the restoration of a damaged cathedral. Furthermore, the parameters of the mortar used as a possible means of repairing a damaged wall were controlled by performing ultrasonic tests. Perez-Gracia et al. [43] combined the UT and GPR techniques in the assessment of the geometry and physical properties of historical columns. The above-mentioned papers have presented many successful applications of combined GPR and UT methods; however, a thorough comparison of these methods by analyzing measurement data recorded along the same traces is rather limited.

The paper presents the results of the integrated ultrasonic testing and ground penetrating radar inspection conducted in St. Nicholas' Church in Gdańsk, Poland. The aim of the study was to present the practical aspects of the application of both techniques in detecting and imaging the underfloor inclusions, such as air voids, brick walls, pipes, rubble and human remains. Experimental measurements of the floor were conducted in the area of both (south and north) aisles, and also around a trial pit. Preliminary investigations were conducted on physical models consisting of two concrete slabs stacked on top of each other and gradually moved apart to simulate a slot of varying thickness, in order to better understand the phenomena of electromagnetic and ultrasonic wave propagation within the air voids and concentrated inclusions. In addition, the numerical simulations of electromagnetic waves were performed to support the interpretation of the GPR results. The analysis of the results obtained allowed concluding that GPR was suitable for the imaging of concentrated inclusions, whereas UT enabled detecting air voids. The presented research revealed the possibilities and limitations of both methods, indicating their complementarity in the context of non-destructive diagnostics of historical buildings.

2. Materials and Methods

2.1. Object of Investigations

Integrated GPR and ultrasonic inspection were conducted in St. Nicholas' Church in Gdańsk, Poland (Google Maps coordinates in a WGS84 system: 54.352206 N, 18.651510 E). This historical object from the 14th century is the only church in the city, which survived World War II without destruction. However, due to the long process of the settlement of the pillars, some damages appeared and progressed in the structural elements. Recently, the process of damage has accelerated, resulting in the serious cracking of vaults, displacement of arches and deformation of the floor. The poor condition of the church led to its closure in November 2018 and undertaking of repair works.

The conducted floor inspection reported in this paper was the first stage of investigations directed to assess the technical state of the church. NDT tests were performed on the floor of two aisles, south and north, as shown in Figure 1. In both aisles, the top floor layer was made of stone tiles measuring approximately 43 cm × 43 cm. In the central part of each aisle, tombstones were laid. The cross-section layers of the floor were identified in a few trial pits. One of the pits located in the north aisle is shown in Figure 2. Based on a visual inspection of the cross-section visible in the trial pit, particular layers in the floor were identified (Figure 3). It was found that the stone tiles laid on the cement mortar layer had a thickness of approximately 2–6 cm. Below was a layer of sand and ground backfill. It was also identified that at some edges, an air gap with a thickness of approximately 1–2 cm was visible below the cement mortar.

(a)

(b)

Figure 1. Photograph of the floor in St. Nicholas' Church in Gdańsk, Poland (Google Maps coordinates in a WGS84 system: 54.352206 N, 18.651510 E): (a) south aisle; (b) north aisle.

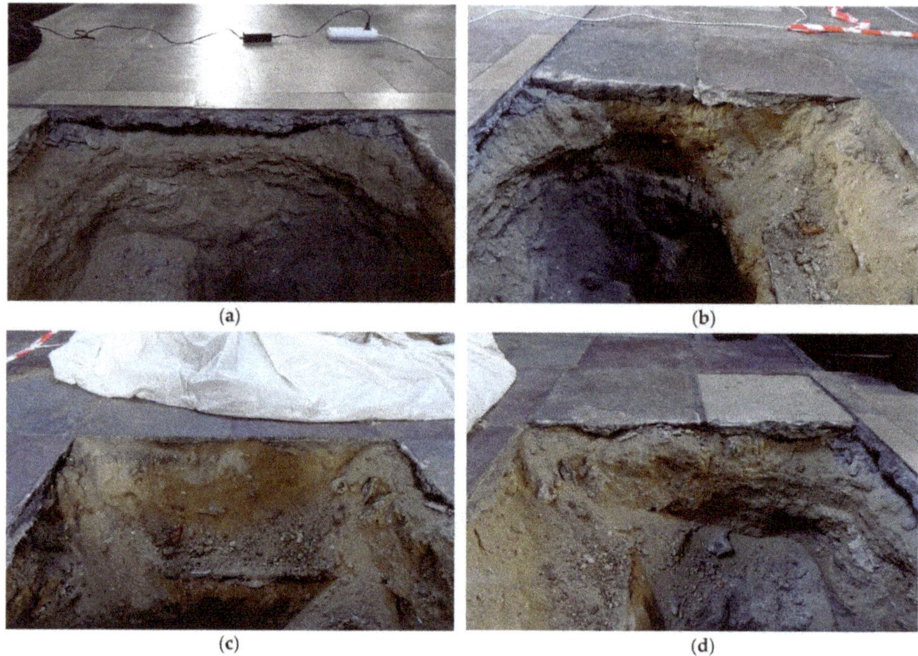

Figure 2. Photographs of the trial pit in the floor: (**a**) edge 1; (**b**) edge 2; (**c**) edge 3; (**d**) edge 4.

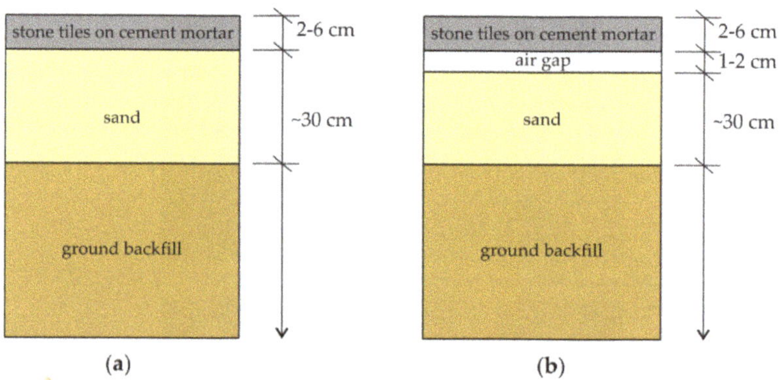

Figure 3. Schematic sketch of floor layers in the trial pit: (**a**) appropriate cross-section (without air gap); (**b**) cross-section with damage (air gap).

Additional investigations were conducted at the laboratory, on two circular concrete slabs. The slabs had a diameter of 49 cm and a thickness of 10 cm. One of the slabs had a table tennis ball with a diameter of 40 mm embedded in its center, at the half height of the slab. During the tests, the slabs were stacked on top of each other and gradually moved apart to simulate air gaps of varying thickness: 1, 4, 20 and 50 mm (see Figure 4).

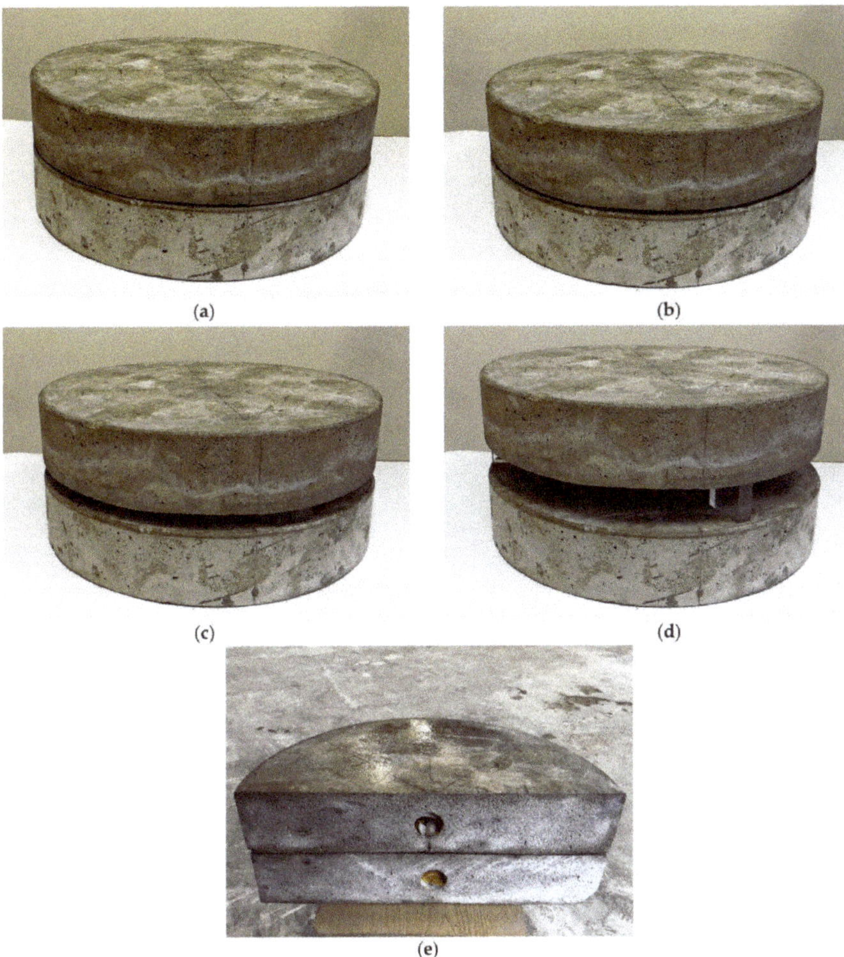

Figure 4. Circular concrete slabs with air gaps of different thickness: (**a**) 1 mm; (**b**) 4 mm; (**c**) 20 mm; (**d**) 50 mm; and (**e**) upper plate after cutting.

2.2. Data Acquisition and Equipment

Two NDT techniques were used for inspecting the floor: ground penetrating radar and ultrasonic testing. In both approaches, a pulse-echo test system was used which comprises both the transmitting (T) and receiving (R) antennas. During the pulse-echo measurements (Figure 5), the antenna is moved along the tested surface and a single time signal (so-called an A-scan) is recorded for a specific position. The assembling of the A-scans gives an image called a B-scan or echogram. As the antenna approaches an element that differs in electrical (in the ground penetrating radar method) or mechanical (in the ultrasound method) properties from the surrounding medium, the time the wave returns to the receiving antenna changes. A reflection is then created on the echogram reflecting the disturbing element. A reflection from any point inclusion (e.g., circular air gap, reinforcing bar) is represented in the B-scan as a hyperbola, while longitudinal inclusions are represented as line patterns.

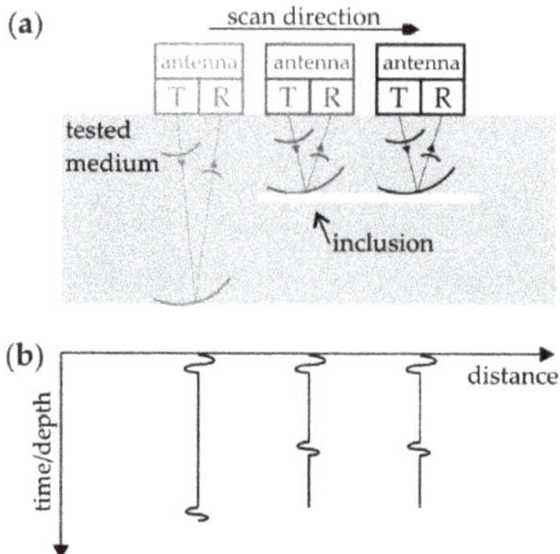

Figure 5. Scheme of measurements performed in pulse-echo mode: (**a**) collection of signals; (**b**) sketch of obtained echogram.

The Aladdin system (IDS GeoRadar, Pisa, Italy) equipped with a 2 GHz antenna was used to perform the GPR measurements. For each trace, the registered time range was 32 ns and the number of captured samples was 1024. The step distance between particular A-scans was 1 cm. The GPR data were registered in K2 FastWave and then processed in GRED HD by the following operations: automatic time zero correction, bandpass filtration in a frequency range of 500–3000 MHz and smoothed gain. Ultrasonic testing was performed using the pulse analyzer Pundit PL-200 (Proceq SA, Schwerzenbach, Switzerland) equipped with a 50 kHz antenna. For each A-scan, 1000 samples were captured with a time step of 1 μs, and the step distance between particular traces was 1 cm.

The GPR inspection of the floor was conducted in both aisles (Figure 6). During the measurements, 147 profiles were traced in the south aisle (A-1 to A-147) and 150 profiles in the north aisle (B-1 to B-150). The distance between particular profiles was 21.5 cm. Some inaccessible areas were omitted during the GPR scanning, so the shape of the scanned area was irregular. Next, four scans were acquired along the edges of the trial pit (denoted as C-1, C-2, C-3 and C-4 in Figure 7a). Finally, one profile was traced on the concrete slabs. Four measurements were made, each for the different thicknesses of the air gap, i.e., D-1 (1 mm), D-2 (4 mm), D-3 (20 mm) and D-4 (50 mm). Additionally, along selected traces, UT measurements were carried out (see Figure 8). The summary of the conducted GPR and UT surveys is given in Table 1.

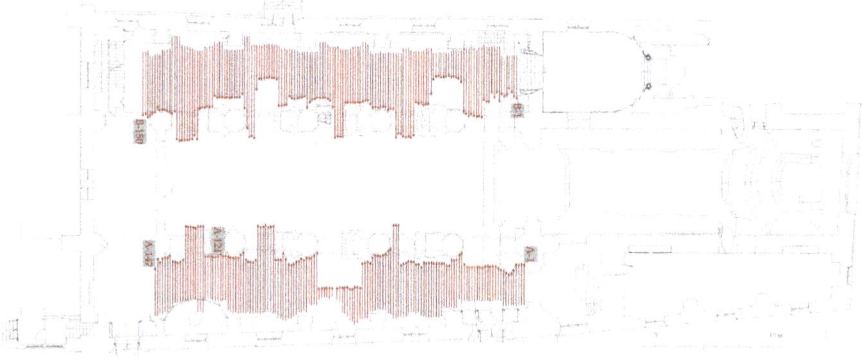

Figure 6. Plane view of the church with ground penetrating radar (GPR) traces in the south (traces A-1 to A-147) and north (traces B-1 to B-150) aisles.

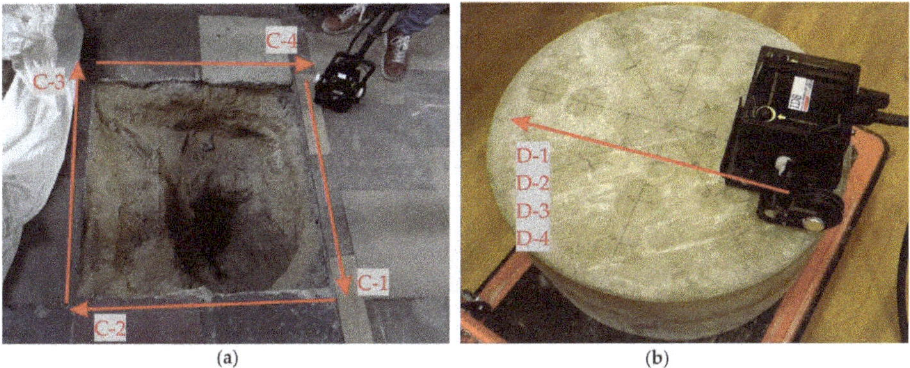

Figure 7. GPR traces registered along (**a**) edges of the trial pit (traces C-1 to C-4) and (**b**) the circular slabs with air gaps of varying thickness (traces D-1 to D-4).

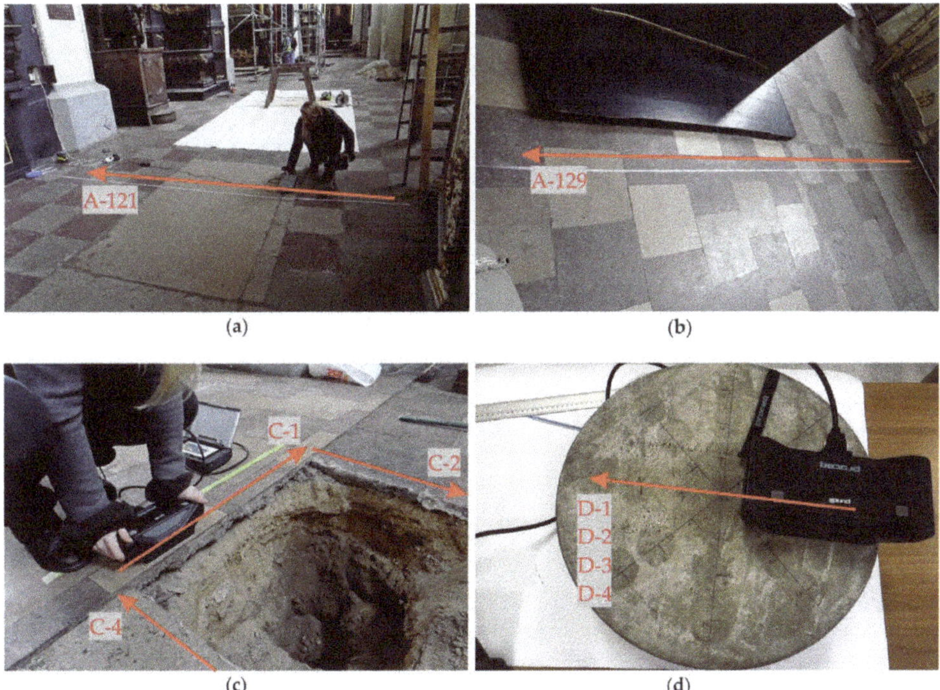

Figure 8. Ultrasonic measurements: (**a,b**) along traces A-121 and A-129 in the south aisle; (**c**) along edges of the trial pit; (**d**) along circular slabs.

Table 1. Scheme of GPR and ultrasonic testing (UT) investigations.

Surveys	GPR Traces	UT Traces
A	A-1 to A-147	A-121 and A-129
B	B-1 to B-150	–
C	C-1 to C-4	C-1 to C-4
D	D-1 to D-4	D-1 to D-4

2.3. Numerical Simulations of Electomagnetic Wave Propagation

In order to better interpret the results obtained from the GPR studies, several numerical models with inclusions expected under the tested floor were prepared. The calculations with the models were performed to give information about the behavior of electromagnetic waves under the influence of anomalies, which can be used to analyze the GPR B-scans of the floor. Numerical modeling of electromagnetic wave propagation was carried out by the finite-difference time-domain (FDTD) method using the gprMax open source software (release 3.1.5) [44]. Two groups of 2D models were created. The first one (models #1) corresponded to the floor, while the second one (models #2) represented the concrete slab. The models were discretized using a 1 mm × 1 mm grid. The time step was selected automatically based on the Courant–Friedrichs–Lewy (CFL) condition. The outer space of the models was restricted by perfectly match layer (PML) absorbing boundary conditions. The excitation signal emitted by the transmitting antenna was the Ricker function with a central frequency of 2 GHz, and the distance between the transmitting and receiving antenna was set as 6 cm, according to the actual distance in the IDS antenna.

The FDTD models of the floor are shown in Figure 9. The models with external dimensions of 2.96 m × 1.12 m were prepared in four variants. The aim of the simulations made on models #1.1–1.4 was to enhance the interpretation of the GPR surveys by analyzing how different a prior known underfloor inclusion influenced the registered echograms. Each model included four stone tiles with dimensions of 43 cm × 4 cm and a tombstone with the dimensions of 120 cm × 15 cm. An air gap with a thickness of 1 cm was inserted under two stone tiles (on the left side of the tombstone). The other two stones were laid directly on the ground (on the right side of the tombstone). Model #1.1 (Figure 9a) included a plain layer of sand under the stone tiles and tombstone. In model #1.2, three walls were inserted to represent underfloor crypts. The walls were made of bricks with dimensions of 6.5 cm × 12 cm and a 1 cm thick mortar. Additional two models (#1.3 and #1.4) comprised of concentrated inclusions in the form of brick rubble. The following values of the electric permittivity were adopted: $\varepsilon_r = 9$ (tiles), $\varepsilon_r = 3$ (sand), $\varepsilon_r = 6$ (brick) and $\varepsilon_r = 4$ (mortar). The conductivity for all materials was set as $\sigma = 0.01$ S/m. A-scans were registered at 280 nodes, starting from 0.08 m and giving the scan length of 2.79 m.

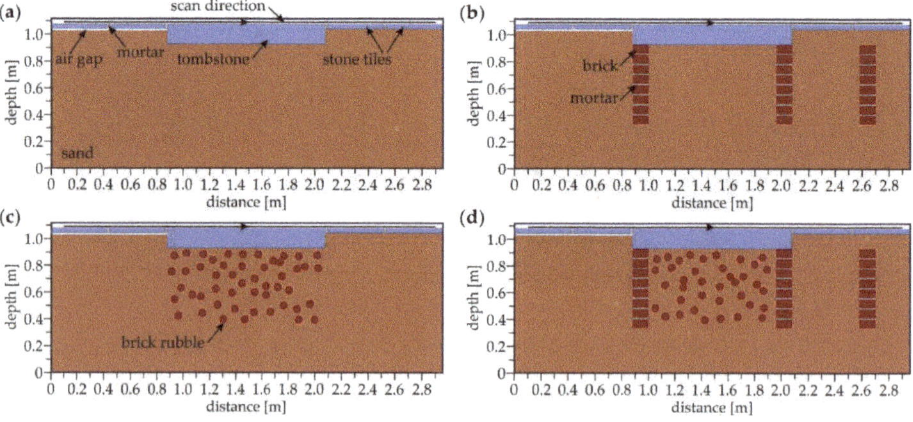

Figure 9. Finite-difference time-domain (FDTD) model of the floor: (**a**) model #1.1; (**b**) model #1.2; (**c**) model #1.3; (**d**) model #1.4.

Figure 10 illustrates the FDTD models of two concrete slabs separated with an air gap. The 2D models with external dimensions of 0.7 m × 0.45 m consisted of two concrete sections with dimensions of 0.49 m × 0.1 m separated by an air gap with a thickness of 1 (model #2.1), 4 (model #2.2), 20 (model #2.3) and 50 mm (model #2.4). In the upper slab, a circular air inclusion with a diameter of 40 mm was inserted. The electric permittivity of concrete $\varepsilon_r = 4$, corresponding to the velocity of electromagnetic waves equal to 15 cm/ns, was determined using the "depth to known reflector" method [45]. The conductivity of concrete was adopted as $\sigma = 0.01$ S/m. During the FDTD simulations, 41 A-scans were registered, giving the scan length of 0.4 m.

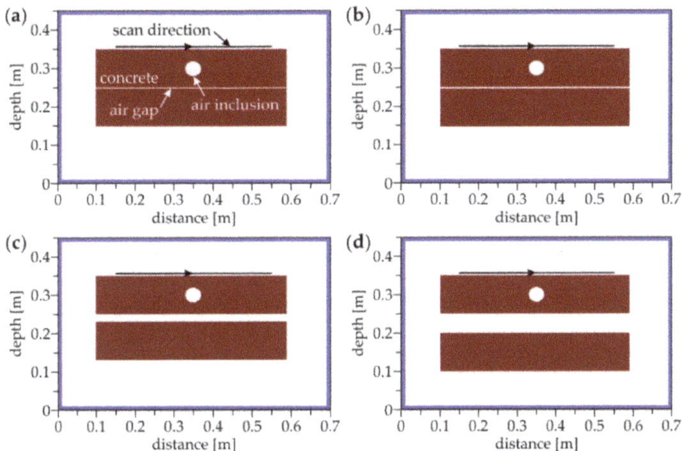

Figure 10. FDTD model of the concrete slabs with an air gap of the thicknesses (**a**) 1 mm (model #2.1); (**b**) 4 mm (model #2.2); (**c**) 20 mm (model #2.3); and (**d**) 50 mm (model #2.4).

3. Results and Discussion

3.1. Preliminary Investigations for Concrete Slabs

The results for the concrete slabs were first discussed as an initial step towards a more detailed analysis of the floor. Figure 11 shows the experimental GPR B-scans for the concrete slabs with different thicknesses of the air gap. The scan length was approximately 0.4 m. The depth axis was determined with the assumption that the electromagnetic wave velocity was equal to 15 cm/ns (based on the "depth to known reflector method" [45]). For the air gap thickness equal to 1 mm (Figure 11a), the half hyperbolas were clearly detected at the depth of 0.1 (orange arrows) and 0.2 m (green arrows), representing the reflections from the bottom faces of the upper and the lower slabs, respectively. This observation indicated the proper assumption of the electromagnetic wave velocity in the analyzed elements. The bottom faces of both slabs were visible as the lines with an intensity different from the adjacent part of the B-scan. The position of the reflection from the upper slab did not change, whereas the lower slab appeared deeper when increasing the air gap thickness (compare Figure 11a–d). For the air gaps with the thicknesses of 4, 20 and 50 mm, the position of the bottom plate was detected at about 20.5, 22 and 24 cm depths, respectively. This change showed the increase in the distance between both slabs and indicated a high compatibility of the obtained numerical results with the real thicknesses of the gaps. The identified gaps thicknesses were slightly different from the actual ones because the depth axis in Figure 11 was calculated for the velocity of the electromagnetic wave in concrete (15 cm/ns). In addition, the line at the depth of 0.1 m transformed into two separate lines, which indicated the opening of the air gap. It should also be mentioned that the line representing the bottom face of the lower slab became less pronounced for the greater distances between the slabs. This remark allowed concluding that the larger the air gap was, the harder it was to detect elements below it. Moreover, other half hyperbolas (denoted by blue arrows) were present at the depth of about 0.05 m, indicating a technological break during the concreting of the upper slab (visible in all B-scans, see Figure 11a–d).

The upper slab was prepared in two stages unlike the lower one, in which no additional perturbations could be seen. A full hyperbola was observed in the center of each B-scan at the depth of about 0.05 m (marked by the red arrow), revealing the presence of the table tennis ball, which was placed in the upper slab between both stages of concreting.

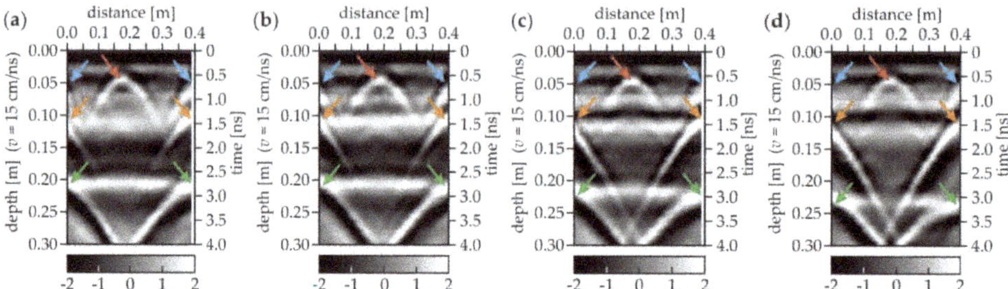

Figure 11. Experimental GPR B-scans for concrete slabs with air gap of thicknesses (**a**) 1 mm (trace D-1); (**b**) 4 mm (trace D-2); (**c**) 20 mm (trace D-3); and (**d**) 50 mm (trace D-4).

The numerical GPR B-scans corresponding with the above-described experimental results are presented in Figure 12. The half hyperbolas marked by the orange and green arrows clearly show the bottom faces of the upper and the lower slabs, respectively. The line denoting the bottom face of the lower slab became less noticeable and appeared deeper when increasing the distance between both slabs (cf. Figure 12a–d). For air gaps with thicknesses of 4, 20 and 50 mm, the position of the bottom plate is detected at about 19, 21 and 22 cm depths, respectively. The opening of the air gap was observed as the shift of the reflection from the bottom face of the lower slab occurred and also by the gradual separation of the single line into two lines. The hyperbola revealing the presence of the table tennis ball (red arrow) was clearly visible at the depth of 0.05 m. However, the technological break was not detectable (it was not modelled). The high agreement of the experimental and numerical results allowed concluding that the modelling of electromagnetic wave propagation was correct. The possibility of detecting air gaps and concentrated inclusions was confirmed.

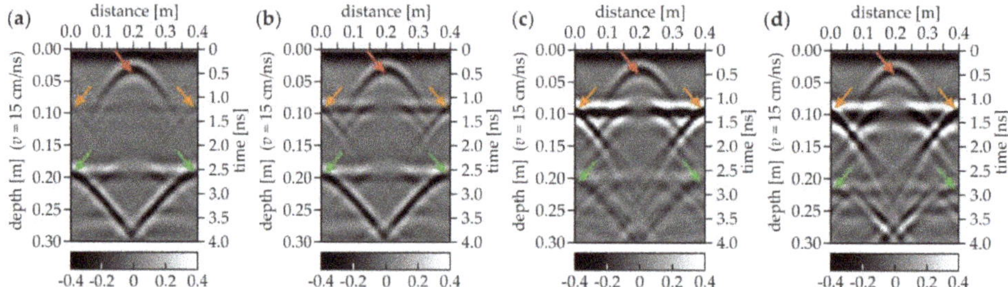

Figure 12. Numerical GPR B-scans for concrete slabs with air gap of thicknesses (**a**) 1 mm (model #2.1); (**b**) 4 mm (model #2.2); (**c**) 20 mm (model #2.3); and (**d**) 50 mm (model #2.4).

The UT B-scans for the analyzed slabs are presented in Figure 13. The ultrasonic pressure wave velocity determined before the main tests was equal to 2055 m/s. The scan length was reduced to 0.3 m due to the dimensions of the used UT antenna. There were no noticeable differences between all scans (Figure 13a–d), thus it could be concluded that the thickness of the air gap did not affect the results obtained. The lines at the depth of about 0.1 m representing the reflections from the bottom face of the upper slab were seen in all B-scans. Moreover, these lines were repeated regularly along the depth axis with the step of about 0.1 m (the slab thickness). Ultrasonic waves did not penetrate into the lower slab, they were fully reflected from the bottom face of the upper slab. The deeper lines were only the multiple reflections from the faces of the upper slab. The intensity of these lines decreased with the depth because of the damping of the ultrasonic waves. It is worth noting that the deeper lines appeared only near the edges of the slab. The presence of the boundaries strengthened the reflections, thus the signals were damped slower at the sides compared with the center of the slab. An important observation is that the table tennis ball was not observed at all. The reason might be the limitation of the used UT antenna. The Pundit PL-200PE instruction states that inclusions with a diameter of at least 30 mm should be detected, however, this condition deals with cylindrical elements. Being a relatively small spherical (concentrated) inclusion, the ball was not possible to be detected, despite the fact that its diameter was greater than minimum. To sum up, the air voids could be successfully detected using the UT technique, but without estimating the thickness. On the other hand, the content of the element below the air gap could not be imaged because the ultrasonic waves were entirely reflected from the air gap. Additionally, small concentrated inclusions could not be detected.

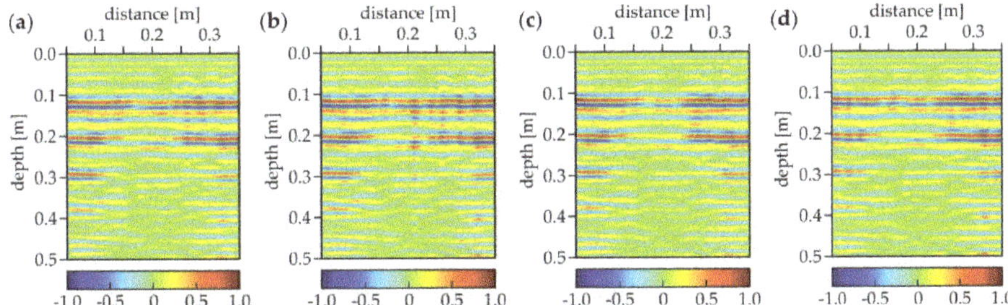

Figure 13. Ultrasonic B-scans for concrete slabs with air gap of thicknesses (**a**) 1 mm (trace D-1); (**b**) 4 mm (trace D-2); (**c**) 20 mm (trace D-3); and (**d**) 50 mm (trace D-4).

3.2. Numerical Models of the Floor

Figure 14 shows the GPR B-scans acquired in the numerical models of the considered floor. Taking into account the complexity of the analyzed medium, the electromagnetic wave velocity was set as constant and equal to 10 cm/ns. The strategy applied did not take into account the heterogeneity factor. Therefore, the results were only approximations of the real subsurface geometry. However, this fact did not disable the proper interpretation of the results. In the model #1.1 (Figure 14a), the positions of the stone tiles were clearly identified based on the line patterns denoting the reflections from the vertical joints between tiles. The half hyperbolas at the ends of the tombstone were visible in the center of the scan. The lower faces of the tiles and the tombstone were located at the depths of 0.04 and 0.15 m, respectively. Some additional reflections with a lower intensity were observed below, indicating the further wave reflections from the analyzed elements (e.g., at the depth of 0.08 m below the tiles and 0.3 m below the tombstone). The difference between the tiles laid with and without the air gap was small, i.e., additional reflections below the tiles located above the air gap were insignificantly stronger. This observation might lead to the hypothesis that the detection of the small air gaps based on the GPR scanning could be problematic. For the model #1.2 (Figure 14b), the conclusions from the identification of the tombstone and the tiles were the same as for model #1.1. However, additional hyperbola patterns appeared, being the reflections from all singular bricks of the walls located below the edges of the tombstone. It is worth noting that the intensity of the hyperbolas decreased with the depth, which is a common relation observed for electromagnetic waves. Similarly, another brick wall was visible at the distance of 2.55 m. It is interesting to note that the walls seemed to be located at different depths. This apparent observation was the result of the assumption that the electromagnetic wave velocity was constant. The wave needed more time to travel though the tombstone (to reach the edge walls) compared with the stone tiles and the sand (above the additional wall). In the model #1.3 (Figure 14c), the additional hyperbolas were observed below the tombstone, proving the presence of the rubble. The great amount of the inclusions made the interpretation of the scan more complicated. It is also worth noting that the deeper hyperbolas had a much lower intensity, thus the identification of inclusions located deeper was straitened. The interpretation of the reflections below the tombstone was far more difficult when considering the simultaneous presence of the rubble and the brick walls (model #1.4, Figure 14d). The superimposition of the hyperbolas denoting the rubble and the singular bricks made them undistinguishable. Therefore, it was not possible to state whether a certain hyperbola denoted the brick or any different kind of inclusion. To sum up, the results obtained from the numerical calculations gave some information about the possibilities and limitations of GPR scanning. The location of the tombstone and the tiles could be determined. The rubble below the tombstone was detectable as well as the brick walls; however, when these elements appeared together, the interpretation was complicated. What is essential is that the air gaps were difficult for the identification.

3.3. Experimental Surveys for the Trial Pit

Figure 15 presents the GPR B-scans collected around the trial pit. As can be seen in Figure 7a, several rectangular tiles formed a joint at one of the pit's sides. Therefore, during the acquisition of the C-1 scan (Figure 15a), the GPR antenna was moved along the joint, thus the bottom faces of the tiles were not clearly imaged. However, the analysis of the upper part of the scan allowed identifying two different rows of tiles; the typical tile (43 cm × 43 cm) row was replaced at the edge of the trial pit with a narrower one (43 cm × 10 cm) laid with the overlap (cf. Figure 2a). The overlap length was about 0.16 m and it could be seen the clearest at the distance of 0.42–0.58 m. The ground under the C-1 scan had many inclusions, resulting in numerous irregularly distributed reflections (cf. the concentrated inclusions in Figure 14c,d). The C-2 scan (Figure 15b) clearly presents the line patterns being the reflections from the joints. The bottom faces of the tiles were identified at dissimilar levels, therefore the tiles (or the mortar layers below them) had different thicknesses. The hyperbola located at the distance of 0.7 m and the depth of 0.18 m denoted the presence of a metallic (steel or aluminum) pipe (cf. Figure 2). The ground on the right side of the scan (around the pipe, at the distance of 0.5–1.0 m) seemed less heterogeneous than the remaining part. This might be caused by the fact that the original ground was removed and replaced with another, more homogeneous one during placing the pipe under the floor. The C-3 scan (Figure 15c) shows the clear image of two tiles, one with the regular bottom face and the second with an unbalanced shape. The ground seemed to be original because it is highly heterogeneous, such as in the C-1 scan. The C-4 scan (Figure 15d) also clearly shows two typical tiles together with the narrow one (located at the distance of 0.88–0.98 m). The shape of the bottom face of all tiles was irregular. The shift was observed in the scan (at the distance of 0.85 m), caused by the slip of the antenna at the joint between the standard and the narrow tile (cf. Figure 2d). The reflection from the pipe was visible at the distance of 0.25 m and the depth of 0.18 m (the same as observed in the C-2 scan). The ground around the pipe was less inhomogeneous (the distance of 0.0–0.5 m) compared with the remaining part. Generally speaking, it was difficult to say whether there were any air gaps under the floor. The subtle difference between the image of the tiles with and without the air gap (observed in the numerical results) was here blurred by the ground inhomogeneities and a signal noise.

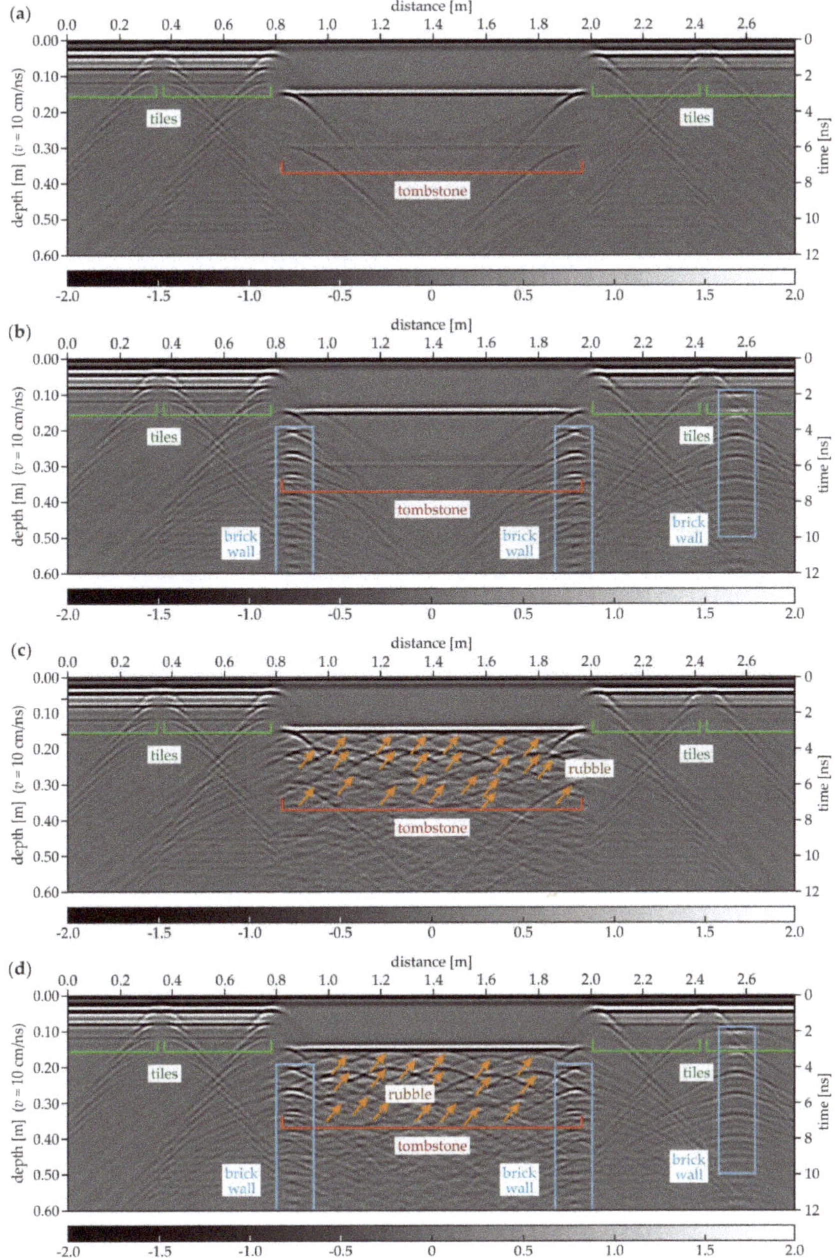

Figure 14. Numerical GPR B-scans for the floor: (**a**) model #1.1; (**b**) model #1.2; (**c**) model #1.3; (**d**) model #1.4.

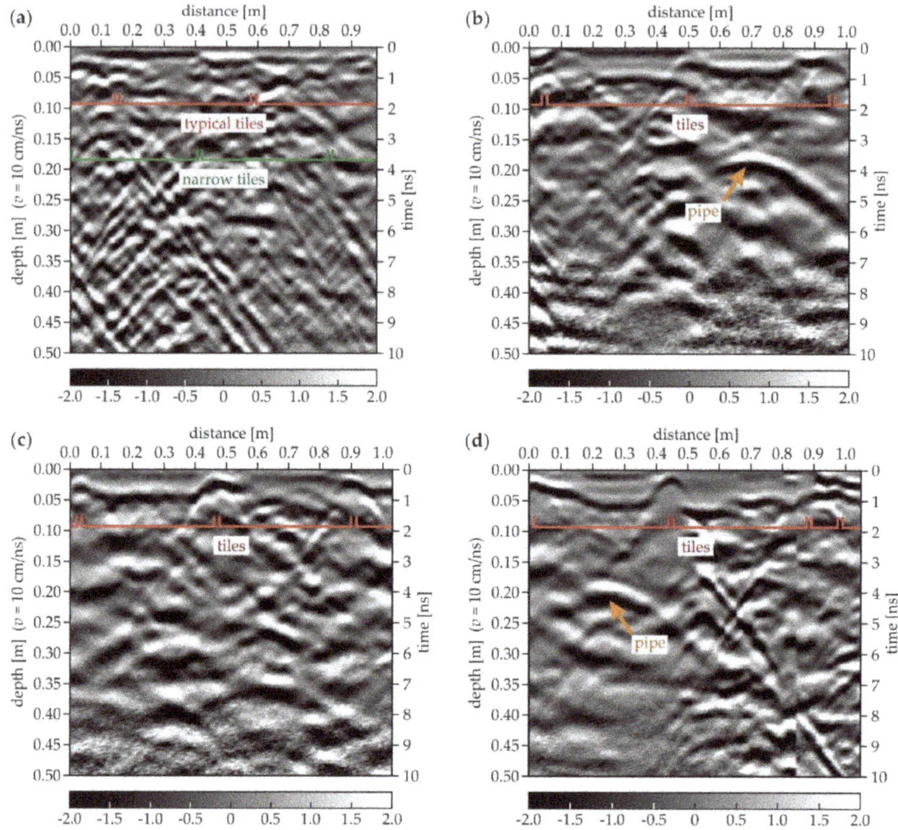

Figure 15. GPR B-scans for the trial pit: (**a**) edge 1 (trace C-1); (**b**) edge 2 (trace C-2); (**c**) edge 3 (trace C-3); (**d**) edge 4 (trace C-4).

Figure 16 contains the UT B-scans corresponding to the above-discussed GPR B-scans. The C-1 scan (Figure 16a) was performed through the narrow tiles, as can be seen in Figure 7a, thus the joints of the standard tiles were not observed. In the top part of the scan (at the depth of about 0.03 m), there was a straight line denoting the regular bottom face of the narrow tiles. However, there was no regularity in the lower part of the scan, where many multiple reflections occurred. Compared with the results for the concrete slabs, it might be stated that the air gap with an irregular shape was present in this area. The non-uniform settlement of the original ground probably led to the appearance of the air gap. The C-2 scan (Figure 16b) shows the image of two different tiles. The right tile, unlike the left one, was clearly visible at the top of the scan; it was possibly removed and then placed again with the use of a different (stronger) mortar. The multiple reflections appeared under the left tile, indicating the presence of the air gap (this tile laid on the original ground). On the contrary, there were no reflections under the right tile. The pipe detected in the GPR scans was not visible, which stays in agreement with the results for the concrete slabs where the UT scans did not reveal the table tennis ball. This observation allowed concluding that concentrated inclusions could not be detected by the UT scanning. It also should be mentioned that, according to the UT antenna instruction, the diameter of the pipe did not exceed 30 mm. In the C-3 scan (Figure 16c), two tiles were also visible. The left one was much more pronounced: it could have been replaced during some renovation works. Additionally, there were multiple reflections under both tiles, indicating the presence of the air gaps (resulting from the ground settlement). However, it needs to be noted that the reflections were less intensive under the right tile because it was weakly identified itself. The possibility of detecting the tiles in the C-4 scan (Figure 16d) was also distinguished: the left one was much better imaged. Although the ground below the left tile was replaced, the multiple reflections indicating the presence of an air gap were visible under both tiles. Like in the C-3 scan, the reflections under the right tile were weaker because it was weakly imaged itself. It also needs to be added that the pipe was not observed like in the C-2 scan.

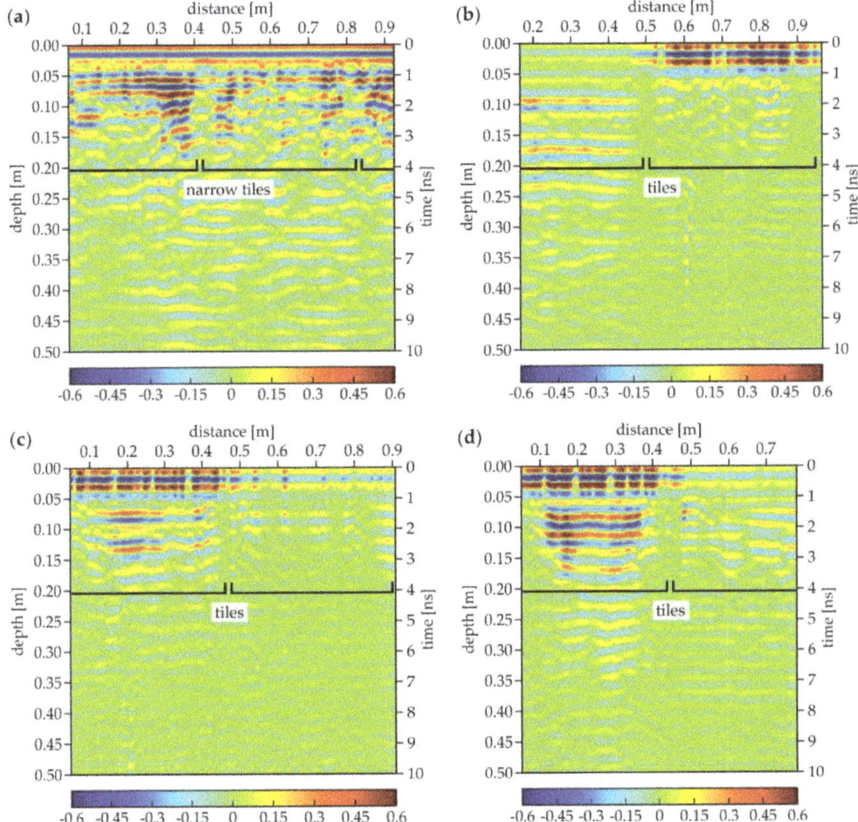

Figure 16. Ultrasonic B-scans for the trial pit: (**a**) edge 1 (trace C-1); (**b**) edge 2 (trace C-2); (**c**) edge 3 (trace C-3); (**d**) edge 4 (trace C-4).

3.4. Experimental Surveys for Floor

Figure 17 presents two examples of the GPR B-scans acquired in the area of the south aisle. The A-19 scan (Figure 17a) clearly reveals the location of the tiles based on the reflections from the tile grouts. The bottom faces of the tiles were identified at different depths, thus they had different thicknesses, varying between 0.04 and 0.09 m. The tombstone with a thickness of approximately 0.16 m could be observed in the center of the scan. An additional reflection denoting the tombstone adornment was also visible at the top of the scan. The ground under the tombstone had many concentrated inclusions, probably being the rubble or the human remains from the original crypts (cf. Figure 14c,d). Aside from this area, inhomogeneities were not observed. It was difficult to state whether there were any brick walls under the edges of the tombstone because the hyperbolas denoting the concentrated inclusions blurred the image. The A-117 scan (Figure 17b) shows the tiles with different thicknesses between 0.03 and 0.08 m. The tombstone was also visible; however, its shape was far more regular compared with the one from the A-19 scan. The ground below the tombstone had no strong inhomogeneities, thus it could be stated that not all the tombstones were laid in the area of the original crypts. Other interesting reflections were the hyperbolas located vertically one above another at the distances of 1.3, 3.4 and 4.2 m. They indicated the presence of three brick walls (cf. Figure 14b,d), that might be the remainders of the original supporting structures of the crypts.

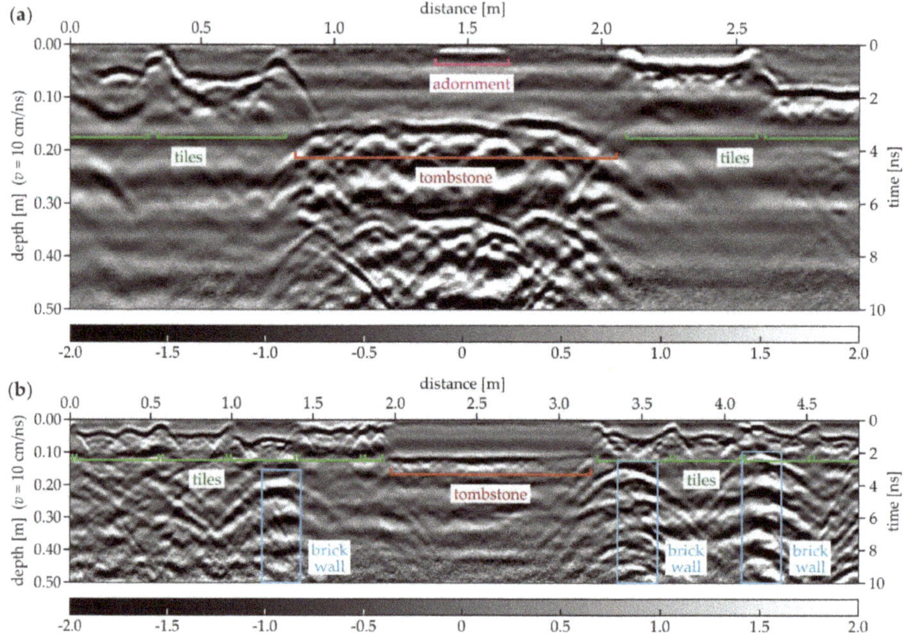

Figure 17. GPR B-scans for the south aisle (survey A): (**a**) A-19; (**b**) A-117.

The next two GPR B-scans from the south aisle are presented together with the corresponding results of the UT scanning. In the A-121 GPR B-scan (Figure 18a), the tiles with different thicknesses (in the range of 0.02–0.04 m) were visible, and also a tombstone with the thickness of about 0.12 m was identified. The ground below the tombstone contained some concentrated inclusions (probably rubble or human remains). The brick wall was present at a distance of about 3.4 m. The UT scan (Figure 18b) clearly revealed the tile pattern, and also the tombstone was observed. The presence of the multiple reflections under each tile allowed stating that there were many air gaps below them, which agreed with the fact that the floor of the south aisle experienced significant settlement. The reflections denoting the air gaps generally did not appear below the tombstone, probably because of its weight. What is also worth noting is that no additional reflections from the ground heterogeneities could be identified. The UT scanning did not detect concentrated inclusions, which were already observed for the pipe in the area of the trial pit and the table tennis ball in the concrete slab. The GPR B-scan for the trace A-129 is presented in Figure 19a. The tiles with different thicknesses (varying between 0.02 and 0.06 m) are clearly detectable. Two tombstones with a thickness of approximately 0.12 m and irregular shape of the bottom face were also present. The ground under both tombstones had many concentrated inclusions unlike aside from this area. The brick wall could be visible at the distance of about 4.3 m; however, the intensity of the reflections from the bricks was not as strong as in the A-121 scan. In the UT scan (Figure 19b), the tiles and the tombstones could be localized. There were no significant reflections under the tombstones, thus the air gaps were not likely to be there, unlike under the most of the tiles, where multiple reflections appeared. This observation corresponded with the settlement of the floor in the area of the south aisle.

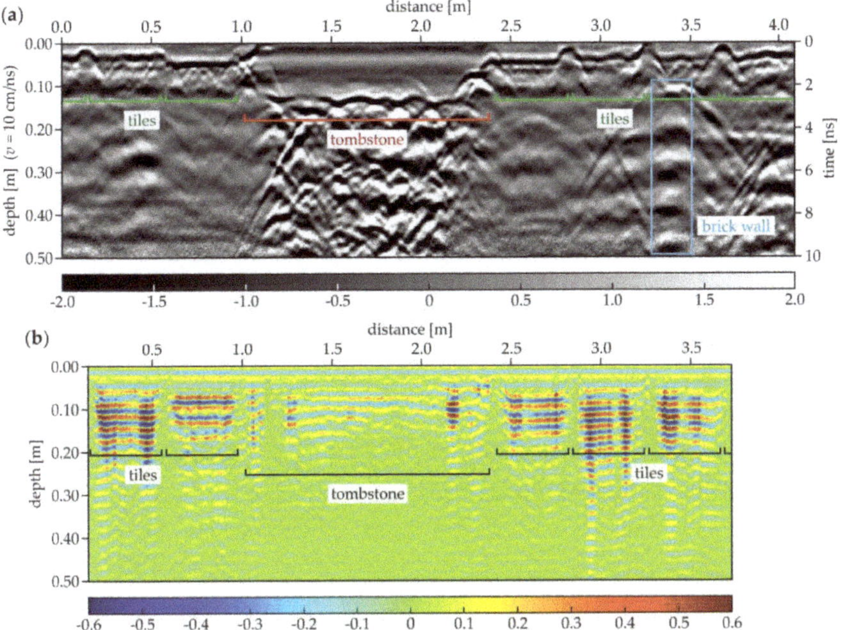

Figure 18. GPR and UT B-scans for trace A-121 (south aisle): (**a**) GPR scan; (**b**) UT scan.

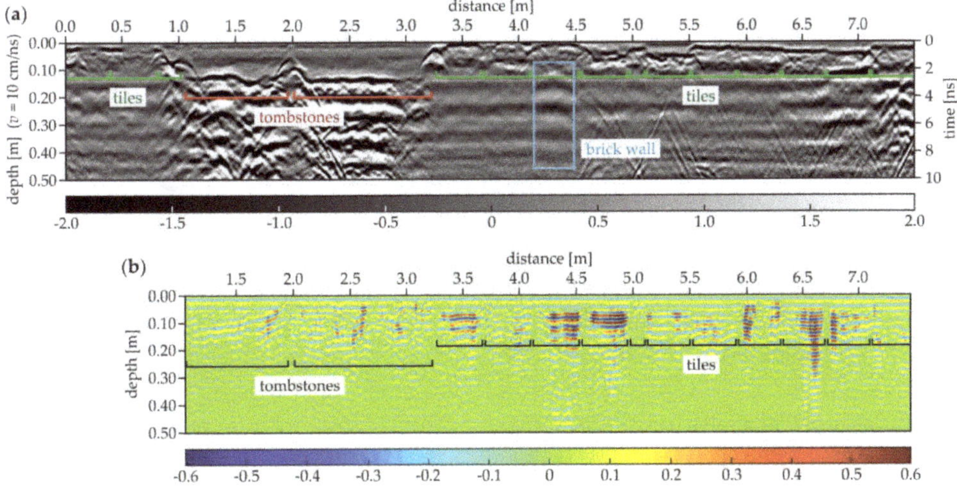

Figure 19. GPR and UT B-scans for trace A-129 (south aisle): (**a**) GPR scan; (**b**) UT scan.

Figure 20 shows the examples of the GPR B-scans for the north aisle. The B-5 scan (Figure 20a) shows the regular pattern of tiles with a constant thickness of 0.03 m. Several evenly spaced hyperbolas were visible at two levels (the depth of 0.11 and 0.18 m) and the distance of 0.0–2.1 m. These reflections represented the reinforcement of the staircase located under the floor. The ground below the tiles aside of the staircase seemed to be devoid of significant heterogeneity. In the B-35 scan (Figure 20b), the tiles had different thicknesses in the range of 0.02–0.06 m. The tombstone had a thickness varying between 0.10 and 0.15 m and the skew shape of the bottom face. The ground below the tombstone had some concentrated inclusions. The B-143 scan (Figure 20c) presents the tiles with the thicknesses between 0.02 and 0.05 m. Two tombstones were also visible, the left one with the thickness of about 0.09 m and the right with the thickness of approximately 0.16 m. The ground under the thinner tombstone did not have inhomogeneities, unlike the ground under the second one, where multiple reflections from the concentrated inclusions occurred. This observation confirms the fact that some of the tombstones were placed aside from the area of the original crypts.

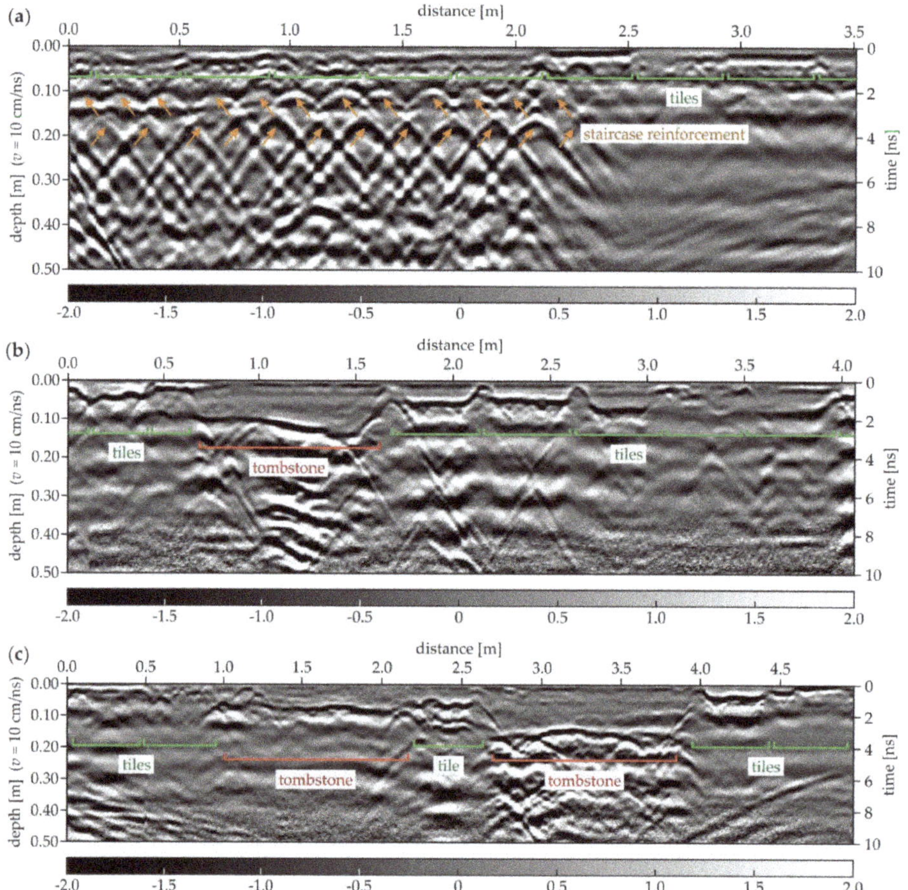

Figure 20. GPR B-scans for the north aisle (survey B): (**a**) B-5; (**b**) B-35; (**c**) B-143.

3.5. GPR Tomographic Imaging of Floor

Figure 21 presents the tomographic images of the south aisle. At the depth of 0.16 m (Figure 21a), the tombstones identified in the above-discussed B-scans of the south aisle were detected. The elements were laid in a straight line along the whole aisle at the width between about 2.75 and 4.25 m. Additionally, the single tombstone was located aside from the main row at the length between 2.5 and 4.5 m (it was seen in the A-129 scan, Figure 19). The distinction of singular tombstones from the main row was not possible; the scanning was performed along the tombstone grouts, thus they could not be imaged. The singular tiles were not detected due to a single direction profiling. The tile grouts in the lengthwise direction (perpendicular to the measurement traces) were identified, unlike the crosswise grouts (oriented in the same direction as the scan traces). At the depth of 4.06 cm (Figure 21b), the imaging of the tombstones was far clearer; however, there was still no possibility to distinguish the singular elements. On the contrary, the individual tiles could be detected; the chosen tomography depth was between the minimum and the maximum thickness of the tiles, thus some of them were visible, whereas other ones could not be observed. At the depth of 12.19 cm (Figure 21c), most of the tombstones were clearly visible. The lower faces of some thicker tombstones were below the tomography depth, unlike the thinner ones, which were not detected at this level. The adornment was detected on the second tombstone from the left. The tiles could not be identified because their bottom faces were located certainly above the actual tomography level. What is more, the brick wall was moderately imaged at the width of about 5.2 m and the length between 2.2 and 13.8 m. The tombstones were still distinguishable at the depth of 23.75 cm (Figure 21d). The inhomogeneities in the ground under certain tombstones were visible, especially on the right side of the tomography. The brick wall discovered at the depth of 12.19 cm was still detectable and seemed to be longer. What is more, another two brick walls were identified: the first one at the width of 4.5 m and the length between 5.5 and 14.0 m, the second at the width of 2.2 m and the length between 6.5 and 15.0 m. All of the three walls were clearly imaged in the A-117 B-scan (see Figure 17b).

Figure 22 shows the tomographic images of the north aisle. Similarly to the results for the south aisle, the tombstones were detected at the depth of 0.16 cm (Figure 22a); however, their configuration was not so regular. The lengthwise tile grouts were also imaged, unlike the crosswise grouts. The adornments were visible on some of the tombstones. At the depth of 4.06 cm, the tombstones were shown far more clearly. Some individual tiles were also observed. At the depth of 12.19 cm, the singular tombstones could be distinguished. Some of them did not seem to be rectangular because of the irregular shape of their bottom faces. It is also worth noting that the staircase reinforcement was visible at the length of 30.0–32.0 m and the width of 5.2–7.8 m. Only the lengthwise bars were noticeable, and the crosswise reinforcement could not be imaged because of the single direction of the GPR scanning.

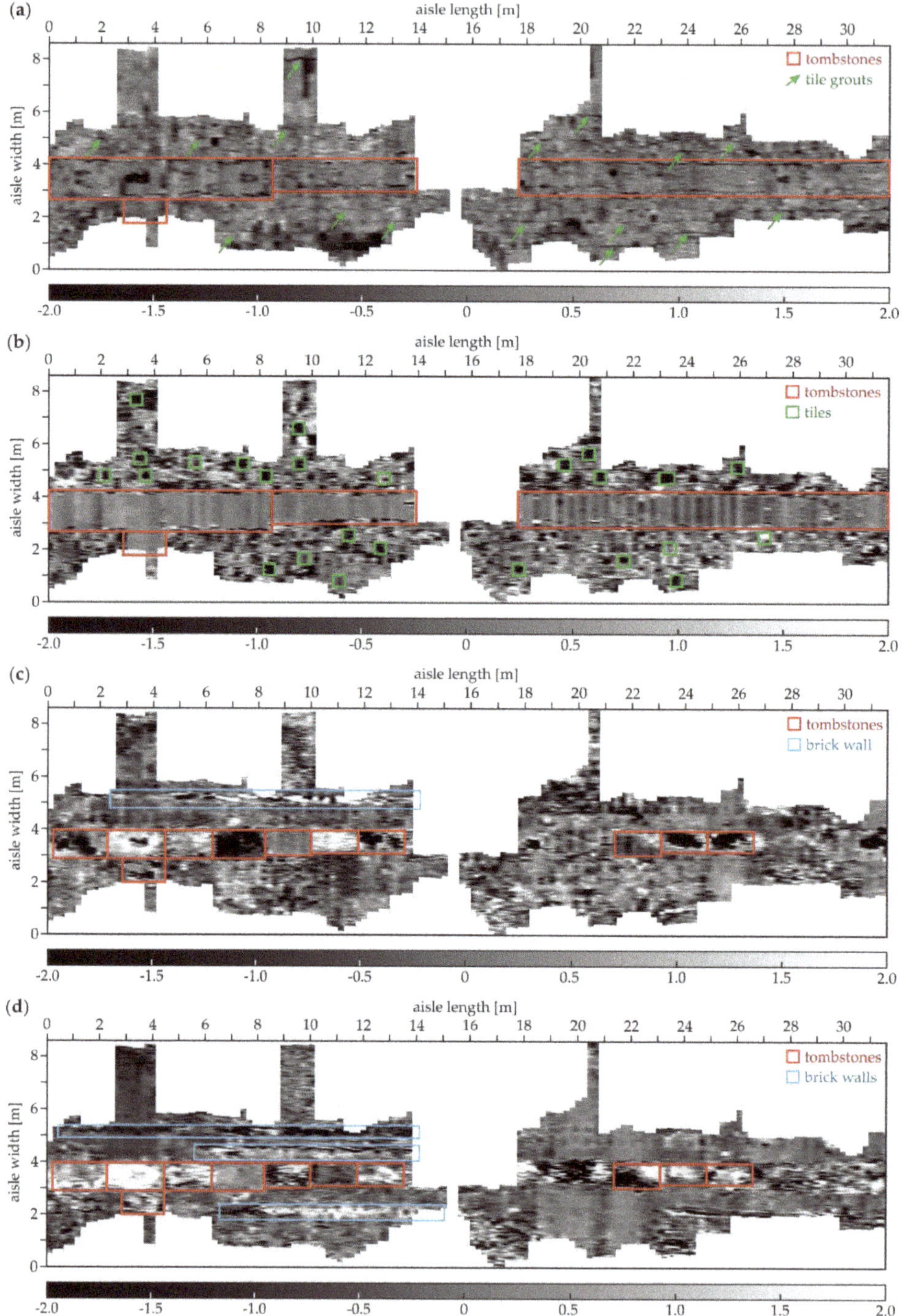

Figure 21. GPR tomographic images of the south aisle (survey A) at different depths: (**a**) 0.16 cm; (**b**) 4.06 cm; (**c**) 12.19 cm; (**d**) 23.75 cm.

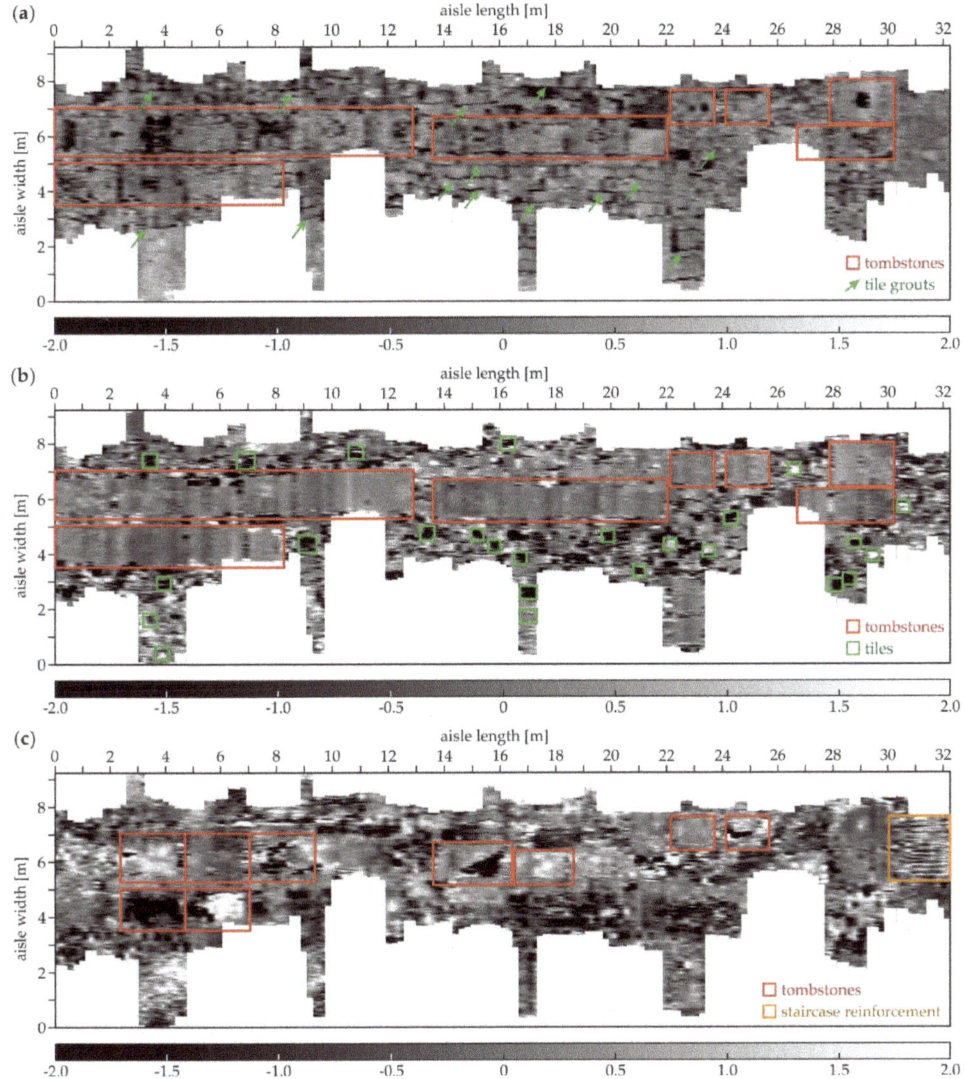

Figure 22. GPR tomographic images of the north aisle (survey B) at different depths: (**a**) 0.16 cm; (**b**) 4.06 cm; (**c**) 12.19 cm.

4. Conclusions

In this study, the integrated GPR and UT inspection was conducted on the floor of a historical church. The performed investigations focused on the detection of air gaps and other anomalies located under the stone tiles and the tombstones. The research was supported by laboratory models illustrating the behavior of electromagnetic and ultrasonic waves. The numerical models were prepared to illustrate the propagation of electromagnetic waves in the medium containing inclusions such as air gaps, brick walls, pipes and brick rubble. The results obtained provided useful information about the possibilities and limitations of the GPR and UT methods.

The possibility of an efficient application of the GPR technique for detecting small concentrated inclusions was confirmed. The table tennis ball was identified in the concrete slab (both in the numerical and experimental results); the pipe was detected in the area of the trial pit; some bricks and concentrated inclusions were observed in the ground under the floor. The larger surface and volume elements (stone tiles, tombstones) were successfully imaged too. The GPR method was also able to show the air gaps as lines (single or double, depending on the air gap thickness). However, the exact imaging of the air voids was possible only for thick layers; in the case of thin air gaps, it was difficult to state whether the lines denoted the air gap or a boundary between two media.

The UT measurements allowed an efficient detecting of the air gaps, independent of their thickness. This resulted from the phenomenon of total reflection of elastic waves at the boundary of the analyzed element. The location of stone tiles and tombstones could be also clearly visible. On the other hand, because of wave reflection, the UT inspection did not allow to detect anything below the air gap. What is more, ultrasonic waves did not identify concentrated inclusions, such as the table tennis ball, the pipe and the ground inhomogeneities, which could be caused by the limitations of the UT antenna used.

The tomographic snapshots provided the overall image of the examined area at a specific depth. This type of imaging could be useful when the exact location of underfloor inclusions in relation to the entire scanned area is the object of interest. However, when the detailed profile of a part of considered structure is important, an analysis of B-scans is more appropriate. For the presented study, a good compliance of both ways of imaging was observed.

To increase the applicability and practicality of non-destructive inspection techniques, the use of integrated GPR and UT methods was recommended. The methods complemented each other, allowing an exclusion of their limitations. The UT technique was efficient at visualizing air gaps of different thicknesses, however, it was not suitable for imaging small inclusions and anything below the air voids. What is also important, the UT measurements were time-consuming. The GPR method successfully detected concentrated inclusions with different sizes; it also allowed inspecting a large area with a relatively low time cost. However, the GPR measurements did not allow detecting air voids.

To summarize, the integrated inspection combining the GPR and UT techniques appeared to be effective for non-destructive diagnostics of underfloor structures in cultural heritage buildings. The proposed approach can be useful for the detection of anomalies laying under the floor, such as air gaps, bricks and pipes, which can appear in historical objects. The complementarity of both methods enables a precise analysis of the tested structure.

Author Contributions: Conceptualization and Methodology, M.R., E.W. and M.Z.; Experimental Investigations, M.R., E.W. and M.Z.; FDTD Calculations, M.R.; Formal Analysis, M.R., E.W. and M.Z.; Visualization, M.R. and E.W.; Writing—Original Draft Preparation, M.R., E.W. and M.Z.; Writing—Review and Editing, M.R., E.W. and M.Z.; Supervision, M.R. All authors have read and agreed to the published version of the manuscript.

Funding: This research received no external funding.

Acknowledgments: The authors would like to thank Michał Osek OP, the Prior of St. Nicholas' Church, for enabling in situ surveys and providing valuable historical information. The support of Jacek Lachowicz during the GPR surveys is gratefully acknowledged. The authors would like also to thank Piotr Samól for providing a plane view of the church (the plane view was used in the background of Figure 6).

Conflicts of Interest: The authors declare no conflict of interest.

References

1. Ziolkowski, P.; Szulwic, J.; Miskiewicz, M. Deformation analysis of a composite bridge during proof loading using point cloud processing. *Sensors* **2018**, *18*, 4332. [CrossRef]
2. Kwiatkowski, J.; Anigacz, W.; Beben, D. Comparison of non-destructive techniques for technological bridge deflection testing. *Materials* **2020**, *13*, 1908. [CrossRef] [PubMed]
3. Miśkiewicz, M.; Pyrzowski, L.; Sobczyk, B. Short and long term measurements in assessment of FRP composite footbridge behavior. *Materials* **2020**, *13*, 525. [CrossRef] [PubMed]

4. Binczyk, M.; Kalitowski, P.; Szulwic, J.; Tysiac, P. Nondestructive testing of the miter gates using various measurement methods. *Sensors* **2020**, *20*, 1749. [CrossRef] [PubMed]

5. Jonaitis, B.; Antonovič, V.; Šneideris, A.; Boris, R.; Zavalis, R. Analysis of physical and mechanical properties of the mortar in the historic retaining wall of the Gediminas Castle Hill (Vilnius, Lithuania). *Materials* **2018**, *12*, 8. [CrossRef] [PubMed]

6. Goodman, D.; Piro, S. *GPR Remote Sensing in Archaeology*; Springer: Berlin/Heidelberg, Germany, 2013; ISBN 9783642318573.

7. Neubauer, W.; Eder-Hinterleitner, A.; Seren, S.; Melichar, P. Georadar in the roman civil town carnuntum Austria: An approach for archaeological interpretation of GPR data. *Archaeol. Prospect.* **2002**, *9*, 135–156. [CrossRef]

8. Zhao, W.; Forte, E.; Fontana, F.; Pipan, M.; Tian, G. GPR imaging and characterization of ancient Roman ruins in the Aquileia Archaeological Park, NE Italy. *Measurement* **2018**, *113*, 161–171. [CrossRef]

9. Beben, D.; Ukleja, J.; Maleska, T.; Anigacz, W. Study on the restoration of a masonry arch viaduct: Numerical analysis and lab tests. *Materials* **2020**, *13*, 1846. [CrossRef]

10. Bajno, D.; Bednarz, L.; Matkowski, Z. Monitoring of thermal and moisture processes in various types of external historical walls. *Materials* **2020**, *13*, 505. [CrossRef]

11. Barilaro, D.; Branca, C.; Gresta, S.; Imposa, S.; Leone, A.; Majolino, D. Ground penetrating radar (G.P.R.) surveys applied to the research of crypts in San Sebastiano's church in Catania (Sicily). *J. Cult. Herit.* **2007**, *8*, 73–76. [CrossRef]

12. Rucka, M.; Wojtczak, E.; Zielińska, M. Interpolation methods in GPR tomographic imaging of linear and volume anomalies for cultural heritage diagnostics. *Measurement* **2020**, *154*, 107494. [CrossRef]

13. Imposa, S.; Mele, G. Ground penetrating radar survey inside the S. Agata cathedral of Catania (eastern Sicily). *Int. J. Archit. Herit.* **2011**, *5*, 188–197. [CrossRef]

14. Yalçıner, C.Ç.; Kurban, Y.C.; Altunel, E. Research using GPR into the cause of cracks and depressions in the floor of the gallery of Hagia Sophia Museum. *Constr. Build. Mater.* **2017**, *139*, 458–466. [CrossRef]

15. Gaber, A.; El-Qady, G.; Khozym, A.; Abdallatif, T.; Kamal, S.A.M. Indirect preservation of egyptian historical sites using 3D GPR survey. *Egypt. J. Remote Sens. Space Sci.* **2018**, *21*, S75–S84. [CrossRef]

16. Imposa, S.; Barone, F.; Gresta, S.; Leone, A. Ground penetrating radar survey finalized to the recovery of the "S. Agata alla Badia" church in Catania (eastern Sicily, Italy). *Environ. Geol.* **2009**, *58*, 889–896. [CrossRef]

17. Pérez-Gracia, V.; Caselles, O.; Clapés, J.; Osorio, R.; Canas, J.A.; Pujades, L.G. Radar exploration applied to historical buildings: A case study of the Marques de Lllió palace, in Barcelona (Spain). *Eng. Fail. Anal.* **2009**, *16*, 1039–1050. [CrossRef]

18. Drahor, M.G.; Berge, M.A.; Öztürk, C. Integrated geophysical surveys for the subsurface mapping of buried structures under and surrounding of the Agios Voukolos Church in İzmir, Turkey. *J. Archaeol. Sci.* **2011**, *38*, 2231–2242. [CrossRef]

19. Pérez-Gracia, V.; Caselles, J.O.; Clapes, J.; Osorio, R.; Martínez, G.; Canas, J.A. Integrated near-surface geophysical survey of the Cathedral of Mallorca. *J. Archaeol. Sci.* **2009**, *36*, 1289–1299. [CrossRef]

20. Tapete, D.; Casagli, N.; Luzi, G.; Fanti, R.; Gigli, G.; Leva, D. Integrating radar and laser-based remote sensing techniques for monitoring structural deformation of archaeological monuments. *J. Archaeol. Sci.* **2013**, *40*, 176–189. [CrossRef]

21. Moropoulou, A.; Labropoulos, K.C.; Delegou, E.T.; Karoglou, M.; Bakolas, A. Non-destructive techniques as a tool for the protection of built cultural heritage. *Constr. Build. Mater.* **2013**, *48*, 1222–1239. [CrossRef]

22. Faella, G.; Frunzio, G.; Guadagnuolo, M.; Donadio, A.; Ferri, L. The Church of the Nativity in Bethlehem: Non-destructive tests for the structural knowledge. *J. Cult. Herit.* **2012**, *13*, e27–e41. [CrossRef]

23. Lachowicz, J.; Rucka, M. 3-D finite-difference time-domain modelling of ground penetrating radar for identification of rebars in complex reinforced concrete structures. *Arch. Civ. Mech. Eng.* **2018**, *18*, 1228–1240. [CrossRef]

24. Drobiec, Ł.; Jasiński, R.; Mazur, W. Accuracy of eddy-current and radar methods used in reinforcement detection. *Materials* **2019**, *12*, 1168. [CrossRef] [PubMed]

25. Orlando, L.; Slob, E. Using multicomponent GPR to monitor cracks in a historical building. *J. Appl. Geophys.* **2009**, *67*, 327–334. [CrossRef]

26. Jiao, L.; Ye, Q.; Cao, X.; Huston, D.; Xia, T. Identifying concrete structure defects in GPR image. *Meas. J. Int. Meas. Confed.* **2020**, *160*, 107839. [CrossRef]

27. Luo, T.X.H.; Lai, W.W.L. GPR pattern recognition of shallow subsurface air voids. *Tunn. Undergr. Space Technol.* **2020**, *99*, 103355. [CrossRef]

28. Johnston, B.; Ruffell, A.; McKinley, J.; Warke, P. Detecting voids within a historical building façade: A comparative study of three high frequency GPR antenna. *J. Cult. Herit.* **2018**, *32*, 117–123. [CrossRef]

29. Rucka, M.; Lachowicz, J.; Zielińska, M. GPR investigation of the strengthening system of a historic masonry tower. *J. Appl. Geophys.* **2016**, *131*, 94–102. [CrossRef]

30. Lachowicz, J.; Rucka, M. Diagnostics of pillars in St. Mary's Church (Gdańsk, Poland) using the GPR method. *Int. J. Archit. Herit.* **2018**, *13*, 1–11. [CrossRef]

31. Fu, Q.; Yan, L.; Ning, T.; Wang, B.; Kasal, B. Interfacial bond behavior between wood chip concrete and engineered timber glued by various adhesives. *Constr. Build. Mater.* **2020**, *238*, 117743. [CrossRef]

32. Ryuzono, K.; Yashiro, S.; Nagai, H.; Toyama, N. Topology optimization-based damage identification using visualized ultrasonic wave propagation. *Materials* **2020**, *13*, 33. [CrossRef] [PubMed]

33. Mousavi, M.; Holloway, D.; Olivier, J.C.; Alavi, A.H.; Gandomi, A.H. A Shannon entropy approach for structural damage identification based on self-powered sensor data. *Eng. Struct.* **2019**, *200*, 109619. [CrossRef]

34. Rucka, M.; Wilde, K. Ultrasound monitoring for evaluation of damage in reinforced concrete. *Bull. Pol. Acad. Sci. Tech. Sci.* **2015**, *63*, 65–75. [CrossRef]

35. Schabowicz, K. Ultrasonic tomography—The latest nondestructive technique for testing concrete members—Description, test methodology, application example. *Arch. Civ. Mech. Eng.* **2014**, *14*, 295–303. [CrossRef]

36. Hola, J.; Sadowski, Ł.; Schabowicz, K. Nondestructive identification of delaminations in concrete floor toppings with acoustic methods. *Autom. Constr.* **2011**, *20*, 799–807. [CrossRef]

37. Cantero-Chinchilla, S.; Chiachío, J.; Chiachío, M.; Chronopoulos, D.; Jones, A. A robust Bayesian methodology for damage localization in plate-like structures using ultrasonic guided-waves. *Mech. Syst. Signal Process.* **2019**, *122*, 192–205. [CrossRef]

38. Cantero-Chinchilla, S.; Chiachío, J.; Chiachío, M.; Chronopoulos, D.; Jones, A. Optimal sensor configuration for ultrasonic guided-wave inspection based on value of information. *Mech. Syst. Signal Process.* **2020**, *135*, 106377. [CrossRef]

39. Zielińska, M.; Rucka, M. Non-Destructive Assessment of Masonry Pillars using Ultrasonic Tomography. *Materials* **2018**, *11*, 2543. [CrossRef]

40. Rucka, M.; Wojtczak, E.; Lachowicz, J. Damage imaging in Lamb wave-based inspection of adhesive joints. *Appl. Sci.* **2018**, *8*, 522. [CrossRef]

41. Guadagnuolo, M.; Faella, G.; Donadio, A.; Ferri, L. Integrated evaluation of the Church of S. Nicola di Mira: Conservation versus safety. *NDT E Int.* **2014**, *68*, 53–65. [CrossRef]

42. Binda, L.; Saisi, A.; Tiraboschi, C.; Valle, S.; Colla, C.; Forde, M. Application of sonic and radar tests on the piers and walls of the Cathedral of Noto. *Constr. Build. Mater.* **2003**, *17*, 613–627. [CrossRef]

43. Pérez-Gracia, V.; Caselles, J.O.; Clapés, J.; Martinez, G.; Osorio, R. Non-destructive analysis in cultural heritage buildings: Evaluating the Mallorca cathedral supporting structures. *NDT E Int.* **2013**, *59*, 40–47. [CrossRef]

44. Warren, C.; Giannopoulos, A.; Giannakis, I. GprMax: Open source software to simulate electromagnetic wave propagation for Ground Penetrating Radar. *Comput. Phys. Commun.* **2016**, *209*, 163–170. [CrossRef]

45. Sham, J.F.C.; Lai, W.W.-L. Development of a new algorithm for accurate estimation of GPR's wave propagation velocity by common-offset survey method. *NDT E Int.* **2016**, *83*, 104–113. [CrossRef]

Article

Detection and Imaging of Debonding in Adhesive Joints of Concrete Beams Strengthened with Steel Plates Using Guided Waves and Weighted Root Mean Square

Erwin Wojtczak *, Magdalena Rucka and Magdalena Knak

Department of Mechanics of Materials and Structures, Faculty of Civil and Environmental Engineering, Gdansk University of Technology, Narutowicza 11/12, 80-233 Gdansk, Poland; magdalena.rucka@pg.edu.pl (M.R.); s168197@student.pg.edu.pl (M.K.)
* Correspondence: erwin.wojtczak@pg.edu.pl; Tel.: +48-58-347-2497

Received: 10 April 2020; Accepted: 6 May 2020; Published: 8 May 2020

Abstract: Strengthening of engineering structures is an important issue, especially for elements subjected to variable loads. In the case of concrete beams or slabs, one of the most popular approaches assumes mounting an external reinforcement in the form of steel or composite elements by structural adhesives. A significant disadvantage of adhesive joints is the lack of access to the adhesive film for visual condition assessment, thus, there is a need for non-destructive diagnostics of these kinds of connections. The aim of this paper was the identification and visualization of defects in adhesive joints between concrete beams and steel plates using the guided wave propagation technique. The initial theoretical and numerical analyses were performed. The experimental wave field was excited and measured by the scanning laser Doppler vibrometry. The collected signals were processed by the weighted root mean square (WRMS) calculation. As a result, 2-D damage maps were obtained. The numerical simulations were performed to corroborate the experimental results. The results showed that the guided waves could be successfully applied in non-destructive diagnostics of adhesive joints between concrete and steel elements. However, the quality of damage visualizations strongly depended on the location of excitation.

Keywords: adhesive joint; concrete beam; guided waves; debonding; damage detection; damage imaging; root mean square

---

1. Introduction

Engineering structures are subjected to various loads, e.g., dead loads, live loads, environmental loads, and other, over the entire service life. In many situations, it may be necessary to increase the load capacity. The main reasons are planned changes in the way a structure is used and carrying out rehabilitation and retrofitting works due to degradation of a structure or its natural aging. To achieve higher load capacity, strengthening is commonly used. Various reinforcement systems such as steel elements (plates, rods, and flat bars) or composite elements (carbon tapes, carbon mats, and other fiber reinforcement polymers) are increasingly used [1–3]. Such externally bonded reinforcement (EBR) is very often connected to a strengthened structure by adhesive bonding [4]. It is worth noticing, that the technical condition of the connection between joined elements has a great influence on the behavior of the whole structure and, if not properly evaluated, may cause an unexpected collapse.

The existing literature presents many examples of the use of adhesives in connections between concrete and steel [4–8] as well as between concrete and fiber reinforced polymeric (FRP) composites [9–17]. Czaderski and Meier [4] proved that the condition of concrete elements strengthened using the EBR technique can be good even after several dozen years of exploitation. However, the monitoring of such structures is crucial because of the possibility of debonding between the substrate and overlay, which is the most common damage type in this type of connections. It should be also noted that debonding is a dangerous defect due to the lack of its visibility, thus the problem of detection of such damage is not a trivial task. The strength of elements reinforced with EBR technique is usually considered as the bond strength between joined elements [15], thus the condition of the adhesive layer is of great importance. Debonding in externally reinforced concrete elements, which is initiated by crack grows, is the aim of the work of many researchers. The reinforcement in the form of steel plates was considered as the protection of concrete beams against failure induced by flexural [6] or shear cracking [5,7,8]. Typical peel-off tests were conducted for concrete beams strengthened with glass fiber reinforced polyester to determine their strength [9]. The influence of carbon fiber reinforced polymers (CFRP) [10], steel reinforced polymers (SRP), and steel reinforced grouts (SRG) [12] on the failure induced by flexural cracking in concrete beams during bending load was also considered. Pre-stressing of CFRP laminates was shown to be the effective method of reduction of debonding in strengthened concrete beams under flexural load [13]. A number of works were conducted based on the shear tests for concrete beams reinforced with SRG [16] and CFRP [17]. The influence of additional CFRP anchorages on the shear strength was also analyzed [14]. The combined influence of normal and shear stresses was investigated for U-shaped CFRP reinforcement covering the bottom and the sides of the beams [11,15].

With the increasing use of adhesive connections to reinforced structures, there is a need to develop efficient techniques of non-destructive detection and imaging of debonding between a substrate and overlay, as for example active thermography [18–22]. A group of the most effective techniques to evaluate the interfacial adhesion consists of the methods utilizing elastic wave propagation. Existing papers have presented numerous successful applications of wave propagation phenomenon for damage identification in civil engineering structures [23–27]. The impact-echo (IE) method, based on the use of low-frequency vibrations, supported with the frequency spectrum analysis and wavelet analysis, was efficiently utilized for the assessment of adhesion in concrete structures [28,29]. Another promising approach, applicable also for the evaluation of adhesive joints, is based on the analysis of wave energy distribution supported with weighted root mean square (WRMS) calculation, e.g., in aluminum [30] and steel connections [31,32]. Recently, ultrasonic waves have also been increasingly used to inspect and evaluate adhesive bonding between concrete elements and externally bonded reinforcement. Castaings et al. [33] analyzed the propagation of a pseudo-Rayleigh wave mode to evaluate the adhesion level between concrete block and glass-epoxy composite plate joined by the epoxy resin, based on the numerical dispersion calculations and experimental measurements. Shen et al. [34] investigated a concrete block strengthened with steel plate joined by epoxy adhesive. Dispersion properties for a three-layer medium were determined by analytical, experimental, and numerical approaches, giving comparable results. Propagation of shear waves was used by Zeng et al. [35] to detect bond slip at the steel–concrete interface in concrete encased I-shape beam under tensile test. Song and Popovics [36] studied the influence of different bonding conditions on dispersion characteristics of fundamental A_0 and S_0 wave modes propagating in the steel-clad concrete beams. They observed that attenuation characteristics of guided waves are sensitive to adhesion level at the steel–concrete interface. Li et al. [37] applied the time-reversal method and continuous wavelet transform to evaluate debonding in the CFRP-reinforced concrete blocks. The analysis of strain distribution of wave propagation in numerical models was used as an attempt to imaging of damaged areas; however the exact size and position of defects were not determined. Zima and Rucka [38] presented experimental and numerical results of condition assessment of adhesive joint between concrete beam and steel plate with various states of degradation. The correlation between the amplitude of peak values in wave signals and bonding quality was observed. The numerical guided wave field represented by acceleration magnitudes was used to visualize debonding areas. Mechanical degradation of adhesive connection in concrete–steel specimens during push-out tests was monitored ultrasonically by Rucka [39]. Changes in the interface slip between adherends were treated as an indicator of approaching failure. Chen et al. [40] proposed detecting interfacial debonding in concrete-filled steel tubes using the transient multichannel analysis of surface waves. The differences in dispersion characteristics allowed determining the presence of defects. Liu et al. [41] monitored debonding between CFRP sheet and concrete beam during three-point bending using the pitch-catch wave-based method. The decrease of signal amplitude was observed in the case of damaged connection compared with an intact one. Experimental strain distribution at different loading levels was used to illustrate the development of damaged areas. Interfacial debonding in steel–concrete and steel-epoxy-concrete plates was quantitatively evaluated by Ke et al. [42]. Comparison of signal spectrograms obtained by the short time Fourier transformation of signals obtained allowed determining the debonding range between joined parts. Yan et al. [43] investigated steel–concrete–steel sandwich beams under bending test. Acoustic emission and electromechanical impedance methods were used to evaluate the cracking process and the structural state of analyzed specimens. Giri et al. [44] analyzed the prediction of a gap in concrete–steel and concrete–aluminum connections using partial least squared regression technique based on the Lamb wave propagation phenomenon. Nonlinear Rayleigh waves were used by Ng et al. [45] for debonding detection in CFRP-retrofitted reinforced concrete blocks. Short time Fourier transformation of wave signals allowed preparing location images of defects with different size. Wang et al. [46] detected the severity and type of damage in concrete columns with CFRP sheeting using Fourier transform and wavelet packet energy analysis. Damage and health indices were used to evaluate concrete cracking and debonding initiation. Huo et al. [47] made a review on monitoring of bond-slip in steel–concrete structures presenting well-known wave-based approaches, including active and passive sensing methods. The reported works present an extensive analysis of debonding detection in steel–concrete composite structures. However, the issue of damage imaging was not deeply considered, except some above-mentioned attempts based on the analysis of the guided wave field represented by the wave acceleration or stress distribution. It should be noted, that the determination of actual location, size, and shape of damaged areas has not been performed, notwithstanding these features have a great impact on the actual strength of adhesive connections.

This paper deals with the condition assessment of concrete beams strengthened with steel plates. The novelty of the study is connected with comprehensive theoretical-numerical-experimental analysis of wave propagation in steel–concrete adhesively bonded specimens and the possibility of debonding imaging using WRMS-based ultrasonic diagnostics. The conducted research was divided into two stages. The initial analysis focused on the characterization of the guided wave propagation in a single-layer and a three-layer media. The main part of the study was directed at the damage identification and imaging in adhesive joints of composite beams with different levels of debonding. The guided waves were sensed and recorded in a number of points on the surface of the steel plate using both contact and non-contact measurement techniques. The signal processing technique based on the WRMS calculation allowed detecting and imaging of the damaged areas. The obtained results indicated that the quality of damage maps strongly depended on the location of excitation.

2. Materials and Methods

2.1. Object of Research

The object of investigations (Figure 1a) was a beam with a square cross section of 100 mm × 100 mm and a length of 500 mm. The beam was made of concrete class C30/37 and strengthened with a steel plate of dimensions 100 mm × 6 mm × 500 mm fixed to the upper surface of the beam by an adhesive film. Five specimens were analyzed (Figure 1b). The first specimen was an intact composite beam (#1) with a full connection between both elements. Three partially damaged beams (#2, #3, and #4) had defects in the form of a debonding, covering a different area of the joint (relative defect surface was 10%, 20%, and 50%, respectively). The last sample was a steel plate (#5), which was laid freely on the beam to simulate a fully debonded joint. The defects were obtained by sticking a Teflon (PTFE) tape onto the appropriate area of the steel plate. To prevent any unintentional defects, the treatment of each bonded elements was performed. The surface of the concrete beam was dedusted and degreased with Loctite-7063 cleaner (Henkel, Düsseldorf, Germany). The steel plate was treated with a fine abrasive paper (grit size 120) and then degreased. The connection between concrete and steel was provided by the two-component epoxy-based structural adhesive Sikadur 30 Normal (Sika, Baar, Switzerland). The specimens were conditioned in room temperature for seven days after gluing. The photograph of prepared specimens with close-ups for selected defects is presented in Figure 2. The thickness of the adhesive film for specimens (#1–#4) was equal to about 2 mm. It was determined after conditioning as the difference between the overall specimen thickness and the sum of beam and plate thicknesses.

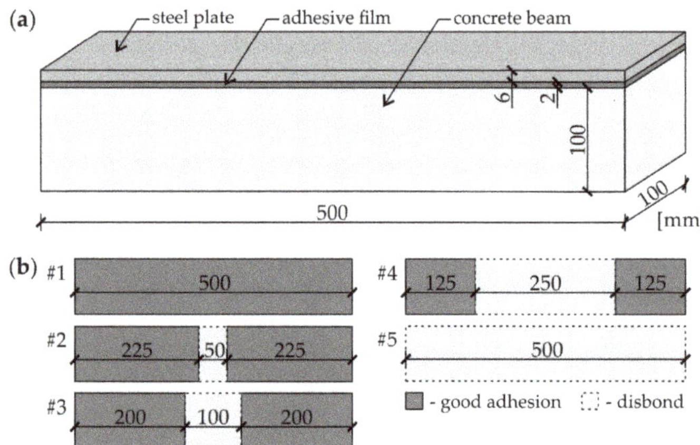

Figure 1. Analyzed specimens: (**a**) geometry of beams and (**b**) variants of defects.

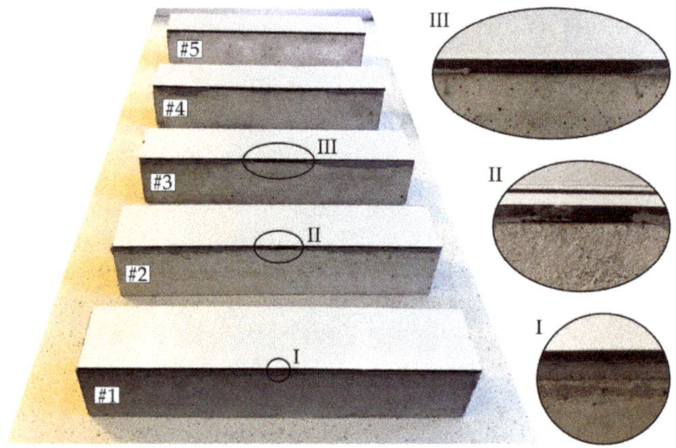

Figure 2. Photograph of prepared specimens (#1–#5) with close-up views of adhesive films in beams #1, #2, and #3.

2.2. Experimental Procedure

The experimental measurements of guided waves were carried out by the scanning laser Doppler vibrometry (SLDV) method using the set-up presented in Figure 3a. The input wave signal was generated by the arbitrary function generator AFG 3022C (Tektronix, Inc., Beaverton, OR, USA) and magnified with the use of the high-voltage amplifier PPA 2000 (EC Electronics, Krakow, Poland). The excitation of the guided wave field was provided by the plate piezoelectric actuators NAC2024 and NAC2025 (Noliac, Kvistgaard, Denmark) bonded to the surface of each examined specimen by the petro wax 080A109 (PCB Piezotronics, Inc., Depew, NY, USA). The signals of propagating waves (out-of-plane velocity components) were collected by the non-contact method using the scanning head of the laser vibrometer PSV-3D-400-M (Polytec GmbH, Berlin, Germany) equipped with the VD-07 velocity decoder. The sampling frequency for each measurement was set to 2.56 MHz. The area of scanning was covered with a retro-reflective sheeting to improve the light backscatter.

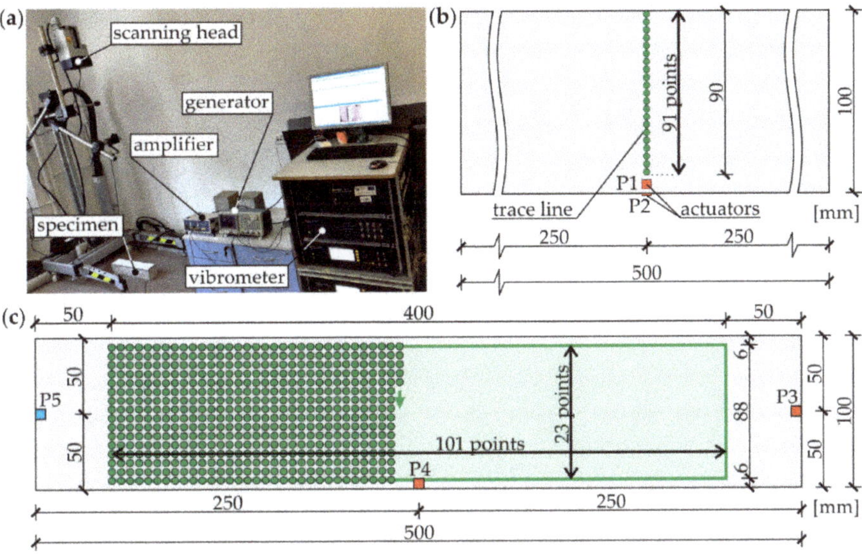

Figure 3. Experimental investigations: (**a**) setup for scanning laser Doppler vibrometry (SLDV) measurements; (**b**) scheme of measurements for dispersion curves determination; and (**c**) scheme for damage identification and visualization measurements.

The samples #1 and #5 were firstly tested to determine dispersion curves in a single-layer and three-layer medium, according to the scheme presented in Figure 3b. For this purpose, the propagation of antisymmetric and symmetric modes was induced by actuators P1 and P2, respectively, both attached to the steel plate. The input signal had the form of a wave packet obtained by the Hanning window modulation of the single-cycle sine function. To obtain a wide frequency spectrum for both excitation types, measurements were conducted five times with the different carrier frequency, equal to 50, 100, 150, 200, and 250 kHz, consecutively. The time-domain wave signals were acquired in 91 points spaced evenly over a straight line with a length of 90 mm, resulting with a resolution equal to 1 mm. The influence of potential reflections of propagating waves at the boundaries, induced by the short path length in relation with beam thickness, was further eliminated by analyzing only the initial part of the collected signals. The main investigations aimed at damage imaging in all beams #1–#5 were conducted with respect to the scheme shown in Figure 3c. Each specimen was tested twice, changing the location of the excitation point (P3 and P4). Excitation signal was a five-peak wave packet obtained from a sinusoidal burst with a central frequency equal to 100 kHz. The signals of propagating waves were collected in 2323 points distributed on the top surface of the steel plate, covering the area of 440 × 88 mm^2. The scanning was conducted point by point in the regular mesh of 23 rows and 101 columns with a resolution of 4 mm in both directions.

Additional measurements with a contact method were performed to identify the damaged specimens. The input signal was identical to the one used for damage imaging with SLDV (five-cycle sine wave packet with a central frequency of 100 kHz). The piezoelectric actuators were attached at the ends of each beam to excite and acquire wave signals in points P3 and P5, respectively (Figure 3c).

The excitation signal was created by the arbitrary waveform generator AFG 3022C (Tektronix, Inc., Beaverton, OR, USA) and then amplified by the high-voltage amplifier A400DI (FLC Electronics AB, Partille, Sweden). The collection of signals was provided by the digital oscilloscope PicoScope 4824 (Pico Technology, St Neots, Great Britain) with a sampling frequency of 20 MHz.

2.3. Numerical Modeling

The numerical simulations of the guided wave propagation in the considered specimens were conducted in Abaqus/Explicit software (ver 6.14, Dassault Systemes, Vélizy-Villacoublay, France) applying the finite element method. All structural elements (steel plate, concrete beam and adhesive film) were assumed to be independent parts bonded rigidly by surface to surface tie connection, ensuring compatibility of translational degrees of freedom at contacting nodes. The materials of modelled parts were supposed to meet the requirements of a homogeneous, isotropic and linearly elastic material model. The mechanical properties of each material are presented in Table 1. The transient dynamic analysis was conducted with the use of the central difference method. Total time of calculations was assumed as 2 ms for each simulation. The propagation of wave was analyzed by considering the behavior of models under the concentrated load with varying amplitude, applied in a specific node of a discretized structure. The results of the analyses were out-of-plane (vertical) velocity signals collected in a certain number of nodes.

Table 1. Material parameters for each material in numerical simulations.

Material	Density ρ (kg/m^3)	Elastic Modulus E (GPa)	Poisson's Ratio ν (–)
concrete	2364.4	49.5	0.12
steel	7822.8	200.3	0.30
adhesive	1611.8	14.9	0.30

Two kinds of models were prepared for different purposes, plane (2-D) and spatial (3-D). Two-dimensional models (example in Figure 4a) were prepared for specimens #1 and #5 to analyze the differences between guided wave fields in a single-layer plate and a three-layer medium. Two-dimensional four-node plane strain elements with reduced integration (CPE4R) were used. The element size was constant and equal to 1×1 mm^2 throughout the whole model. This value satisfies the requirement of the appropriate mapping of wave behavior (at least 10 nodes for the length of the shortest considered wave) presented in [48]. The calculations were conducted with a fixed time step with a value of 10^{-7} s that meets the recommendation of at least 20 integration points per cycle of the wave with the highest frequency of interest [49]. For initial analysis, the wave was excited at one end of the plate (point R1) and its signal was collected at another end (point R5) to determine the time-of-flight (TOF) of the wave through the length of the specimen. In the same simulation, guided wave field was saved for the whole model. The excitation had the form of a concentrated force varying in time in accordance with a 5-cycle tone burst with a center frequency of 10, 23, and 100 kHz (three different simulations). The plane models were also used to determine the numerical dispersion curves, similarly as in the experimental investigations (cf. Figure 3b). The input signal was applied in points R1 and R2 (independently) as a wave packet modulated from a single-cycle sine function with a different frequency (50, 100, 150, 200, and 250 kHz), resulting with ten consecutive simulations. The guided wave responses were obtained at 101 points distributed along the line with a length of 100 mm, giving 1 mm spacing between each point. It is worth noting that the measurement path was oriented perpendicularly to the one in experimental investigations. However, the results of both approaches could be compared, as long the materials used were modeled as isotropic and only the initial parts of signals were further analyzed. Three-dimensional models (example in Figure 4b) were prepared for specimens #1–#5 to verify the results of experimental measurements. Eight-node linear brick elements with reduced integration (C3D8R) and a global size of 2 mm were applied. The integration step was 2×10^{-7} s. Both element size and time step met the requirements from [48,49], described above. The scheme of excitation and collection of signals was identical with the experimental measurements (cf. Figure 3c). The wave (Hanning windowed five-cycle sine function with a carrier frequency of 100 kHz) was excited in point R3 or R4 and collected in 2323 points covering the area of 88×400 mm^2.

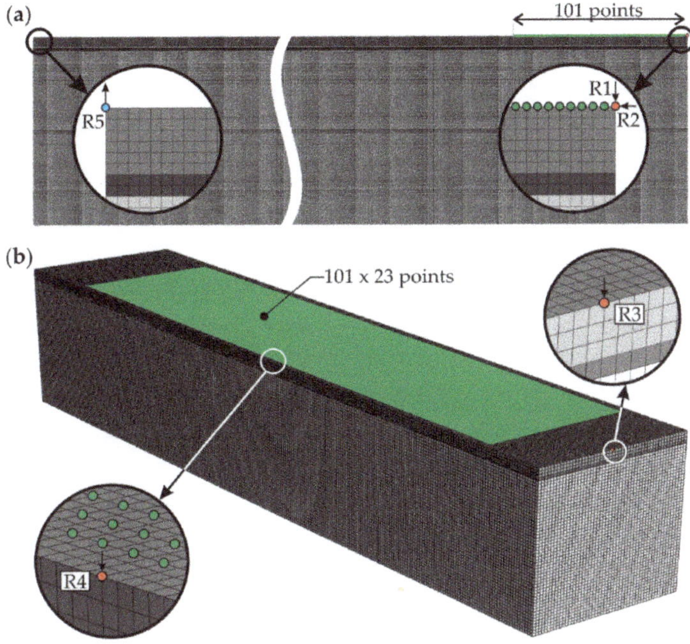

Figure 4. Numerical models for specimen #1: (**a**) 2-D plane model and (**b**) 3-D spatial model.

2.4. Signal Processing with Weighted Root Mean Square

The direct analysis of the excited guided wave field usually does not give useful information about the shape, position and size of defects occurring in an examined structure. The utility of the results can be enhanced by further signal processing. The simple and widely used approach is based on the calculation of the effective value, also called the root mean square (RMS). For damage identification and imaging, it is advantageous to use the weighted variant of RMS (WRMS) that decreases the impact of the incident wave on the results obtained. The WRMS for a discrete time domain signal $s_r = s(t_r)$ consisted of N values sampled with a constant Δt interval can be calculated with respect to the formula [31,50–53]:

$$ WRMS = \sqrt{\frac{1}{N} \sum_{r=1}^{N} w_r s_r^2} \tag{1} $$

in which w_r is a so-called weighting factor, defined as the function of the number of consecutive sample r as:

$$ w_r = r^m, \ m \geq 0, \ r = 1, 2, \ldots, N \tag{2} $$

where m is a non-negative power of the weighting factor. For $m = 0$ the above formulas describe the simple root mean square (without weighting). According to the literature cited above, the value of m was usually set arbitrary, however, it can be concluded that efficient results can be obtained for the higher values of m, e.g., $m = 1$ (linear weighting factor) or $m = 2$ (square variant). Further increasing of power m does not enhance the results in a significant manner.

2.5. Dispersion Curve Estimation with Matrix Pencil Method

The wavenumber–frequency relations can be calculated based on a set of wave propagation signals collected in a number of points spaced evenly along a measurement line. One of the popular methods is the use of the two-dimensional fast Fourier transform (2D-FFT [48,54,55]). However, the curves obtained are in the form of 2D maps which can be problematic when comparing different results. To solve this problem, a simple and robust algorithm called the Matrix Pencil (MP) technique [56–58] can be used. The method is based on the calculation of the one-dimensional fast Fourier transform (FFT). Its advantage is a good elimination of noise effects on the obtained results. Let us have the set of time-domain signals measured in the m points distributed along z axis with the interval Δz. The Fourier coefficients $X(z,\omega)$ for each signal $x(z,t)$ can be calculated with respect to the formula:

$$X(z,\omega) \;=\; \int\limits_{-\infty}^{\infty} x(z,t)e^{-i\omega t}dt \tag{3}$$

where ω is a circular frequency. Then, the wavenumbers are estimated using the forward/backward averaging technique. The sequence x_n of m complex values at each fixed frequency ω_0 is assumed to be the sum of p complex exponentials representing wavenumbers, as follows:

$$x_n \;=\; x_n(\omega_0) \;=\; X(z_n,\omega_0) \;=\; \sum_{j=1}^{p} a_j e^{-ik_j z}, \qquad n \;=\; 1,2,\ldots,m. \tag{4}$$

In the above formula, p is the model order, so-called pencil parameter that should be chosen based on the unknown number of searched signal modes q and satisfy the limitation:

$$q \leq p \leq m - q. \tag{5}$$

The algorithm requires the construction of a parent Hankel matrix $\mathbf{H}$ ($m - p$ by $p + 1$) from the sequence x_n as presented below:

$$\mathbf{H} \;=\; \begin{bmatrix} x_1 & x_2 & \cdots & x_{p+1} \\ x_2 & x_3 & \cdots & x_{p+2} \\ \vdots & \vdots & \ddots & \vdots \\ x_{m-p} & x_{m-p+1} & \cdots & x_m \end{bmatrix}. \tag{6}$$

Two submatrices of $\mathbf{H}$ need to be created, the $\mathbf{H}_0$ formed by deleting the first column of $\mathbf{H}$ and $\mathbf{H}_1$ formed by deleting the last column of $\mathbf{H}$. The sets of forward λ_f and backward λ_b exponential estimates can be calculated by solving the two eigenvalue problems, respectively:

$$\left(\mathbf{H}_0^+ \mathbf{H}_1 - \lambda_f \mathbf{I}\right)e \;=\; 0, \quad \left(\mathbf{H}_1^+ \mathbf{H}_0 - \lambda_b \mathbf{I}\right)e \;=\; 0 \tag{7}$$

where $(.)^+$ indicates the Moore-Penrose generalized inverse and $\mathbf{I}$ is the identity matrix. The sets of forward k_f and backward k_b complex wavenumbers are computed as follows:

$$k_f \;=\; \frac{\ln \lambda_f}{i\Delta z}, \quad k_b \;=\; -\frac{\ln \lambda_b}{i\Delta z}. \tag{8}$$

The final wavenumber k values can be calculated by the averaging technique presented in [56]. Firstly, the sets k_f and k_b are sorted and matched in pairs in the order of sorting. If the relative difference between two values of a certain pair is within a specified tolerance l, the pair is averaged arithmetically, if not, the pair is discarded.

3. Results and Discussion

3.1. Wave Propagation in Single-Layer and Multi-Layer Media

The first part of this section aimed at comparing the characteristics of guided waves propagating in a single layer plate (simulating the steel plate #5) and a two-layer plate (steel-adhesive) on a concrete half-space (three-layer medium modeling steel-adhesive-concrete composite beam #1). The materials for all the layers were assumed to be homogeneous, isotropic and linearly elastic (according to the material parameters presented in Table 1). Dispersion curves were calculated using a theoretical, numerical and experimental approach. The theoretical dispersion curves for the single-layer plate #5 were obtained in the way of solving well-known Rayleigh–Lamb equations in Matlab®. In the case of the three-layer medium, the theoretical curves were calculated using an open-source toolbox called Elastic-Matrix [59], based on the partial-wave method. Although the proposed algorithm is stated to be inaccurate for leaky cases, the appropriate solution can be obtained by selecting appropriate calculation parameters. The experimental and finite element method (FEM) numerical results were attained from the collected time-domain wave signals (91 and 101 signals, respectively) using the Matrix Pencil Method [56–58]. The pencil parameter and tolerance were chosen heuristically and equal to $p = 16$, $l = 1.4\%$ for the experimental analysis and $p = 13$, $l = 2.2\%$ for the numerical one. The determination of optimal values of these parameters requires further work; however, this aspect is not essential for the usefulness of results obtained in this research. The value of p was greater in the experimental calculation because the signal noise required computing more estimates, however, the tolerance needed to be more rigorous to exclude false results. The quality of results was enhanced by the authorial script deleting 'lonely' points. The complete wavenumber–frequency spectra were obtained by superimposing the dispersion curves calculated for the individual measurements with different location of the actuator and the excitation frequency.

The dispersion curves obtained from the three approaches are presented in Figure 5. The wavenumber–frequency relations for specimen #1 (Figure 5a) reveal the fact, that two flexural (A_0, A_1) and one longitudinal (S_0) modes can propagate in the frequency range up to 400 kHz. It is essential to note, that the fundamental S_0 mode does not appear in the range up to 122 kHz and the A_1 mode is present for the frequencies above 260 kHz, thus only one mode can propagate for the assumed excitation frequency equal to 100 kHz. The comparison of the three approaches allowed concluding that the overall agreement between the numerical, experimental and theoretical results was very good. However, the higher mode A_1 was not identified in the experimental approach, probably due to the setup limitations. It is also important to note that neither experimental nor numerical calculations detected A_0 in the range up to 60 kHz, what could have been caused by the strict tolerance level. However, the increase of l would significantly decrease the legibility of the curves obtained. The analysis of dispersion curves for specimen #5 (Figure 5b) leads to similar conclusions. Compared with the results for beam #1, the S_0 mode can propagate in the whole assumed frequency range, such as A_0. The dispersion relations for both specimens for the frequencies above 150 kHz are comparable. The important differences are observable in the lower frequency range. To expose these dissimilarities, the theoretical dispersion curves for group velocity were prepared for both #1 and #5 specimens (Figure 5c). No significant difference can be seen for the A_1 curve. The shapes of A_0 and S_0 curves are also very similar for the higher frequency range. As mentioned above, S_0 mode is not present in the lower frequency range. To be noticed, it is important to analyze the shape of the A_0 mode, because for the damage imaging, measurements the waves were excited perpendicularly to the plate surface and the out-of-plane velocity signals were collected. It is interesting that in the frequency range up to about 23 kHz it propagates with higher velocity in the specimen #1, whereas above this limit, the velocity of propagation is higher in specimen #5.

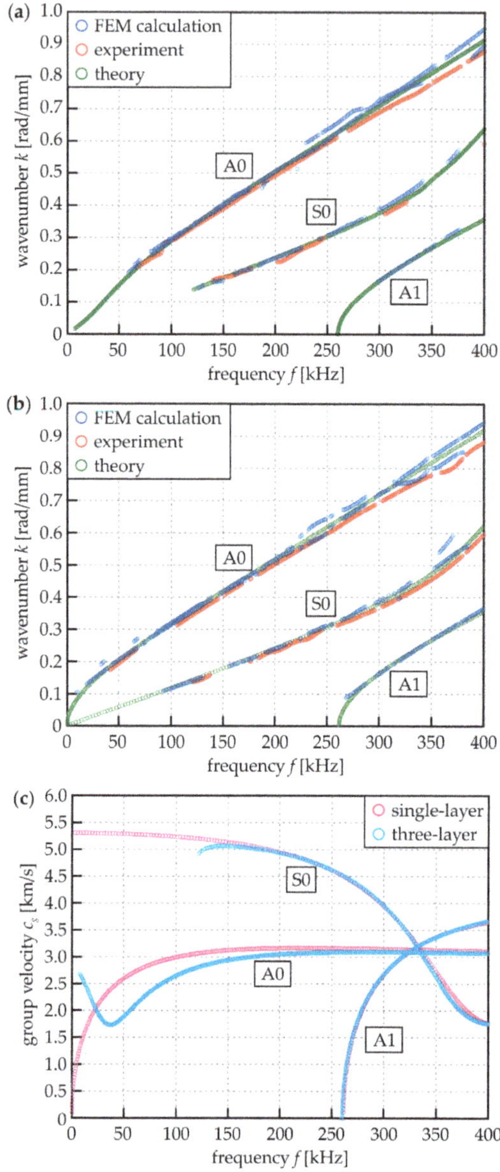

Figure 5. Dispersion curves for analyzed specimens: (**a**) wavenumber vs. frequency for three-layer plate #1; (**b**) wavenumber vs. frequency for single-layer steel plate #5; and (**c**) theoretical group velocity vs. frequency for three-layer plate #1 and single-layer plate #5.

The simple FEM numerical simulations were performed to check whether this dispersion relationship is actually observable. Figure 6 shows the time signals of propagating waves in specimens #1 and #5 obtained from the numerical models in point R5 as the results of R1 excitation (cf. Figure 4a). It is clearly visible, (especially in close-ups) that the wave packet related to the A_0 mode appears earlier in the signal from specimen #1 for the excitation frequency equal to 10 kHz (Figure 6a). If the excitation frequency is equal to 23 kHz, no difference in time of flight (TOF) of the first wave packet is observable (Figure 6b). The further increasing of the frequency value leads to the situation when the wave propagates faster in specimen #5 (Figure 6c). The TOF values for both models were calculated based on the dispersion curves (theoretical TOF) and are presented in Table 2, together with the values got from signals recorded. The relative differences between theoretical and numerical TOFs do not exceed a few percent, which indicates on the good compliance between dispersion curves and FEM simulations.

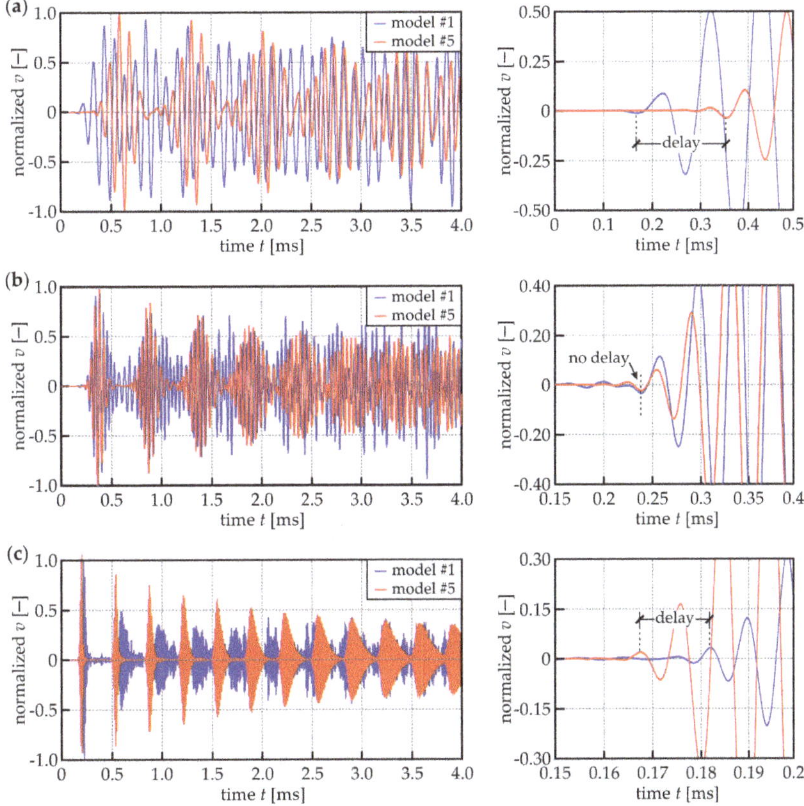

Figure 6. Normalized velocity signals (whole signals and zoom) registered at point R5 in numerical models of specimens #1 and #5 for different excitation frequencies: (**a**) 10 kHz; (**b**) 23 kHz; and (**c**) 100 kHz.

Table 2. Time of flight of the wave packet based on the dispersion curves and the wave signals.

Frequency f (kHz)	Theoretical TOF (#1) (µs)	Numerical TOF (#1) (µs)	Theoretical TOF (#5) (µs)	Numerical TOF (#5) (µs)
10	193	177	344	358
23	246	258	247	255
100	189	183	167	168

In addition to the analysis of the time signals, the guided wave fields representing the propagation of A_0 mode for specimens #1 and #5 for a specific time instance 0.17 ms are shown in Figure 7. It is visible that the incident wave with the frequency of 10 kHz propagates faster in beam #1 compared with plate #5 (Figure 7a), what stays in the agreement with the comparison of dispersion curves. What is also important, the wave propagates through the three-layer medium as a single waveguide, although the wave was excited only in the steel plate. It is possible because the A_0 wavelength is large enough (271 mm), compared to the overall sample thickness (108 mm). The length of the A_0 mode in beam #1 is significantly smaller (74 mm), but much greater than the thickness of the plate (6 mm). For the frequency equal to 23 kHz (Figure 7b) the wave packets in both specimens propagate with similar velocity. The distribution of the wave field trough the thickness of specimen #1 is not uniform, the amplitude decreases with the depth because the wavelength (107 mm) is slightly smaller than the specimen thickness. The interesting situation takes place when the excitation frequency increases to 100 kHz (Figure 7c). The wave propagation velocity in specimen #1 is smaller compared with specimen #5 (what is confirmed by the dispersion curves), but the wave propagates mainly in the steel plate and adhesive film. The reason is that the wavelength is now much smaller (22 mm) than the overall specimen thickness. For the concrete part, only the Rayleigh waves can be observed propagating at the interface between concrete and adhesive and the bottom boundary. The length of the wave propagating in plate #5 is comparable with the one present in the beam #1 and equal to 20 mm, which is also clearly visible in the snapshot. The analysis of the FEM simulations illustrating in Figure 7 allows concluding that the differences in the dispersion curves for both, three-layer and single-layer media are less pronounced for the higher frequency range (above about 120 kHz), because in this case, the wave propagates mainly in the steel plate, passing by the influence of the concrete part. Thus, the damage visualization in analyzed specimens would not be successful in this case. On the other hand, lower frequencies result in poorer image resolution. As the compromise, the excitation frequency equal to 100 kHz was chosen for further investigations aimed at damage detection and imaging.

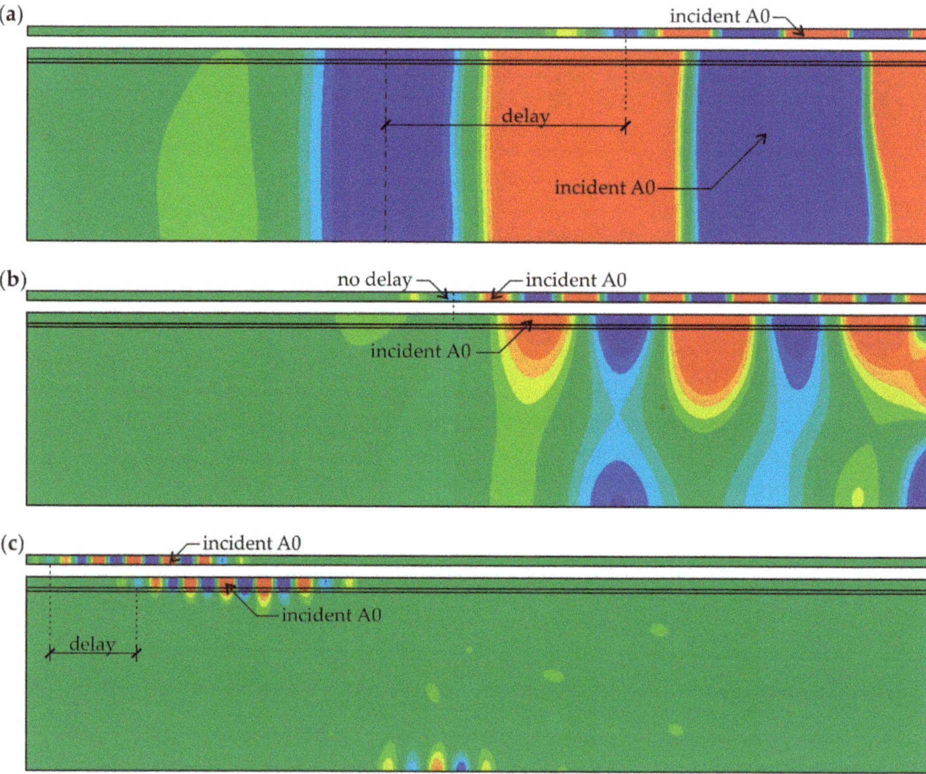

Figure 7. Numerical wave fields in 2-D finite element method (FEM) models of specimens #1 and #5 ($t = 0.17$ ms) for different excitation frequencies: (**a**) 10 kHz; (**b**) 23 kHz; and (**c**) 100 kHz.

3.2. WRMS-Based Damage Identification and Imaging

The non-destructive diagnostics of specimens #1–#5 was conducted in two stages: identification and imaging. The aim of the damage identification was to detect, whether there are any defects in the adhesive joint between the concrete beam and the steel plate. The root mean squares values were calculated for the time signals excited by the actuator located in point P5 and measured by the sensor located in point P3 (see Figure 3c). This means that the wave propagated through the whole length of the specimen. The calculations were performed for the time of averaging assumed as $t_a = 1$ ms, with a linear weighting factor ($m = 1$, $w_r = r$). The results for all specimens are presented in Figure 8a. It is clearly visible that the WRMS value for specimen #5 (single-layer steel plate) is significantly higher than in others. In the case of specimen #5, the point of excitation was located on the free steel plate, characterized by the relatively small material damping. For specimens #1–#4, the wave was excited on the area where the steel plate had a good connection with the adhesive layer and concrete part (i.e., materials with higher damping) what resulted in the instantaneous leak and dissipation of energy in the bottom part of the specimen. For this reason, a close-up of the chart for the specimens #1–#4 is shown in Figure 8b. The correlation between WRMS value and the size of the defect is clearly visible, with the greater the damaged area, the higher WRMS value. This relationship occurs because the wave energy is magnified by the multiple reflections of propagating waves at the boundaries of defects. It is also worth noticing that the WRMS increased significantly between intact #1 and damaged #2 specimen (71%) but the further growth of defect surface did not result in the significant growth of WRMS. The damaged area of specimen #4 was 2.5 times bigger than in specimen #3 but the growths in comparison with #1 were equal to 116% and 129%, respectively. Concluding, the comparative analysis of WRMS values gave the answer about the presence of the defect, however, it did not provide the information about the size and location of damaged areas.

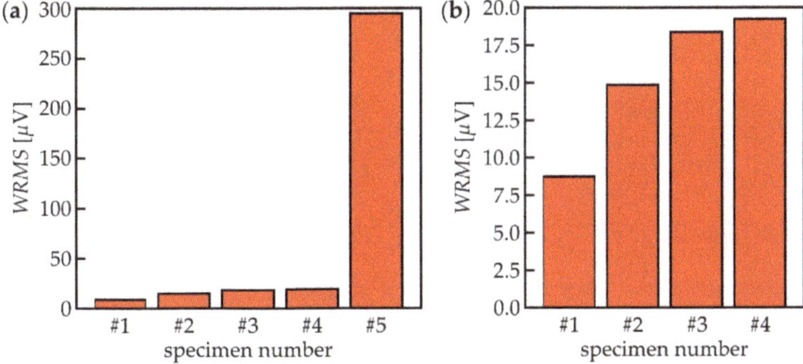

Figure 8. Weighted root mean square (WRMS) values of experimental signals collected in P5 point: (**a**) specimens #1–#5 and (**b**) specimens #1–#4.

The second stage, i.e., damage imaging, was performed to identify the exact position and size of the damaged area of each specimen. The WRMS map in the form of the two-dimensional defect representation over the scanned area was prepared based on the experimental investigations and numerical simulations on 3-D FEM models (see Figures 3c and 4b). The parameters of the calculations were assumed to be the same as for damage identification stage ($t_a = 1$ ms, $w_r = r$). The results for specimens #1–#5 are presented below. Figure 9 shows the WRMS maps for the intact composite beam (specimen #1). Regardless of the location of the excitation point, the visualizations reveal that no internal defects were detected inside the analyzed adhesive joint. The values in each experimental and numerical map are concentrated over a certain value. The horizontal lines with magnified values in images for excitation P3 (Figure 9a), such as the vertical lines for excitation P4 (Figure 9b) are the effects of the specimen and the load symmetry. Some interesting conclusions can be derived from the analysis of the results for specimen #2 with a partial debonding (Figure 10). It is clearly visible that the map obtained for the excitation applied in point P3 (Figure 10a) results in the moderate defect imaging when compared with the excitation in point P4 (Figure 10b). The reason is that the wave excited in point P3 (in the area of good adhesion) strongly leaks into concrete, so the wave energy is damped significantly before meeting the boundaries of the defect, thus it has no possibility to be clearly visualized. It was already possible to conclude in the stage of damage identification (the high WRMS value for specimen #5 indicated on this effect, see Figure 8a). However, the damaged area is somehow detectable and it is characterized by slightly smaller WRMS values than the remaining part of the sample. The energy of the wave excited by the actuator P4 is concentrated in the damaged area and the damage map has a high quality. The properly boded areas have much lower WRMS values because the waves transmitted to the three-layer medium are highly damped.

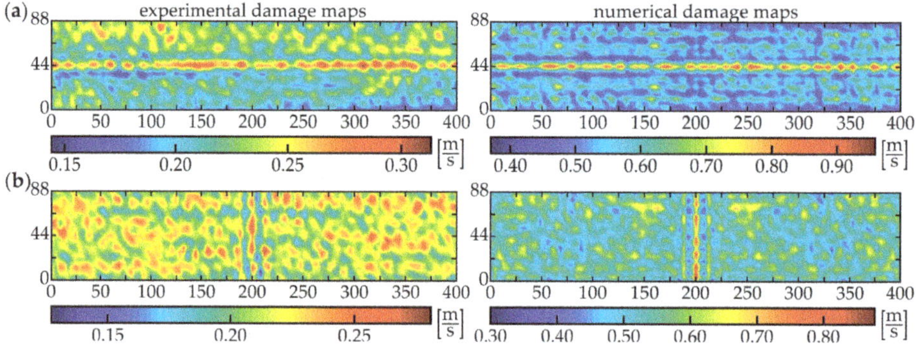

Figure 9. Experimental and numerical WRMS damage maps for specimen #1: (**a**) Excitation P3 and (**b**) Excitation P4.

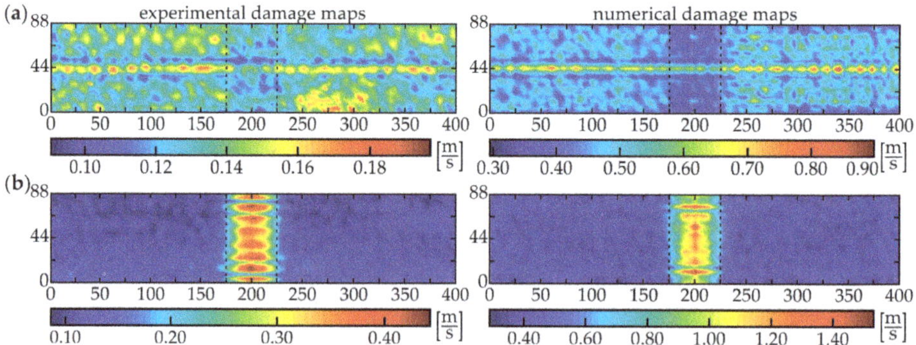

Figure 10. Experimental and numerical WRMS damage maps for specimen #2: (**a**) excitation P3 and (**b**) excitation P4.

A good agreement of experimental and numerical results is observed, however, the quality of the numerical visualization for the excitation in point R3 is richer compared with the experimental one. This may be caused by the noise present in the experimental signals. Similar conclusions can be derived from the analysis of damage maps of scanning for the two other partially damaged beams: #3 (Figure 11) and #4 (Figure 12). The WRMS maps for the fully debonded joint #5 (Figure 13) are similar to the ones obtained for the intact composite beam #1 (cf. Figure 9). This convergence allows stating that the damage visualization is possible only when the area of scanning covers both areas of good adhesion and debonding. The only difference between both specimens is that the value of WRMS is significantly higher for the steel plate (see color bars in Figures 9 and 13).

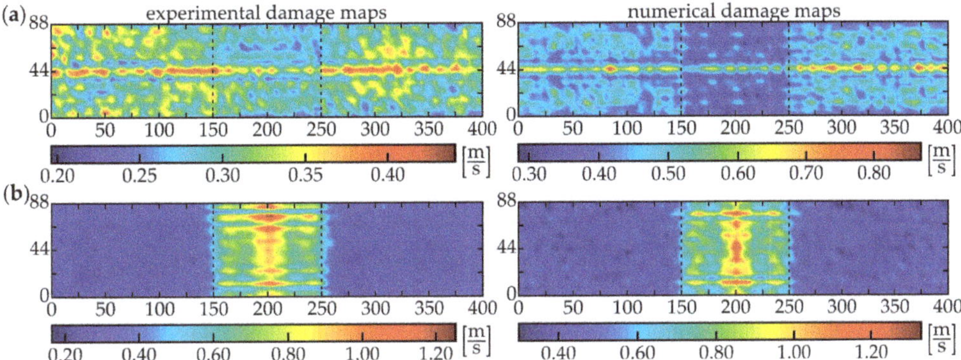

Figure 11. Experimental and numerical WRMS damage maps for specimen #3: (**a**) excitation P3 and (**b**) excitation P4.

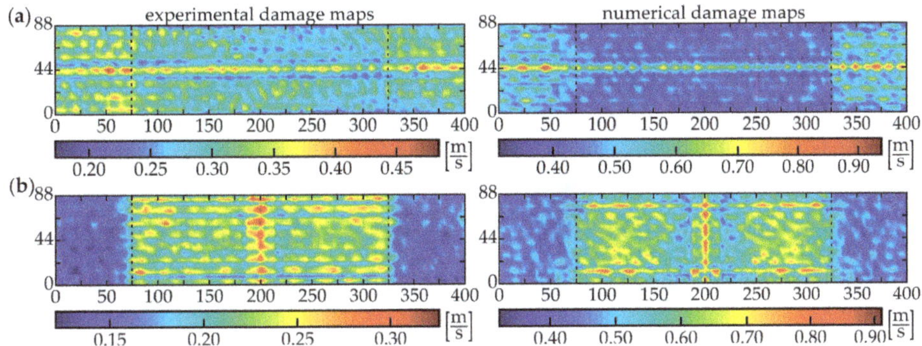

Figure 12. Experimental and numerical WRMS damage maps for specimen #4: (**a**) excitation P3 and (**b**) excitation P4.

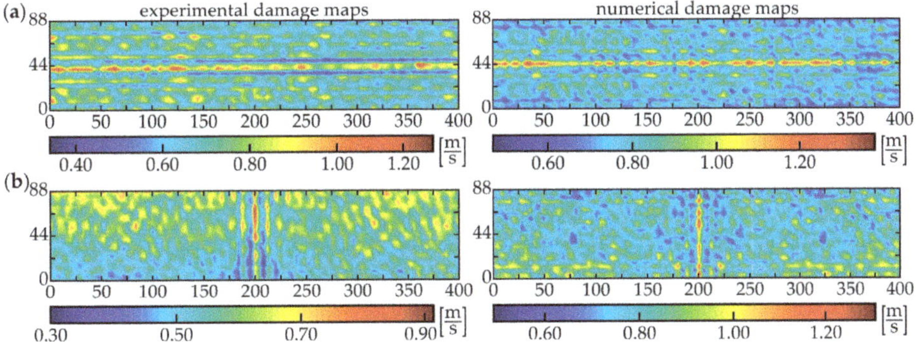

Figure 13. Experimental and numerical WRMS damage maps for specimen #5: (**a**) excitation P3 and (**b**) excitation P4.

4. Conclusions

This paper describes a WRMS-based ultrasonic non-destructive testing approach based on the scanning laser Doppler vibrometry technique. The damage identification and imaging in composite steel–concrete beams were performed. Based on the results of the experimental investigations evaluated by numerical FEM simulations and supported by theoretical analysis, the following final remarks can be formulated.

- The appropriate choice of the excitation frequency is an important issue for the effectiveness of proposed technique. The wave characteristics for a three-layer composite beam and a single-layer steel plate (simulating debonding) were similar for a frequency range above about 120 kHz, thus the damage imaging will not be effective in a higher frequency range. However, the lower frequencies were also not effective because of the decrease in image resolution.
- The comparable analysis of WRMS values calculated for the single wave signals allowed the initial detection of damaged specimens. However, the actual defect size, shape, and position were indeterminable at this stage.
- The main factor affecting the efficiency of damage imaging was the phenomenon of wave leakage and dissipation of the energy, therefore, the position of the excitation point was crucial in the context of debonding detection. The scanning with the actuator placed directly on the damaged area (single-layer steel plate), where the wave leakage did not occur, gave more valuable results. However, the results obtained for the excitation in the area of good adhesion (three-layer composite beam) were also useful, despite the more intensive wave leakage into the concrete beam.
- The visualization of defects was possible only when both debonded and properly connected areas were covered by the scanning region. If the image did not have areas with significantly different values, it was not clear whether the entire analyzed area was well-bonded or debonded.

The leading conclusion of the presented research is that the guided wave propagation method supported with WRMS calculation can be successfully applied for damaged imaging in steel–concrete composite beams. However, considering the complexity of guided wave propagation phenomenon, it is important to perform some initial analysis, including dispersion curves determination and FEM numerical calculations. Further works will be directed to develop the proposed method to be applicable for non-destructive diagnostics of real-scale engineering structures.

Author Contributions: Conceptualization and Methodology, E.W. and M.R.; Experimental Investigations, M.R. and E.W.; FEM Calculations, E.W. and M.K.; Formal Analysis, E.W.; Visualization, E.W.; Writing—Original Draft Preparation, E.W. and M.K.; Writing—Review and Editing, MR.; Supervision, M.R.; All authors have read and agreed to the published version of the manuscript.

Funding: This research received no external funding.

Acknowledgments: Abaqus calculations were carried out at the Academic Computer Centre in Gdańsk.

Conflicts of Interest: The authors declare no conflict of interest.

References

1. Zhao, X.L.; Zhang, L. State-of-the-art review on FRP strengthened steel structures. *Eng. Struct.* **2007**, *29*, 1808–1823. [CrossRef]
2. De Lorenzis, L.; Teng, J.G. Near-surface mounted FRP reinforcement: An emerging technique for strengthening structures. *Compos. Part B Eng.* **2007**, *38*, 119–143. [CrossRef]
3. Teng, J.G.; Yu, T.; Fernando, D. Strengthening of steel structures with fiber-reinforced polymer composites. *J. Constr. Steel Res.* **2012**, *78*, 131–143. [CrossRef]
4. Czaderski, C.; Meier, U. EBR strengthening technique for concrete, long-term behaviour and historical survey. *Polymers* **2018**, *10*, 77. [CrossRef]
5. Barnes, R.A.; Mays, G.C. The transfer of stress through a steel to concrete adhesive bond. *Int. J. Adhes. Adhes.* **2001**, *21*, 495–502. [CrossRef]

6.   Ali, M.S.M.; Oehlers, D.J.; Bradford, M.A. Debonding of steel plates adhesively bonded to the compression faces of RC beams. *Constr. Build. Mater.* **2005**, *19*, 413–422.

7.   Bez Batti, M.M.; do Vale Silva, B.; Piccinini, Â.C.; dos Santos Godinho, D.; Antunes, E.G.P. Experimental analysis of the strengthening of reinforced concrete beams in shear using steel plates. *Infrastructures* **2018**, *3*, 52. [CrossRef]

8.   Alam, M.A.; Onik, S.A.; Mustapha, K.N. Crack based bond strength model of externally bonded steel plate and CFRP laminate to predict debonding failure of shear strengthened RC beams. *J. Build. Eng.* **2020**, *27*, 100943. [CrossRef]

9.   Giurgiutiu, V.; Lyons, J.; Petrou, M.; Laub, D.; Whitley, S. Fracture mechanics testing of the bond between composite overlays and a concrete substrate. *J. Adhes. Sci. Technol.* **2001**, *15*, 1351–1371. [CrossRef]

10.   Schilde, K.; Seim, W. Experimental and numerical investigations of bond between CFRP and concrete. *Constr. Build. Mater.* **2007**, *21*, 709–726. [CrossRef]

11.   Pan, J.; Leung, C.K.Y.; Luo, M. Effect of multiple secondary cracks on FRP debonding from the substrate of reinforced concrete beams. *Constr. Build. Mater.* **2010**, *24*, 2507–2516. [CrossRef]

12.   Napoli, A.; Realfonzo, R. Reinforced concrete beams strengthened with SRP/SRG systems: Experimental investigation. *Constr. Build. Mater.* **2015**, *93*, 654–677. [CrossRef]

13.   Gao, P.; Gu, X.; Mosallam, A.S. Flexural behavior of preloaded reinforced concrete beams strengthened by prestressed CFRP laminates. *Compos. Struct.* **2016**, *157*, 33–50. [CrossRef]

14.   Mertoğlu, Ç.; Anil, Ö.; Durucan, C. Bond slip behavior of anchored CFRP strips on concrete surfaces. *Constr. Build. Mater.* **2016**, *123*, 553–564. [CrossRef]

15.   Akroush, N.; Almahallawi, T.; Seif, M.; Sayed-Ahmed, E.Y. CFRP shear strengthening of reinforced concrete beams in zones of combined shear and normal stresses. *Compos. Struct.* **2017**, *162*, 47–53. [CrossRef]

16.   Ascione, F.; Lamberti, M.; Napoli, A.; Realfonzo, R. Experimental bond behavior of Steel Reinforced Grout systems for strengthening concrete elements. *Constr. Build. Mater.* **2020**, *232*, 117105. [CrossRef]

17.   Zhang, P.; Lei, D.; Ren, Q.; He, J.; Shen, H.; Yang, Z. Experimental and numerical investigation of debonding process of the FRP plate-concrete interface. *Constr. Build. Mater.* **2020**, *235*, 117457. [CrossRef]

18.   Lai, W.L.; Lee, K.K.; Kou, S.C.; Poon, C.S.; Tsang, W.F. A study of full-field debond behaviour and durability of CFRP-concrete composite beams by pulsed infrared thermography (IRT). *NDT E Int.* **2012**, *52*, 112–121. [CrossRef]

19.   Tashan, J.; Al-Mahaidi, R. Bond defect detection using PTT IRT in concrete structures strengthened with different CFRP systems. *Compos. Struct.* **2014**, *111*, 13–19. [CrossRef]

20.   Yi, Q.; Tian, G.Y.; Yilmaz, B.; Malekmohammadi, H.; Laureti, S.; Ricci, M.; Jasiuniene, E. Evaluation of debonding in CFRP-epoxy adhesive single-lap joints using eddy current pulse-compression thermography. *Compos. Part B Eng.* **2019**, *178*. [CrossRef]

21.   Yazdani, N.; Beneberu, E.; Riad, M. Nondestructive Evaluation of FRP-Concrete Interface Bond due to Surface Defects. *Adv. Civ. Eng.* **2019**, *2019*. [CrossRef]

22.   Gu, J.C.; Unjoh, S.; Naito, H. Detectability of delamination regions using infrared thermography in concrete members strengthened by CFRP jacketing. *Compos. Struct.* **2020**, *245*, 112328. [CrossRef]

23.   Shiotani, T.; Momoki, S.; Chai, H.; Aggelis, D.G. Elastic wave validation of large concrete structures repaired by means of cement grouting. *Constr. Build. Mater.* **2009**, *23*, 2647–2652. [CrossRef]

24.   Rucka, M.; Wilde, K. Ultrasound monitoring for evaluation of damage in reinforced concrete. *Bull. Polish Acad. Sci. Tech. Sci.* **2015**, *63*, 65–75. [CrossRef]

25.   Choi, H.; Ham, Y.; Popovics, J.S. Integrated visualization for reinforced concrete using ultrasonic tomography and image-based 3-D reconstruction. *Constr. Build. Mater.* **2016**, *123*, 384–393. [CrossRef]

26.   Zielińska, M.; Rucka, M. Non-Destructive Assessment of Masonry Pillars using Ultrasonic Tomography. *Materials* **2018**, *11*, 2543. [CrossRef]

27.   Słoński, M.; Schabowicz, K.; Krawczyk, E. Detection of Flaws in Concrete Using Ultrasonic Tomography and Convolutional Neural Networks. *Materials* **2020**, *13*, 1557. [CrossRef]

28.   Garbacz, A.; Piotrowski, T.; Courard, L.; Kwaśniewski, L. On the evaluation of interface quality in concrete repair system by means of impact-echo signal analysis. *Constr. Build. Mater.* **2017**, *134*, 311–323. [CrossRef]

29.   Sadowski, Ł.; Hoła, J.; Czarnecki, S. Non-destructive neural identification of the bond between concrete layers in existing elements. *Constr. Build. Mater.* **2016**, *127*, 49–58. [CrossRef]

30. Marks, R.; Clarke, A.; Featherston, C.; Paget, C.; Pullin, R. Lamb Wave Interaction with Adhesively Bonded Stiffeners and Disbonds Using 3D Vibrometry. *Appl. Sci.* **2016**, *6*, 12. [CrossRef]

31. Rucka, M.; Wojtczak, E.; Lachowicz, J. Damage Imaging in Lamb Wave-Based Inspection of Adhesive Joints. *Appl. Sci.* **2018**, *8*, 522. [CrossRef]

32. Wojtczak, E.; Rucka, M. Wave frequency effects on damage imaging in adhesive joints using lamb waves and RMS. *Materials* **2019**, *12*, 1842. [CrossRef] [PubMed]

33. Castaings, M.; Hosten, B.; François, D. The sensitivity of surface guided modes to the bond quality between a concrete block and a composite plate. *Ultrasonics* **2004**, *42*, 1067–1071. [CrossRef] [PubMed]

34. Shen, Y.; Hirose, S.; Yamaguchi, Y. Dispersion of ultrasonic surface waves in a steel-epoxy-concrete bonding layered medium based on analytical, experimental, and numerical study. *Case Stud. Nondestruct. Test. Eval.* **2014**, *2*, 49–63. [CrossRef]

35. Zeng, L.; Parvasi, S.M.; Kong, Q.; Huo, L.; Lim, I.; Li, M.; Song, G. Bond slip detection of concrete-encased composite structure using shear wave based active sensing approach. *Smart Mater. Struct.* **2015**, *24*, 125026. [CrossRef]

36. Song, H.; Popovics, J.S. Characterization of steel–concrete interface bonding conditions using attenuation characteristics of guided waves. *Cem. Concr. Compos.* **2017**, *83*, 111–124. [CrossRef]

37. Li, J.; Lu, Y.; Guan, R.; Qu, W. Guided waves for debonding identification in CFRP-reinforced concrete beams. *Constr. Build. Mater.* **2017**, *131*, 388–399. [CrossRef]

38. Zima, B.; Rucka, M. Guided wave propagation for assessment of adhesive bonding between steel and concrete. *Procedia Eng.* **2017**, *199*, 2300–2305. [CrossRef]

39. Rucka, M. Failure Monitoring and Condition Assessment of Steel–concrete Adhesive Connection Using Ultrasonic Waves. *Appl. Sci.* **2018**, *8*, 320. [CrossRef]

40. Chen, H.; Xu, B.; Wang, J.; Luan, L.; Zhou, T.; Nie, X.; Mo, Y.L. Interfacial debonding detection for rectangular cfst using the masw method and its physical mechanism analysis at the meso-level. *Sensors* **2019**, *19*, 2778. [CrossRef]

41. Liu, S.; Sun, W.; Jing, H.; Dong, Z. Debonding Detection and Monitoring for CFRP Reinforced Concrete Beams Using Pizeoceramic Sensors. *Materials* **2019**, *12*, 2150. [CrossRef]

42. Ke, Y.T.; Cheng, C.C.; Lin, Y.C.; Huang, C.L.; Hsu, K.T. Quantitative assessment of bonding between steel plate and reinforced concrete structure using dispersive characteristics of lamb waves. *NDT E Int.* **2019**, *102*, 311–321. [CrossRef]

43. Yan, J.; Zhou, W.; Zhang, X.; Lin, Y. Interface monitoring of steel–concrete–steel sandwich structures using piezoelectric transducers. *Nucl. Eng. Technol.* **2019**, *51*, 1132–1141. [CrossRef]

44. Giri, P.; Mishra, S.; Clark, S.M.; Samali, B. Detection of gaps in concrete–metal composite structures based on the feature extraction method using piezoelectric transducers. *Sensors* **2019**, *19*, 1769. [CrossRef] [PubMed]

45. Ng, C.T.; Mohseni, H.; Lam, H.F. Debonding detection in CFRP-retrofitted reinforced concrete structures using nonlinear Rayleigh wave. *Mech. Syst. Signal. Process.* **2019**, *125*, 245–256. [CrossRef]

46. Wang, Y.; Li, X.; Li, J.; Wang, Q.; Xu, B.; Deng, J. Debonding damage detection of the CFRP-concrete interface based on piezoelectric ceramics by the wave-based method. *Constr. Build. Mater.* **2019**, *210*, 514–524. [CrossRef]

47. Huo, L.; Cheng, H.; Kong, Q.; Chen, X. Bond-slip monitoring of concrete structures using smart sensors—A review. *Sensors* **2019**, *19*, 1231. [CrossRef]

48. Alleyne, D.; Cawley, P. A two-dimensional Fourier transform method for the measurement of propagating multimode signals. *J. Acoust. Soc. Am.* **1991**, *89*, 1159–1168. [CrossRef]

49. Moser, F.; Jacobs, L.J.; Qu, J. Modeling elastic wave propagation in waveguides with the finite element method. *NDT E Int.* **1999**, *32*, 225–234. [CrossRef]

50. Żak, A.; Radzieński, M.; Krawczuk, M.; Ostachowicz, W. Damage detection strategies based on propagation of guided elastic waves. *Smart Mater. Struct.* **2012**, *21*, 035024. [CrossRef]

51. Lee, C.; Zhang, A.; Yu, B.; Park, S. Comparison study between RMS and edge detection image processing algorithms for a pulsed laser UWPI (Ultrasonic wave propagation imaging)-based NDT technique. *Sensors* **2017**, *17*, 1224. [CrossRef] [PubMed]

52. Pieczonka, Ł.; Ambroziński, Ł.; Staszewski, W.J.; Barnoncel, D.; Pérès, P. Damage detection in composite panels based on mode-converted Lamb waves sensed using 3D laser scanning vibrometer. *Opt. Lasers Eng.* **2017**, *99*, 80–87. [CrossRef]

53. Kudela, P.; Wandowski, T.; Malinowski, P.; Ostachowicz, W. Application of scanning laser Doppler vibrometry for delamination detection in composite structures. *Opt. Lasers Eng.* **2016**, *99*, 46–57. [CrossRef]

54. Harb, M.S.; Yuan, F.G. A rapid, fully non-contact, hybrid system for generating Lamb wave dispersion curves. *Ultrasonics* **2015**, *61*, 62–70. [CrossRef] [PubMed]

55. Gauthier, C.; Galy, J.; Ech-Cherif El-Kettani, M.; Leduc, D.; Izbicki, J.L. Evaluation of epoxy crosslinking using ultrasonic Lamb waves. *Int. J. Adhes. Adhes.* **2018**, *80*, 1–6. [CrossRef]

56. Ekstrom, M.P. Dispersion Estimation from Borehole Acoustic Arrays Using a Modified Matrix Pencil Algorithm. *IEEE Proc. ASILOMAR-29* **1996**.

57. Mazzotti, M.; Bartoli, I.; Castellazzi, G.; Marzani, A. Computation of leaky guided waves dispersion spectrum using vibroacoustic analyses and the Matrix Pencil Method: A validation study for immersed rectangular waveguides. *Ultrasonics* **2014**, *54*, 1895–1898. [CrossRef]

58. Chang, C.Y.; Yuan, F.G. Extraction of guided wave dispersion curve in isotropic and anisotropic materials by Matrix Pencil method. *Ultrasonics* **2018**, *89*, 143–154. [CrossRef]

59. Ramasawmy, D.R.; Cox, B.T.; Treeby, B.E. ElasticMatrix: A MATLAB toolbox for anisotropic elastic wave propagation in layered media. *SoftwareX* **2020**, *11*, 100397. [CrossRef]

Article

Evaluation of Ferromagnetic Steel Hardness Based on an Analysis of the Barkhausen Noise Number of Events

Maciej Roskosz [1,*], **Krzysztof Fryczowski** [2] **and Krzysztof Schabowicz** [3]

[1] Faculty of Mechanical Engineering and Robotics, AGH University of Science and Technology, aleja Mickiewicza 30, 30-059 Kraków, Poland

[2] Faculty of Energy and Environmental Engineering, Silesian University of Technology, ul. Akademicka 2A, 44-100 Gliwice, Poland; krzysztof.fryczowski@polsl.pl

[3] Faculty of Civil Engineering, Wrocław University of Technology, Wybrzeże Wyspiańskiego 27, 50-370 Wrocław, Poland; k.schabowicz@pwr.edu.pl

* Correspondence: mroskosz@agh.edu.pl

Received: 6 April 2020; Accepted: 27 April 2020; Published: 29 April 2020

Abstract: Measurements are carried out of the Barkhausen noise (BN) and hardness on specimens where changes in hardness were obtained due to strain hardening (S235 and DC01 steels) and due to thermochemical treatment (AMS 6414 steel). A method is presented of processing the recorded BN signal to extract diagnostic information. The BN number of events is selected as the signal characteristic property to develop relevant correlations. A new methodology is presented for the development of correlations between the Barkhausen noise number of events and hardness. A possibility is indicated of developing correlations with a high R2 determination coefficient. The method limitations are specified.

Keywords: hardness; Barkhausen noise; number of events; non-destructive testing inverse problem

1. Introduction

Hardness is a property that cannot be defined unequivocally. In fact, many other properties are measured in hardness tests. The tests can be classified depending on the material resistance observed during the testing. Static methods of hardness measurement consist of pushing the indenter into the tested material beyond the yield point to cause plastic strains. In such methods, hardness can be defined as the measure of the material strength to plastic strains arising due to the indenter pressing force.

During a traditional hardness measurement, the hardness testing machine indenter causes plastic strain to the surface domain where the measurement is performed. For some elements, such damage to the surface is unacceptable. For others, plastic deformation of the surface is not an operating problem but a diagnostic one: it is impossible to repeat the hardness test in exactly the same location under the same conditions. These obstacles can be overcome using so-called inverse problems of non-destructive testing, where mechanical properties are determined based on other physical quantities.

In the testing presented herein, this physical quantity is the Barkhausen noise (BN), which is created due to a time-variable impact of a magnetic field on a ferromagnetic. The variable magnetic field causes the stochastic displacement of domain walls, which involves a sudden change in magnetization detectable by the measuring sensor. More details about the theory of the Barkhausen noise can be found in several models [1–3].

The Barkhausen noise is used as a diagnostic signal to determine the material stress-and-strain state [4–7], the state of the microstructure [8], defects in surface treatment [9], and hardness [5,9–15].

Ortega-Labra et al. [4] presented a novel system for measuring stress applied to a pipe section using the rotational continuous magnetic Barkhausen noise (MBN). The presented results show that the shape and the amplitude of the MBN angular dependence change with an increase in applied stress. The outcome reveals that the applied stress value could be estimated using the amplitude of the angular dependence of the measured MBN signal. Additionally, the shape of the angular dependence is related to the applied stress direction. This suggests that using a proper procedure, the direction of the applied stress could also be estimated based on the MBN.

Liu et al. [5] presented the application of a multifunctional magnetic testing technique in the quantitative prediction of tensile stress, surface hardness, and the case depth in medium-carbon steel rods. Simultaneous measurements were carried out on the MBN signal, the tangential magnetic field, and the magnetic hysteresis curve in tensioned steel rods with different depths of the surface-hardened layer. A total of 16 feature parameters were extracted from four types of signal patterns as candidate inputs of the back-propagation neural network. Very promising results were obtained, where the prediction model error was less than 5.3% for tensile stress, 5% for the case depth, and 0.22% for surface hardness.

In [6], skewness is utilized as a feature for applied stress detection and compared with the root-mean-square (RMS) peak value and peak position, which are the most common features of the MBN signal. Different levels of compressive and tensile stresses are applied to En36 steel specimens to evaluate the new feature effectiveness. Its non-linear behavior and sensitivity are compared with the RMS peak value and peak position. The domain wall energy and the eddy current damping are discussed as the cause of the phenomenon.

In [9], He et al. presented a review of non-destructive testing methods for the metallic grinding burn detection. One of them makes use of measurements of the magnetic Barkhausen noise (MBN) and the magnetoelastic acoustic emission (MAE). It is found that the BN method itself produces only relative results that need to be compared with calibration blocks to define certain limits for rejection. Several aspects for further investigation are defined, such as the relationship between the BN signal multiparameters and the measured material properties (stress state, hardness), the impact factors of BN measurements including microstructural domain walls, and macroscopic operations, such as the influence of the holding force, the contact angle between the probe and the workpiece, and the probe moving speed.

Wilson et al. [10] studied P9 and T22 steel specimens subjected to different kinds of heat treatment used to modify the microstructure. Magnetic hysteresis loops and the Barkhausen noise were measured and compared to the Vickers hardness number. It is demonstrated that a rise in hardness involves a rise in coercivity, a change in the location of the maximum value of the Barkhausen noise envelope, and a drop in magnetic permeability.

Moorthy et al. [11] examined the impact of a post-hardening change in the specimen structure on the Barkhausen noise envelope on the carburized layer of steel EN36. Several frequencies of the excitation field were applied, and a frequency analysis of the measured signal was conducted in different ranges. It is proved that the envelope maximum value correlates with the hardness measured in the layer to the depth of 100 μm.

The aim of the studies presented in [12] was to use the Barkhausen noise to characterize hardness in steel AISI 4140 subjected to the Jominy test. Various types of time-frequency representations were used: the spectrogram, the Wigner–Ville distribution, the Capongram, the ARgram, the wavelet transform scalogram, and the Mellingram. It is proved that due to the Barkhausen noise non-stationary character, time-frequency representations are a rich source of information. A method is proposed of extracting information from these representations by creating an algorithm reducing the number of variables from 3 (time, frequency, amplitude) to 2 (time or frequency and amplitude).

In [13], O'Sullivan et al. measured magnetoacoustic emission (MAE) and the Barkhausen noise (BN) to determine the properties of AISI 430 steel specimens subjected to heat treatment to change their microstructure. Parameters such as the RMS value of the Barkhausen noise and the magnetoacoustic emission absolute value proposed by the authors were compared e.g., to the Vickers hardness number.

It is shown that a rise in plastic strain causes a drop in the BN and MAE and that the parameters are linearly dependent on the Vickers hardness number.

Franco et al. [14] present relations between the Barkhausen noise parameters and Rockwell hardness using SAE 4140 and SAE 6150 steel specimens subjected to the Jominy hardenability test. The Barkhausen noise is described using the maximum value, the location, and the RMS value of the envelope. A linear relation is found between all parameters of the Barkhausen noise and Rockwell hardness.

In [15], many properties of the Barkhausen noise are analyzed, and a method is proposed of their selection and mutual complementation for the purposes of hardness evaluation. An algorithm is presented for the analysis of the signal main properties. Combined with the method of the analysis of the correlations between them, the algorithm is used to select the optimal properties. Multidimensional linear regression with selected properties is applied, and a statistical linear model of merging BN properties is built to forecast the material hardness.

This paper presents the results of testing aiming to develop diagnostic correlations between hardness and the Barkhausen noise total number of events NoE_{TOT} BN. A good correlation between the two quantities would enable periodic and multiple measurements of hardness in the same location, which cannot be done using traditional hardness measurement methods. Two cases of differences in the material hardness are considered. One is a change in hardness due to cold plastic strain and the other is the change caused by thermochemical treatment.

NoE_{TOT} is selected as the diagnostic signal property to be analyzed because the authors already obtained promising results from their previous preliminary research on hardness evaluation [16,17] and from their comprehensive studies on the assessment of the state of active stresses in steel elements [7]. The following quantities are analyzed in [7,16,17]: the BN RMS value of voltage (U_{RMS}) [18–20], the BN energy (E_{BN}) [20,21], characteristic parameters related to the BN signal envelope [22,23], and the distribution of the BN total number of events (NoE_{TOT} BN) [24]. The best determination coefficients R^2 and the smallest standard deviations are obtained for the NoE_{TOT} BN. Another issue, not taken up in this paper, is the analysis of the intensity of changes in the features describing the Barkhausen noise that occur due to changes in hardness or in the stress-and-strain state. The signal feature that changes the most intensely is the one best suited to enable the development of a diagnostic correlation. The issue is related to digital processing of the signal on the one hand, and to analysis of the variability of the signal individual features on the other. Such analyses are presented for example in [25], where a comparison is made between changes in the Barkhausen noise profile skewness and the noise signal RMS value caused by active compressive and tensile stresses. Within certain stress ranges, the sensitivity of skewness is higher than that of the RMS value, which makes the former a better diagnostic signal.

Among others, the Barkhausen noise depends on the state of the material microstructure and on the material stress state. The impacts can be separated from each other assuming that the two factors are not changed at the same time. Changes in hardness due to strain hardening are accompanied by changes in internal (residual) stresses. Similarly, changes in hardness due to thermochemical treatment involve, apart from changes in the specimen microstructure, changes in the level of residual stresses. In the presented testing, the impact of internal stresses was not taken into consideration in the analysis of results. In consequence, the developed diagnostic correlations are less universal because they are related to the structural properties of specimens.

2. Measuring Apparatus

The Barkhausen noise was measured using the MEB4-C system made by the Mag-Lab s.c. company (Gdańsk, Poland). This is an independent system of measuring, analyzing, and recording the Barkhausen noise (BN) voltage signals. The system diagram is shown in Figure 1. The measurements were performed using a surface measuring head with systems of the electromagnet, Barkhausen noise detection, and magnetization control. The head diagram is presented in Figure 2. The system enables configuring settings in the range of the magnetizing current amplitudes, the rate of changes in the magnetizing current (the magnetizing current frequency), and the measured signal amplification.

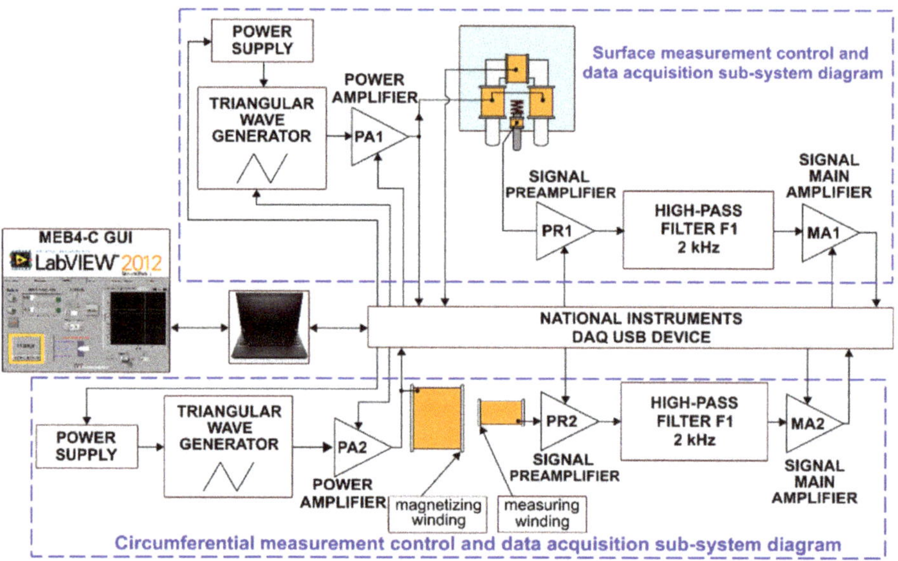

Figure 1. Schematic diagram of the MEB4-C system.

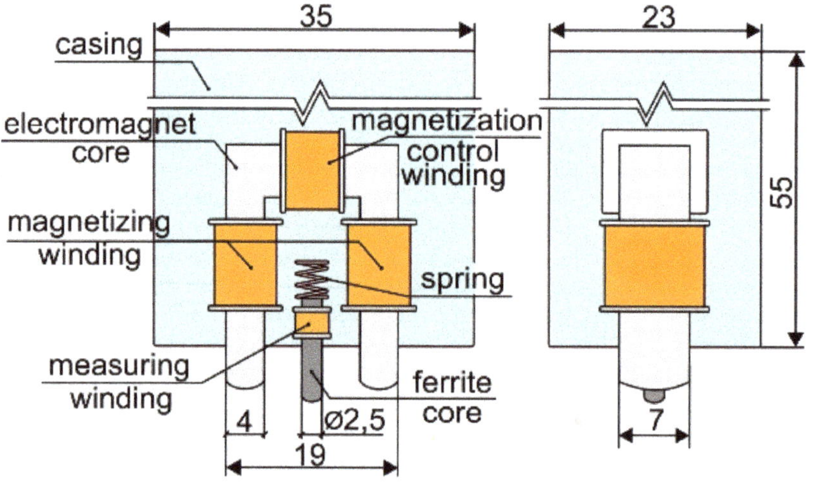

Figure 2. Surface measuring probe.

3. BN Signal Analysis

The basic unit of information is a single cycle of changes in magnetization. The cycle is presented in detail in Figure 3. Two halves of such a full cycle can be distinguished: one in which the magnetization current intensity diminishes (marked as $I\searrow$) and the other in which the current intensity rises (marked as $I\nearrow$).

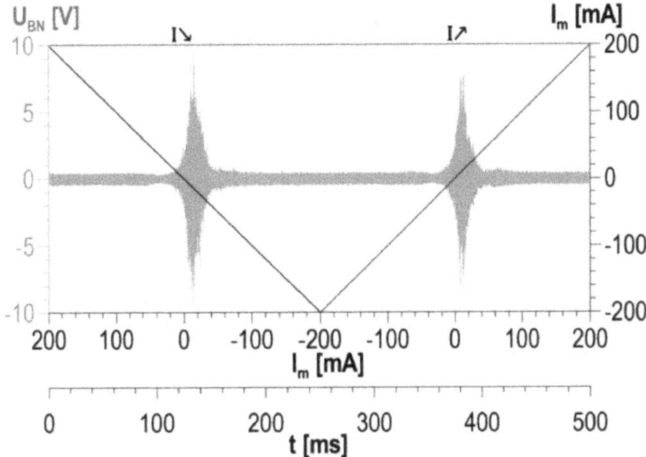

Figure 3. Detailed description of a single cycle of changes in magnetization.

To obtain information from a single cycle of changes in magnetization, the signal is processed in different ways. The processing method produces values of individual properties of the Barkhausen noise. For a multiple number of cycles of changes in magnetization, a multiple number is obtained of values of the quantity for which a statistical analysis has to be carried out. The analysis gives the average value of the quantity and the value of standard deviation being a variability measure of the BN signal analyzed properties.

The measured signal of the Barkhausen noise is a set of voltage pulses, among which so-called events can be distinguished. It is assumed that for a set value of threshold voltage U_g, the event occurrence is determined based on three subsequent signal samples with a value exceeding the threshold voltage, where the value of the second signal is higher than the value of the first and third (cf. Figure 4). Detected events are summed up, which gives the distribution of the total number of events—NoE_{TOT}—depending on threshold voltage U_g (cf. the example presented in Figure 5). The threshold voltage-dependent NoE_{TOT} distribution is obtained for each cycle of changes in magnetization. In the case of a multiple number of cycles of changes in magnetization, averaged NoE_{TOT} distributions are determined together with statistical features, such as standard deviation.

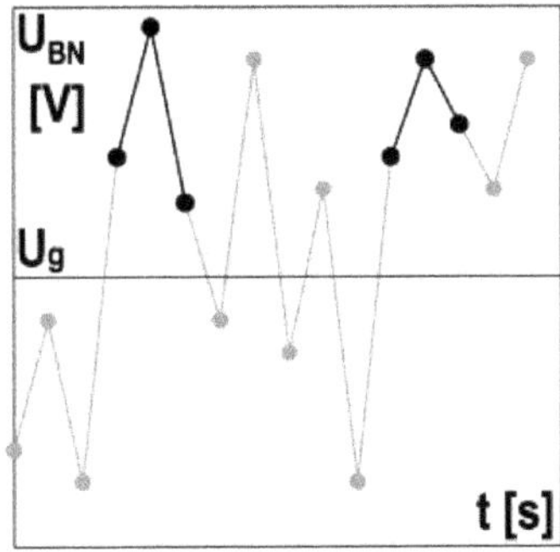

Figure 4. Method of event determination above the set value of threshold voltage U_g.

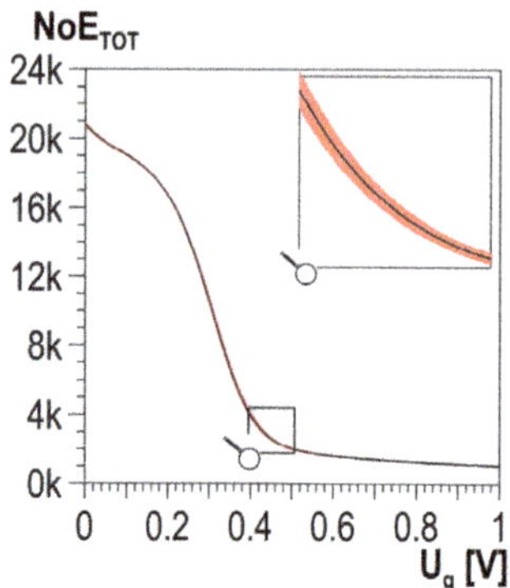

Figure 5. Averaged distribution of the total number of events (NoE_{TOT}) with standard deviation values (marked in red).

The relation between the number of events (NoE_{TOT}) and the corresponding values of hardness were analyzed for individual values of discrimination voltage U_{g_j} using a linear regression model. The analysis resulted in the value of determination coefficient R^2 for each value of discrimination voltage U_{g_j}. The value reflects the quality of the goodness of fit of the linear model developed based on input data. A value of determination coefficient R^2 equal or higher than 0.8 proves a good fit of the linear regression model.

Additionally, the resolution coefficient ($RE_{COF.}$) was defined. The coefficient makes it possible to find the discrimination voltage value for which the best resolution is obtained, where the resolution is defined for a given value of discrimination voltage U_{g_j} as the quotient of the difference between the maximum and the minimum value of the obtained total number of events ($NoE_{TOTmax}(U_{g_j})$ and $NoE_{TOTmin}(U_{g_j})$, respectively) and the sum of the standard deviation values $SD_i(U_{g_j})$ corresponding to individual points of the hardness measurement.

$$RE(U_{g_j}) = \frac{NoE_{TOTmax}(U_{g_j}) - NoE_{TOTmin}(U_{g_j})}{\sum_i^n SD_i(U_{g_j})} \tag{1}$$

To normalize the resolution coefficient (RE) results with the determination coefficient (R^2) varying in interval <0,1>, the normalized coefficient of resolution ($RE_{COF.}$) was introduced. The coefficient is defined as:

$$RE_{COF.}(U_{g_j}) = \frac{RE(U_{g_j})}{max(RE(U_{g_j}))}. \tag{2}$$

The best correlation occurs for the discrimination voltage for which maximum values of R^2 and RE_{COF} are obtained.

4. Testing of Specimens with Plastic Strain

Cold plastic strain increases dislocation density (by 4 to 5 orders of magnitude) [26], which involves the material hardening due to the decrease in distances between dislocations and the increasing forces of their interaction. Due to the smaller distance, the dislocations block each other, and their further displacement (further deformation of the material) is only possible if a higher external stress is applied; the phenomenon is referred to as strain hardening. A plastically deformed material is characterized by increased strength properties (hardness, yield point) with simultaneously decreased plastic properties (elongation, narrowing) [26].

4.1. Testing Details

The tests were performed on plate specimens made of S235 steel (according to Standard EN 10025) and DC01 steel (according to Standard EN 10130). The steel grades properties are listed in Table 1.

Table 1. Steel chemical composition and strength properties.

	\multicolumn{6}{c}{Chemical Composition Mass Fraction [%]}						Strength Properties	
	C	**Mn**	**P**	**S**	**N**	**Cu**	**Re [MPa]**	**Rm [MPa]**
S235JR	0.21	1.4	0.035	0.035	0.012	0.55	235	410
DC01	≤0.12	≤0.60	≤0.045	≤0.045	-	-	140–280	270–410

The specimens were loaded on a Galdabini Sun 10P tensile testing machine (Galdabini, Cardano al Campo (VA), Italy). They were repeatedly subjected to uniaxial tension in such a manner that with every loading cycle, there was a slight increment in plastic strain. When the ultimate strength value was exceeded, a narrowing started to form in the specimen. The loading process was discontinued the moment that the narrowing was visible enough to detect through a visual inspection. Due to the material hardening, a change in hardness occurs in the plastically deformed area. Non-uniform plastic strain leads to the creation of a specific hardness profile on the specimen surface.

For the measuring apparatus described in Section 2, the details of the configurations used during the BN measurements are presented in Table 2.

Table 2. Details of measuring configurations.

Configuration	S235 Steel	DC01 Steel
sampling frequency f_p [kHz]	800	800
magnetizing current amplitude [mA]	150	200
magnetizing current frequency [Hz]	2.02	2.02
signal amplification [dB]	21	35
number of cycles of changes in magnetization	5	5

At the preliminary stage of the testing, measurements were carried out changing the magnetization direction every 30°; angles 0° and 180° correspond to the load direction. Example results are presented in Figure 6. They confirm the literature data [27,28] that prove that the signal maximum and minimum values are obtained for the maximum strain direction and for the normal direction, respectively.

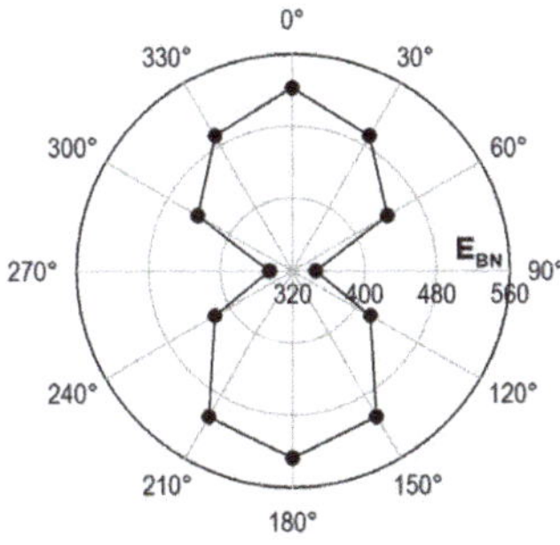

Figure 6. Polar diagram illustrating average Barkhausen noise energy E_{BN} in the place of the specimen narrowing (0–180°—direction parallel to the tensile load direction, 90–270°—direction normal to the tensile load direction).

In the engineering practice, diagnostic correlations should be independent of the directions of principal residual stresses in the material, which are usually unknown. One proposal for a simplified solution is the absolute value of the Barkhausen noise properties:

$$V = \sqrt{V_X^2 + V_Y^2} \tag{3}$$

where V_X and V_Y, respectively, are the property values for two directions of the magnetizing field application (e.g., E_{BN} or NoE_{TOT}) for which the most extreme values are obtained.

Considering the above, the Barkhausen noise was measured only in two directions perpendicular to each other, i.e., in direction X, parallel to the load direction, and in direction Y, normal to the load direction.

After the Barkhausen noise was measured in the same measuring points, HV5 hardness measurements were performed using the Krautkramer TIV hardness tester (General Electric, Boston, MA, USA). The measurements were carried out along a measurement line being the specimen axis of symmetry. Example testing results for plastically deformed specimens made of S235 and DC01 steel are shown in Figures 7 and 8, respectively.

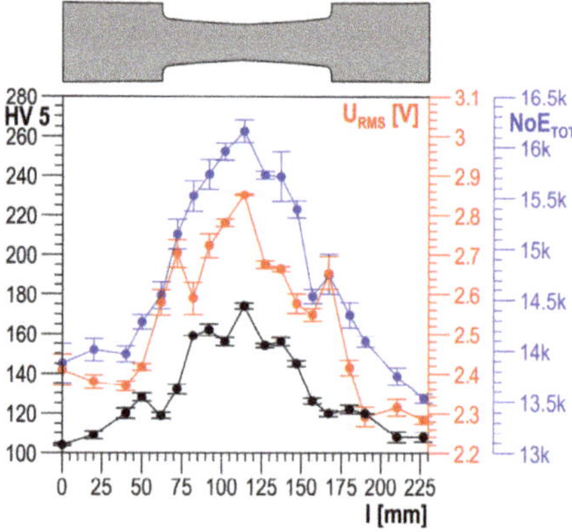

Figure 7. Distribution of HV5, root mean square (RMS), and NoE_{TOT} along the S235 steel specimen length.

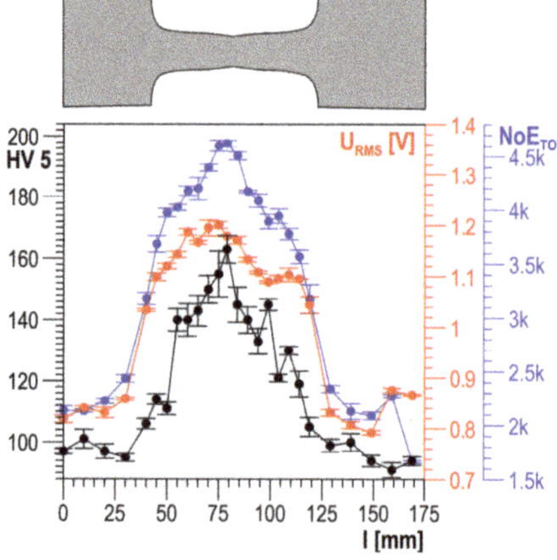

Figure 8. Distribution of HV5, RMS, and NoE_{TOT} along the DC01 steel specimen length.

4.2. Analysis of Results

According to the measurement methodology presented in Section 3, the analysis covers the results obtained in two directions perpendicular to each other and their absolute value. The distributions of determination coefficient R^2 of the linear correlation between *HV5* hardness and the total number of events NoE_{TOT} and of resolution coefficient RE_{COF} as a function of threshold voltage U_g for S235 steel are presented in Figure 9a–c for the parallel direction, the normal direction, and the absolute value, respectively. In all three charts (Figure 9a–c), the curves illustrating the history of determination coefficient R^2, after the initial phase of a clear increase (in the range of U_g from 0 to about 0.4 V), look quite similar; for higher voltage values, there are no essential changes. The distributions of coefficient RE_{COF} are characterized by a global maximum falling on U_g values ranging from 0.5 to 0.6 V. These maxima decide the selection of the discrimination voltage for which the linear correlations between NoE_{TOT} and HV5 are developed and presented in Figure 10a for the parallel direction, in Figure 10b for the normal direction, and in Figure 10c for the absolute value. The values of determination coefficient R^2 differ considerably for the two directions and total about 0.7 for the parallel direction (which is too low for practical use) and about 0.92 for the normal direction. However, in the analyzed case, the impact of the parallel direction of magnetization on the value of the linear correlation determination coefficient R^2 for the absolute value ($R^2 = 0.914$) is very slight, and the correlation can be used in practice.

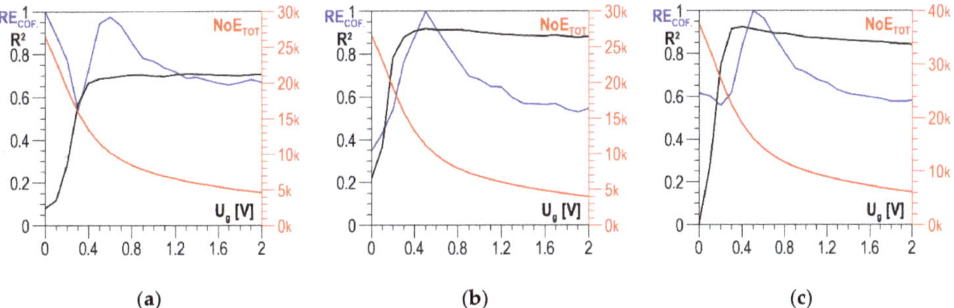

Figure 9. Averaged NoE_{TOT} distributions and distributions of coefficients RE_{COF} and R^2 as a function of threshold voltage U_g—S235 steel (**a**) parallel direction; (**b**) normal direction; (**c**) absolute value.

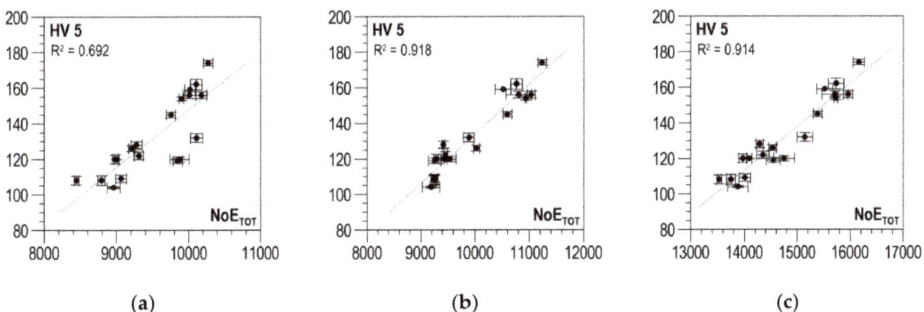

Figure 10. Linear correlations between NoE_{TOT} and *HV5* hardness—S235 steel (**a**) parallel direction, $U_g = 0.6$ V; (**b**) normal direction, $U_g = 0.5$ V; (**c**) absolute value, $U_g = 0.5$ V.

The distributions of determination coefficient R^2 of the linear correlation between *HV5* hardness and the total number of events NoE_{TOT} and of resolution coefficient RE_{COF} as a function of threshold voltage U_g for DC01 steel are presented in Figure 11a–c for the parallel direction, the normal direction, and the absolute value, respectively. In all three charts below, the curves illustrating the history of determination coefficient R^2, after the initial phase of a clear increase (in the range of U_g from 0 to about 0.5 V), look quite similar; for higher voltage values, there are no essential changes. A characteristic feature of the RE_{COF} distributions is that the first local maximum falls within the range of voltage U_g variability where the determination coefficient R^2 takes low values. It is only the values of the second local maximum that decide the selection of the discrimination voltage for which the linear correlations between NoE_{TOT} and $HV5$ are developed and presented in Figure 12a for the parallel direction, in Figure 12b for the normal direction, and in Figure 12c for the absolute value.

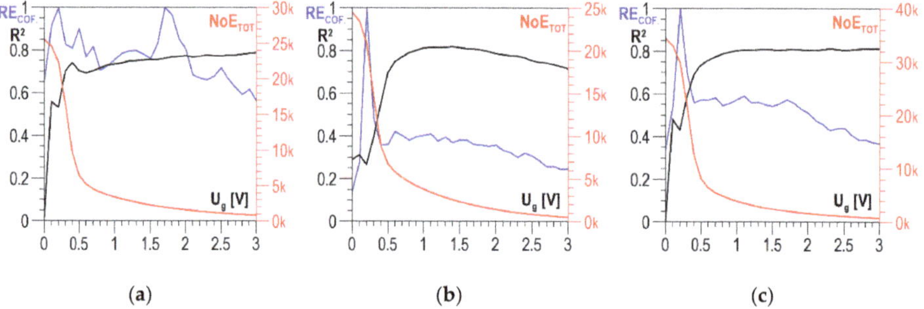

Figure 11. Averaged NoE_{TOT} distributions and distributions of coefficients RE_{COF} and R^2 as a function of threshold voltage U_g—DC01 steel (**a**) parallel direction; (**b**) normal direction; (**c**) absolute value

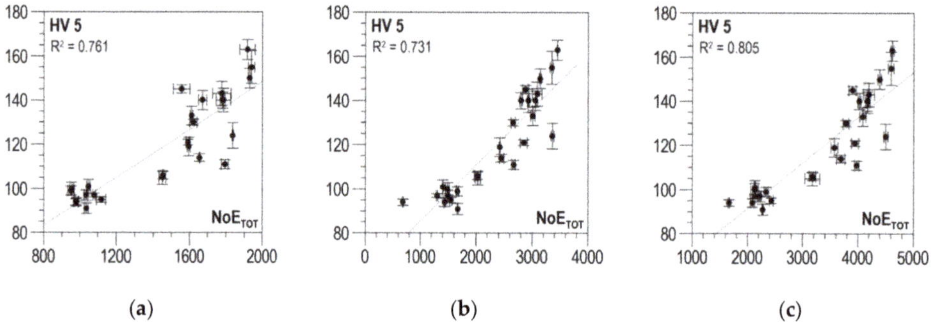

Figure 12. Linear correlations between NoE_{TOT} and HV5 hardness—DC01 steel (**a**) parallel direction, U_g = 1.7 V; (**b**) normal direction, U_g = 1.1 V; (**c**) absolute value, U_g = 1.1 V.

The values of determination coefficient R^2 for the two directions of magnetization vary from about 0.73 to about 0.8, which is too low to use in practice. Finding the absolute value improves the situation only slightly, raising coefficient R^2 for optimal discrimination voltage U_g to about 0.8, which is still just on the limit of possible practical use.

In the presented results of the testing, higher NoE_{TOT} values correspond to higher hardness (higher degree of plastic strain), which is also observed for the values of other properties of the Barkhausen noise (Figures 7 and 8) [7,16,17]. However, this is not a permanent trend that occurs for every material. S235JRG2 steel is investigated in [29], and it is found that a rise in the plastic strain degree involves a decrease in the BN U_{RMS} value measured in the load direction and an increase in the value determined in the normal direction. It is also found that the main reason for the observed changes is the number of dislocations causing non-homogeneous microdeformations. The stress field around dislocations interacts with all suitably oriented displacements of domain walls. If the number of displacements is reduced, the values of the BN properties are reduced, too. The authors of [29] state that this makes it possible to account for the drop in the BN property value measured in the load direction, but it still cannot explain the rise of the value in the normal direction. A similar trend of changes is observed in [13]. In [30], Fe–2%Si alloy is investigated using an enveloping coil, which allows the assumption that the magnetization direction is the same as the direction of the load. It is found that a rise in plastic strain involves a rise in the integrals of U_b voltage. This trend is opposite to the one observed in [29]. The results of the testing of plastically deformed Armco iron presented in [31] show a non-uniform trend of changes in U_b voltage integrals. After an initial increase with a rise in plastic strain, if the strain value exceeds about 10%, a further rise in strain involves a decrease in the values. Another essential factor is additionally found in [32]: the impact of the specimen loading history (the method of achieving plastic strain). In the case of specimens for which plastic strain was achieved in a single loading cycle, the BN properties decreased with a rise in plastic strain. For specimens where subsequent plastic strain values were achieved gradually to the value of about 1%, the values of the BN properties increased. A further rise in strain caused a gradual drop in the values of the BN properties. However, the drop never reached the level measured for specimens where strain was achieved in a single loading cycle.

The analysis of the phenomena of interaction between a change in the degree of plastic strain and dislocation density is a complex process that requires many more microscopic and macroscopic studies on the changes that arise in the domain structure of polycrystalline ferromagnetic materials due to a complex state of stresses and strains.

5. Testing of Specimens Subjected to Thermochemical Treatment

5.1. Testing Details

The testing was carried out on a set of three specimens made of AMS 6414 steel subjected to carburization in an atmosphere with a different concentration of carbon and then to hardening [33]. The process resulted in specimens with a different degree of hardness and a constant thickness of the carburized layer of 1.2 mm. The surface of the specimens was prepared for the testing by grinding. The core of the specimens has a martensitic structure. As for the layers, an increase in martensite refinement can be observed with a rise in hardness. This is because the number of locations of martensite nucleation during the hardening process increases with a rise in the content of carbon in the carburized layer. As a result, the growth of martensite is reduced. Moreover, cementite precipitates can be found in the hardest layer [33].

The Barkhausen noise initial measurements were performed so that an angular distribution of properties could be obtained, as presented in Figure 6. The results were analyzed, and the directions of magnetization parallel with and normal to the direction of grinding traces were adopted as the measurement directions (Figure 13). Measurements were carried out for three configurations of the measuring device settings, the details of which are listed in Table 3.

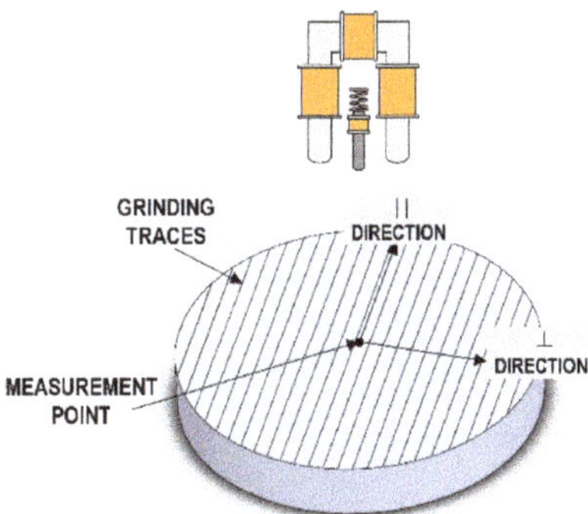

Figure 13. Details of the Barkhausen noise testing performed on specimens made of AMS 6414 steel.

Table 3. Details of the measuring configuration.

Configuration	Configuration C1	Configuration C2	Configuration C3
sampling frequency f_p [kHz]	800	800	800
magnetizing current amplitude [mA]	200	200	200
magnetizing current frequency [Hz]	2.02	5.31	10.4
signal amplification [dB]	50	50	25
number of cycles of changes in magnetization	5	15	25

Microhardness on the surface of carburized specimens was measured using a Zwick Roell microhardness tester. The measurement results are listed in Table 4.

Table 4. *HV 0.5* hardness measurement results.

Specimen	HV 5 Measurement Results										HV 0.5 Average Value	Standard Deviation
1	757	749	746	768	748	757	740	757	757	753	753.2	7.40
2	815	800	809	789	801	809	803	814	815	803	805.8	7.82
3	834	827	827	821	851	846	821	853	846	836	836.2	11.53

5.2. Analysis of Results

Averaged NoE_{TOT} distributions were obtained using the data-analysis procedure described in Section 3 for three measuring configurations and two magnetization directions perpendicular to each other. The absolute value was also determined according to Relation (2). For Configuration C1, distributions of R^2 as a function of U_g, after strong oscillations in the U_g range from 0 to about 0.4 V, stabilize, taking values close to 1 for the parallel direction (Figure 14a) and for the absolute value (Figure 14c), and values oscillating between 0.9 and 1 for the normal direction (Figure 14b). The selection of discrimination voltage U_g, for which the correlations presented in Figure 15a–c were developed, is decided by the global maximum of the RE_{COF} distribution. In the case under analysis, the maximum of voltage U_g is about 0.8 V. The obtained values of determination coefficient R^2 take values higher than 0.97, which proves a very good correlation.

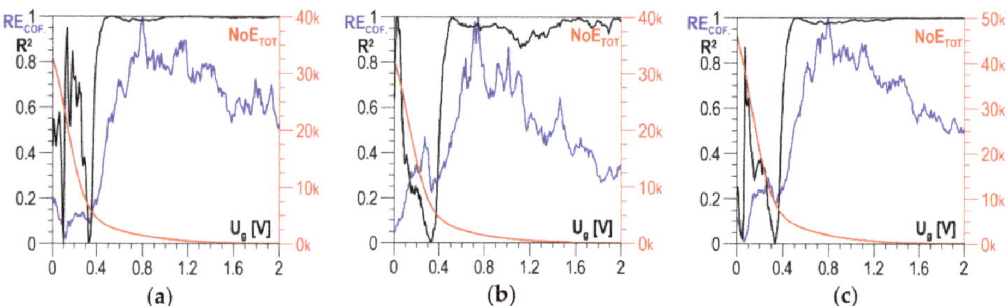

Figure 14. Averaged NoE_{TOT} distributions and distributions of coefficients RE_{COF} and R^2 as a function of threshold voltage U_g—Configuration C1 (**a**) parallel direction; (**b**) normal direction; (**c**) absolute value.

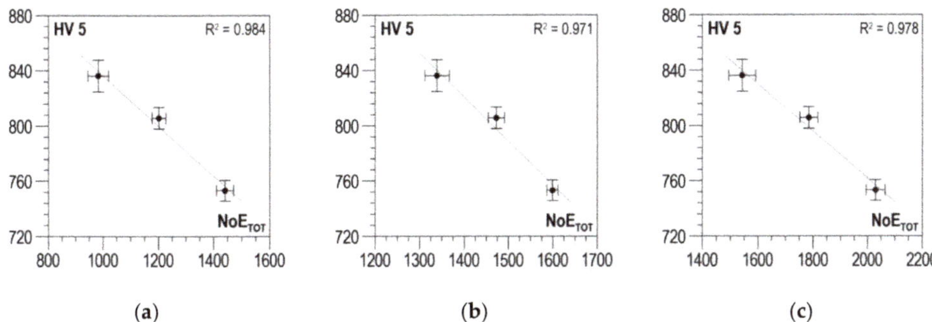

Figure 15. Linear correlations between NoE_{TOT} and $HV5$ hardness—Configuration C1 (**a**) parallel direction, $U_g = 0.8$ V; (**b**) normal direction, $U_g = 0.74$ V; (**c**) absolute value, $U_g = 0.79$ V.

For Configuration C2, the distributions of R^2 depending on U_g (Figure 16a–c) are characterized by considerable variability with two local maxima reaching the value of 1. The RE_{COF} distributions also demonstrate high variability and have a global maximum that decides the value of discrimination voltage U_g for which the correlations shown in Figure 17a–c were developed. For Configuration C2, the voltage values are higher compared to Configuration C1 (from 1.12 to 1.22 V). The obtained values of determination coefficient R^2 take lower values compared to Configuration C1. They are included in the range from about 0.89 to about 0.96, which proves a good correlation.

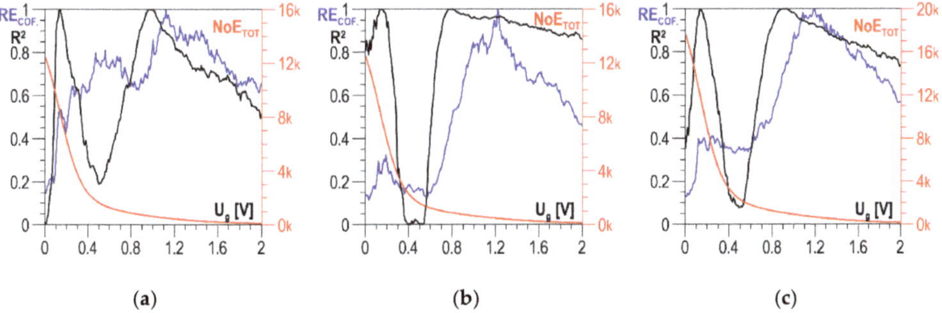

Figure 16. Averaged NoE_{TOT} distributions and distributions of coefficients RE_{COF} and R^2 as a function of threshold voltage U_g—Configuration C2 (**a**) parallel direction; (**b**) normal direction; (**c**) absolute value.

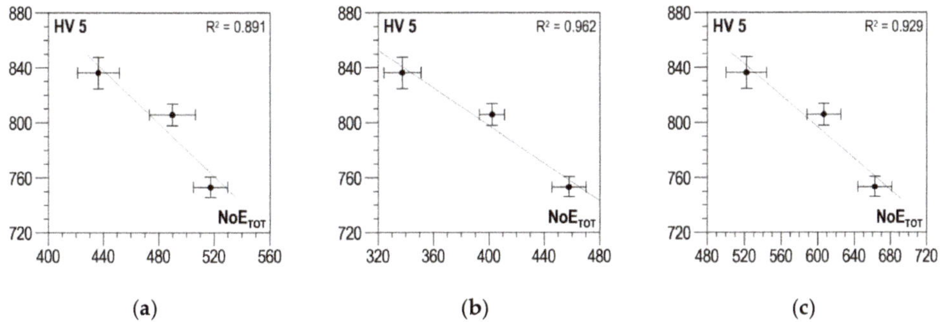

Figure 17. Linear correlations between NoE_{TOT} and $HV5$ hardness—Configuration C2 (**a**) parallel direction, $U_g = 1.12$ V; (**b**) normal direction, $U_g = 1.22$ V; (**c**) absolute value, $U_g = 1.19$ V.

For Configuration C3, the determination coefficient in the U_g-dependent distributions of R^2 (Figure 18a–c) takes values higher than 0.95 for the U_g range from 0.1 to 0.4 V. This is also the range for which the NoE_{TOT} values are higher than 0.

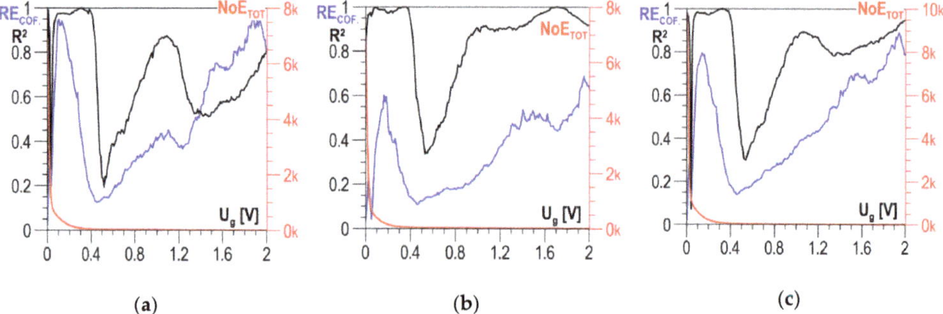

Figure 18. NoE_{TOT} distributions and distributions of coefficients RE_{COF} and R^2 as a function of threshold voltage U_g—Configuration C3 (**a**) parallel direction; (**b**) normal direction; (**c**) absolute value.

The RE_{COF} distributions for this voltage range have a local maximum that decides the selection of discrimination voltage U_g (from 0.1 to 0.16 V) for which the correlations presented in Figure 19a–c were developed. The values of determination coefficient R^2 are included in the range from about 0.97 to about 0.987, which proves a very good correlation.

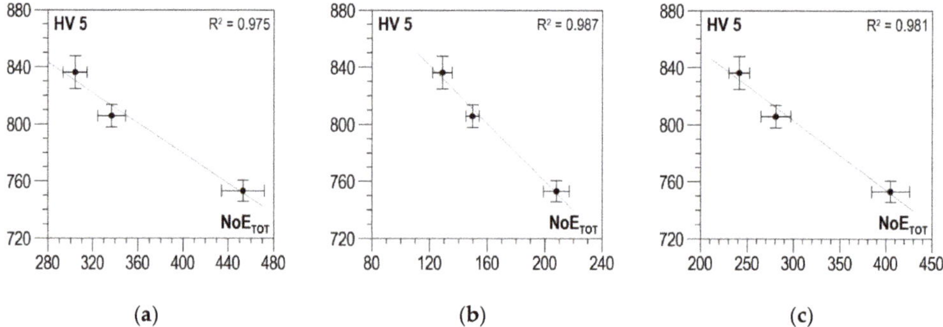

Figure 19. Linear correlations between NoE_{TOT} and $HV5$ hardness—Configuration C3 (**a**) parallel direction, $U_g = 0.1$ V; (**b**) normal direction, $U_g = 0.16$ V; (**c**) absolute value, $U_g = 0.14$ V.

In each of the configurations under analysis, a rise in hardness involves a drop in the number of events (NoE_{TOT}). For the tested specimens, martensite refinement increases with a rise in hardness, which means that a smaller number of events (NoE_{TOT}) corresponds to a structure with a smaller grain size. Similar results are obtained in [34,35], whereas completely opposite relations are found in [8]. This proves that the Barkhausen noise affects not only the grain size but also the nature of grain boundaries and their mutual orientation [35]. Moreover, for a specific configuration of factors affecting the Barkhausen noise, a repeatability of results is achieved. The electromagnetic phenomena occurring in the microscale due to the effect of the size of the measuring converters make use of diagnostic relations developed based on macroscale parameters (averaged information on the material state from the macro area). Compared to the conclusions presented in [36], the diagnostic information obtained in this manner is both repeatable and reliable.

6. Summary and Conclusions

The Barkhausen noise and *HV* hardness were tested in specimens subjected to plastic strain and thermochemical treatment. The measurement data were processed numerically, which resulted in distributions of the total number of events (NoE_{TOT}) as a function of discrimination voltage U_g. Next, the linear correlation determination coefficient (R^2) and the resolution coefficient (RE_{COF}) introduced by the authors were analyzed in parallel to find the optimal correlation. The obtained results prove the great potential of using the Barkhausen noise total number of events (NoE_{TOT} BN) for the development of correlations enabling hardness determination. Distributions of the total number of events (NoE_{TOT}) in selected intervals of threshold voltage U_g make it possible to develop correlations characterized by a high determination coefficient R^2 in the case of changes in hardness due to both strain hardening and thermochemical treatment.

A new methodology was developed for the determination of diagnostic correlations to solve inverse problems of non-destructive testing consisting of hardness evaluation based on the Barkhausen noise total number of events (NoE_{TOT} BN). However, it has to be emphasized that if these correlations are to be applied, factors that affect the diagnostic signal, such as the surface shape and roughness or the element edge impact, must be taken into account. The factors are described in detail in [7,17]. The sampling frequency can also play a significant role, because some pulses could overlap in the time scale.

Compared to classic static and dynamic methods of hardness evaluation using an indenter, the developed method of measuring hardness based on the total number of events (NoE_{TOT}) of the Barkhausen noise causes no plastic strain on the tested surface and makes it possible to perform hardness measurements in the same place repeatedly.

Additionally, it is possible to utilize other properties of the Barkhausen noise to develop multi-property correlations [15], where each property will have a specific weight established based on an analysis of the determination coefficient (R^2) and the resolution coefficient introduced by the authors (RE_{COF}).

The authors, as well as other scientists [37], are aware of the deficit in basic research on the relations between changes in a broadly understood state of the material and changes in electromagnetic properties, both at the micro and macroscopic scale. Another important research issue that emerged while analyzing the testing results is the impact of the plastic strain process (whether it proceeds continuously or in stages) on changes in parameters of the Barkhausen noise quantitative description. The results of such research would enable more confident actions in inverse problems of non-destructive testing using electromagnetic phenomena.

Author Contributions: conceptualization, M.R. and K.F.; methodology, K.F.; software, K.F.; validation, M.R. and K.F.; investigation, M.R. and K.F.; resources, K.F.; data curation, K.F.; writing—original draft preparation, M.R., K.F. and K.S.; writing—review and editing, M.R., K.F. and K.S.; visualization, K.F.; supervision, M.R. and K.S. All authors have read and agreed to the published version of the manuscript.

Funding: This research received no external funding.

Conflicts of Interest: The authors declare no conflict of interest.

References

1. Kypris, O.; Nlebedim, I.C.; Jiles, D.C. Measuring stress variation with depth using Barkhausen signals. *J. Magn. Magn. Mater.* **2016**, *407*, 377–395. [CrossRef]
2. Alessandro, B.; Beatrice, C.; Bertotti, G.; Montorsi, A. Domain-wall dynamics and Barkhausen effect in metallic ferromagnetic materials. I. Theory. *J. Appl. Phys.* **1998**, *68*, 2901–2907. [CrossRef]
3. Sipahi, L.B.; Jiles, D.C.; Chandler, D. Comprehensive analysis of Barkhausen emission spectra using pulse height analysis, frequency spectrum, and pulse wave form analysis. *J. Appl. Phys.* **1993**, *73*, 5623–5625. [CrossRef]

4. Ortega-Labra, O.; Le Manh, T.; Martinez-Ortiz, P.; Hallen, J.M.; Perez-Benitez, J.A. A novel system for non-destructive evaluation of surface stress in pipelines using rotational continuous magnetic Barkhausen noise. *Meas. J. Int. Meas. Confed.* **2019**, *136*, 761–774. [CrossRef]

5. Liu, X.; Shang, W.; He, C.; Zhang, R.; Wu, B. Simultaneous quantitative prediction of tensile stress, surface hardness and case depth in medium carbon steel rods based on multifunctional magnetic testing techniques. *Meas. J. Int. Meas. Confed.* **2018**, *128*, 455–463. [CrossRef]

6. Ding, S.; Tian, G.; Moorthy, V.; Wang, P. New feature extraction for applied stress detection on ferromagnetic material using magnetic Barkhausen noise. *Meas. J. Int. Meas. Confed.* **2015**, *73*, 515–519. [CrossRef]

7. Roskosz, M.; Fryczowski, K. Magnetic methods of characterization of active stresses in steel elements. *J. Magn. Magn. Mater.* **2020**, *499*, 166272. [CrossRef]

8. Pal'a, J.; Bydžovský, J. Barkhausen noise as a function of grain size in non-oriented FeSi steel. *Meas. J. Int. Meas. Confed.* **2013**, *46*, 866–870. [CrossRef]

9. He, B.; Wei, C.; Ding, S.; Shi, Z. A survey of methods for detecting metallic grinding burn. *Meas. J. Int. Meas. Confed.* **2019**, *134*, 426–439. [CrossRef]

10. Wilson, J.W.; Karimian, N.; Liu, J.; Yin, W.; Davis, C.L.; Peyton, A.J. Measurement of the magnetic properties of P9 and T22 steel taken from service in power station. *J. Magn. Magn. Mater.* **2014**, *360*, 52–58. [CrossRef]

11. Moorthy, V.; Shaw, B.A.; Evans, J.T. Evaluation of tempering induced changes in the hardness profile of case-carburised EN36 steel using magnetic Barkhausen noise analysis. *NDT E Int.* **2003**, *36*, 43–49. [CrossRef]

12. Padovese, L.R.; Martin, N.; Millioz, F. Time–frequency and time-scale analysis of Barkhausen noise signals. *Proc. Inst. Mech. Eng. Part G J. Aerosp. Eng.* **2009**, *223*, 577–588. [CrossRef]

13. O'Sullivan, D.; Cotterell, M.; Tanner, D.A.; Meszaros, I. Characterisation of ferritic stainless steel by Barkhausen techniques. *NDT E Int.* **2004**, *37*, 489–496. [CrossRef]

14. Franco, F.A.; González, M.F.R.; De Campos, M.F.; Padovese, L.R. Relation between magnetic barkhausen noise and hardness for jominy quench tests in SAE 4140 and 6150 steels. *J. Nondestruct. Eval.* **2013**, *32*, 93–103. [CrossRef]

15. Ding, S.; Tian, G.; Sutthaweekul, R. Non-destructive hardness prediction for 18CrNiMo7-6 steel based on feature selection and fusion of Magnetic Barkhausen Noise. *NDT E Int.* **2019**, *107*, 102138. [CrossRef]

16. Fryczowski, K.; Roskosz, M.; Kukla, D.; Szwed, M. Using of Barkhausen noise in hardness evaluation of carburized and inductively tempered layers on AMS 6414 steel. *Weld. Technol. Rev.* **2017**, *89*, 70–75.

17. Fryczowski, K. The Evaluation of Material Condition Using Magnetic Non-Destructive Methods. Ph.D. Thesis, Silesian University of Technology, Faculty of Energy and Environmental Engineering, Gliwice, Poland, 2018.

18. Padovese, L.R.; Martin, N. Analysing MBN signals of different materials by time-frequency methods. In Proceedings of the Fifth International Conference on Condition Monitoring and Machinery Failure Preventaion Technologies, Edinburgh, UK, 1–6 June 2008; pp. 1–12.

19. Sorsa, A.; Santa-Aho, S.; Vippola, M.; Lepistö, T.; Leiviskä, K. Utilization of frequency-domain information of Barkhausen noise signal in quantitative prediction of material properties. *AIP Conf. Proc.* **2014**, *1581*, 1256–1263.

20. Liu, T.; Kikuchi, H.; Kamada, Y.; Ara, K.; Kobayashi, S.; Takahashi, S. Comprehensive analysis of Barkhausen noise properties in the cold rolled mild steel. *J. Magn. Magn. Mater.* **2007**, *310*, 2006–2008. [CrossRef]

21. Gurruchaga, K.; Martínez -De-Guerenu, A.; Soto, M.; Arizti, F. Magnetic barkhausen noise for characterization of recovery and recrystallization. *IEEE Trans. Magn.* **2010**, *46*, 513–516. [CrossRef]

22. Vashista, M.; Moorthy, V. On the shape of the magnetic Barkhausen noise profile for better revelation of the effect of microstructures on the magnetisation process in ferritic steels. *J. Magn. Magn. Mater.* **2015**, *393*, 584–592. [CrossRef]

23. Moorthy, V.; Shaw, B.A.; Hopkins, P. Magnetic Barkhausen emission technique for detecting the overstressing during bending fatigue in case-carburised En36 steel. *NDT E Int.* **2005**, *38*, 159–166. [CrossRef]

24. Vashista, M.; Paul, S. Novel processing of barkhausen noise signal for assessment of residual stress in surface ground components exhibiting poor magnetic response. *J. Magn. Magn. Mater.* **2011**, *323*, 2579–2584. [CrossRef]

25. Ding, S.; Tian, G.Y.; Dobmann, G.; Wang, P. Analysis of domain wall dynamics based on skewness of magnetic Barkhausen noise for applied stress determination. *J. Magn. Magn. Mater.* **2017**, *421*, 225–229. [CrossRef]

26. Przybyłowicz, K. *Structural Aspects of Metal Deformation*; Wydawnictwo Naukowe PWN: Warsaw, Poland, 2002; ISBN 978-83-01-19851-0.

27. Martínez-Ortiz, P.; Pérez-Benítez, J.A.; Espina-Hernández, J.H.; Caleyo, F.; Mehboob, N.; Grössinger, R.; Hallen, J.M. Influence of the maximum applied magnetic field on the angular dependence of Magnetic Barkhausen Noise in API5L steels. *J. Magn. Magn. Mater.* **2016**, *401*, 108–115. [CrossRef]

28. Dhar, A.; Clapham, L.; Atherton, D.L. Influence of uniaxial plastic deformation on magnetic Barkhausen noise in steel. *NDT E Int.* **2001**, *34*, 507–514. [CrossRef]

29. Lindgren, M.; Lepistö, T. Effect of prestraining on Barkhausen noise vs. stress relation. *NDT E Int.* **2001**, *34*, 337–344. [CrossRef]

30. Piotrowski, L.; Augustyniak, B.; Chmielewski, M.; Landgraf, F.J.G.; Sablik, M.J. Impact of plastic deformation on magnetoacoustic properties of Fe-2%Si alloy. *NDT E Int.* **2009**, *42*, 92–96. [CrossRef]

31. Piotrowski, L.; Augustyniak, B.; Chmielewski, M.; Hristoforou, E.; Kosmas, K. Evaluation of barkhausen noise and magnetoacoustic emission signals properties for plastically deformed armco iron. *IEEE Trans. Magn.* **2010**, *46*, 239–242. [CrossRef]

32. He, M.; Matsumoto, T.; Uchimoto, T.; Takagi, T.; Chen, H.; Xie, S.; Chen, Z. Caution to apply magnetic Barkhausen noise method to nondestructive evaluation of plastic deformation in some ferromagnetic materials. *Chinese J. Mech. Eng. (English Ed.)* **2019**, *32*, 1–6. [CrossRef]

33. Kukla, D.; Piotrowski, L.; Szwed, M. Evaluation of hardness layers carburizing and induction hardened the steel AMS 6414 using magnetic non-destructive techniques. *Weld. Technol. Rev.* **2016**, *88*, 46–50.

34. Ranjan, R.; Rastogi, P.K. Magnetic Magnetic properties of decarburized steels: An investigation of the effects of grain size carbon content. *IEEE Trans. Magn.* **1987**, *23*, 1869–1876. [CrossRef]

35. Yamaura, S.; Furuya, Y.; Watanabe, T. The effect of grain boundary microstructure on Barkhausen noise in ferromagnetic materials. *Acta Mater.* **2001**, *49*, 3019–3027. [CrossRef]

36. Zeng, K.; Tian, G.; Liu, J.; Gao, B.; Qiu, F. Repeatability and stability study of residual magnetic field for domain wall characterization. *J. Magn. Magn. Mater.* **2019**, *485*, 391–400. [CrossRef]

37. Sonntag, N.; Skrotzki, B.; Stegemann, R.; Löwe, P.; Kreutzbruck, M. The role of surface topography on deformation- induced magnetization under inhomogeneous elastic-plastic deformation. *Materials* **2018**, *11*, 1518. [CrossRef] [PubMed]

Article

FEM-Based Wave Propagation Modelling for SHM: Certain Numerical Issues in 1D Structures

Magdalena Palacz *, Arkadiusz Żak and Marek Krawczuk

Department of Mechatronics and High Voltage Engineering, Faculty of Electrical and Control Engineering, Gdańsk University of Technology, Narutowicza 11/12, 80-233 Gdańsk, Poland; arkadiusz.zak@pg.edu.pl (A.Ż.); marek.krawczuk@pg.edu.pl (M.K.)
* Correspondence: mpalacz@pg.edu.pl; Tel.: +48-58-347-25-08

Received: 2 April 2020; Accepted: 24 April 2020; Published: 28 April 2020

Abstract: The numerical modelling of structural elements is an important aspect of modern diagnostic systems. However, the process of numerical implementation requires advanced levels of consideration of multiple aspects. Important issues of that process are the positive and negative aspects of the methods applied. Therefore the aim of this article is to familiarise the reader with the most important aspects related to the process of numerical modelling of one-dimensional problems related to the phenomena of the propagation of elastic waves and their application for damage detection purposes.

Keywords: wave propagation; damage detection; numerical modelling

1. Introduction

Structural condition monitoring has been a subject of great interest for many leading research groups for last several decades. It has also become more and more popular among engineers, who put a lot of effort into the reduction of maintenance costs [1]. As technical data for further analysis, various structural parameters and their changes have been considered in the case of damage detection algorithms [2,3]. This includes structural parameters obtained from: modal analysis [4–9], impedance measurements [10,11], ultrasonic inspection [12–15], wave propagation [16–30], and many others [31–35]. A simple graph is presented in Figure 1 to illustrate in numerals the most popular parameters used in damage detection algorithms. One can see that modal analysis and wave propagation are *the leaders* among many scientific and industrial interests, and the number of publications devoted to these two subjects is relatively large in comparison to ultrasonic inspection and impedance based algorithms. In the authors' opinion, the simplicity of measurements, equipment availability, and sensitivity to structural changes are the reasons for this *popularity*. Nowadays, numerical calculations based on customised numerical/computational models are ideal alternatives to actual measurements. According to the literature, numerical modelling may be considered as one of the most important aspects of structural condition monitoring [36,37]. In the following figure, Figure 2, a summary of the number of publications devoted to numerical modelling of one-dimensional structural elements, devoted to modelling of wave propagation phenomena, and finally devoted to the use of various numerical models for SHM (structural health monitoring) purposes, has been presented. In every case the number of publications increases, which is definitely a sign that the subject of this paper is of interest to the scientific society.

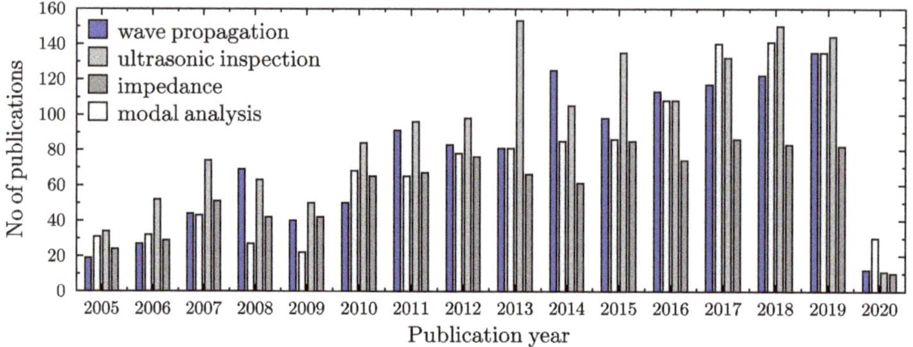

Figure 1. Graphic summary of the number of publications, related with structural parameters used for structural health monitoring (SHM) systems. Web of Science (26 March 2020).

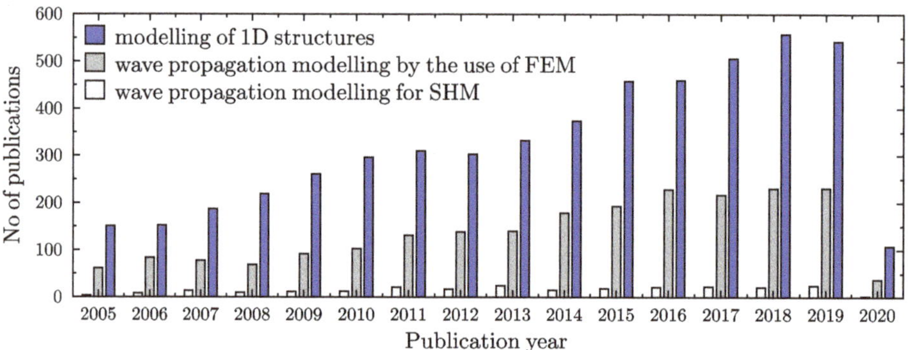

Figure 2. Graphic summary of the number of publications, related with modelling of 1D elements. Web of Science (26 March 2020).

The purpose of this article is to draw the attention of the research community to some aspects of numerical modelling that may have a significant impact on the correctness of the results of numerical computations; i.e., aspects that may lead to significant errors and misinterpretation of results. The aim of the article is also to draw attention to the problems of numerical nature, which are immanent to modelling. The authors illustrate this in the case of simple one-dimensional structural elements (rods and beams) for various discretisation methods known from the literature.

2. Technical Background

As mentioned in Introduction, proper numerical modelling of designed structures has become, nowadays, a key aspect of modern engineering, and it usually covers the total product life cycle starting from its design, current state analysis, and damage prognosis for life-time assessment. Changes in propagating waves have been one of the most popular physical measures used for damage detection purposes [38–44]. For their proper numerical analysis it is extremely important to employ reliable numerical tools [45–50]. For modelling of the wave propagation phenomenon, various commonly known computer methods have been in use. However, the classical finite element method (FEM) may be numerically inefficient, since in the process of wave propagation many higher order vibrational modes participate in the motion. For this reason very dense FE meshes, resulting from very small wavelengths participating in vibrational movements, should be applied because the size of FEs should be comparable to the wavelength of the shortest signal component. This procedure significantly increases the computational size of analysed problems, and directs researchers to other, more efficient numerical methods [51–53]. At this point, we would like to strongly emphasise that in the current paper we do not discuss the numerical consequences of taking into account such physical phenomena as wave dispersion and/or attenuation, as we instead concentrate on numerical properties of particular computational approaches.

For several decades, for wave propagation modelling purposes, the time domain spectral finite element method (TDSFEM), proposed by Patera [54], has been successfully applied [36,38,45,55–61]. It is a computational technique, which combines the properties of approximating polynomials of spectral methods and the same approach as the FEM to divide the analysed area into finite elements. The finite element properties of the method are represented by the fact that for every simple geometrical object, particular points (called nodes) with certain approximating functions (called shape functions or node functions) are defined. These functions describe the distributions of analysed physical quantities within elements and along their boundaries. Non-uniform distribution of nodes within a single element is the result of the distance between zeros of certain polynomials. This allows one to avoid the Runge phenomenon; i.e., large oscillations of approximating polynomials near the structural/element edges [62]. This fact is central for employing high-order polynomials in this method, which is impossible in the case of equally distributed nodes in classical finite element methods [36].

The process of building and solving a numerical model by the use of TDSFEM is similar to FEM and consists of the following steps:

- Division of the analysed structure into a finite number of geometrically simple elements, called spectral finite elements with a certain number of characteristic points called nodes. The spectral finite elements are connected together in a finite number of nodes located at their edges. The number of nodes in the element indicates a selection of the function used for description of the distribution of the physical quantities within the spectral finite elements, depending on their node values. These functions are called node functions or shape functions—Lobatto, Chebyshew, or Laguerre polynomials.
- Transformation of the ordinary or differential equations describing the analysed physical phenomenon to equations of the spectral finite element method. This transformation may be a weak formulation of the method, where there is a weighted residual method applied or a strong formulation, where there the method of minimising the variation functional of the phenomenon is applied. The aforementioned equations, being the problem description, are composed at the level of individual elements and are called local equations, whereas the transformations mentioned correspond to the characteristic matrices of the elements, which are derived. At this step the element matrices are aggregated to form the global characteristic matrices.
- Implementation of boundary conditions.
- Starting the solution process with the appropriate numerical method, leading to obtaining values of sought physical quantities in nodes of individual elements.

For more detailed information about TDSFEM, procedures can be found in [36]. Implementation of TDSFEM with numerous examples of sophisticated models applied to different structural elements have been presented by [42], wherein numerical models for wave propagation with dispersion effects have been described. The models have been experimentally verified and applied for damage detection analysis.

3. Numerical Considerations

In order to study the reliability of numerical models applied for wave propagation modelling, various computational investigations have been carried out. Changes in parameters, such as their accuracy or convergence, have been analysed, as they may affect the results of such simulations directly.

3.1. Rod Structure

As a numerical example, a uniform structural element of a circular cross-section has been selected, as presented in Figure 3. It has been assumed that the length of the element is $L = 2000$ mm, while its diameter is $d = 50$ mm. It has also been assumed that the element is made out of aluminium alloy (elastic modulus $E = 67.5$ GPa, Poisson ratio $\nu = 0.33$, material density $\rho = 2700$ kg/m^3). It has been assumed that the element is modelled by the use of elementary rod theory. For this type of element, the analytical solutions are commonly known. Therefore, it was easy to validate the correctness of the determined dynamic parameters; i.e., the natural frequencies and modes of natural vibrations. Several cases has been analysed for different approximation functions, in particular:

- Two-node element, Chebyshev polynomial, $p = 1$, 300 finite elements, 301 DOFs;
- Four-node element, Chebyshev polynomial, $p = 3$, 100 finite elements, 301 DOFs;
- Six-node element, Chebyshev polynomial, $p = 5$, 60 finite elements, 301 DOFs;
- Two-node element, Hermite polynomial, $p = 3$, 150 finite elements, 302 DOFs;
- Two-node element, Hermite polynomial, $p = 5$, 100 finite elements, 303 DOFs;
- B-spline polynomial, $p = 3$, 300 nodes, 302 DOFs.

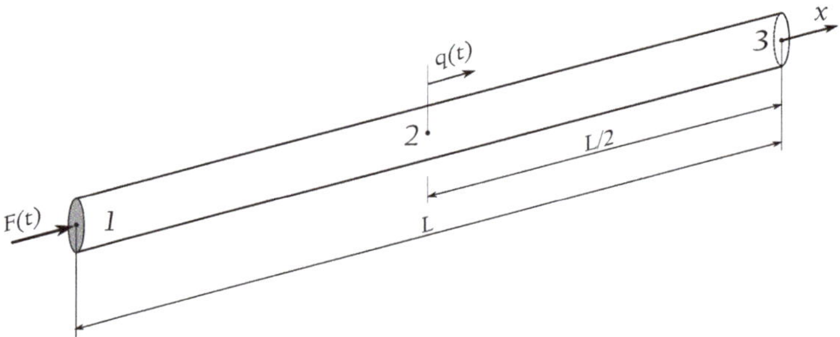

Figure 3. Geometry of a uniform aluminium rod.

It should be stressed that the application of Hermite and B-spline approximation polynomials enforces the continuity of the strain and stress fields. In the case of Hermite polynomials of the third degree, $p = 3$, this only comprises the first derivatives of the displacement fields (i.e., the strain and stress fields remain continuous but not necessarily smooth). In the case of Hermite polynomials of the fifth degree, $p = 5$, and for B-spline approximation polynomials of the third degree, $p = 3$; the first and second derivatives of the displacement fields are continuous (i.e., strain and stress fields remain continuous and smooth). It should be also added that in the the case of numerical analysis, three different types of boundary conditions have been employed. Firstly, for the analysis of natural frequencies and mode shapes, the clamped-clamped boundary conditions for rods and simply-supported boundary conditions for beams have been used. Secondly, for the analysis of wave propagation, the free–free type of boundary conditions have always been in use, despite the type of structural element.

Computational analysis of structural dynamics requires appropriate modelling of dynamic responses, which are directly linked with accurate modelling of structural natural frequencies and mode shapes. Figure 4 shows the results of numerical calculations of natural frequencies for the same rod (Figure 3) modelled by the listed numerical models. The letter N stands for the number of elements used in each analysed case. On the basis of the results presented, it may be concluded that some numerical models show certain discontinuities in their frequency spectra. This phenomenon is caused by the discontinuity of strain and stress fields expressed by spatial derivatives of the approximation functions [63]. That is particularly well visible in the case of Chebyshev approximation polynomials; however, the frequency spectra related to the use of Hermite or B-spline approximation polynomials seem free of such discontinuities.

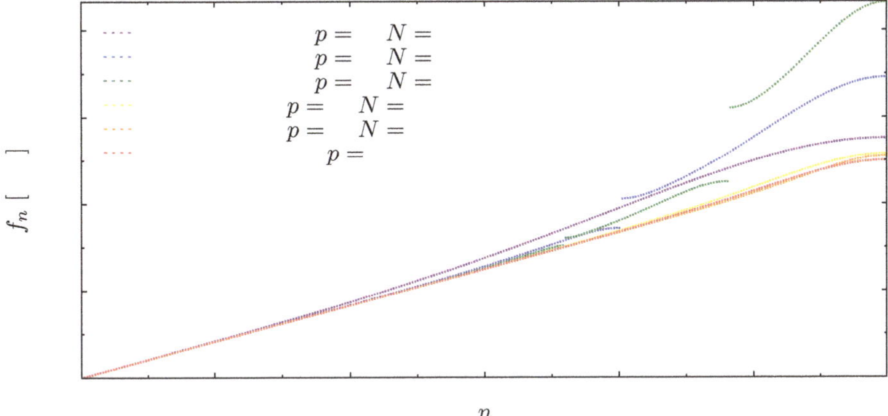

Figure 4. Natural frequencies of the rod calculated based on different approximation polynomials, in the case of the rof with fixed ends.

In order to determine the extent to which the presence of such frequency gaps influence the correctness of the results obtained, the following diagrams (Figure 5) have been presented to illustrate the changes in the relative error of the rod natural frequency spectra. It is clearly seen that the use of a higher approximation Chebyshev polynomial results in a significant increase in the relative errors of frequency spectra corresponding to higher frequencies. On the other hand, for the same order of approximation polynomials, but in the case of Hermite polynomials, results in the increase of the relative error are on a level comparable to the error present in calculations obtained by the use of other approximation functions. Moreover, a small change in the relative frequency error may be noticed for all approximation functions analysed. The value of this error seems to be dependent not only on the smoothness of the approximation polynomials used, but the type of node distribution: non-equidistant (Chebyshev $p = 3$ or Chebyshev $p = 5$) or equidistant (Hermite $p = 3$, Hermite $p = 5$, or B-spline $p = 3$).

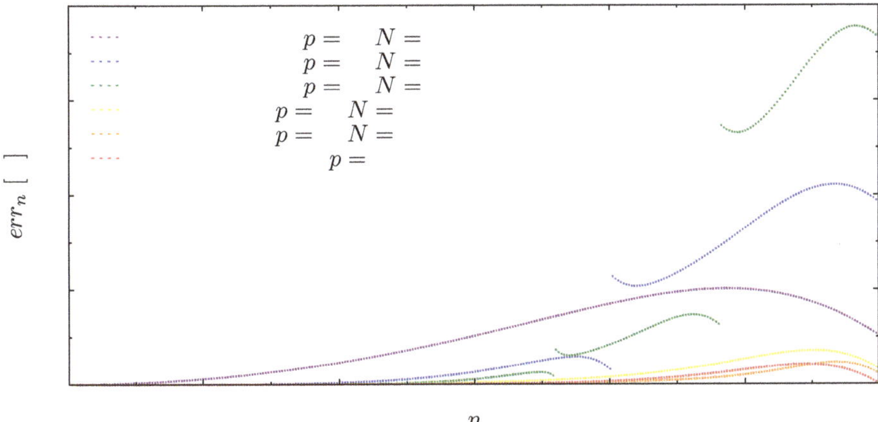

Figure 5. Relative errors of natural frequencies of the rod calculations based on different approximation polynomials, in the case of the rof with fixed ends.

In the analysis of the dynamics of the examined structure, properly represented modes of natural vibrations, obtained numerically, are of great importance. In order to verify the correctness of the results obtained by the proposed numerical models, the errors in the representations of calculated modes of natural vibrations have been determined in comparison to the modes known from analytical solutions for the elementary rod theory, as seen in Figure 6. The fitness value equal to 1 indicates the maximum degree of fitness (i.e., the coefficient of determination) of the determined modes of natural vibrations, while values smaller than 1 indicate inaccurate fitness. It can be seen from the diagrams presented in Figure 6 that in each case considered, the fitness decreases for higher frequencies. Moreover, there are such approximation polynomials (i.e., Chebyshev and Hermite), for which there are strict boundaries of total incompatibility of the determined modes. These discontinuities appear at natural frequency numbers that are multiples of the numbers of finite elements of numerical models [63]. Based on the results obtained, it can be concluded that the correctness of numerical calculations in the dynamics of the rod under investigation is significantly determined by the type of approximation polynomials. The appearance of the first frequency gap limits the usable part of the frequency spectrum in a much more significant manner than the observable decrease of the fitness for higher frequencies. Only in the cases of Chebyshev polynomials of the first degree (linear shape functions) $p = 1$, and B-spline approximation polynomials of the third degree $p = 3$, do the fitness lines remain smooth. In the first case out of these two, however, the values of fitness decrease for higher frequencies much faster; thus, the usable part of the frequency spectrum is smaller.

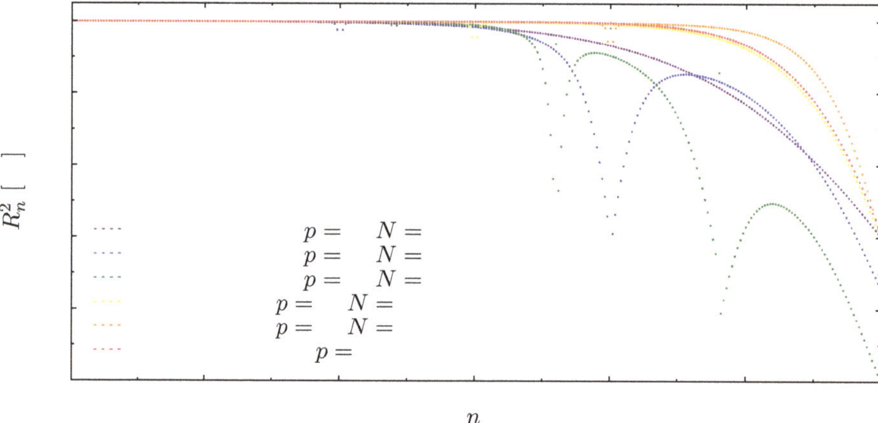

Figure 6. Fitness of natural vibration modes of the rod calculated based on different approximation polynomials, in the case of the rof with fixed ends.

Wave Propagation Analysis

Changes in the propagation of elastic waves have been successfully used in various systems for assessing the technical conditions of mechanical structures. For that purpose, numerical simulations may be an alternative method with which to avoid unexpected costs of maintenance. Modelling of elastic wave propagation requires the use of numerical models, which ensure the correct representation of the structure dynamics, especially for higher frequencies, in order to avoid any signal losses or distortions. This is why scientific research on the development of numerical methods that enable the analysis of this phenomenon has been very popular among various groups in the world.

For this purpose, simulations have been carried out to verify the developed numerical models in regard to their application to the analysis of the elastic wave propagation phenomenon. Similar to the described analysis of dynamic parameters, rod models have been developed by the use of elementary theory and the same approximation polynomials have been studied. An example of analysed excitation signals is shown in Figure 7, which illustrates the time signals and their normalised power spectral densities (*psd*). In both cases (the carrier frequency $f_c = 75$ kHz and $f_c = 150$ kHz) it was a sinusoidal signal modulated by a Hanning window. The signal carrier frequencies f_c have been selected to visualise the effect of the periodicity of various numerical models on the correctness of the results obtained. The value of the carrier frequency $f_c = 75$ kHz is outside any spectrum discontinuities, whereas the carrier frequency $f_c = 150$ kHz is located near such a discontinuity. It should be also noticed that in each case of the excitation signals, a certain range of frequencies $< f_c - 2f_m, f_c + 2f_m >$ (f_m—modulation frequency equal to f_c/m, m—modulation equal to 10) has been simulated. For the carrier frequency of 75 kHz signal excited frequencies fall into the frequency range starting from 60 kHz and ending at 90 kHz, whereas in the case of the carrier frequency of 150 kHz, signal excited frequencies fall into the frequency range starting from 120 kHz and ending at 180 kHz. In both cases, the excitation amplitude was 1 N. Figure 7 shows the time and power spectra of the excitation signals.

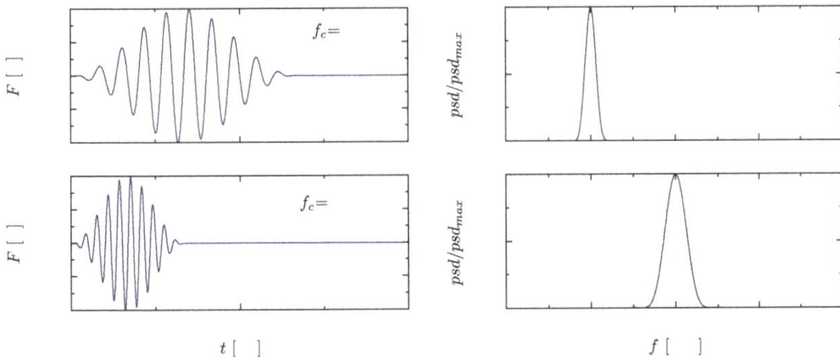

Figure 7. Excitation signals for two different carrier frequencies f_c in the time domain (**left**) and their power density spectra in the frequency domain (**right**).

In the following figures the examples of patters of propagating elastic waves have been shown. Thus, Figure 8 illustrates the changes registered for the rod under investigation in the case in which the rod is excited by a signal of the carrier frequency of 75 kHz. Illustrated signals have been registered at the excited end of the rod. The time of analysis has been set to $0.8\ ms + 2/f_m$ and it was divided into 2^{13} equal time steps. For each signal, an appropriate time window is marked, within which the propagating wave packet should be located at the end of the analysis. As a solution method for the equations of motion, the Newmark method has been chosen ($\alpha = 0.5$, $\beta = 0.25$). Material damping has been neglected. It is worth mentioning that the selected carrier frequency of 75 kHz is beyond the boundaries of frequency spectrum discontinuity areas. As it can be noticed in the presented figure, the obtained patterns of propagating elastic waves, except the one obtained for Chebyshev polynomials of the first order $p = 1$, seem to be correct; i.e., there is no visible influence of the periodicity of numerical models used on the representation of the calculated patterns of elastic waves.

The next example (Figure 9) demonstrates similar changes, but registered for the excitation signal of the carrier frequency equal to 150 kHz. This carrier frequency is located close to the frequency band only slightly visible in the diagrams in Figures 5 and 6. A narrower time window is directly related to the shorter excitation time. In the discussed case it can be clearly noticed that a higher frequency of the excitation signal definitely requires reliable numerical models. In all of the numerical models based on the use of Chebyshev approximation polynomials, the registered signals have fallen outside the time window. One can see at least two examples of the registered signals, which may indicate the discursive nature of the wave propagation phenomena. However, this is very much misleading because the applied rod theory is non-dispersive [63]. The signals obtained by the use of numerical models based on the other approximation polynomials are represented in a correct manner.

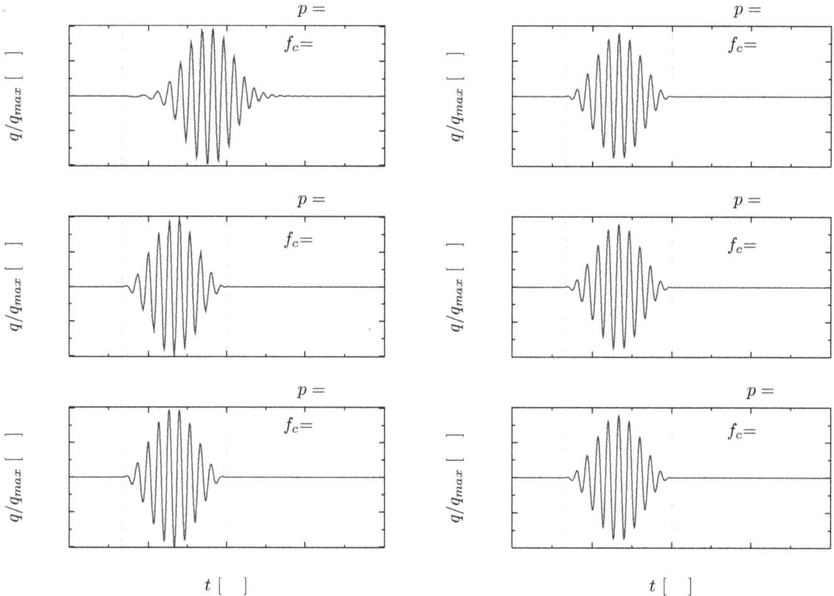

Figure 8. Wave propagation patterns of the rod calculated based on different approximation polynomials for the carrier frequency $f_c = 75$ kHz, in the case of the rod of free ends.

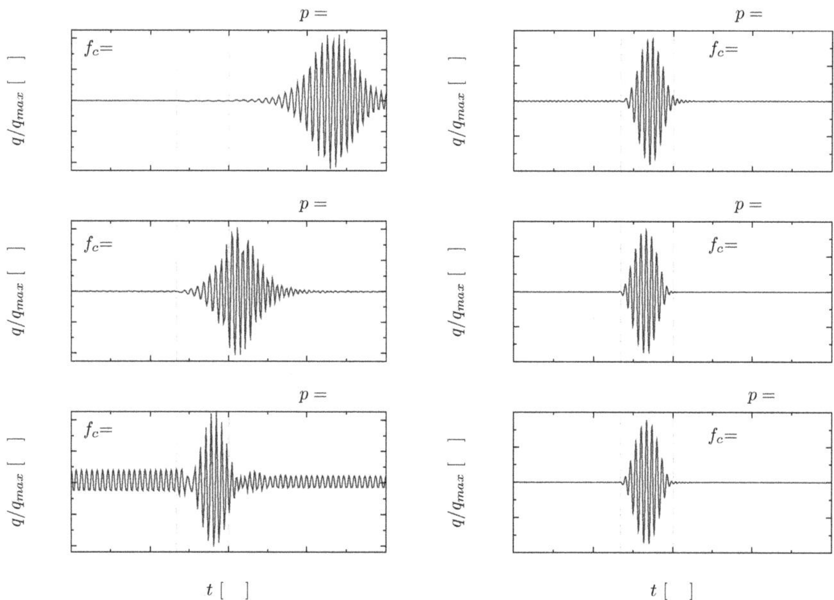

Figure 9. Wave propagation patterns of the rod calculated based on different approximation polynomials for the carrier frequency $f_c = 150$ kHz, in the case of the rod of free ends.

3.2. Beam Structure

It should be emphasised that the results of numerical investigations presented so far have been directly related to the propagation of longitudinal elastic waves in rod structural elements. However, the conclusions drawn can be easily generalised and extended onto other types of structural elements, such as beams, plates, shells, or solid elements. As the next numerical example, a beam of a uniform cross-section has been selected, as presented in Figure 3. All geometrical and material properties, as well as the form and type of boundary conditions and excitation signals were the same as in the case of the analysed rod element. The modelled beam has been divided into the same number of spectral finite elements. The elementary theory for beams has been employed in this analysis.

Figure 10 presents the results of natural frequency calculations of the beam modelled by means of several numerical models using different approximation polynomials: Hermite polynomials of the third order $p = 3$, Hermite polynomials of the fifth order $p = 5$, and B-spline approximation polynomials of the third order $p = 3$. Again, the letter N stands for the number of finite elements used in each analysed case. On the basis of the results presented, it may be concluded that numerical models based on Hermite approximation polynomials show certain discontinuities in their frequency spectra. This phenomenon is particularly visible near the boundaries of frequency bands related to the total number of finite elements used in the analysis. As before, these are the so-called frequency band gaps and their appearance is a direct proof of the periodical nature of the numerical models under investigation [63].

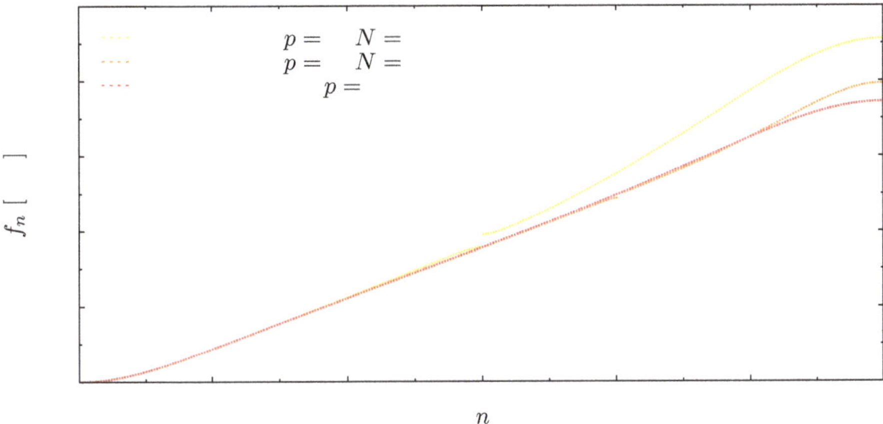

Figure 10. Natural frequencies of the beam calculated based on different approximation polynomials, in the case of a beam with simply-supported ends.

In order to determine the extent to which such frequency gaps influence the correctness of the results obtained, the following diagrams (Figure 11) have been presented to illustrate changes in the relative error of the beam's natural frequency spectrum. It is clearly seen that Hermite polynomials of the third order $p = 3$ result in significant increases in the relative errors in the parts of the frequency spectra corresponding to higher natural frequencies. On the other hand, Hermite polynomials of the fifth degree $p = 5$ result in an increase in the relative error that reaches only 10%. Moreover, the minimal changes in the relative frequency error are noticed for B-spline approximation, for the third degree of approximation polynomials $p = 3$. It can be summarised that the nature of the chances in the relative frequency error appears as independent of the order of approximation polynomials or node distribution—non-equidistant or equidistant—but it is dependent on the order of the smoothness of the displacement field in relation to the order of approximation polynomials.

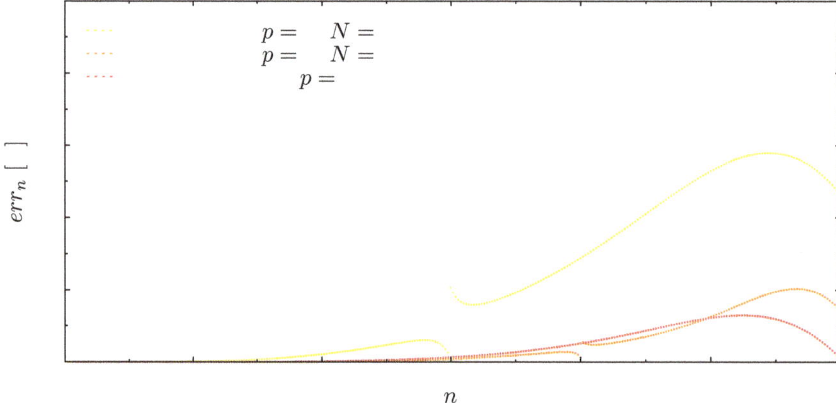

Figure 11. Relative errors of natural frequencies of the beam calculated based on different approximation polynomials, in the case of a beam with simply-supported ends.

Additionally, in the case of beam elements, in order to verify the correctness of the results achieved by the proposed numerical models, errors in the representation of the mode shapes have been determined (Figure 12). Again, the fitness value equal to 1 indicates the maximum degree of fitness (i.e., the coefficient of determination) of the determined modes of natural vibrations, while values smaller than 1 indicate inaccurate fitness.

It can be seen from the diagrams presented in Figure 12 that in each case considered the fitness decreases for higher frequencies. Moreover, in both Hermite approximation polynomials there are are strict boundaries of total incompatibility of the determined modes. These discontinuities appear at natural frequency numbers that are multiples of the number of finite elements of numerical models. As before, in the case of the beam under investigation it should be emphasised that the correctness of numerical calculations is significantly determined by both the number of finite elements and the degree of the polynomial approximation function, as these parameters determine the extent of the usable part of the frequency spectrum that is up as the first frequency band gap. Only in the case of B-spline approximation polynomials of the third degree $p = 3$ are there no visible frequency band gaps in the frequency spectra, and the line representing fitness of the calculated modes remains smooth. Additionally, in this case the usable part of the frequency spectrum is the greatest.

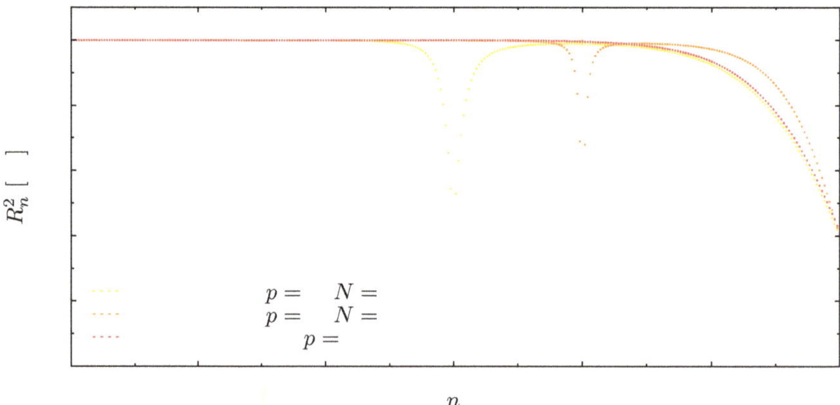

Figure 12. Fitness of natural vibration modes of the beam calculated based on different approximation polynomials, in the case of a beam with simply-supported ends.

Wave Propagation Analysis

In order to illustrate the changes in the patterns of propagating flexural waves in the aluminium beam element, certain results have already been presented in Figure 13. The excitation signals used have been presented in Figure 7. Similarly to the examples discussed above related to the wave propagation analysis in the rod element, the signal carrier frequencies f_c have been selected to take into account the effect of the periodicity of the numerical model on the correctness of the results obtained. Since the group and the phase velocities are different in the case of the elementary theory of the beam employed by the authors, some signal dispersion should be observed. As a consequence, the group speed of propagating waves also depends on the frequency, as it is equal to $c_g = 4854.1$ m/s for $f_c = 150$ kHz and $c_g = 3432.3$ m/s for $f_c = 75$ kHz. Therefore the time of the analysis have been adjusted accordingly in each particular case.

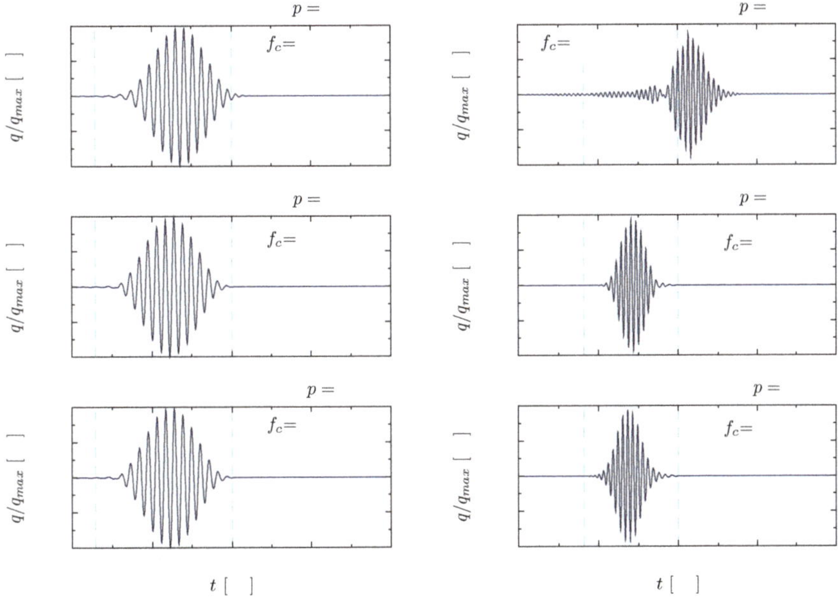

Figure 13. Wave propagation patterns of the beam calculated based on different approximation polynomials for the carrier frequencies $f_c = 75$ kHz and $f_c = 150$ kHz, in the case of a beam with free ends.

The last of the presented diagrams illustrates the changes in the patters of propagating waves in relation to the type of assumed approximation polynomials employed by numerical modes of the beam. The results presented in Figure 13 concern both excitation frequencies $f_c = 75$ and $f_c = 150$ kHz. Based on the results obtained, it can be concluded that in the case of the carrier frequency of 75 kHz the result fits quite well to the selected time window. On the other hand, in the case of the analysis carried out for the signal of the carrier frequency equal to 150 kHz, the influence of the periodic nature of the numerical model on the correctness of the obtained results is clearly visible. Signal frequency components near the frequency band gap are proportionally misrepresented in both aspects: their propagation speed related to increased values of natural frequencies and distorted shapes of modes of natural vibrations.

4. Discussion and Conclusions

The numerical studies carried out by the authors concerned dynamic responses of simple one-dimensional structures. The analysed responses covered natural frequencies and modes of natural vibrations, and the accuracy of their representations. The study aimed at the investigation of the influence of this accuracy on wave propagation responses in two one-dimensional structures, those being: a rod modelled according to the elementary theory of rods and a beam modelled according to the elementary theory of beams. Despite the fact that those structural elements are one-dimensional and the employed theories of their dynamic behaviour are the simplest available, the conclusions presented below remain valid for other structural elements. They also apply to two and three-dimensional structural elements of complex geometries and material properties as long as their numerical representations have the features typical of periodic structures, i.e., characterised by a large number of finite elements of the same or similar size, which is a typical for wave propagation analysis.

The conclusions based on the results obtained by the authors and presented in the current work can be summarised in the following way:

- Numerical models based on the use of finite elements may be thought of as representing periodic structures of certain properties as long as the they include a large number of finite elements of the same or similar size.
- The periodic nature of these models results from the level of the discontinuity of the displacement fields and the order of approximation polynomials employed to build appropriate finite elements, and manifests by the presence of so-called frequency band gaps in calculated frequency spectra.
- As a result of that, the frequency spectra are divided into a number of regions separated by these frequency band gaps, which effectively limits the usable parts of available frequency spectra, based on which the calculated dynamics responses remain unaffected by the periodic nature of discrete numerical models used.
- The number of regions is correlated with the order of approximation polynomials and the level of continuity of the displacement field (1 corresponds to the continuity of the displacement field, 2 to the continuity of the strain and stress fields, and 3 to the continuity of their derivatives).
- The biggest number of such regions was observed in the case of Chebyshev approximation polynomials and the continuity of the displacement field only, which is typical for classical FEM and TDSFEM. The only exceptions are the approximation polynomials of the first degree $p = 1$, where there are no visible frequency band gaps in the calculated spectrum and where the order of approximation polynomials is the same as the level of the continuity of the displacement field. However, the application of such approximation polynomials is characterised by the greatest average errors.
- Application of other types of approximation polynomials, such as Hermite polynomials, leads to smooth frequency spectra only in the case of rods. For calculated beam elements, the use of Hermite approximation polynomials does not improve the situation. In their cases, the orders of approximation polynomials $p = 3$ and $p = 5$ are greater than the level of the continuity of the displacement fields, which are equal to 2 and 3, respectively.
- Only in the case of B-spline approximation polynomials of the third degree $p = 3$ is the observed behaviour different, and the calculated characteristics remain smooth. In this case the order of approximation polynomials is equal to the level of the continuity of the displacement field. As a result the entire spectrum is free of frequency band gaps.
- The influence of frequency band gaps is typically associated to the upper part of the calculated frequency spectra; however, it may significantly influence the representation of modal responses in much lower parts of these spectra.
- This influence may lead to significant numerical errors, as a result of which calculated wave propagation responses may be misrepresented and possess artificial features; for example, they may suggest the presence of strong damping or dispersion.

- As a consequence it is strongly recommended by the authors, prior any wave propagation analysis, that one performs a thorough analysis of natural frequencies and modes of natural vibrations in order to recognise the regions affected by the periodicity of numerical models employed.
- The features discussed in this paper may even more profoundly influence dynamic responses of two-dimensional and three-dimensional structures. This is the subject of the authors' future research.

Author Contributions: Concept, methodology, validation, formal analysis, investigation, resources, data analysis, writing—original draft preparation, M.P.; writing—review and editing, A.Ż.; visualisation, M.P.; supervision, A.Ż., M.K., and M.P. All authors have read and agreed to the published version of the manuscript.

Funding: This research received no external funding.

Acknowledgments: The authors would like to gratefully acknowledge the support of the Academic Computer Centre in Gdańsk, the provider of the software used for the research done by the author and described in this paper.

Conflicts of Interest: The authors declare no conflict of interest.

Abbreviations

The following abbreviations are used in this manuscript:

SHM	Structural health monitoring
FEM	Finite element method
TDSFEM	Time domain spectral finite element method

References

1. Boller, C.; Chang, F.; Fujino, Y. (Eds.) *Encyclopedia of Structural Health Monitoring*; John Wiley and Sons: Hoboken, NJ, USA, 2009.
2. Kessler, S.; Spearing, S.; Soutis, C. Damage detection in composite materials using Lamb wave methods. *Smart Mater. Struct. Struct.* **2002**, *11*, 269–278. [CrossRef]
3. Inman, D.; Farrar, C.; Lopes, V., Jr.; Steffen, V., Jr. (Eds.) *Damage Prognosis for Aerospace, Civil and Mechanical Systems*; John Wiley and Sons: Hoboken, NJ, USA, 2005.
4. Farrar, C.; Doebling, S. *An Overview of Modal-Based Damage Identification Methods*; Technical Report; Los Alamos National Laboratory: Los Alamos, NM, USA, 1997.
5. Doebling, S.; Farrar, C.; Prime, M. A Summary Review of Vibration-based Damage Identification Methods. *Shock Vib. Digest* **1998**, *30*, 91–105. [CrossRef]
6. Israr, A.; Cartmell, M.; Krawczuk, M.; Ostachowicz, W.; Manoach, E.; Trendafilova, I.; Shishkina, E.; Palacz, M. On approximate anatytical solutions for vibrations in cracked plates. *Appl. Mech. Mater.* **2006**, *5–6*, 315–322. [CrossRef]
7. Raja, S.; Prathima Adya, H.; Viswanath, S. Analysis of Piezoelectric Composite Beam and Plate with Multiple Delaminations. *Struct. Health Monit.* **2006**, *5*, 255–266. [CrossRef]
8. Wang, J.; Qiao, P. Improved Damage Detection for Beam-type Structures using a Unigorm Load Surface. *Struct. Health Monit.* **2007**, *6*, 99–110. [CrossRef]
9. Sinou, J. chapter A Review of Damage Detection and Health Monitoring of Mechanical Systems from Changes in the Measurement of Linear and Non-linear Vibrations. In *Mechanical Vibrations: Measurement, Effects and Control*; Nova Science Publishers: Hauppauge, NY, USA, 2009; pp. 643–702.
10. Giurgiutiu, V.; Zagrai, A. Damage Detection in Thin Plates and Aerospace Structures with the Electro-Mechanical Impedance Method. *Struct. Health Monit.* **2005**, *4*, 99–118. [CrossRef]
11. Dhakal, D.; Neupane, K.; Thapa, C.H.; Ramanjaneyulu, G. Different techniques of structural health monitoring. *Int. J. Civ. Struct. Infrastruct. Eng. Res. Dev.* **2013**, *3*, 55–66.
12. Ludwig, R.; Lord, W. Afbeams-element formulation for the study of ultrasonic NDT systems. *IEEE Trans. Ultrason. Ferroelectr. Frequency Control* **1988**, *35*, 809–820. [CrossRef]
13. Kishore, N.; Sridhar, I.; Iyengar, N. Finite element modelling of the scattering of ultrasonic waves by isolated flaws. *NDT E Int.* **2000**, *33*, 297–305. [CrossRef]
14. Rizzo, P.; Lanza di Scalea, F. Feature Extraction for Defect Detection in Strands by Guided Ultrasonic Waves. *Struct. Health Monit.* **2006**, *5*, 297–308. [CrossRef]

15. Broda, D.; Staszewski, W.; Martowicz, A.; Uhl, T.; Silberschmidt, V. Modelling of non-linear crack-wave interactions for damage detection based on ultrasound—A review. *J. Sound Vib.* **2014**, *333*, 1097–1118. [CrossRef]

16. Su, Z.; Ye, L. Fundamental Lamb Mode-based Delamination Detection for CF/EP Composite Laminates Using Distributed Piezoelectrics. *Struct. Health Monit.* **2004**, *3*, 43–68. [CrossRef]

17. Mal, A.; Ricci, F.; Banerjee, S.; Shih, F. A Conceptual Structural Health Monitoring System based on Vibration and Wave Propagation. *Struct. Health Monit.* **2005**, *4*, 283–293. [CrossRef]

18. Monnier, T. Lamb Waves-based Impact Damage Monitoring of a Stiffened Aircraft Panel using Piezoelectric Transducers. *J. Intell. Mater. Syst. Struct.* **2006**, *17*, 411–421. [CrossRef]

19. Lestari, W.; Qiao, P. Application of Wave Propagation Analysis for Damage Identification in Composite Laminated Beams. *Compos. Mater.* **2005**, *39*, 1967–1984. [CrossRef]

20. Su, Z.; Ye, L.; Lu, Y. Guided Lamb waves for identification of damage in composite structures: A review. *J. Sound Vib.* **2006**, *295*, 753–780. [CrossRef]

21. Raghavan, A.; Cesnik, C. Review of Guided-wave Structural Health Monitoring. *Shock Vib. Digest* **2007**, *39*, 91–114. [CrossRef]

22. Grabowska, J.; Palacz, M.; Krawczuk, M.; Ostachowicz, W.; Trendafilova, I.; Manoach, E.; Cartmell, M. Wavelet analysis for damage identification in composite structures. *Key Eng. Mater.* **2007**, *347*, 253–258. [CrossRef]

23. Grabowska, J.; Palacz, M.; Krawczuk, M. Damage identification by wavelet analysis. *Mech. Syst. Signal Process.* **2008**, *22*, 1623–1635. [CrossRef]

24. Ng, C.; Veidt, M.; Lam, H. Guided wave damage characterisation in beams utilising probabilistic optimisation. *Eng. Struct.* **2009**, *31*, 2842–2850. [CrossRef]

25. Joglekar, D.M.; Mitra, M. Nonlinear analysis of flexural wave propagation through 1D waveguides with a breathing crack. *J. Sound Vib.* **2015**, *344*, 242–257. [CrossRef]

26. Mitra, M.; Gopalakrishnan, S. Guided wave based structural health monitoring: A review. *Smart Mater. Struct.* **2016**, *25*, 1–28. [CrossRef]

27. Nazeer, N.; Ratassepp, M.; Fan, Z. Damage detection in bent plates using shear horizontal guided waves. *Ultrasonics* **2017**, *75*, 155–163. [CrossRef] [PubMed]

28. Yu, Y.; Yan, N. Numerical Study on Guided Wave Propagation in Wood Utility Poles: Finite Element Modelling and Parametric Sensitivity Analysis. *Appl. Sci.* **2017**, *7*, 1063.

29. Martinez, M.; Pant, S.; Yanishevsky, M.; Backman, D. Residual stress effects of a fatigue crack on guided Lamb waves. *Smart Mater. Struct.* **2017**, *26*, 1–16. [CrossRef]

30. Kudela, P.; Radzieński, M.; Ostachowicz, W.; Yang, Z. Structural Health Monitoring system based on a concept of Lamb wave focusing by the piezoelectric array. *Mech. Syst. Signal Process.* **2018**, *108*, 21–23. [CrossRef]

31. Murayama, H.; Kageyama, K.; Naruse, H.; Shimada, A.; Uzawa, K. Application of Fiber-Optic Distributed Sensors to Health Monitoring for Full-Scale Composite Structures. *J. Intell. Mater. Syst. Struct.* **2003**, *14*, 3–13. [CrossRef]

32. Giurgiutiu, V.; Cuc, C. Embedded Non-destructive Evaluation for Structural Health Monitoring, Damage Detection, and Failure Prevention. *Shock Vib. Digest* **2005**, *37*, 83–105. [CrossRef]

33. Montalvao, D.; Maia, N.; Ribeiro, A. A Review of Vibration-based Structural Health Monitoring with Special Emphasis on Composite Materials. *Shock Vib. Digest* **2006**, *38(4)*, 295–324. [CrossRef]

34. Reda Taha, M.; Noureldin, A.; Lucero, J.; Baca, T. Wavelet Transform for Structural Health Monitoring: A Compendium of Uses and Features. *Struct. Health Monit.* **2006**, *5*, 267–295. [CrossRef]

35. Nichols, J.; Trickey, S.; Seaver, M.; Moniz, L. Use of Fiber-optic Strain Sensors and Holder Exponents for Detecting and Localizing Damage in an Experimental Plate Structure. *J. Intell. Mater. Syst. Struct.* **2007**, *18*, 51–67. [CrossRef]

36. Ostachowicz, W.; Kudela, P.; Krawczuk, M.; Żak, A. *Guided Waves in Structures for SHM: The Time-Domain Spectral Element Method*; Wiley & Sons: West Sussex, UK, 2012.

37. Hall, S. The effective management and use of structural health data. In Proceedings of the 2nd International Workshop on Structural Health Monitoring, Stanford, CA, USA, 8–10 September 1999; pp. 265–275.

38. de Basabe, J.; Sen, M. A comparison of finite-difference and spectral-element methods for elastic wave propagation in media with a fluid-solid interface. *Geophys. J. Int.* **2015**, *200*, 278–298. [CrossRef]

39. Chakraborty, A.; Gopalakrishnan, S. A spectrally formulated finite element for wave propagation analysis in functionally graded beams. *Int. J. Solids Struct.* **2003**, *40*, 2421–2448. [CrossRef]

40. Chakraborty, A.; Gopalakrishnan, S. A spectral finite element model for wave propagation analysis in laminated composite plate. *J. Vib. Acoust.* **2006**, *128*, 477–488. [CrossRef]

41. Palacz, M.; Krawczuk, M. Analysis of longitudinal wave propagation in a cracked rod by the spectral element method. *Comput. Struct.* **2002**, *80*, 1809–1816. [CrossRef]

42. Rucka, M. *Guided Wave Propagation in Structures. Modelling, Experimental Studies adn Application to Damage Detection*; Politechnika Gdańska: Gdańsk, Poland, 2011.

43. Żak, A.; Krawczuk, M. Certain numerical issues of wave propagation modelling in rods by the Spectral Finite Element Method. *Finite Elements Anal. Des.* **2011**, *47*, 1036–1046. [CrossRef]

44. Żak, A.; Krawczuk, M. Assessment of flexural beam behaviour theories used for dynamics and wave propagation problems. *J. Sound Vib.* **2012**, *331*, 5715–5731. [CrossRef]

45. Faccioli, E.; Maggio, F.; Paolucci, R.; Quarteroni, A. 2D and 3D elastic wave propagation by a pseudo-spectral domain decomposition method. *J. Seismol.* **1997**, *1*, 237–251. [CrossRef]

46. Ham, S.; Bathe, K. A finite element method enriched for wave propagation problems. *Comput. Struct.* **2012**, *94–95*, 1–12. [CrossRef]

47. Chakraborty, A.; Gopalakrishnan, S. A higher-order spectral element for wave propagation analysis in functionally graded materials. *Acta Mech.* **2004**, *172*, 17–43. [CrossRef]

48. Joglekar, D.M.; Mitra, M. Analysis of flexural wave propagation through beams with a breathing crack using wavelet spectral finite element method. *Mech Syst. Signal Process.* **2016**, *76–77*, 576–591. [CrossRef]

49. Komijani, M.; Gracie, R. An enriched finite element model for wave propagation in fractured media. *Finite Elements Anal. Des.* **2017**, *125*, 14–23. [CrossRef]

50. Wang, Y.; Hao, H.; Zhu, X.; Ou, J. Spectral Element Modelling of Wave Propagation with Boundary and Structural Discontinuity Reflections. *Adv. Struct. Eng.* **2012**, *15*, 855–870. [CrossRef]

51. Simonetti, F.; Cawley, P. On the nature of shear horizontal wave propagation in elastic plates coated with viscoelastic materials. *Proc. R. Soc. A* **2004**, *204*, 2197–2221. [CrossRef]

52. Ajith, V.; Gopalakrishnan, S. Wave propagation in a porous composite beam: Porosity determination, location and quantification. *Int. J. Solids Struct.* **2013**, *50*, 556–569. [CrossRef]

53. Ahmida, K.; Arruda, J. On the relation between complex modes and wave propagation phenomena. *J. Sound Vib.* **2002**, *255*, 663–684. [CrossRef]

54. Patera, A. A Spectral Element Method for Fluid Dynamics: Laminar Flow in a Channel Expansion. *J. Comput. Phys.* **1984**, *54*, 468–488. [CrossRef]

55. Bottero, A.; Cristini, P.; Komatitsch, D. An axisymmetric time-domain spectral-element method for full-wave simulations: Application to ocean acoustics. *J. Acoust. Soc. Am.* **2016**, *140*, 3520–3530. [CrossRef]

56. Doliński, L.; Krawczuk, M.; Palacz, M.; Żak, A. Spectral finite element method in condition monitoring and damage detection. In Proceedings of the 8th European Workshop on Structural Health Monitoring, EWSHM 2016, Bilbao, Spain, 5–8 July 2016; Volume 2, pp. 1172–1181.

57. Komatitsch, D.; Vilotte, J.; Vai, R.; Castillo-Covarrubias, J.; Sanchez-Sesma, F. The spectral element method for elastic wave equations-application to 2-D and 3-D seismic problems. *Int. J. Numer. Methods Eng.* **1999**, *45*, 1139–1164. [CrossRef]

58. Ha, S.; Chang, F.K. Optimizing a spectral element for modeling PZT-induced Lamb wave propagation in thin plates. *Smart Mater. Struct.* **2010**, *19*, 015015. [CrossRef]

59. Rucka, M. Modelling of in-plane wave propagation in a plate using spectral element method and Kane-Mindlin theory with application to damage detection. *Arch. Appl. Mech.* **2011**, *81*, 1877–1888. [CrossRef]

60. Żak, A.; Krawczuk, M. A higher order transversely deformable shell-type spectral finite element for dynamic analysis of isotropic structures. *Finite Elements Anal. Des.* **2018**, *142*, 17–29. [CrossRef]

61. Palacz, M. Spectral Methods for Modelling of Wave Propagation in Structures in Terms of Damage Detection—A Review. *Appl. Sci.* **2018**, *8*, 1124. [CrossRef]

62. Boyd, J. *Chebyshew and Fourier Spectral Methods*; Dover Publications, Inc.: Miniola, NY, USA, 2000.

63. Żak, A.; Krawczuk, M.; Palacz, M. Periodic Properties of 1D FE Discrete Models in High Frequency Dynamics. *Math. Probl. Eng.* **2016**, *2016*, 1–15. [CrossRef]

Article

Comparison of Non-Destructive Techniques for Technological Bridge Deflection Testing

Jacek Kwiatkowski [1], Wojciech Anigacz [2] and Damian Beben [2,*]

[1] J&L Consulting Jacek Kwiatkowski, 45-594 Opole, Poland; j.kwiatkowski@jl-consulting.pl
[2] Faculty of Civil Engineering and Architecture, Opole University of Technology, 45-758 Opole, Poland; w.anigacz@po.edu.pl
* Correspondence: d.beben@po.edu.pl; Tel.: +48-77-449-8739

Received: 18 March 2020; Accepted: 15 April 2020; Published: 18 April 2020

Abstract: This paper presents a comparison and assessment of usefulness of various measuring techniques (terrestrial laser scanning (TLS), tachymetry, photogrammetry) applied to establish the behavior of a suspension bridge under different load scenarios. The applied techniques were examined on the bridge with a 165 m span. The tested structure works as the technological bridge for a belt conveyor linking a lime mine and cement plant. The testing range consisted of conducting the non-contact measuring of the bridge and cable displacements under dynamic loads (during the belt conveyor movement—normal service loads) and static loads (while stopped). Tachymetric surveys were carried out using a precise total station (to obtain the reference data). A Canon 750D digital camera was applied in the photogrammetry technique. FARO Focus 3D and Trimble TX8 scanners were employed for the TLS measuring. The obtained results are especially important for bridge inspectors and managers who can use the non-contact measurements of serviced structures.

Keywords: laser scanning; photogrammetry; tachymetry; bridge deflection; non-contact measurement

1. Introduction

Generally, dynamic and static examinations are carried out to verify the bridge structure behavior versus design assumptions or to determine the actual load-carrying capacity of the bridge in connection to planned repair works. Conventional empirical experiments are normally conducted using gauges to measure accelerations, strains, displacements, and temperatures, as well as using wireless and geodesic techniques [1–5]. However, mounting the above-mentioned gauges requires the close contact with the examined construction. In some cases, there may be a problem with access to the testing objects, frequently needing the use of platforms, scaffolding, alpinists, cranes, etc. Consequently, this is the main motivation to examine and evaluate the chosen measuring techniques designed for non-contact observation on the bridge deflections.

Useful methods for observing the bridge deflections with interferometry, computer vision, digital image processing, a laser Doppler vibrometer, a Robotic Total Station, a GPS, and a video deflectometer are presented by Beben [6], Lee and Shinozuka [7], Meng et al. [8], Moschas and Stiros [9], Nassif et al. [10], Olaszek [11], Pan et al. [12], Pieraccini et al. [13], Pieraccini and Miccinesi [14], Psimulis and Stiros [15], Spencer et al. [16] and Zhang et al. [17]. Currently, some researchers are using new technology to monitor bridge behavior, such as for example the smartphone [18]. Most researchers, for example, Anigacz and Kokocinska-Pakiet [1], Schofield and Breanch [19], and Yu et al. [20], prefer a traditional way to do the object testing; i.e., they set a reference measuring grid both on the examined structure and outside. It requires signalizing the tested points on the structure. This way is completely appropriate but requires preparatory work in the field.

It should be emphasized that the latest achievements in measurement technologies allow obtaining a several million points set (called a point cloud) of the analyzed object. The point cloud can be obtained indirectly (photogrammetry) and directly (spatial terrestrial laser scanning (TLS)), as well by using both methods together. This allows receiving complete data about the object shape [21].

Jauregui et al. [22] reported an analysis of bridge displacements using the close-range photogrammetry method. The obtained results were compared to the total station readings and finite element analysis. Jiang et al. [23] presented the latest state-of-the-art on the close-range photogrammetry technique, including monitoring the bridge behavior and changes in its shape. However, the distance from the camera to the tested bridge was quite small. Yu et al. [20] described the latest measuring techniques and compared them to each other (Table 1). Nevertheless, the given data related to the TLS method is overestimated. This is because achieving precision up to 1 mm is practically impossible for the large-span bridges or limited to a short distance from the scanner to the bridge (in the case of small bridges). Yoon et al. [24] applied the TLS method to receive the crucial tunnel data.

Table 1. A set of displacement measuring techniques.

Measuring Techniques	Accuracy	Test Type	Bridge Class	Test Costs
Inclination	<cm	Static/dynamic	large rigid frame	low
Leveling	<mm	Static	no limits	low
Interferometry	<mm	Static/dynamic	no limits	expensive
Global Positioning System	cm	Static/dynamic	large span	average
Terrestrial laser scanning	<mm	Static	medium or small span	expensive
Connection pipe	<mm	Static	no limits	expensive
Electro-optical imaging	mm/cm	Static/dynamic	large span	average *
Computer vision	mm	Static/dynamic	large span	average
Tachymetric	mm	Static/dynamic	no limits	low

* Depends on used instruments.

In a typical survey with a conventional measurement instrument, the creation of a measuring grid is related to the transfer of potential errors to the observed points. Beshr [25] and Osada [26] proposed methods to reduce such errors. However, it should be emphasized that these problems did not appear in an analyzed situation. This is because the photogrammetric and TLS examinations were conducted from the same position. The connecting points of the individual scans constituted white reference spheres and measuring targets, as well as clearly visible bridge details.

The main aim of the research was to examine and evaluate the measurement possibilities for determining the bridge deflections under various load scenarios. The preliminary research for this bridge is given in [27]. Information about bridge deflections, obtained during normal operating conditions, is crucial from a structural safety point of view. In addition, the adequacy of the design principles may be assessed using data on the bridge behavior under load. Furthermore, in the case of old bridges, it is important to know that the admissible deflections of the load-bearing structure are not exceeded. One of the main purposes of the study was to minimize the preliminary activities on the examined bridge. Therefore, it was resolved to apply white reference spheres, surveying targets and well-visible bridge details as the measuring points. In the tachymetry method (as the reference method), several surveying targets were also attached on the transverse beams of the bridge. For this reason, two measuring positions were constantly maintained throughout the test. This allows a direct comparison of results of the test sessions conducted using various methods.

In the present paper, a suspension bridge with a 165 m span was tested under static and dynamic loading. In the tested bridge, typical deflection sensors could not be used because the main span of the bridge is located over a large river. Therefore, the photogrammetry and TLS methods were applied, and the obtained results were compared to a classic measuring technique (tachymetry—used as a reference method). Until now, most photogrammetric measurements were conducted at a relatively short distance from the tested structures. In this case, a greater distance from the camera to the bridge (ca. 120 m) was examined. The laser scanning results were also presented in the form of color-calibrated maps of deviations to present bridge deflections and the most strained elements (important for the bridge inspectors). This method of presenting results is new and highly useful because it supplies important information on deformation of the bridge elements. The most important results of the bridge deflections using applied methods were given and compared to each other and with appropriate values from the bridge standards. In addition, the benefits and drawbacks from each applied technique were also characterized.

2. Description of The Bridge Structure

The examined object was a technological suspension bridge over the Odra River (Southern Poland) designed for joining a marl open-pit mine with a cement plant (Figure 1). Generally, the bridge includes five spans, the main span over the river (165 m) and four spans over the floodplain. The steel transport gallery (together with the belt conveyor) is suspended on two sets of carrying cables (six wires in each set) that are supported on two towers (in the form of portals) and anchored to special reinforced concrete blocks. Each wire in the set has a 48 mm diameter. The height of each tower is 26.50 m, and the total bridge length is 285.00 m. The floodplain sections of the bridge consist of self-supporting trusses with a 30.00 m span. To guarantee lateral stability of the bridge in the case of increased lateral wind action, the bottom parts of the bridge deck are also secured by two cables intersecting at one-third of the main bridge span. An orthotropic plate, made of a 5 mm thick flat sheet, and 50 mm channel bar cross stiffeners spaced at 0.50 m are placed directly on the longitudinal beams. The outermost segments of the river span are set on the lowest beam, concentrating the poles of the pylon. Connections between the hangers, the cables, and the structure of the bridge were designed by placing them in proper steel connectors. The bridge towers rest on a reinforced concrete foundation through the steel articulated bearings. The cross-section of the pole consists of two 550 mm I-sections, connected with batten plates and bridging comprising two I-sections, 500 mm each. The sets of wires are joined to the bridge tower using the head and bolted joints in such a way as to allow limited cable movement on the tower top. The greatest lateral movement on the bearings was anticipated at both sides of the bridge entrance, while tangential bearings were applied in the towers.

Figure 1. Side view of the bridge from the upper water side.

The examined bridge was designed to carry the belt conveyor situated on the truss gallery (for limestone transport). The mass of the limestone output transported via the bridge was in the 900 to 1100 tons per hour range, and the velocity of the belt conveyor was about 7 km per hour. The bridge was loaded using evenly distributed loads (output of limestone). The quantity of limestone was automatically controlled by the feeder and was approximately constant. The bridge was recently refurbished by building a new composite (carbon fiber reinforced polymer) deck (instead of a steel orthotropic deck), changing a broken hanger, and eliminating tower cracks. Before the tests described here, no bridge damages were observed. Considering the national principles concerning evaluating bridge conditions, the structural rating was satisfactory.

3. Methodology of the Experimental Study

3.1. General Remarks

Generally, survey-engineering uses a wide scope of measuring techniques; therefore, it is important to select the most suitable one for bridge deflection testing, especially since the structure span is 165 m.

Therefore, the critical determinants, affecting the choice of measuring technique, are the bridge shape, its parameters, and possible access to the bridge (for mounting the deflection sensors). Deflections of this bridge were also measured by Rabiega et al. [28] using a laser rangefinder. The results obtained were unsatisfactory due to the belt conveyor vibration, affecting the receiver and transmitter.

The tested bridge was monitored using three independent measuring techniques: tachymetry, TLS, and photogrammetry. The measurement range of each applied method was the same. Thus, it can be assumed that those methods are mutually verified. It should be added that the measuring network with benchmarks was situated outside the testing bridge. In the analyzed case, the most important factors are the bridge length and the spatial nature of the bridge (the truss). The bridge length is important because the accuracy of determining the position of the measuring points falls with increasing sight length. It should be also added that before the main bridge testing, a preliminary examination of the selected methods was conducted to choose the most appropriate measuring shields, white spheres, and to determine the proper distances from instruments to the bridge. Additionally, various environmental conditions were also checked. The field testing was conducted after 40 years of bridge operation and a year after the bridge repair. Bridge deflections were recorded in three load scenarios:

(1) A: Static loading (the reference point for other test results, belt conveyor with limestone, without motion),
(2) B: Dynamic loading (the belt conveyor was running but it was empty, without limestone),
(3) C: Dynamic loading, i.e., normal operation (the belt conveyor with limestone was in motion).

Each load scenario (A, B, and C) lasted approximately 1 h, which gave 3 h for the full load cycle. In total, bridge testing took 12 h (four load cycles). The tachymetry method is a classical measurement method (one of the most accurate), and in this test, it was treated as the reference method. The bridge deflections were observed and compared to each other based on these techniques. To receive the reference stage for each applied method, the initial measurements were carried out under the static loading (without the limestone). All results obtained were referred to this stage.

Environmental conditions can have a significant impact on the accuracy of the laser scanning and photogrammetric measurements. Producers of measuring equipment (laser scanners) introduce limitations to the control systems of these devices to prevent measurements when external conditions exceed their values. The most commonly considered factors are vibrations, temperature, and atmospheric pressure. The instrument's internal software makes corrections automatically due to the refraction and curvature of the Earth. Some measuring instruments have the option of introducing additional factors, e.g., humidity. In summary, laser scanners have a zero and one environmental security system. This means that for the level of external factors set by the producer (or user), the instrument works correctly and after exceeding them, the instrument turns off. Factors affecting the measurement accuracy also include dust, fog, unfavorable lighting, and precipitation such as rain or snow. In addition, the impact of these factors may change over time. To check the influence of the selected environmental conditions, preliminary tests were conducted during the rain and high humidity (>80%). The obtained results were unsatisfactory due to the lack of repeatability of results (they deviated far above the standard error). Therefore, the authors tried to minimize the impact of external conditions on the accuracy of measurements by making them in the most favorable weather conditions.

3.2. The Tachymetry Technique

The tachymetry technique is related to creation of the spatial polar coordinates, i.e., vertical and horizontal angles, and the distances from the given reference points. Currently, total stations are theoretically available for measurements with a precision up to 1 mm and for measuring directions of 0.15 mgon (0.5′). However, considering practical experiences, the received accuracy is commonly significantly smaller, i.e., 1–5 mm for the distance measuring and 0.9–3 mgon (3–10′) for the direction measuring. In the analyzed example, to decrease the length of sight (the distance between the measuring points and the total station position), the positions of the total stations were located on both banks of the river (Figure 2). Thus, the length of sight did not exceed 100 m. Notwithstanding the significant discrepancy between the accuracy in the field and lab conditions, the tachymetry technique still seems to be the most precise way to determine the point location. This technique is distinguished by the discretion of measuring. A Leica TC2002 (Heerbrugg, Switzerland) total station was used for both tests. Several measuring points were selected (targets were placed on the transverse beams) for observing the bridge deflections (Figure 3). The bridge deflections were computed using the method proposed by Anigacz and Kokocinska-Pakiet [1].

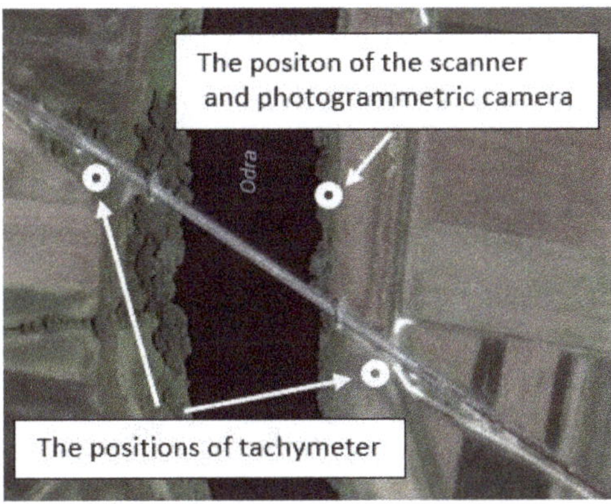

Figure 2. Top view on locations of total stations, laser scanner, and digital camera.

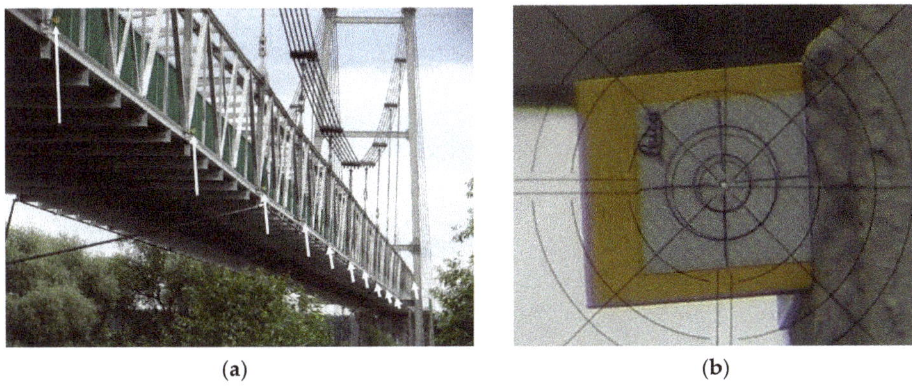

(**a**) (**b**)

Figure 3. The examined bridge structure seen through the total station: (**a**) distribution of the measurement shields on the transport gallery, (**b**) view of the measurement shield through the telescope of the total station.

The tachymetric measurements, as the most accurate measurements, were taken as the reference results for the other testing techniques. The received bridge deflection results have been confirmed by measuring from the second position of the total station. To distinguish the observed bridge deflections, the two-sided measuring shields were applied, which allowed measuring the distance from both sides of the bridge. The measuring shields were installed on the transverse beams of the bridge deck (Figure 3a). Thanks to this location of measurement points, it was possible to determine the deflection at 1/4, 1/2, and 3/4 of the span. A view through the telescope of the total station is presented in Figure 3b. The hole in the center of the shield has a 5 mm diameter. Based on multiple measurements of the total station-shield set used, a 1–2 mm real accuracy was obtained.

3.3. The Photogrammetry Technique

The photogrammetry measuring method appeared almost simultaneously with the invention of photography. The last few decades have seen the development of digital photogrammetry, which has almost completely replaced analogue photogrammetry in engineering applications. The development of analytical methods has enabled the use of non-metric cameras in engineering applications. The non-metric cameras do not have to strictly meet the requirements for metric cameras, which makes them much cheaper.

In the described case, a Canon 750D digital camera (Oita, Japan) with an 85 mm fixed focus lens was used to make the photograms. The camera was situated on a GigaPano turntable to automatically take the sequences of pictures. The distance from the camera to the tested bridge was 120 m. To verify the assumed method of photogrammetric measurements, a laboratory test was conducted. In the presented case, the lens distortion (with an 85 mm fixed focus lens) was determined based on a photo of 420 mm × 297 mm graph paper on which an ideal grid of squares was applied in AutoCAD (Version 23.0) [29].

The key parameter for photogrammetric measurements is a barrel distortion (Figure 4). This figure shows the image of a square grid deformed by the barrel distortion (red) and the standard square grid (black). Figure 5 shows an image of the camera matrix center. The standard grid of squares (blue) almost perfectly coincides with the symmetry axes of the cropped graph paper. This means that for 1 pixel resolution, any lens errors are imperceptible.

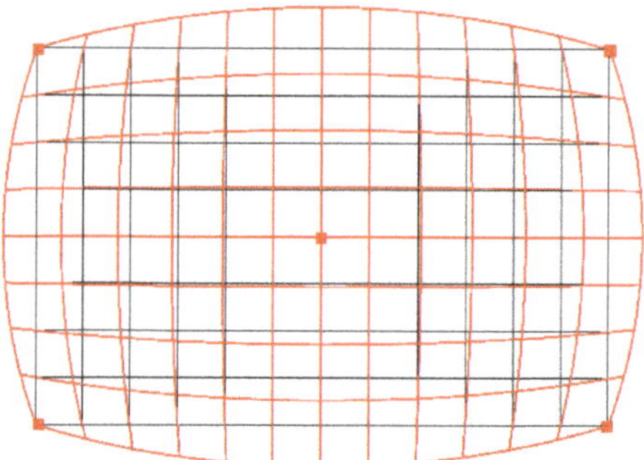

Figure 4. The barrel distortion (red) against the background of a standard grid of squares (black).

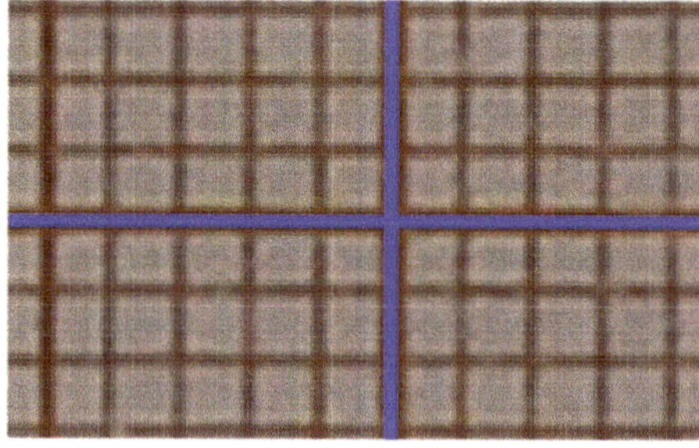

Figure 5. Camera matrix center on the background of the standard square grid (blue).

To verify the quality of the lens, differences (in pixels) between the theoretical (ideal) square grid (created in AutoCAD) and superimposed on the actual image of the square grid (obtained using the tested lens) were measured. The comparison was made in the middle of the outer sides of the frame, where the distortion should be the highest. The basis for determining the amount of distortion was taking a frontal photograph of a sheet of graph paper. Then, the corner image in the pixels was superimposed (shifting along the horizontal line) on the image halfway along the side. The number of pixels extending (in the vertical) beyond the theoretical frame (marked by the blue horizontal line—Figure 6) determines the level of barrel distortion. For the optical system used, the maximum distortion on the upper and lower edges of the frame was 4 pixels (Figure 6a), which means the percentage deformation (2 × 4)/4000 = 0.002, i.e., 0.2%. In other words, the length of the vertical line in the center of the frame is 4008 pixels instead of 4000 pixels, which gives an error of 0.2% ((4008–4000)/4000 = 0.002). One square corresponds to one pixel (Figure 6a). In the analyzed case, the Canon 750D camera matrix has a 22.3 × 14.9 mm physical size and 6000 × 4000 in pixels. The same was done for the shorter edges of the frame, i.e., shifting the corner image along the blue vertical line to the middle of the edge length shows the level of barrel distortion on the short side of the frame. The maximum distortion was 3 pixels (Figure 6b), which means that the percentage deformation is (2 × 3)/6000 = 0.001, i.e., 0.1%. The length of the horizontal line in the center of the frame instead of is 6006 pixels, 6000 pixels, which gives an error of 0.1% ((6006−6000)/6000 = 0.001).

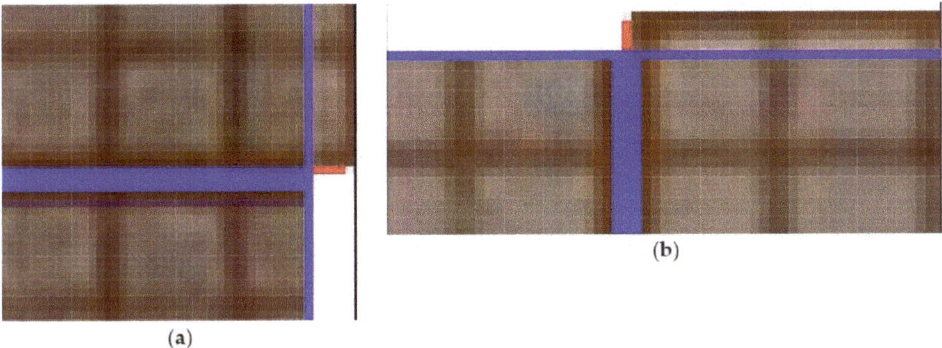

(a)

(b)

Figure 6. Maximum barrel distortion for: (**a**) the upper, (**b**) right edge of the frame, recorded in the middle of its length.

In summary of the lab tests, the registered deformation of the camera-lens system used allows for computerized adjustment with an accuracy of one pixel (Figure 6). Only the central part of the frame (2000/1333 pixels), constituting 1/9 of the entire surface of the frame, was used for the measurement. This is due to use of stitching covering 1/3 of the frame. For the 2000/1333 pixel frame field used, the distortion does not exceed 1 pixel, which is within the interpretational error limit. Therefore, for the expected measurement accuracy and the inability to register deformations in the central part with an accuracy below 1 pixel, it is justified to abandon the calibration procedure.

The AutoPano Giga 4.0 program was applied to process the photograms. The Photoshop CC program (Version 19.0) was applied to calibrate the produced photogram. This was done using a linear rescaling to determine the size of a single pixel.

To receive the bridge deflections at the stage of the dynamic tests, the digital camera was positioned so that it could take a series of photographs at an assumed time, e.g., five photograms per second. As a result, the bridge deflection distributions were measured. To rescale the bridge photograms to the fixed pixel size, the distinctive bridge elements were taken (i.e., the distance between transverse beams of the bridge deck), which were tested earlier using the total station (Figure 3a). The supplementary measurement components, i.e., white reference spheres (Figure 7a) and measuring square shields (Figure 7b), were mounted on the bridge. The photogrammetry technique is characterized by various benefits, such as short measuring time, spatial visualization, and relatively low-cost equipment (digital camera). The results of photogrammetric measurements are dependent mainly on the following determinants:

(1) Digital camera matrix resolution,
(2) Optical system quality (a decrease of the spherical deviation, resolution),
(3) Program capacity to process the photograms, e.g., joining the individual images.

(a)   (b)

Figure 7. The supplementary measuring components fixed on the structure: (**a**) sphere, (**b**) measuring square shield (checker).

3.4. The TLS Technique

The TLS technique permits receiving a quasi-constant, three-dimensional presentation of the observed objects. The precision of the measuring effects depends mainly on the sight length, i.e., the distances between the elements of the tested object and the scanner. The number of the TLS applications is growing rapidly [21,22,30]. TLS measurements are related to determining the distance based on a difference between the reflected and emitted laser beam. Data on the obtained distances and the reflected signals are transformed into coordinates in real time. The operation principles are described in detail by Rodriguez [31].

FARO Focus 3D (Lake Mary, FL, USA) and Trimble TX8 scanners (Sunnyvale, CA, USA) (Figure 8) were applied for the bridge deflection testing. These scanners measure almost 1 million points per second and allow building a cloud consisting of even a billion points. The manufacturer declares that the measurement precision of the FARO Focus 3D scanner is on the level of ±2 mm, while the angular resolution is equal to 0.16 mrad. The maximum scanning range is up to 130 m. The maximum scanning range of the Trimble TX8 scanner is 120 m, but in favorable conditions, it may even extend to 340 m. The laser wavelength is 1.5 μm, and the scanning frequency is 1 MHz. The declared measurement accuracy is ±2 mm, and the angular resolution is 0.07 mrad. A comparison of catalogue data shows that the Trimble TX8 scanner has twice the angular resolution as the FARO Focus 3D scanner. Anigacz et al. [32] applied the FARO laser scanner during the first attempt to check the applicability of the TLS technology to bridge deflection testing. However, the results indicate that the FARO laser scanner used did not allow for obtaining accurate and satisfactory data. The quality of the scanning results of the bridge deflections was close to the maximum range of the scanner (120 m). For this reason, the quality of scans (density of the points) was too weak to obtain the bridge deflections with satisfactory accuracy.

Figure 8. Side view of the bridge measurement sets for laser scanning and photogrammetry.

The Trimble TX8 scanner allows obtaining the point clouds for the unloaded and loaded state clouds with a 5.7 mm resolution at a 30 m distance. This gave about 500 million measurement points for a single scan. The unloaded state was registered on five measuring spheres to the loaded state with accuracy under 1 mm using Trimble RealWorks Advanced Plant software version 10.2 [33]. Visualization of the deviation cloud between the unloaded and loaded state was obtained through an authoritative optimization approach that was then used for comparative analysis in Trimble RealWorks Advanced Plant 10.2. The software permits for the autointerpolation of the center of the measuring shields and spheres, which were also used in the photogrammetry technique (Figure 7).

It is important that measuring scanning sessions were compatible with each other; therefore, succeeding scans should be matched to selected bridge points. This approach significantly reduces the measurement time on the structure.

When TLS is used, the compared point clouds are recorded in a common coordinate system, or both are georeferenced if a global reference system exists. In the analyzed case, one point cloud represents the bridge state without load and the other cloud represents the state with a static load. There are two cases of point identification, i.e., characteristic points of the structure and points signaled by markers recognized by the software (e.g., checker, spheres). It should be emphasized that much better accuracy in determining the location of points is obtained when they are signaled (as in the tested bridge).

In this case, the results of laser scanning were additionally presented in the form of color deviation maps. The generation of deviation maps consists of comparing a georeferenced point cloud with a 3D model. Based on differences in the distance between the 3D model and the point cloud, the system allows you to give colors to the cloud with a given RGB value depending on the reference distance (i.e., the surface of the 3D model). In the analyzed case, navy blue was used for 0–4 mm deformations, green-yellow was used for 24–32 mm deformations, and red was used for 40–50 mm deformations. The color-defined deviation maps can be indicated by the operator performing the study.

4. Analysis of Bridge Deflections

From a practical point of view, it is important to know the maximum bridge deflections. Therefore, the authors decided to analyze the bridge behavior at selected points located at 1/4, 1/2, and 3/4 of the main bridge span. In total, 10 node points situated on the bridge deck were analyzed during the measurements (Figure 3a). In addition, two bridge towers were also observed. It should be emphasized that measurements using the tachymetry technique (the reference method) were conducted over a longer period of time. Nevertheless, the authors tried to record the data at a similar time as for other methods. The possible differences in the bridge deflections can be related to the insufficient synchronization of tests carried out with various techniques.

Figure 9 presents an example of the obtained bridge scans. The maximum deflection results for the selected bridge points are presented in Table 2. Other maximum deflections in each measurement series were similar to those presented in Table 2 (the maximum difference did not exceed 1–3% for considered load scenarios). Therefore, it can be assumed that the results obtained were relatively convergent in individual load cycles.

Figure 9. Bridge deflections under static loading (A) in chosen points received from the TLS technique (Trimble scanner).

Table 2. Results of the maximum recorded bridge deflections (in mm).

Measuring Technique	Load Scenarios	Bridge Deflection at the Span of:		
		1/4	1/2	3/4
Tachymetry (reference method)	A	24.0	49.1	37.7
	B	0.3	2.1	1.7
	C	19.7	35.4	24.1
Photogrammetry	A	18	60	46
	B	0	3	2
	C	15	43	30
TLS (FARO Focus)	A	a	a	a
TLS (Trimble TX8)	A	19	49	38
	B	b	b	b
	C	b	b	b

[a] Former TLS results [32] were considered uncertain (the quality of point cloud was weak); [b] bridge deflections under dynamic loading are not possible to extract at this stage (the next research are still required); A, B, and C mean the load scenarios according to static loading (with limestone), dynamic loading (without the limestone), and dynamic loading (with limestone), respectively.

The obtained results also show the bridge deflections for various load scenarios (A, B, and C), and it is easy to observe the differences obtained between applied measuring techniques. The presented results were calculated in relation to the dead load of the bridge (reference stage: static loading without the limestone). Maximum bridge deflections were 60 mm and 49 mm for the photogrammetry and TLS (Trimble scanner), respectively, which means that the TLS results are quite consistent with the reference method (tachymetry); in contrast, the maximum bridge displacements obtained from the photogrammetry technique are higher by 18–20% than those obtained from the tachymetry and TLS techniques. These differences are related to the pixel size of the applied digital camera (4 mm) in the photogrammetry technique. Generally, the maximum deflections were observed in the middle of the main span of the bridge. The bridge deflections at other points (at 1/4 and 3/4 of the bridge span) were much lower.

The obtained results expressly confirmed that the bridge deflections under the static loading (load scenario A) were greater than those received under dynamic loading (load scenario C). The obtained disparities differed depending on the bridge measuring points (17–37%). This is due to load scenario A, where the load (limestone) can be considered as a linear load, and it was positioned stationary on the bridge. The measurement under this load lasted at least 1 h. The bridge deflections according to the load scenario B (dynamic loading without limestone) were at a low level.

Generally, the bridge deflections received from the TLS and tachymetry techniques under the static loading (A) are quite close to each other (Table 2). The TLS method for the bridge deflection measurements requires application of an adequate scanner, i.e., Trimble TX8. The results are convergent with the reference: the tachymetry method. Using the FARO Focus 3D scanner was less precise. It was the result of the too-large distance between the apparatus and the bridge and too-small angular resolution of a laser scanner (point density). In addition, the fact that the suspension bridge has an openwork structure also affects poor results (scans of larger density than ones obtained from the FARO scanner are required). To verify the test, the same scanner (FARO) was used to conduct inventory work on a historic (the oldest in Europe) suspension cast-iron bridge in Ozimek on the Mala Panew river (Poland). The tests were conducted at a much shorter distance between the scanner and the bridge, i.e., 40 m. The recorded point cloud density allowed identifying the elements (details) of the bridge, and fully satisfactory bridge inventory results were obtained [32].

Additionally, the TLS results can be presented using a calibrated map of colors (Figure 10). This method of presenting the TLS result is helpful for determining the most strenuous bridge elements. In addition, it indicates the bridge elements (in colors) where the deformations can be exceeded due to damages (which are not visible from the bridge deck). In general, incorrect behavior of some bridge elements can be detected. With this information, the bridge inspector can pay more attention to these elements during inspections. Red indicates the largest deflections and navy blue indicates the smallest. Maximum deflections of 49 mm were obtained in the middle of the main bridge span (they were recorded in suspension cables and the upper parts of the transport gallery). Additionally, the deformations of the bridge pylons are small (Figure 10). However, due to the resolution of a single screen, Figure 10 presents an indicative image that shows the tendency of displacements of the entire bridge structure. For a more accurate reading of displacements, in software working with point clouds, it is necessary to enlarge the given bridge element (Figure 11) or to measure the location of the same point in two clouds, one representing different states of the structure and one that is georeferenced. Figure 11 shows the details of the bridge deflections under the static loads (load scenario A) in the middle part of the structure obtained using a Trimble TX8 scanner. This also presents the highest deflection of bridge elements (red—cables and upper elements of the truss).

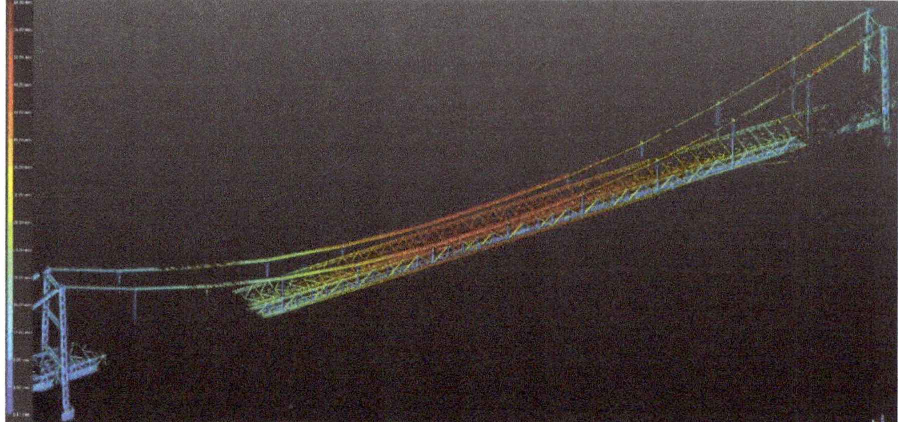

Figure 10. Bridge displacements under load scenario A shown by means of color deviation maps (Trimble TX8 scanner).

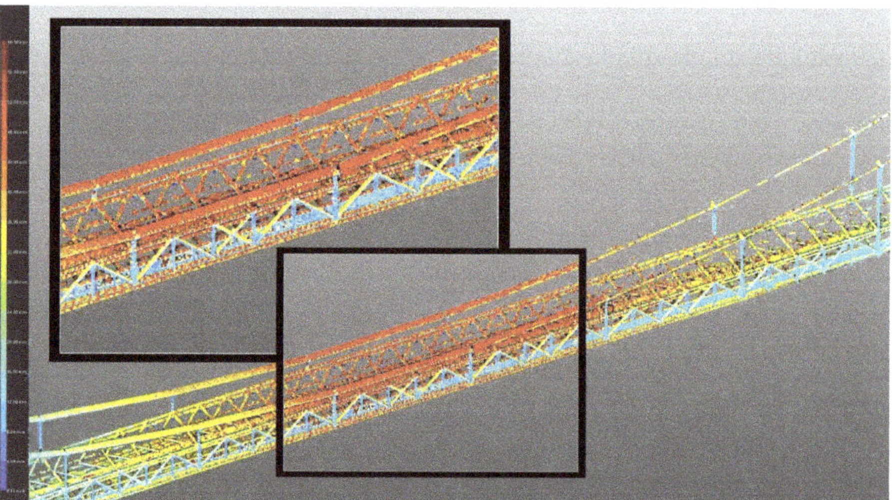

Figure 11. View of bridge deflection details (in colors) using a Trimble TX8 scanner.

Considering the TLS measuring sessions and the TLS results analysis, it is not possible to distinguish the bridge deflections under the dynamic loading. This is related to relatively rapidly changing bridge deflections in relation to the rotation speed of the laser scanner. Further research is needed on this topic. The authors believe that using the "in-line scanning" function can give interesting results. Then, multiple one-line scans (for example, near the middle of the bridge span) may allow determining the bridge deflections under the dynamic loading.

Figure 12 shows the chosen bridge deflections received from the photogrammetry technique under the static loading (load scenario A). Well-identified points of the bridge structure were chosen for the measurements. Generally, the processing of the received photogrammetric results was based on overlapping the subsequent photograms (from the various load scenarios). For example, the first photogram came from the static loading scenario (without the limestone), and the next photogram displays two photograms superimposed (with and without the loads). On the images superimposed (presented in the small windows), one pixel has been rescaled to 10 mm. Next, the number of pixels of the tested bridge components moving to each other was determined (similar to in the lab tests mentioned earlier).

Figure 12. Bridge deflections under static loading (A) obtained from the photogrammetry.

The deflection of the bridge section (crossbeam) under the dynamic loading (load scenario C) using the photogrammetry technique is presented in Figure 13. The first photogram (Figure 13a) was made when the belt conveyor was stopped (without the load), and the next one (Figure 13b) presents the image with two photograms superimposed (static loading and dynamic with the limestone (influence of the service dynamic loads)). As during the static loading (A), Figure 13b presents the superimposed images in which one pixel means 10 mm. As a result, the change in the bridge deflections under load scenario C (effect of normal bridge operation) was obtained. The results obtained indicate that the long distance from the digital camera to the bridge (ca. 120 m) did not adversely affect the quality of the bridge deflection readings. It is important to choose the appropriate camera focus, which means that the pixel size should be as large as possible, e.g., 2 mm. A larger scale (larger camera's focus) of the photograms may permit obtaining detailed data on the observed bridge elements, e.g., cracks, corrosion, excessive distortions, etc. This can be useful for bridge managers and inspectors.

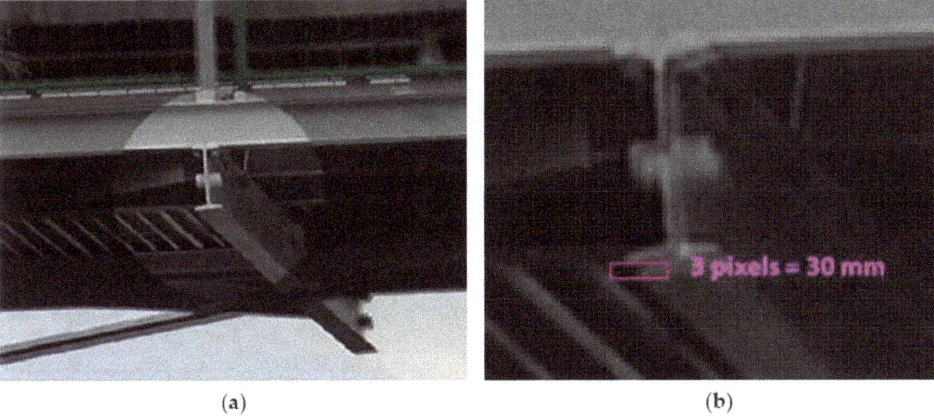

(a)   (b)

Figure 13. The analyzed detail in 3/4 of the bridge span: (**a**) without load, (**b**) images superimposed (without and with load) showing the scale of the bridge deflection under dynamic loads (load scenario C).

The obtained measurement errors in the case of the tested bridge are consistent with those reported by the manufacturers of the applied instruments. They were confirmed at reference points for multiple measurements of the measured quantity. The real measurement errors for the tested bridge were as follows: 2 mm for the tachymeter, 1 pixel (4 mm) for photogrammetry, and 2 mm for the TLS method.

The tested structure constitutes the technological bridge (without any specific requirements); therefore, the obtained results were compared to the regulations for railway and road bridges. The deflection limit of the steel railway bridges, in accordance with the bridge standard EN 1991-2 Eurocode [34], was calculated using the formula $l/15v-400$, where v is the maximum speed and l is the bridge span. The measured maximum deflection under the static loading in the middle of the main span is in the upper limit of permissible displacements (56 mm). However, considering the admissible deflections provided for the steel road bridges (PN-82/S-10052 [35]), the measured maximum test deflections are considerably below the admissible value ($l/500 = 330$ mm). To recap, the tested bridge (after repair) fulfills requirements for the steel road bridges.

It would be best to measure the bridge deflections with a few total stations and digital cameras at the same time. Then, it would be possible to control the selected points using various techniques. As a result, accurate bridge deflections would be achieved. This approach would allow a complete measuring synchronization, but this way is practically impossible to conduct. This is due to the need to monitor a dozen points on the bridge at the same time. The total station and scanner need some time to take the measurements. The full synchronization is possible only for photogrammetry method and requires using several cameras that would be simultaneously activated by radio waves.

5. Conclusions

Considering the data gained from the bridge deflection testing under various load situations, carried out with using the TLS, photogrammetry, and tachymetry technologies, these final conclusions can be formulated:

(1) Tachymetry and photogrammetry techniques allow obtaining the bridge deflection under the static and dynamic load scenarios. However, the photogrammetry method gave overestimated deflections by 18% compared to the reference method: tachymetry. The long distance from the digital camera to the bridge (ca. 120 m) did not adversely affect the obtained deflection results.

(2) At the present time, the TLS technique enables obtaining bridge deflections during the static loading with almost identical precision as for tachymetry (reference method). Nevertheless, more experimental research is required for the TLS technology, especially for determining bridge behavior under rapidly changing dynamic loads.

(3) The elaborated TLS technology is particularly useful for topologically complex structures such as truss bridges. Automatically generating colors using calibrated deviation maps of the differences between scans (e.g., without and with load) allows selecting the points and elements of the bridge to which inspectors should pay more attention. Therefore, this method can also be helpful in bridge management.

(4) The greatest bridge deflection amounted to ca. cca 49 mm and it was identified in the center of the bridge span. This deflection was measured using the reference method (tachymetry) and photogrammetry. The highest bridge deflections, generated during the normal operation (load scenario C), should be defined as comparatively small in comparison to static loading (load scenario A). The bridge deflections received from the static loading are greater than those caused by dynamic loading. The received maximum deflection was considerably smaller than the permissible value for the road bridges; simultaneously, it was in the upper limit of permissible deflection in the regulations for the railway bridges.

(5) Taking into account all the analyzed measuring technologies, the TLS is the costliest; however, it allows presenting the data using colored maps. The photogrammetry technique permits conducting the testing of the bridges and other structures relatively cheaply and with satisfactory

accuracy. For static loading, the TLS technique is better than the photogrammetry method; however, in the case of dynamic loading, it is the opposite. It would be best if both methods complement each other.

Author Contributions: Conceptualization, W.A. and D.B.; methodology, J.K., W.A., and D.B.; software, J.K.; validation, W.A. and D.B.; formal analysis, J.K., W.A., and D.B.; investigation, J.K. and W.A.; resources, D.B.; data curation, J.K., W.A., and D.B.; writing—original draft preparation, W.A. and D.B.; writing—review and editing, W.A. and D.B.; visualization, J.K.; supervision, W.A. and D.B.; project administration, W.A. and D.B.; funding acquisition, J.K., W.A., and D.B. All authors have read and agreed to the published version of the manuscript.

Funding: This research received no external funding. The APC was funded by Opole University of Technology.

Conflicts of Interest: The authors declare no conflict of interest.

References

1. Anigacz, W.; Kokocinska-Pakiet, E. Long-term monitoring of earth dam settlement. In *Life-Cycle of Structural Systems: Design, Assessment, Maintenance and Management*; Furuta, H., Frangopol, D.M., Akiyama, M., Eds.; Taylor & Francis: London, UK, 2014; pp. 195–199.

2. Beben, D. Corrugated steel plate (CSP) culvert response to service train loads. *J. Perform. Constr. Facil.* **2014**, *28*, 376–390. [CrossRef]

3. Lantsoght, E.O.L.; van der Veen, C.; de Boer, A.; Hordijk, D.A. State-of-the-art on load testing of concrete bridges. *Eng. Struct.* **2017**, *150*, 231–241. [CrossRef]

4. Maguire, M.; Roberts-Wollmann, C.; Cousins, T. Live-load testing and long-term monitoring of the Varina-Enon bridge: Investigating thermal distress. *J. Bridge Eng.* **2018**, *23*, 04018003. [CrossRef]

5. Yang, Y.B.; Xu, H.; Zhang, B.; Xiong, F.; Wang, Z.L. Measuring bridge frequencies by a test vehicle in non-moving and moving states. *Eng. Struct.* **2020**, *203*, 109859. [CrossRef]

6. Beben, D. Application of the interferometric radar for dynamic tests of corrugated steel plate (CSP) culvert. *NDT E Int.* **2011**, *44*, 405–412. [CrossRef]

7. Lee, J.J.; Shinozuka, M. A vision-based system for remote sensing of bridge displacement. *NDT E Int.* **2006**, *39*, 425–431. [CrossRef]

8. Meng, X.; Dodson, A.H.; Roberts, G.W. Detecting bridge dynamics with GPS and triaxial accelerometers. *Eng. Struct.* **2007**, *29*, 3178–3184. [CrossRef]

9. Moschas, F.; Stiros, S.C. Three-dimensional dynamic deflections and natural frequencies of a stiff footbridge based on measurements of collocated sensors. *Struct. Cont. Health Monit.* **2014**, *21*, 23–42. [CrossRef]

10. Nassif, H.H.; Gindy, M.; Davis, J. Comparison of laser Doppler vibrometer with contact sensors for monitoring bridge deflection and vibration. *NDT E Int.* **2005**, *38*, 213–218. [CrossRef]

11. Olaszek, P. Investigation of the dynamic characteristic of bridge structures using a computer vision method. *Measurement* **1999**, *25*, 227–236. [CrossRef]

12. Pan, B.; Tian, L.; Song, X. Real-time, non-contact and targetless measurement of vertical deflection of bridges using off-axis digital image correlation. *NDT E Int.* **2016**, *79*, 73–80. [CrossRef]

13. Pieraccini, M.; Parrini, F.; Fratini, M.; Atzeni, C.; Spinelli, P.; Micheloni, M. Static and dynamic testing of bridges through microwave interferometry. *NDT E Int.* **2007**, *40*, 208–214. [CrossRef]

14. Pieraccini, M.; Miccinesi, L. An interferometric MIMO radar for bridge monitoring. *IEEE Geosci. Remote Sens. Lett.* **2019**, *16*, 1383–1387. [CrossRef]

15. Psimoulis, P.A.; Stiros, S.C. Measuring Deflections of a Short-Span Railway Bridge Using a Robotic Total Station. *J. Bridge Eng.* **2013**, *18*, 182–185. [CrossRef]

16. Spencer, B.F.; Hoskere, V.; Narazaki, Y. Advances in computer vision-based civil infrastructure inspection and monitoring. *Engineering* **2019**, *5*, 199–222. [CrossRef]

17. Zhang, Q.; Ma, C.; Meng, X.; Xie, Y.; Psimoulis, P.; Wu, L.; Yue, Q.; Dai, X. Galileo Augmenting GPS Single-Frequency Single-Epoch Precise Positioning with Baseline Constrain for Bridge Dynamic Monitoring. *Remote Sens.* **2019**, *11*, 438. [CrossRef]

18. Zhao, X.; Han, R.; Yu, Y.; Hu, W.; Jiao, D.; Mao, X.; Li, M.; Ou, J. Smartphone-based mobile testing technique for quick bridge cable-force measurement. *J. Bridge Eng.* **2017**, *22*, 06016012. [CrossRef]

19. Schofield, W.; Breach, M. *Engineering Surveying*, 6th ed.; Elsevier: Amsterdam, The Netherlands, 2007.

20. Yu, Y.; Liu, H.; Li, D.; Mao, X.; Ou, J. Bridge deflection measurement using wireless mems inclination sensor systems. *Int. J. Smart Sens. Intell. Syst.* **2013**, *6*, 38–57. [CrossRef]

21. Kwiatkowski, J.L. Construction Life Cycles Monitoring of Industrial Installations Verified by Hybrid Bulk Measurement Techniques. Ph.D. Thesis, Opole University of Technology, Opole, Poland, May 2014.

22. Jauregui, D.V.; White, K.R.; Woodward, C.B.; Leitch, K.R. Noncontact photogrammetric measurement of vertical bridge deflection. *J. Bridge Eng.* **2003**, *8*, 212–222. [CrossRef]

23. Jiang, R.; Jauregui, D.V.; White, K.R. Close-range photogrammetry applications in bridge measurement: Literature review. *Measurement* **2008**, *41*, 823–834. [CrossRef]

24. Yoon, J.-S.; Sagong, M.; Lee, J.S.; Lee, K.-S. Feature extraction of a concrete tunnel liner from 3D laser scanning data. *NDT E Int.* **2009**, *42*, 97–105. [CrossRef]

25. Beshr, A.A.A. Structural deformation monitoring and analysis of highway bridge using accurate geodetic techniques. *Engineering* **2015**, *7*, 488–498. [CrossRef]

26. Osada, E. *Geodesy*, 2nd ed.; Wroclaw University of Science and Technology Press: Wroclaw, Poland, 2002.

27. Anigacz, W.; Beben, D.; Kwiatkowski, J. Displacements monitoring of suspension bridge using geodetic techniques. In *Experimental Vibration Analysis of Civil Structures*; Conte, J.P., Astroza, R., Benzoni, G., Feltrin, G., Loh, K.J., Moaveni, B., Eds.; Springer: Berlin, Germany, 2018; pp. 331–342.

28. Rabiega, J.; Lorenc, W.; Kozuch, M.P.; Kuzawa, M.J.; Lakis, L.; Watroba, P.; Kostiuk, R. Technical condition and repair as well as dynamic tests and monitoring of the multi-span technological bridge over the Odra River. In *Large Multi-Span Bridges: Design, Construction Technologies, Monitoring*; Biliszczuk, J., Ed.; Lower Silesian Educational Publishing House: Wroclaw, Poland, 2016; pp. 487–496.

29. AutoDesk Knowledge Network. 2019. Available online: https://knowledge.autodesk.com/support/recap/learn?sort=score (accessed on 25 August 2019).

30. Berenyi, A.; Lovas, T.; Barsi, A.; Dunai, L. Potential of terrestrial laser scanning in load test measurements of bridges. *Period Polytech.* **2009**, *53*, 25–33. [CrossRef]

31. Rodriguez, J.A.M. *Laser Scanner Technology*; IntechOpen: London, UK, 2012.

32. Anigacz, W.; Beben, D.; Kwiatkowski, J. Analysis of suitability of new measurement technologies to examine changes in geometric parameters of a suspension bridge. In *Maintenance, Monitoring, Safety, Risk and Resilience of Bridges and Bridge Network*; Bittencourt, T.N., Frangopol, D.M., Beck, A., Eds.; Taylor & Francis: London, UK, 2016; pp. 2204–2211.

33. Trimble Realworks 10.2 User Guide. 2017. Available online: https://www.trimble.com/3d-laser-scanning/realworks.aspx (accessed on 12 October 2019).

34. EN 1991-2 Eurocode 1. *Actions on Structures. Part. 2: Movable Loads on Bridges*; European Committee for Standardization: Brussels, Belgium, 2003.

35. PN-82/S-10052. *Bridge Structures. Steel Structures. Design*; Polish Committee of Standardization: Warsaw, Poland, 1982.

 *materials* 

Article

Methodology for Controlling the Technological Process of Executing Floors Made of Cement-Based Materials

Łukasz Sadowski *, Anna Hoła and Jerzy Hoła

Department of Building Engineering, Wroclaw University of Science and Technology, Wroclaw, Wybrzeże Wyspiańskiego 27, 50-370 Wroclaw, Poland; anna.hola@pwr.edu.pl (A.H.); jerzy.hola@pwr.edu.pl (J.H.)
* Correspondence: lukasz.sadowski@pwr.edu.pl

Received: 22 January 2020; Accepted: 17 February 2020; Published: 20 February 2020

Abstract: The article presents original complex methodology for the effective control of the entire process of executing floors made of cement-based materials. This methodology has been lacking in literature so far. The methodology was developed on the basis of many years of the authors' experience, which was acquired when diagnosing the technical condition of such floors. The methodology was preceded by a synthetic summary of the most important technological and technical requirements for floors made of cement-based materials. It was also enriched with a discussion of the problem documented by sample research results showing the state that may be the result of disregarding and not performing the necessary control activities.

Keywords: methodology; floors; cement-based materials; casting; forming; non-destructive and semi-destructive tests

1. Introduction

Currently, floors are commonly made of cement-based materials. They are used in civil engineering for various purposes, e.g., as the top layer of the floor in residential and industrial buildings, in public buildings, and in multi-car garages [1]. Floors made of cement-based materials can also be a substrate for the finishing layer of other materials, such as epoxy resin coatings, ceramic tiles, parquet, etc. [2–4]. The substrate for floors is usually concrete or a reinforced concrete ceiling, but can also often be foamed polystyrene, which is a layer of thermal or sound insulation, which was presented in Figure 1.

In order to make floors made of cement-based materials without faults, it is necessary to comply with the technological and technical requirements that are contained in the standards and set by the designer [5–10]. However, this is not the only condition that needs to be met. It is first necessary to ensure by the technical supervision that the conditions necessary to make a decision about the commencement of flooring are met, and then to monitor the entire implementation process on an ongoing basis. In the final phase, strength requirements should be checked before the floor is commissioned. It should be stated that in the case of strength parameters, not only the flexural and compressive strength should be monitored. It should also test the compressive strength along the thickness of the floor. The non-destructive ultrasonic method with the use of exponential heads is irreplaceable for this purpose. Ultrasounds are still very popular in civil engineering (for example for damage detection, as presented in [11]). The problem of lower compressive strength in the upper zone of floors, which is related to the horizontal direction of their concreting, is described, among others, in [12,13]. The subsurface tensile strength and abrasion resistance should also be controlled, which depends on the compressive strength in the upper zone of the floor. This upper zone of the floor is, after all, subjected to direct operational effects, and therefore maintaining an adequate compressive strength and subsurface tensile strength in this particular zone is very important for durability and safety of use [14–17]. This is very important when the floor is the final finishing layer. If the floor is laid directly on a concrete substrate or a reinforced concrete ceiling, the pull-off adhesion between the floor and the substrate also requires control, which is due to the fact that the durability of the exploited floor is dependent on this parameter [18–20].

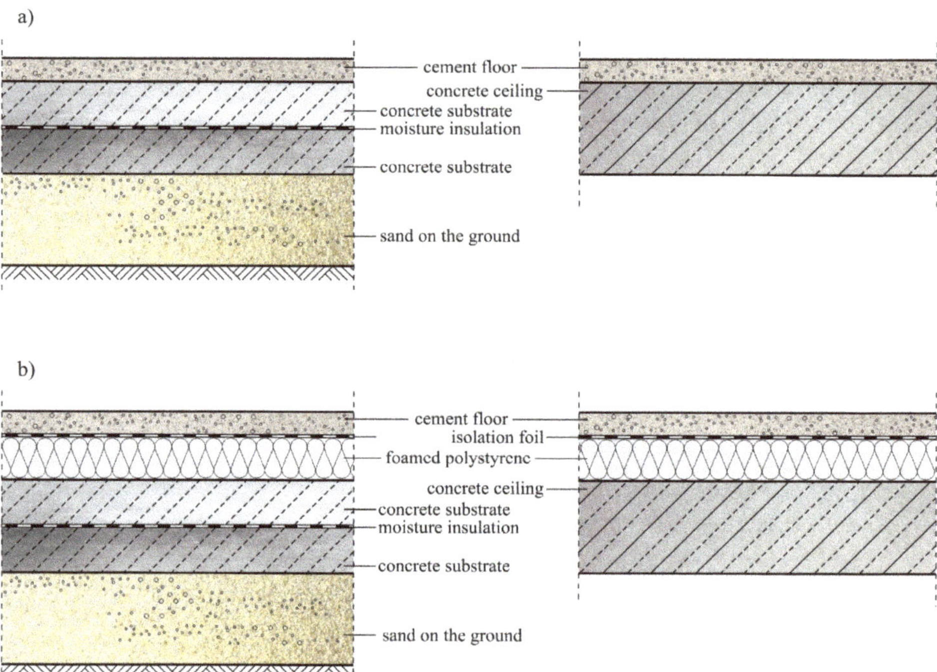

Figure 1. Applied variants of various floor systems made on: (**a**) a concrete substrate or a reinforced concrete ceiling: (**b**) a foamed polystyrene laid on a concrete substrate or reinforced concrete ceiling.

In order for the entire flooring process to be correct, its control should be effective. Comprehensive control can guarantee this. This control should be complex and methodological. However, such a methodology is missing in the literature. According to the authors, the consequence of this can be seen in the very frequent cases of cement floors being executed badly, regardless of whether they are made of ready-mix cement mortars or prepared directly at the construction site [21–24]. In construction practice, there are known cases of removing defective floors and then re-executing them again [25–29].

Considering the above, the main goal assumed in this article is to develop—based on many years of experience acquired by the authors when testing the technical condition of floors, the quality of which has been questioned by users—an original and complex methodology for the effective control of the entire process of executing floors made of cement-based materials. The developed methodology is preceded in the article by a list of more important requirements for such floors. These are not only the result of a failure to comply with the relevant requirements, but above all the incorrect control of this technological process, which can be seen to be unmethodical, ineffective and not random. The article has been enriched with a discussion comparing and contrasting the current state and standards of quality control of floor performance with comprehensive control of the developed methodology. Its purpose is to show that the practical application of the proposed methodology will effectively eliminate many implementation problems and thus improve the quality of floors made of cement-based materials.

2. A Synthetic Summary of the Major Requirements for Floors Made of Cement-Based Materials

In the case of floors made of cement-based materials, they can be made of ready-mix cement mortars, or prepared directly at the construction site. Usually, fine aggregate with a grain size of up to 2 mm (e.g., quartz sand) is used to make them. In order to improve their workability, the admixtures in the form of plasticizers are used. The use of plasticizers depends on whether the mixture is laid by hand or by pumping. Table 1 presents a synthetic summary of the major technological and technical requirements for the execution of floors made of cement-based materials. These requirements are the same, regardless of whether the floors are made of ready-mix cement mortars or prepared directly on site.

Table 1. A synthetic summary of the major technological and technical requirements for the execution of floors made of cement-based materials.

Important Technological and Technical Requirements		
Floor thickness	-	Compliant with the design, usually 40–60 mm.
Beginning of the execution of the floor	- - - - -	After building windows and doors in the building, In the absence of so-called drafts that overdry the cement-based material, Without the possibility of intense sunlight or heating of the cement-based material, At ambient and substrate temperatures between +10 and +25 °C, At a relative humidity of 65%–95%.
Preparation of the mixture	Mechanical mixing: - -	without homogenizing the mixture, with homogenizing the mixture according to standards in the mix manufacturer's requirements.
Unevenness of the surface of a concrete substrate or reinforced concrete ceiling	-	It is required not to exceed the upper and lower deviations, recorded in [30–35].
Corrugation of insulation foil laid under the floor	-	It is not allowed.
The deviation of the surface from the horizontal plane	-	It is required not to exceed the upper and lower deviations, recorded in [30–35].
Execution of peripheral expansion joints	- -	For full thickness of floors, made of non-absorbent elastic foam, Minimum joint width 7 mm.
Execution of cut expansion joints	- - - - -	Maximum field size should be 36 m^2, Maximum length of dilated field should be 6 m, The maximum proportions of the sides of the dilated field 1:1.5, The minimum depth of cut expansion joints from 1/3 to 1/2 of the floor thickness, at right angles, Joint widths from 3 to 5 mm.
Deadline for making cut-in expansion joints	-	In the first 24 h after execution of the floor.
Compressive strength	-	According to the project; usually at least 20 MPa.
Flexural strength	-	According to the project; usually at least 5 MPa.
Subsurface tensile strength	-	According to the project; usually at least 1.5 MPa [15].
Abrasion resistance	-	According to the project, usually a maximum of 22 cm^3
Cracks	-	They are not allowed [7].
Unevenness of the surface of a floor	-	It is required not to exceed the upper and lower deviations, recorded in [30–35].

3. Methodology of Controlling the Technological Process of Executing Floors made of Cement-Based Materials

The developed methodology for controlling the technological process of executing floors made of cement-based materials is described below and graphically presented in Figures 2–5. Figure 2 presents the general, and Figures 3–5 the detailed scheme of this methodology.

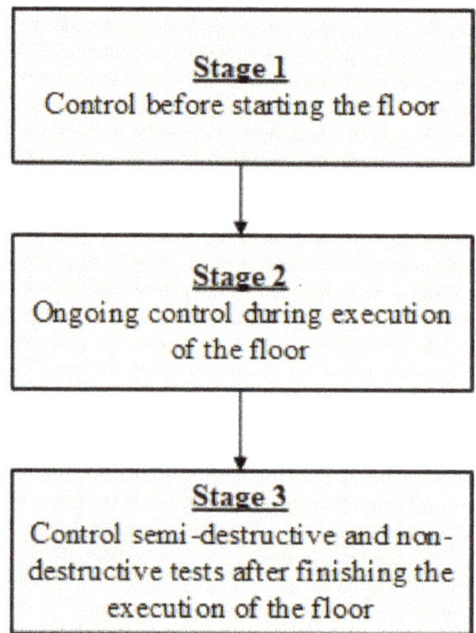

Figure 2. General diagram illustrating the developed methodology.

In Stage 1, which was presented in Figure 3, the conditions that need to be met must be checked in order to make the decision to start executing the floor.

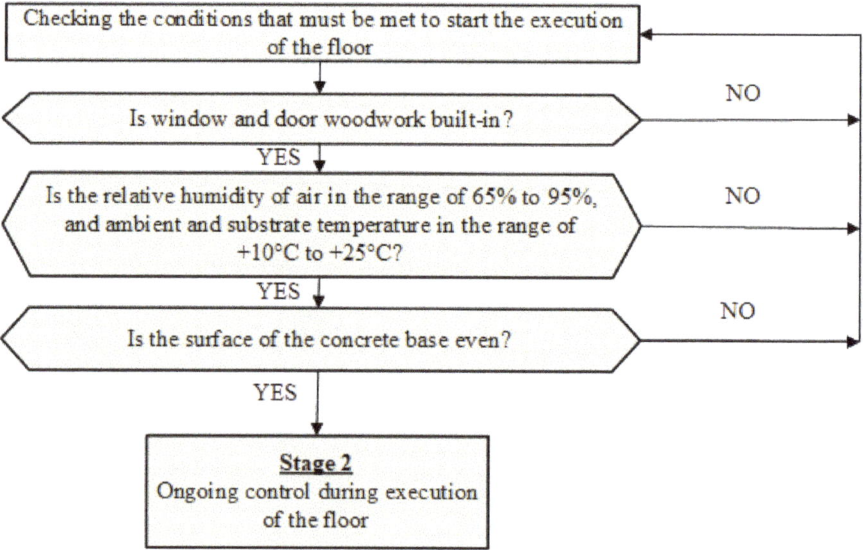

Figure 3. Methodology of controlling the technological process of executing floors made of cement-based materials—Stage 1 (control before starting the floor).

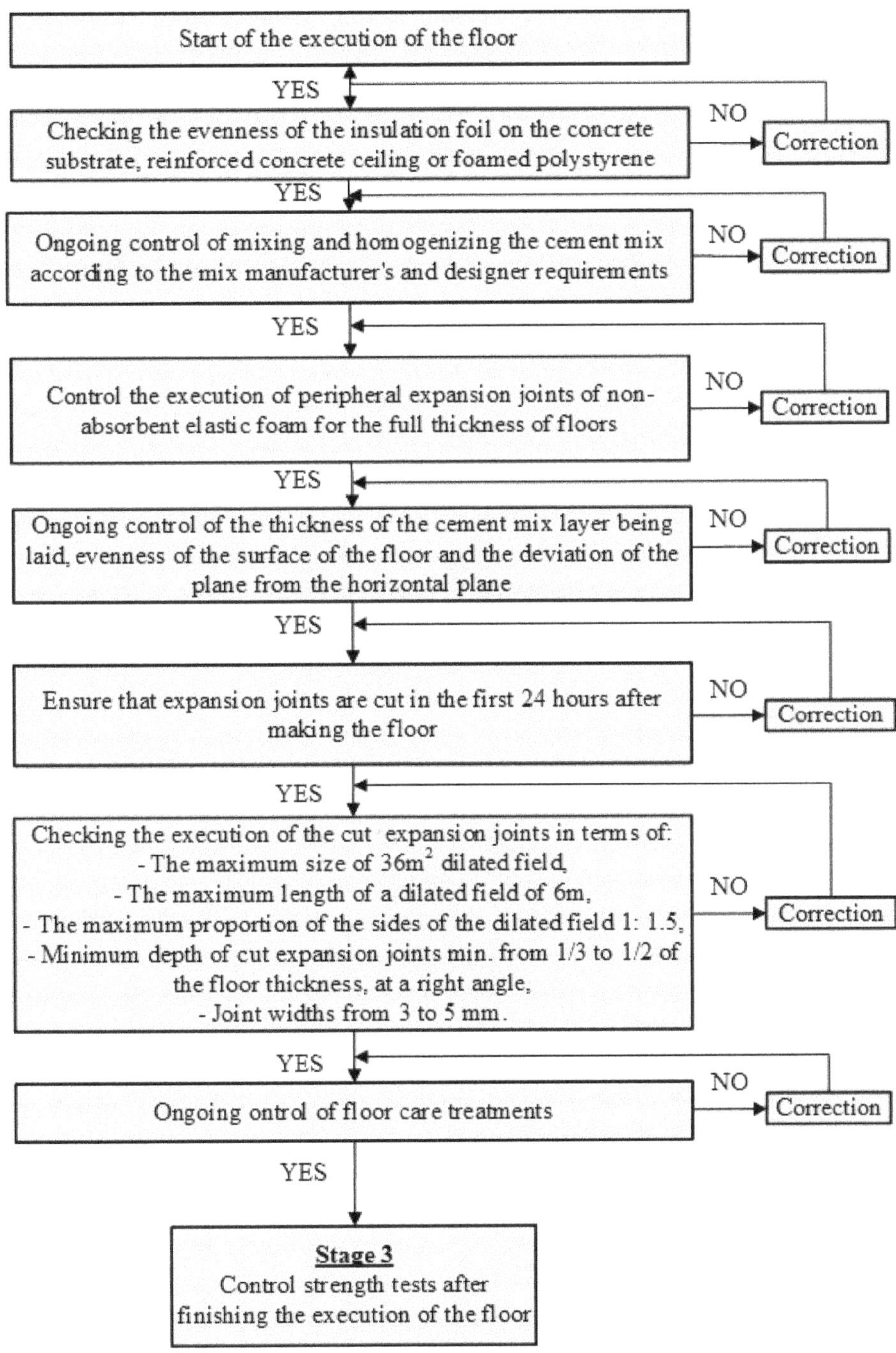

Figure 4. Methodology of controlling the technological process of executing floors made of cement-based materials—Stage 2 (ongoing control during execution of the floor).

Stage 1 begins from checking whether the window and door joinery has been built into the building. After ensuring this condition, the relative humidity of the air in the building and the air and substrate temperature on which the floor will be laid are in the required ranges [7]. The next step is to check the evenness of the upper surface of the concrete base. Unevenness of the upper surface of the concrete base is permissible, but the upper and lower deviations need to be maintained according to [30–35]. If these conditions are met, insulating foil can be laid on the substrate. The 2nd stage then begins, in which the ongoing process of floor control takes place (Figure 4).

As it can be seen from Figure 4 the first step of Stage 2 is to check the evenness of the insulation foil on the surface of the concrete substrate, reinforced concrete ceiling or foamed polystyrene. In this step, it should be borne in mind that corrugation of the insulation foil laid under the floor is not allowed. The next step is to control the mixing and homogenization of the cement mix according to the mix manufacturer's requirements. When laying the cement mix, the thickness of the mortar layer and the evenness of the top surface of the floor and its deviation from the horizontal plane should be checked. Peripheral expansion joints are then checked, which should be made of non-absorbent flexible foam, usually with a minimum thickness of 7 mm for the full thickness of the floors. Unevenness of the upper floor surface is only permissible if the upper and lower deviations are maintained, according to [30–35]. The next step is to make sure that expansion joints are cut in the first 24 h after executing the floor. Ongoing floor care, consisting of moisturizing it, should then be carried out. Then proceed to the control strength tests after the floor (stage 3).

Stage 3 consists of carrying out control strength tests of the floor, without which there should be no final acceptance and commissioning (Figure 5). This stage requires a slightly broader and more detailed explanation. As illustrated in Figure 5, the first step in Stage 3 is to check the subsurface tensile strength of the floor using the semi-destructive pull-off method in randomly made places, 28 days after concreting, and in accordance with EN 1542 [36]. The pull-off tests should begin with selecting representative measuring places and preparing the surface. Then, in these places, an incision should be made in the floor at least 15 mm deep where a steel measuring disc with a diameter of 50 mm is then attached with glue. This disc should then be pulled-off from the floor (Figure 6) and the value of the subsurface tensile strength should be determined in accordance with EN 1542 [36].

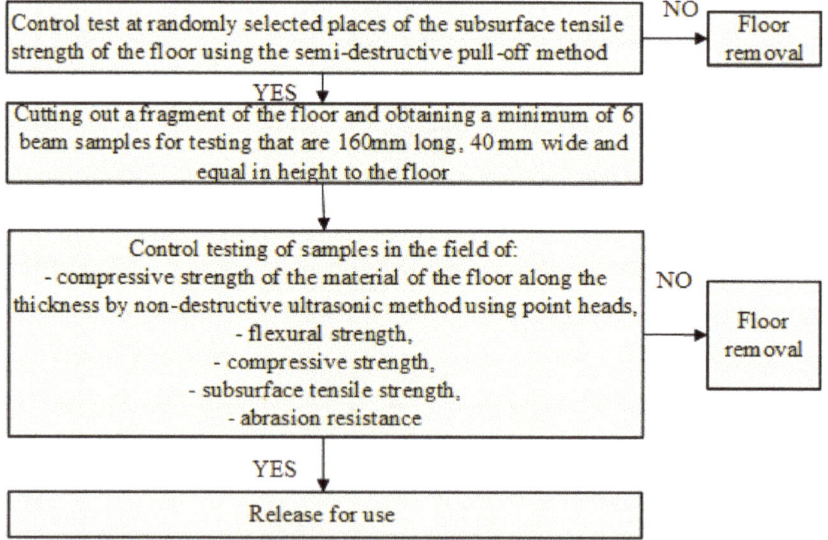

Figure 5. Methodology of controlling the technological process of executing floors made of cement-based materials—Stage 3 (control strength tests after finishing the execution of the floor).

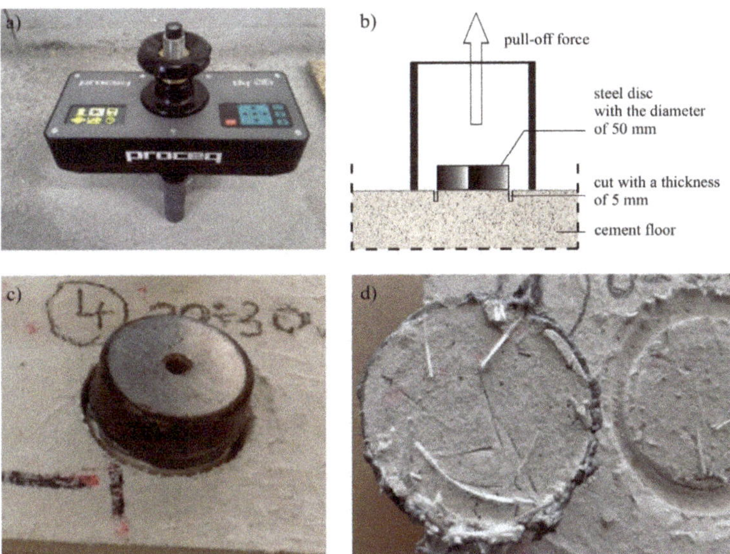

Figure 6. Control of the subsurface tensile strength of the floor using the pull-off method: (**a**) pull-off apparatus; (**b**) scheme of the method; (**c**) view of the glued steel disc; (**d**) view of the floor surface after the pull-off tests.

A fragment of the floor, with such a size that at least six beam samples with a length of 160 mm, a width of 40 mm and a height equal to the floor thickness can be obtained, should then be cut off in order to check both the compressive strength of the cement mortar along the thickness and the flexural and compressive strength of the floor. The test samples should first be subjected to ultrasonic testing along the direction of concreting the floor. For ultrasonic testing, it is proposed to use an ultrasonic probe that has special exponential heads with a frequency of 40 kHz and point contact with the tested surface. According to [37] this frequency is optimum for testing cement floor samples with a dimension of $40 \times 40 \times 160$ mm^3. The measuring points should be applied to the lateral surfaces of the samples at a spacing of 5 mm in three rows (Figure 7).

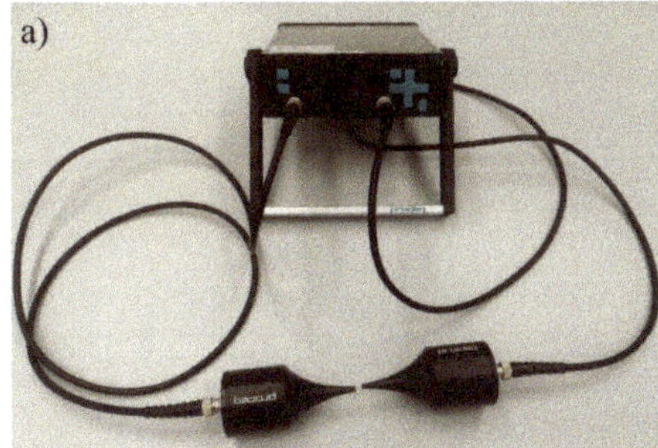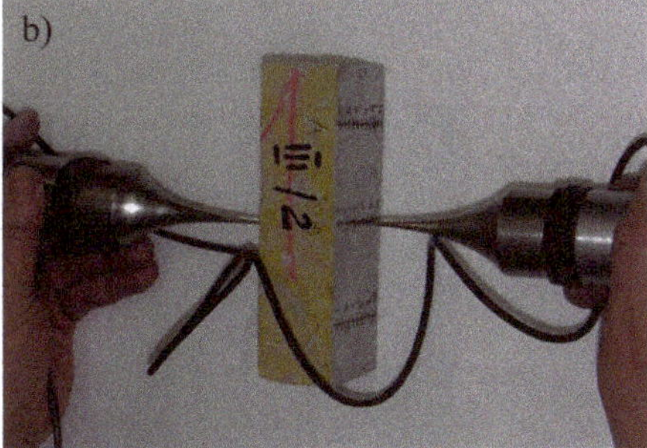

Figure 7. Example view: (**a**) head and probe for ultrasonic testing; (**b**) floor sample tested using the ultrasonic method.

After performing the ultrasonic tests, the lower zone of the beam samples should be cut so that their thickness is 40 mm, and then subjected to flexural and compressive strength tests. On the basis of the conducted ultrasonic tests, a correlation relationship (or possibly a hypothetical relationship from the literature) between the velocity of the longitudinal ultrasonic wave and the compressive strength of cement mortar should be developed for the tested floors. This relationship will be used to identify the course of compressive strength along the thickness of the tested floor. The compressive strength in the upper zone should not be less than 10% when compared to the strength in the middle zone. However, the flexural and compressive strength should not be less than the values for the floor that were specified by the designer. After meeting the above strength conditions, the floor can be approved for use. In the event of non-compliance with strength conditions, the floor, after consulting the designer, should be either allowed to be used, or removed.

4. Discussion of the Problem

As already emphasized in the introduction, the literature lacks a comprehensive methodology for controlling the process of making cement floors, which step by step would enforce subsequent control activities. This section only draws attention to those control activities, which, if neglected or omitted, result in serious flooring problems that significantly affect their final quality.

It should be started that the importance of control measures to be taken before flooring process is being started. These activities, included in Stage 1 of the developed methodology, are key to achieving a good final result.

So, currently the important problem is that it is allowed to make floors in buildings without built-in window and door joinery. This promotes the formation of drafts and intense sunlight drying of freshly laid cement mortar [38]. Due consideration is not given to the thermal and humidity conditions that should be provided in the facility, namely the required air and substrate temperature in the range from +10 to +25 °C and relative humidity in the range from 65% to 95%. Cement mortar built into the floor is not cared for by moistening it with water. These are the main reasons for the low surface compressive strength of the upper floor zone and large differences in this strength along the thickness, which is very often demonstrated on the basis of samples taken from the floors tested by the ultrasonic method. An example illustrating the above is Figure 5, which presents examples of the authors' results of the course of the longitudinal velocity of the ultrasonic wave c_l along the thickness h of the cement mortar, and the course of compressive strength f_m of the cement mortar along the thickness h of the cement floor determined on this basis. These tests were performed by a non-destructive ultrasonic method with exponential heads with a frequency of 40 kHz.

Figure 8b indicates the mean compressive strength f_{m1} of 20 MPa, which is required by the designer and declared by the cement mortar manufacturer, and also the average strength f_m obtained from the tests and shown with a vertical line. Figure 8 shows that within the subsurface zone of the floor up to a depth of about 15 mm from the upper floor surface, the compressive strength f_m was much lower than the strength f_{m1} required by the designer and declared by the manufacturer, which disqualified this floor. The occurrence of such large differences in strength along the floor thickness could not be seen in aggregate segregation, which is known and described in the literature [39–45].

An example is also Table 2, which shows how significantly the strength parameters of floors might differ, for which the control activities preceding its execution were neglected in relation to the requirements contained in Table 1.

The evenness of the upper surface of the concrete base was not checked, including the occurrence of so-called local "humps", especially interfering with the leveling of the polystyrene arrangement, which, combined with the lack of ongoing control of the thickness of the mixture being laid, resulted in obtaining a floor thickness that was not in accordance with the design. Numerous studies of the authors show that the thickness of floors is either smaller or larger in relation to the design, which is illustrated in Figure 9. Too low a thickness promotes cracking of the floor during use, whereas too large a thickness limits the floor's load-bearing capacity.

Table 2. Sample results of strength tests carried out for one of the cement floors tested in accordance with [46] (own study based on the data provided [25]).

Summary of Average Strength Values	
Subsurface tensile strength f_h (MPa):	
- Obtained on the basis of research	0.55 ± 0.21
- Required according to Table 1	1.5
Compressive strength (MPa):	
- Obtained on the basis of research	15.1 ± 4.5
- Required according to Table 1	20.0
Flexural strength (MPa):	
- Obtained on the basis of research	2.3 ± 1.0
- Required according to Table 1	5.0

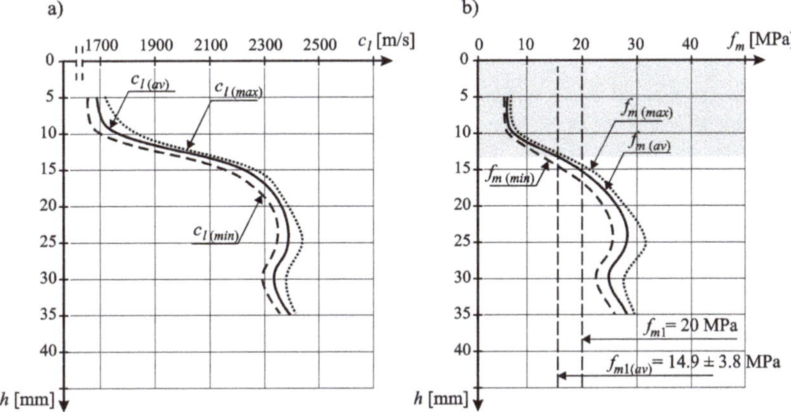

Figure 8. An example of the course: (**a**) the longitudinal velocity of the ultrasonic wave c_l along the thickness h and (**b**) the compressive strength f_m of the cement mortar along thickness h.

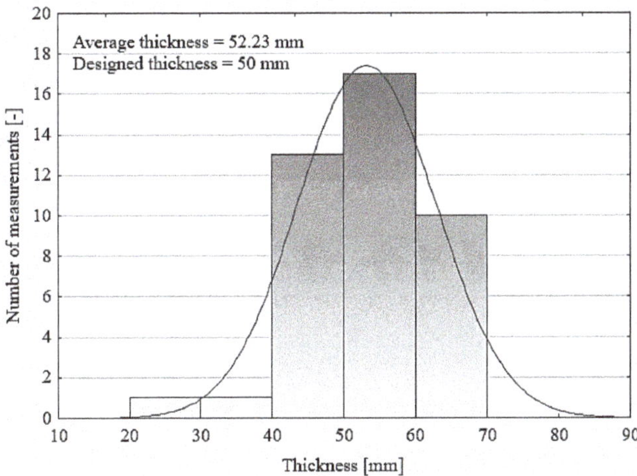

Figure 9. The thickness of the floors in comparison with the designed thickness (based on the data presented in [25]).

The problem of controlling the evenness of the insulation foil laying on the surface of a concrete substrate or polystyrene was underestimated. As a result of the corrugation of insulation foil laid under the floor on a foamed polystyrene surface, there were irregular grooved recesses on the bottom surface of the floor. As shown in Figure 10, these grooves locally reduced the thickness of the floor by a few, or even several millimeters, which promoted cracking of the floor during use.

Figure 10. Example view: (**a**) the bottom surface of the floor, with grooved recesses constituting the "imprint" of the corrugated foil and (**b**) measuring the depth of the grooved cavities.

In the developed methodology for controlling the process of making cement floors, in Stage 2, very much attention was paid to checking the expansion joints. Lack of such control, or sporadic controls, contribute to the incorrect execution of expansion joints, or their lack and consequent cracking of floors (for example Figure 11). Incorrect circumferential separating of floors from load-bearing walls results in a lack of continuity of expansion joints, floors being too small in width, joints being too shallow or the filling of joints with a material other than the non-absorbent flexible foam used for this purpose, e.g., multilayer cardboard.

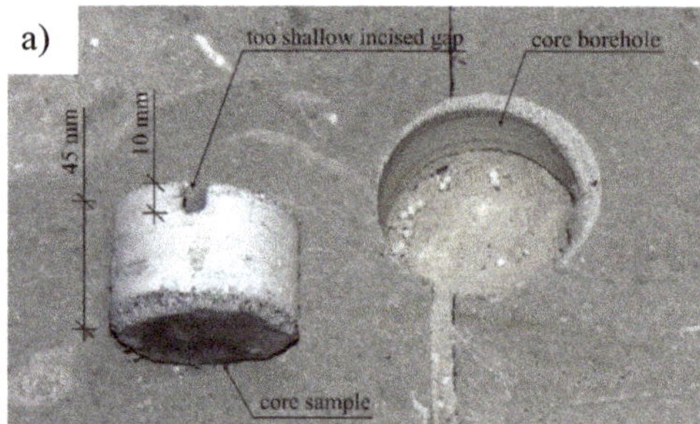

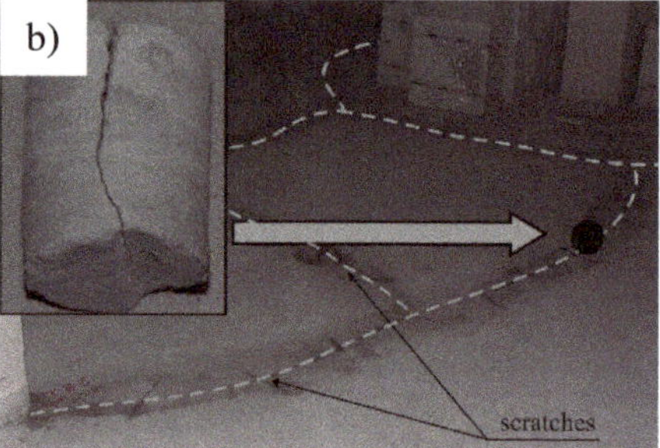

Figure 11. An exemplary view: (**a**) of a core borehole made in the floor, and also a core sample showing an expansion joint that is cut too shallow and (**b**) cracks on the surface of floors, along with an approximation of the cut core sample documenting the depth of one of the cracks.

Whereas a common disadvantage are expansion joints that are incorrectly cut, or cut at the wrong time, in cement floors. Such errors cause the dimensions of the dilated fields to be too large, the cuts to be too shallow, or the gaps to be cut too late (Figure 11). The developed methodology clearly indicates the necessary scope of control, which should exclude the problem of incorrectly made expansion joints.

In the methodology developed to control the process of making cement floors, in Stage 3, a lot of attention was paid to control strength tests after flooring. Currently, the compressive and flexural strength is being tested. At present, subsurface tensile strength, abrasion resistance and compressive strength of a cement mortar along the thickness are very rarely tested. Lack of these tests results in less durability or loosening of the applied overlay.

5. Conclusions

The article presents the original and effective methodology of controlling the technological process of executing floors made of cement-based materials, which there is a lack of in the literature. This methodology was developed for the entire process of the cement floor technology, which is also missing in the literature. The methodology also indicated the need to use necessary semi-destructive and non-destructive methods for such control.

The methodology was developed on the basis of many years of experience of the authors, which was acquired during the examination of the technical condition of floors in various construction objects. This methodology has three control stages. Both the individual stages and the methodology as a whole indicate or even force, step by step, consecutive control activities to be carried out during flooring. In the discussion of the problem, attention was drawn to those selected control activities, which are currently being neglected and which result in serious implementation problems. This was documented by sample research results, clearly showing the negative effects of such omissions on the quality of the floors made.

The use in construction practice of the proposed comprehensive methodology for controlling the process of flooring made of cement-based materials should have a positive effect on the current state. It should contribute to the effective elimination of many serious implementation problems and, as a consequence, significantly improve the quality and durability of floors.

Author Contributions: Conceptualization, Ł.S. and J.H.; methodology L.S., J.H.; validation, L.S. and A.H.; writing—original draft preparation, Ł.S. and A.H.; writing—review and editing, J.H.; visualization, Ł.S. and A.H.; supervision, J.H. All authors have read and agreed to the published version of the manuscript.

Funding: This research received funding from the subsidy no. 82010003902 intended for the Department of Building Engineering at Wrocław University of Science and Technology and funded by the Ministry of Science and Higher Education of Poland (MNiSW).

Conflicts of Interest: The authors declare no conflict of interest.

References

1. Williamson, N.I.H. Concrete floors for warehousing and distribution facilities: Fit for purpose? *Proc. Inst. Civ. Eng. Struct. Build.* **1996**, *116*, 235–243. [CrossRef]
2. Tarr, S.M.; Farny, J.A. *Concrete Floors on Ground*; Portland Cement Association: Skokie, IL, USA, 2008.
3. Marais, L.R.; Perrie, B.D. *Concrete Industrial Floors on the Ground*; Portland Cement Institute: Skokie, IL, USA, 1993.
4. Pye, P.; Harrison, H.W. *Floors & Flooring: Performance, Diagnosis, Maintenance, Repair and the Avoidance of Defects*; BRE Building Elements Series: London, UK, 1997.
5. Deshpande, H.; Joshi, R.; Bangar, P. Design Considerations for Reinforced Concrete Flat Slab Floor System. *Int. J. Sci. Eng. Res.* **2014**, *2229*, 5518.
6. Ransom, W.H. *Building Failures: Diagnosis and Avoidance*; Routledge: London, UK, 2002.
7. Hajduk, P. *Projektowanie Podłóg Przemysłowych*; Wydawnictwo Naukowe PWNL: Warszawa, Poland, 2013.
8. Czarnecki, L.; Emmons, P. *Naprawa i Ochrona Konstrukcji Betonowych*; Polski Cement: Kraków, Poland, 2007.
9. Bissonnette, B.; Courard, L.; Garbacz, A. *Concrete Surface Engineering*; CRC Press: Boca Raton, FL, USA, 2015.
10. Garber, G. *Design and Construction of Concrete Floors*; Elsevier: London, UK, 2006.
11. Rucka, M.; Wilde, K. Experimental Study on Ultrasonic Monitoring of Splitting Failure in Reinforced Concrete. *J. Nondestruct. Eval.* **2013**, *32*, 372–383. [CrossRef]
12. Canbaz, M.; Topçu, I.B.; Ateşin, Ö. Effect of admixture ratio and aggregate type on self-leveling screed properties. *Constr. Build. Mater.* **2016**, *116*, 321–325. [CrossRef]
13. Stawiski, B. *Ultradźwiękowe Badania Betonów i Zapraw Głowicami Punktowymi*; Oficyna Wydawnicza Politechniki Wrocławskiej: Wrocław, Poland, 2009.

14. Chaplin, R. Abrasion resistant concrete floors. In *Advances in Concrete Slab Technology*; Elsevier: Pergamon, Turkey, 1980; pp. 532–543.

15. Szymanowski, J. Evaluation of the Adhesion between Overlays and Substrates in Concrete Floors: Literature Survey, Recent Non-Destructive and Semi-Destructive Testing Methods, and Research Gaps. *Buildings* **2019**, *9*, 203. [CrossRef]

16. Cattell, D. *Specialist Floor Finishes: Design and Installation*; Routledge: London, UK, 2003.

17. Collepardi, M.; Troli, R.; Bressan, M.; Liberatore, F.; Sforza, G. Crack-free concrete for outside industrial floors in the absence of wet curing and contraction joints. *Cem. Concr. Compos.* **2008**, *30*, 887–891. [CrossRef]

18. Sadowski, Ł. *Adhesion in Layered Cement Composites*; Springer: Cham, Switzerland, 2019.

19. Sadowski, Ł.; Żak, A.; Hoła, J. Multi-sensor evaluation of the concrete within the interlayer bond with regard to pull-off adhesion. *Arch. Civ. Mech. Eng.* **2018**, *18*, 573–582. [CrossRef]

20. Sadowski, Ł.; Hoła, J.; Czarnecki, S.; Wang, D. Pull-off adhesion prediction of variable thick overlay to the substrate. *Autom. Constr.* **2018**, *85*, 10–23. [CrossRef]

21. Nowobilski, T.; Hoła, B. Flooring in housing building—An overview of material solutions (in Polish). *Builder* **2017**, *21*, 84–87.

22. Hoła, J.; Sadowski, Ł.; Nowacki, A. Analysis of the causes of cracks in marble slabs in a large-surface floor of a representative commercial facility. *Eng. Fail. Anal.* **2019**, *97*, 1–9. [CrossRef]

23. Gajzler, M. Supporting of repairs processes of concrete industrial floors. *Eng. Struct. Technol.* **2016**, *8*, 8–14. [CrossRef]

24. Bissonnette, B.; Attiogbe, E.K.; Miltenberger, M.; Fortin, C. Drying Shrinkage, Curling and Joint Opening of Slabs on Ground. *ACI Mater. J.* **2007**, *104*, 259–266.

25. Hoła, J.; Sadowski, Ł.; Hoła, A. The effect of failure to comply with technological and technical requirements on the condition of newly built cement mortar floors. *J. Mater. Des. Appl.* **2019**, *233*, 268–275. [CrossRef]

26. Delatte, N. Using Failure Case Studies in Civil Engineering Education. *Forensic Eng.* **2000**, 430–440.

27. Tang, P.; Huber, D.; Akinci, B. Characterization of Laser Scanners and Algorithms for Detecting Flatness Defects on Concrete Surfaces. *J. Comput. Civ. Eng.* **2011**, *25*, 31–42. [CrossRef]

28. Bissonnette, B.; Courard, L.; Beushausen, H.; Fowler, D.; Trevino, M.; Vaysburd, A. Recommendations for the repair, the lining or the strengthening of concrete slabs or pavements with bonded cement-based material overlays. *Mater. Struct.* **2013**, *46*, 481–494. [CrossRef]

29. Douglas, J.; McEwen, I. Defects diagnosis—A case study involving chemical analysis. *Constr. Build. Mater.* **1998**, *12*, 259–267. [CrossRef]

30. *Toleranzen im Hochbau Bauwerke*; DIN 18202: Berlin Germany, 2013.

31. Loprencipe, G.; Cantisani, G. Evaluation methods for improving surface geometry of concrete floors: A case study. *Case Stud. Struct. Eng.* **2015**, *4*, 14–25. [CrossRef]

32. Walsh, K.D.; Bashford, H.H.; Mason, B.C.A. State of Practice of Residential Floor Slab Flatness. *J. Perform. Constr. Facil.* **2001**, *15*, 127–134. [CrossRef]

33. Face, A. Floor flatness and levelness: The F-number system. *Constr. Specif.* **1987**, *40*, 124–132.

34. *Posadzki z Betonu i Zaprawy Cementowej. Wymagania i Badania Techniczne przy Odbiorze*; PN-62/B-10145: Warszawa, Poland, 1962.

35. *Standard Test Method for Determining FF Floor Flatness and FL Floor Levelness Numbers*; ASTM 1155M-96: West Conshohocken, PA, USA, 2001.

36. *Wyroby i Systemy do Ochrony i Napraw Konstrukcji Betonowych-Metody Badań-Pomiar Przyczepności Przez Odrywanie*; PN-EN 1542: Warszawa, Poland, 2004.

37. Gudra, T.; Stawiski, B. Non-destructive strength characterization of concrete using surface waves. *Ndt E Int.* **2000**, *33*, 1–6. [CrossRef]

38. Anderberg, A.; Wadsö, L. Drying and Hydration of Cement Based Self-Leveling Flooring Compounds. *Dry. Technol.* **2007**, *25*, 1995–2003. [CrossRef]

39. Navarrete, I.; Lopez, M. Understanding the relationship between the segregation of concrete and coarse aggregate density and size. *Constr. Build. Mater.* **2017**, *149*, 741–748. [CrossRef]

40. Gao, X.; Zhang, J.; Su, Y. Influence of vibration-induced segregation on mechanical property and chloride ion permeability of concrete with variable rheological performance. *Constr. Build. Mater.* **2019**, *194*, 32–41. [CrossRef]

41. Safawi, M.I.; Iwaki, I.; Miura, T. The segregation tendency in the vibration of high fluidity concrete. *Cem. Concr. Res.* **2004**, *34*, 219–226. [CrossRef]
42. Navarrete, I.; Lopez, M. Estimating the segregation of concrete based on mixture design and vibratory energy. *Constr. Build. Mater.* **2016**, *122*, 384–390. [CrossRef]
43. Josserand, L.; Coussy, O.; De Larrard, F. Bleeding of concrete as an ageing consolidation process. *Cem. Concr. Res.* **2006**, *36*, 1603–1608. [CrossRef]
44. Topçu, I.B.; Elgün, V.B. Influence of concrete properties on bleeding and evaporation. *Cem. Concr. Res.* **2004**, *34*, 275–281. [CrossRef]
45. Hoshino, M. Relation between bleeding, coarse aggregate, and specimen height of concrete. *Mater. J.* **1989**, *86*, 185–190.
46. *Metody Badania Materiałów na Podkłady Podłogowe-Część 2: Oznaczanie Wytrzymałości na Zginanie i ściskanie*; EN 13892-2: Warszawa, Poland, 2004.

 *materials* 

Article

The Use of Dijkstra's Algorithm in Assessing the Correctness of Imaging Brittle Damage in Concrete Beams by Means of Ultrasonic Transmission Tomography

Zbigniew Perkowski * and Karolina Tatara

Faculty of Civil Engineering and Architecture, Department of Physics of Materials, Opole University of Technology, Katowicka 48, 45-061 Opole, Poland; k.tatara@po.edu.pl
* Correspondence: z.perkowski@po.edu.pl

Received: 21 December 2019; Accepted: 21 January 2020; Published: 23 January 2020

Abstract: The accuracy of transmission ultrasonic tomography for the detection of brittle damage in concrete beams can be effectively supported by the graph theory and, in particular, by Dijkstra's algorithm. It allows determining real paths of the fastest ultrasonic wave propagation in concrete containing localized elastically degraded zones at any stage of their evolution. This work confronts this type of approach with results that can be obtained from non-local isotropic damage mechanics. On this basis, the authors developed a method of reducing errors in tomographic reconstruction of longitudinal wave velocity maps which are caused by using the simplifying assumptions of straightness of the fastest wave propagation paths. The method is based on the appropriate elongation of measured propagation times of the wave transmitted between opposite sending-receiving transducers if the actual propagation paths deviate from straight lines. Thanks to this, the mathematical apparatus used typically in the tomography, in which the straightness of the fastest paths is assumed, can be still used. The work considers also the aspect of using fictitious wave sending-receiving points in ultrasonic tomography for which wave propagation times are calculated by interpolation of measured ones. The considerations are supported by experimental research conducted on laboratory reinforced concrete (RC) beams in the test of three-point bending and a prefabricated damaged RC beam.

Keywords: non-destructive testing; ultrasonic tomography; graph theory; concrete; damage mechanics; elastic degradation; damage parameter; internal length; experimental research

1. Introduction

Concrete is one of the most commonly used materials in civil structures. From a scientific and technical point of view, it is a subject of interest both at the stage of designing a recipe, manufacturing various types of elements and during its operation. Ensuring safe and long-term use of concrete structures and elements requires, among other things, appropriate diagnostics. It can use destructive testing (e.g., by testing the strength of drilled cores [1,2]), semi-destructive testing (e.g., pull-out [1,3], pull-off [4,5] methods) and non-destructive testing (NDT, e.g., using sclerometer tests [1,3,6], thermal imaging techniques [7,8], analysis of natural frequencies [9,10], stereological investigations [11], acoustic emission [12,13], X-ray tomography [12], ground-penetrating radars [7,14,15], ultrasound [1,3,7,15–25] including ultrasound tomography [15,19–25]). The choice of method depends on the material characteristics that we want or are able to measure. All three types of tests are widely known, but especially NDT, thanks to the introduction of a number of modern measurement techniques in the building industry and intensive research, is becoming more and more popular and reliable. Particularly interesting in this area are ultrasound techniques which use at their basis typical phenomena associated with wave motion physics—e.g., reflection, diffraction, attenuation, change of propagation velocity depending on changes in stiffness and density of the medium. The simplest method in the case of concrete structures is an assessment of the velocity of longitudinal waves between selected points of the tested element—the lower the velocity, the lower the stiffness of concrete and its quality [16]. However, it concerns the average speed measurement on the section between the ultrasonic transducers. An interesting, practical case of this type of analysis is article [18] where ultrasonic measurements carried out on a river dam were verified by means of visual inspection of cores taken from it. In the literature, there are also analyses concerning the change of time and intensity of ultrasound wave passing through the area of concrete where a single crack builds up [17]. The use of such research, however, requires in advance knowledge of where such a defect may develop in an element. On the other hand, tomographic methods are deprived of this type of inconvenience, where only a set of transceiver converters suitable for concrete is required and external access to the tested element on one side in a reflective mode (e.g., Reference [21]) or with access on two or more sides in a transmission mode (e.g., Reference [15,20,22–25]). In the latter view, the state of the material is most often shown indirectly by means of reconstructed maps of the propagation velocity of a selected type of ultrasound wave. For the sake of convenience, longitudinal waves are usually selected for this purpose in the unambiguous interpretation of measurements, as they move the fastest and are not dispersed. It should be emphasized that very accurate maps of cracks in the concrete structure can be made, on the other hand, using X-ray tomography [12]; however, currently, due to the cost of the equipment and the possibility of its use, it is practically impossible to use it directly on real building structures in field research. It is also worth mentioning at this point that, from the point of view of developing mathematical foundations for tomography, its beginning dates back to 1917, when Johann Radon proposed a solution to the problem of reconstruction of the shape of an object on the basis of its projections [26].

The tomographic imaging concrete elements available in the literature focus mainly on the identification of defects with much lower acoustic resistance than the surrounding concrete: e.g., artificially introduced defects, for research purposes, in the form of inclusions from foamed polystyrene [21–23], expanded polypropylene [25], prisms from cracked concrete [23] or air-filled pipes [21], cavities in defectively injected pipes for placing prestressing cables [15], areas strongly cracked as a result of excessive loads [20,22,24] or freeze-thaw cycles [19]. Therefore, the first goal that the authors set for themselves in this paper was to carry out an analysis of the extent to which it is possible to detect brittle defects in concrete beams starting from the early stage of their development, when microcracks do not yet form defects capable of effective reflection of waves or their significant slowing down. For this purpose, the methodology of damage mechanics was applied in terms of one of the most recognized concrete models in this field, formulated by Chaboche [27] and Mazars [28], and, in the non-local terms, developed by Pijaudier-Cabot [29–31]. Then, depending on the degree of brittle damage described in a "fuzzy" way by the damage parameter, it is possible to model the development of a localized decrease in material stiffness and the associated reduction in the speed of sound waves in concrete. This fact can also be used in the tomographic assessment of concrete [32,33]. For this reason, in order to reliably calculate the distributions of drop in stiffness and changes in the velocity of the longitudinal wave around the forming crack, the authors of the paper proposed an effective way of identifying the parameters of a non-local model of brittle damage evolution using experimental data from [34]. These data were then used in computer simulations of tomographic identification of this type of defect in various phases of its formation and were confronted with the results of our own experimental research.

Another important aspect of ultrasound tomography measurements is its accuracy which may be affected by the diffraction of waves when passing through and around areas of different acoustic resistance than the rest of the medium. It is commonly assumed in order to significantly simplify calculations that paths of the fastest wave propagation are rectilinear (e.g., Reference [15,19–24,35]) which is then called rays. This introduces disturbances in tomographic reconstructions when the actual paths differ strongly from the geometry assumed so far. The studies available in the literature show that this assumption does not cause any disturbances in the location of the damaged areas [21–24]. However, the obtained values of wave propagation velocities on the reconstructed maps differ significantly from the actual values at high levels of concrete degradation [32,33] or inclusions with significantly different acoustic resistance from the matrix [25]. In this case, it should be taken into account that according to the Fermat principle, wave disturbance travels from one point to another such a path that needs minimum or maximum time, or the same in comparison to other, adjacent paths [36], which determines the course of its fastest propagation. As stated in Reference [25], the first attempt to consider the Fermat principle in ultrasound imaging was made by Johnson et al. [37]. They proposed the use of the ray-tracing technique which in the case of concrete structures was first adopted in works [38,39]. An alternative to this type of approach is the use of graph theory, in particular Dijkstra's algorithm [40]. If we apply it in the analyzed issue to a full graph the nodes of which will cover the studied area, and the weights of the edges will be equal to the time of the wave passing through them, then in this way we can approximately determine the shape of paths of the fastest sound propagation and the time needed for the wave to travel along them. The results will be the more accurate the denser the network of nodes. This very interesting concept, inspired by the works of various authors and applied to seismic waves, was developed in Moser's work [41] in 1991 (as in Reference [25]). Extensive studies in this field with examples of calculations using experimental data from concrete cubes with inclusions from expanded polypropylene have been presented in Reference [25]. A proposal to use this methodology in the ultrasonic testing of structural elements was also presented by the authors in Reference [33,42]. In the light of the quoted information, it can be noted that so far there is no analysis in the literature concerning this type of problem in the case of tomographic assessment of the damage evolution described according to the concept of damage mechanics. Therefore, the second aim of this article was to present in this area appropriate numerical analyses with the use of Dijkstra's algorithm in determining shape of the paths of the fastest ultrasound wave propagation. On this basis, the author's own method of improving the accuracy of tomographic calculations was formulated because of the inconsistency of the assumptions with reality in the case of using straight-line rays to approximate the geometry of the fastest propagation paths. For this purpose, it is proposed to not interfere with the assumption of straightness of ultrasound pulse pathways but, on the other hand, to properly scale the measured times of their propagation between the assumed transceiver points elongating them in the case of rays that pass through elastically degraded areas. The advantage of such an approach is that it does not complicate well established mathematical methods (e.g., Reference [35,43–46]) which have been implemented in tomographic imaging techniques.

The paper also raises the practical aspect of conducting such measurements by analyzing in the light of the presented arguments how the accuracy of results may be affected by the introduction of the so-called fictitious transceiver points for which the times of propagation of ultrasound waves are interpolated on the basis of measurements from the real points. In this respect, the authors were inspired by work [47] where this approach was presented in the study of moisture distribution in walls. It may significantly reduce the number of points used and the labor intensity in real measurement situations, which is particularly important in the case where fewer ultrasonic transducers are available.

Due to the scope of studies within the framework of the presented calculation examples and experiments, the authors limited themselves to the case of concrete beams with elastically degraded zones perpendicular to the beam axis. All the calculations made in the article were done with the use of the authors' computer programs written in the MATLAB software environment.

2. Ultrasonic Transmission Tomography

The considerations presented in the paper concern the case of concrete beams that contain elastically degraded zones (in the form of grouped micro-cracks or cracks) running across the entire cross-section—e.g., as a result of simple bending or tension. Therefore, tomographic analyses were narrowed down to the identification of brittle damage in a flat longitudinal section of beam which will also be the plane of symmetry of this element, focusing on the assessment of changes in the speed of ultrasonic waves due to a local change in the stiffness of the cracked material. For this reason, in all computational examples and experimental studies presented in the paper, a system of opposite transmitting/receiving points of ultrasound waves in the transmission mode was used (Figure 1). Longitudinal waves have been selected as non-dispersive and the fastest of all ultrasonic wave types for the study. This requires, however, that their length should be small enough in relation to the dimensions of the cross-section of the element to be effectively generated (practical recommendations in this respect can be found in, e.g., Reference [48]). On the other hand, it limits the distance between the intended transmitting/receiving points because of the attenuation of the longitudinal waves and the angle at which they may propagate from the transmitting point in an effective manner in reception because of the amplitude distribution. In the latter case, on the basis of individual experiences of the authors, the angle of inclination of diagonal rays to a bar axis was limited to range from 45° to 135°; in Reference [33], a numerical simulation of the process of propagation of ultrasounds during its excitation by the transducer was performed where it was shown that the amplitudes of longitudinal waves are negligible in contrast to transverse waves outside this angular range, which may result in incorrect and increased propagation time reading.

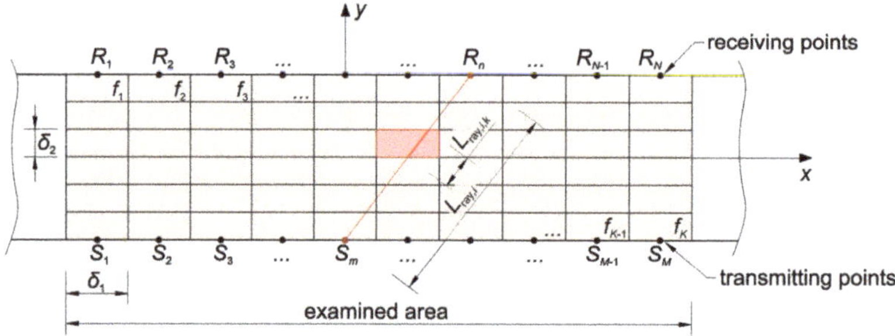

Figure 1. Scheme of a cell system, transmitting/receiving points, and rays in a plane area examined tomographically.

From the mathematical point of view, by indirect visualization of the material structure by means of maps of distributions of quantities characterizing the propagation of ultrasound waves, it is necessary to build an appropriate system of equations (e.g., Reference [35]). The plane problem assumes that the reconstructed image consists of a finite number of plane cells which, in the examined area, is separated by an orthogonal grid of a step of $\delta_1 \times \delta_2$ (Figure 1). The function is searched for in an approximate way:

$$f = c_L^{-1}(x, y), \tag{1}$$

where: c_L—longitudinal wave velocity (m/s); x, y—spatial variables (m). For this purpose, it is assumed that f in each cell takes a constant value of f_k where $k = 1, 2, \ldots, K$.

In the problem, the longitudinal wave propagation times between the selected sending (S_m) and receiving (R_n) points must be given where $m = 1, 2, \ldots, M; n = 1, 2, \ldots, N$. It is assumed to simplify considerations that the fastest propagation path between these points can be modeled as a straight line, which in tomography is referred to as a ray. The propagation time of $t_{ray,i}$ (s) over the i-th ray, connecting points S_m and R_n, can be calculated from the integral:

$$t_{ray,i} = \int_{S_m}^{R_n} f(l_i) dl_i, \tag{2}$$

where: l_i—variable describing the position on the i-th ray (m); $i = 1, 2, \ldots, I$. Considering that averaged values f are being sought in the cellular areas, we can write that:

$$t_{\text{ray},i} = \sum_{k=1}^{K} L_{\text{ray},i,k} f_k \rightarrow \mathbf{A}\mathbf{x} = \mathbf{b}, \tag{3}$$

where: $L_{\text{ray},i,k}$—parts of the length of the i-th ray falling on the k-th cell (m) (if the ray does not pass through cell k, then $L_{\text{ray},i,k} = 0$), $\mathbf{A} = \left[L_{\text{ray},i,k}\right]_{I \times K}$ (m), $\mathbf{x} = \left[f_1, f_2, \ldots, f_K\right]^{\text{T}}$ (s/m), $\mathbf{b} = \left[t_{\text{ray},1}, t_{\text{ray},2}, \ldots, t_{\text{ray},I}\right]^{\text{T}}$ (s). The above is the definition of a system of equations in relation to the value of f_k, which is used to determine the velocity distribution of ultrasound wave propagation in tomography. After its solution, we get that:

$$c_{\text{L tom},k} = \frac{1}{f_k}, \tag{4}$$

where $c_{\text{L tom},k}$—mean velocity of the longitudinal wave in the k-th cell (m/s) (in the sense of the presented method). Please note that I (the number of rays used) must not be less than K (the number of the assumed cells) and the rays must evenly cover the test area. The system of equations formulated in such a way is an ill-conditioned one, which forces the use of iterative techniques of its solving. The basic method in this respect is an algorithm developed by the Polish mathematician Stefan Kaczmarz (1937). In 1970, Gordon and his collaborators, working on the application of this technique in medicine, rediscovered this method and named it Algebraic Reconstruction Technique (ART) [43]. It was this one that was used in the first in the world computed tomography scanner constructed by Hounsfield in 1972 [44]. On the basis of the Kaczmarz algorithm, many other methods were developed. Currently, the literature distinguishes three basic ways of imaging: the aforementioned ART, Simultaneous Iterative Reconstruction Technique (SIRT), and Simultaneous Algebraic Reconstruction Technique (SART) which is a combination linking the advantages of the ART and SIRT methods [35,46]. For this purpose, the Tikhonov regularization method can also be used in the method of least squares (e.g., Reference [24]).

In this article, all tomographic images presented below were solved with the use of a randomized Kaczmarz algorithm. The final result was taken as $\mathbf{x}_{\text{mean},q}$, i.e., the mean of q independently obtained solutions of equation system (3). Hence:

$$\mathbf{x}_{\text{mean},q} = \frac{1}{q} \sum_{r=1}^{q} \mathbf{x}_r. \tag{5}$$

In a given solution $\mathbf{x}_r$, its subsequent approximations were made by projecting the previous approximation in a direction perpendicular to the randomly selected straight lines defined by the equations of system (3), but so that each of these lines would be used only once. The starting point for the iteration of each $\mathbf{x}_r$ was the vector:

$$\mathbf{x}_0 = \left[c_{\text{L ref}}^{-1}, c_{\text{L ref}}^{-1}, \ldots, c_{\text{L ref}}^{-1}\right]_{1 \times K}^{\text{T}}, \tag{6}$$

where $c_{\text{L ref}}$—reference value of the longitudinal wave velocity (m/s). The number q of the averaged solutions of equation system (3) were selected so that the condition was met:

$$\left|\frac{e_{g,q} - e_{g,q-1}}{e_{g,q}}\right| < 10^{-5} \text{ and } e_{g,q} = \frac{\left\|\mathbf{A}\mathbf{x}_{\text{mean},q} - \mathbf{b}\right\|}{\left\|\mathbf{b}\right\|}. \tag{7}$$

As mentioned in the introduction, another inconvenience of the presented method of tomographic imaging is the fact that ultrasonic waves are diffracted when avoiding areas with different acoustic resistance, so that the real paths of the fastest propagation are curvilinear. This is one of the basic sources of inaccuracy of the presented approach if there are sub-areas with significantly different acoustic resistances in the tested concrete element in relation to the rest of the element [25,32,33]. These issues will be discussed in the context of the following calculation examples and experimental studies concerning the detection of elastically degraded concrete zones.

3. Localized Elastic Degradation in Concrete—Crack Model in Concrete According to the Damage Mechanics

In concrete, due to its brittleness, the evolution of damage occurs in particular at the action of tensile stresses [27–31]. The process begins with the formation of microcracks which, with further increase in stress, grow into localized damage zones and cracks visible to the naked eye. For this reason, the presence of such zones at the final stage can be easily detected visually, as they contain cracks of the order of tenths of a millimeter in width. On the other hand, they are not visible at an early stage of development, and what is important, the initial microcracks usually do not occur in random places of concrete structure. When the cementitious material structure develops, the high and low internal cohesion zones can be distinguished. Especially, the places with low cohesion are the ones where cracks begin to grow because, in such places, even a small amount of released energy causes their propagation. They form mainly around the micro-pores or near the phase separation surfaces [11]. From the analyses presented in this respect within the damage mechanics, it is known that this process obviously leads to elastic degradation of concrete, which means a local decrease in material stiffness [27–31,49]. This phenomenon gives grounds for detection and control of this type of damage by ultrasound tomography if the spatial distribution of propagation velocity of a selected type of mechanical waves (e.g., Reference [24,33]) is assessed within such a framework. That is why, in the article, the preliminary calculations were oriented on determining the spatial distribution of concrete stiffness change in the case of localized damage under tension. The necessary information was obtained in this way in order to analyze the propagation of ultrasound waves in a localized elastically degraded concrete zone which evolution leads finally to forming macro-cracks. The calculations were performed using the assumptions of one of the most popular models in this respect, introduced by Chaboche [27] and Mazars [28], which takes into account the weakening of the material due to the isotropic damage accumulation and which was developed by Pijaudier-Cabot [29–31] in non-local terms. In a uniaxial tensile state, the model assumes the following relationship between stress and deformation (using the principle of strain equivalence):

$$\sigma = (1 - D)E_0\varepsilon, \tag{8}$$

where: σ—tensile normal stress (Pa); ε—normal strain [-] E_0—Young's modulus in undamaged material (Pa); D—damage parameter [-]. Due to thermodynamic limitations of the process, the following must be satisfied:

$$\text{if } f_D = \bar{\varepsilon} - \kappa = 0 \text{ and } \dot{\bar{\varepsilon}} > 0, \text{ then}$$
$$D = 1 - (1 - A_t)\frac{\kappa_0}{\kappa} - A_t \exp(-B_t(\kappa - \kappa_0)) \text{ and } \dot{\kappa} = \dot{\bar{\varepsilon}}, \tag{9}$$

$$\text{if } f_D = \bar{\varepsilon} - \kappa < 0 \text{ or } \dot{\bar{\varepsilon}} \le 0, \text{ then}$$
$$\dot{D} = 0 \text{ and } \dot{\kappa} = 0, \tag{10}$$

where: A_t, B_t—material parameters [-]; f_D—load function [-]; κ—variable describing the process of material weakening [-]; κ_0—initial value of the variable κ [-]; $\bar{\varepsilon}$—non-local equivalent tensile strain [-]. To simplify the problem, if we assume that pure tension occurs in a concrete element with its axis described by variable x (m), the non-local equivalent strain at point x will be defined as:

$$\bar{\varepsilon}(x) = \frac{1}{V_r} \int_{s_{(1)}}^{s_{(2)}} \tilde{\varepsilon}(s)\, \psi(x - s)A(s)\, ds, \tag{11}$$

$$V_r(x) = \int_{s_{(1)}}^{s_{(2)}} \psi(x - s)A(s)\, ds\ , \tag{12}$$

$$\psi(x - s) = \exp\left(\frac{-4(x - s)^2}{l_c^2}\right), \tag{13}$$

$$\tilde{\varepsilon} = \sqrt{\sum_{i=1}^{3} \langle \varepsilon_i \rangle^2}, \tag{14}$$

where: $\widetilde{\varepsilon}$—local equivalent tensile strain [-]; ε_i—local principal strain [-]; V_r—representative volume of the material (m³), s—variable describing the position along axis x (m); $s_{(1)}$, $s_{(2)}$—starting and ending point of the considered element (m); A—cross-sectional area of the element (m²); l_c—the characteristic dimension of the non-locality or the so-called internal length (m); ψ—weight function [-]; $\langle\ldots\rangle$—Macauley's operator. Variability of function ψ is adopted in the model identical to the normal distribution with standard deviation $\frac{l_c}{2\sqrt{2}}$. It also means that the range of the non-locality practically ceases to be significant above the distance l_c because $\psi(x - s = l_c) = 1.83\cdot10^{-3}$ (Figure 2). In addition, in the case of uniaxial tension, $\widetilde{\varepsilon} = \varepsilon$ where ε is the normal strain along the axis of the beam, κ_0 will be equal to this deformation at the moment of initiation of the damage evolution, and κ_0 can reach a maximum value of f_t/E_0 where f_t is the tensile strength. In the incremental version needed for the numerical analysis of the problem, the stress and damage evolution Equations (8)–(10) take the following forms:

$$\begin{cases} \text{if } f_D = \widetilde{\varepsilon} - \kappa = 0 \text{ and } \Delta\widetilde{\varepsilon} > 0, \text{ then} \\ \Delta\sigma = (1-D)E_0\Delta\varepsilon - \Delta D E_0\varepsilon \\ \Delta D = \frac{\partial D}{\partial\widetilde{\varepsilon}}\Delta\widetilde{\varepsilon} = \frac{\partial D}{\partial\widetilde{\varepsilon}} \frac{1}{V_r} \int_{s_{(1)}}^{s_{(2)}} \Delta\widetilde{\varepsilon}(s)\,\psi(x-s)A(s)\,ds \\ \Delta\kappa = \Delta\widetilde{\varepsilon} \end{cases} \tag{15}$$

$$\begin{cases} \text{if } f_D = \widetilde{\varepsilon} - \kappa < 0 \text{ or } \Delta\widetilde{\varepsilon} \le 0, \text{ then} \\ \Delta\sigma = (1-D)E_0\Delta\varepsilon \\ \Delta D = 0 \\ \Delta\kappa = 0 \end{cases} \tag{16}$$

where: $\Delta\ldots$—the finite increment of a given quantity.

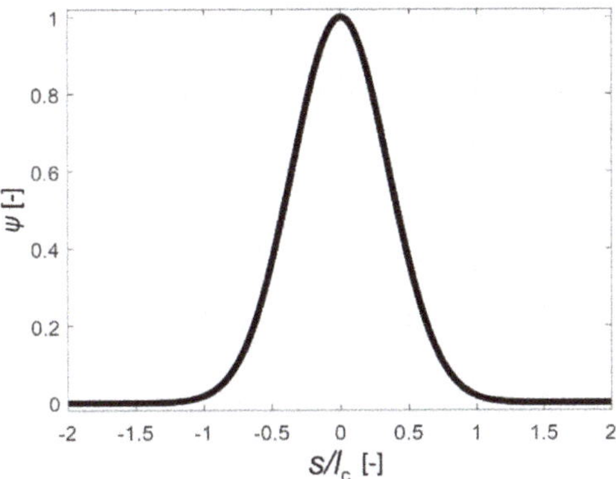

Figure 2. Distribution of the weight function ψ for $x = 0$.

A separate issue related to the use of the discussed model is the adoption of appropriate material parameters that will enable the most accurate capture of the real behavior of concrete. The authors in paper [31] give the approximate typical ranges of these parameters in case of concrete of moderate strength, i.e., $E_0 = 30$–$40\,\text{GPa}$, $A_t = 0.7$–1.2, $B_t = 10^4$–$5\cdot10^4$, $\kappa_0 = 10^{-4}$, and $l_c = 3$–$5\,d_{a\,max}$ where $d_{a\,max}$ is the maximum size of the aggregate. However, so far, there are no exhaustive items in the literature devoted to research and formulation of automatic optimization calculation techniques allowing for precise estimation of all parameters mentioned above for a given type of concrete due to the length l_c. It is assessed using mainly the scale effect and the model is calibrated by a manual trail-and-error method (e.g., Reference [31,49,50]). For this purpose, a suitable coefficient inverse problem is formulated in this article which uses illustratively the data from [34]. These tests concern the uniaxial tensile testing of a series of "dog bone" shaped concrete specimens of mean compressive strength $f_{cm,cube} = 50\,\text{MPa}$ and $d_{a\,max} = 8\,\text{mm}$. The scheme of specimens is shown in Figure 3, and the experimental load-elongation dependencies (P_{exp}-u_{exp}) of a selected series of specimens of an overall length of 30 cm within a range of up to 50 µm are shown in Figure 4a. This relation was obtained on the basis of digitalization of graphs presented in Reference [34], and due to their quality by entering the curve in the middle area formed by the envelope of all several curves measured on the specimens of the length of 30 cm. For calculations, data from this series of samples were selected from all 7.5 cm, 15 cm, 30 cm, 60 cm, 120 cm, and 240 cm length series, which were tested in Reference [34], because in their case the relatively highest number of measurements was obtained with possibly small scatter of measured tensile strength. On the other hand, samples with a length of 7.5 cm were characterized by the largest scatter of measurement results on the basis of which the authors of [34] concluded that the width of their smallest section is smaller than the length l_c and it can be larger. For this reason, its value was suggested as 6–7 $d_{a\ max}$. The elongation measurement base was 0.6 times the notch length in the specimen (Figure 3). It should be noted at this point that a very interesting analysis of the development of elastically degraded zones based on the data from work [34] was also presented in Reference [50] but without formulating a coefficient inverse problem.

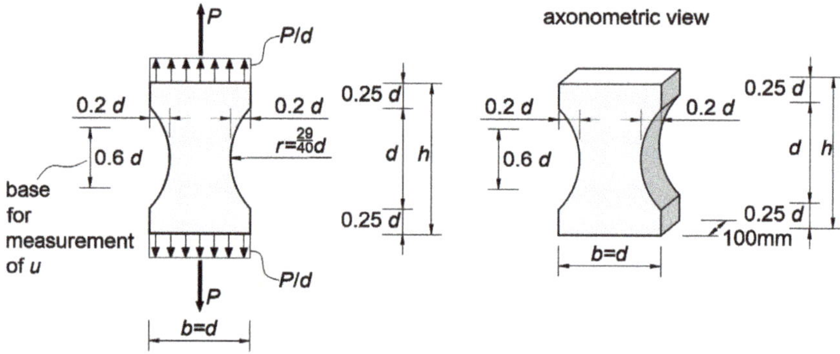

Figure 3. Shape of concrete specimens for the tension tests (based on Reference [34]).

The tests included in Reference [34] were also selected for analysis due to the fact that the shape of the specimens enables, in tension, the location of the brittle damage zone in their middle area, leading to the formation of a single macro-crack when the load capacity is exhausted. In the considerations, for the sake of simplification, a bar in tension of variable cross-section was used as a model of specimens, according to their geometrical features. This choice was dictated by the need to carry out the calculation procedures described below in an acceptable time frame due to the computer hardware available (PC with a processor 3.06 GHz and RAM 8 GB—time of solving one task approximately 2 min). The boundary problem analyzed was solved using the finite element method (FEM) in an incremental version using physical Equations (15),(16). Two-node bar finite elements with 2 degrees of freedom and constant and averaged over the length physical and geometric features (with one integration point in the middle of the element) were used. One hundred and twenty-one finite elements were adopted on the axis of the specimen of the length of 30 cm. In addition, the load increments of the specimen were assumed to be continuously elongated during the calculation, which meant switching the force sign at the transition to the weakening phase. The force increment, in turn, was selected in each step so that the increments in elongation of the measuring base and normal strain in each of the elements would not exceed, respectively, the values of 0.15 µm and 1.25×10^{-6}. Taking into account the above mentioned conditions of calculations allowed to obtain a satisfactory convergence of the solution. Further increase of the number of elements and decrease of permissible increments of strain and displacements did not significantly affect the result (Figure 4b).

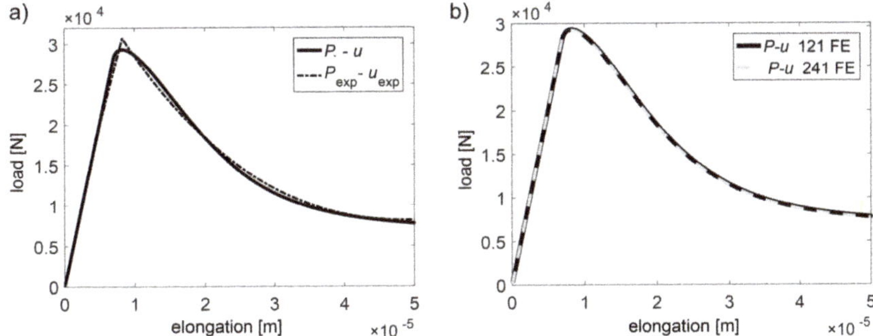

Figure 4. (**a**) Comparison of relations P_{exp}-u_{exp} and P-u using 121 finite elements. (**b**) Comparison of relations P-u using 121 and 241 finite elements.

First, E_0 was estimated by adjusting the slope of the initial straight-line section to the measuring point $\left[P_{exp}, u_{exp}\right] = [20\text{ kN}, 5\text{ }\mu\text{m}]$ in the load-elongation relation (P-u) obtained from the model in the elastic range. The calculus procedure in this case was realized using ordered domain search. Hence, it was determined that $E_0 = 38.29$ GPa. Other parameters, i.e., A_t, B_t, κ_0, and l_c were estimated by minimizing the objective function:

$$F(\mathbf{p}) = \sum_{i=1}^{n}\left(P_i(\mathbf{p}) - P_{exp,i}\right)^2, \tag{17}$$

where: $P_{exp,i}$—specimen loads from the experiment (N) determined in the weakening phase (Figure 4a) for elongations at equidistant intervals starting from $u_{exp} = 8.25$ μm (for $P_{exp} = P_{max} = 30.72$ kN—i.e., maximum load) to $u_{exp} = 50$ μm (for $P_{exp} = 8.18$ kN); P_i—equivalents $P_{exp,i}$ determined on the basis of the assumed model (N); $\mathbf{p} = [p_1, p_2, p_3, p_4]$—vector of variables corresponding to the parameters searched for, i.e., A_t, B_t, κ_0, and l_c, respectively. On that basis, it was assumed that:

$$\arg\min F(\mathbf{p}) = [A_t,\ B_t, \kappa_0, l_c]. \tag{18}$$

The calculations in formula (17) assumed $n = 31$. The minimization of the implicit function (17) was carried out in three stages. In the first stage, a genetic algorithm was used on a population of 20 parameter vectors $\mathbf{p}$ in 20 selections. The vector components of the first population were drawn in preselected intervals: p_1 of $[0.7,\ 1.2]$, p_2 of $[10^4, 5 \times 10^4]$, p_3 of $[5 \times 10^{-4}, P_{max}/(E_0 A_{min})]$ and p_4 of $[3\, d_{a\ max}, 11\, d_{a\ max}]$, where A_{min} [m^2] is the minimum cross-sectional area of the specimen at the center of its length; hence, $P_{max}/A_{min} = f_t$. It should be noted that the upper limit of interval selected for p_4 is greater than that given in Reference [31,34]. The authors assumed finally the value 11 $d_{a\ max}$ as the first smallest one for which the genetic algorithm did not estimate placing the minimum point of F right next to the upper limit of p_4. Subsequent generations of the population were created as follows: $\mathbf{p}$ with the lowest value of F passed unconditionally to the next generation, the next 10 new vectors $\mathbf{p}$ were obtained by arithmetic crossover of randomly selected $\mathbf{p}$ from the group of the first 14 vectors of the old generation after their ordering from the lowest to the highest value of F. The set of the last nine new vectors $\mathbf{p}$ were drawn in the same way as in the first generation. This procedure was repeated independently five times. In the second stage, the same procedure as in the first stage was followed, however, changing the draw intervals of vector $\mathbf{p}$ components. The boundaries of them were defined from 0.8 to 1.2 of the the values of the $\mathbf{p}$ vector components for which the lowest value of F was obtained in stage one. In the third and final stage, an orderly search of the domain of acceptable solutions was performed in the vicinity of the point defined by the vector $\mathbf{p}$ components with the lowest F-value found in the second stage. The search interval was selected from 0.99 to 1.01 of the values of the parameters of this vector. If the minimum F in this area was at the boundary of any interval, the procedure was repeated where the point with the current minimum value F determined the midpoint of the intervals in the next step. In this way, the following values of the parameters of the concrete damage evolution equations were obtained: $A_t = 8.01 \times 10^{-1}$, $B_t = 1.95 \times 10^4$, $\kappa_0 = 5.64 \times 10^{-5}$ and $l_c = 112$ mm while the objective function F reached the value 6.07×10^6 N^2. This outcome corresponds to the global mean square relative error $\sqrt{(\sum_{i=1}^{n} (P_i(\mathbf{p}) - P_{\exp,i})^2 / \sum_{i=1}^{n} P_{\exp,i}^2)} = 0.027$ which expresses matching the experimental curve to the theoretical one in the weakening phase. In Figure 4a, the experimental curve and model one for the determined parameters were compared. Figure 4b also shows the course of the model curve after increasing the number of elements to 241 and reducing the permissible increments of measurement base elongation and normal strain to the value of 0.075 µm and 6.25×10^{-7}, respectively, to confirm the correctness of the calculations from the point of view of ensuring the convergence of the solution. In this case, the mean square relative difference between the solutions in the weakening phase with dense and rare discretization amounted to $\sqrt{(\sum_{i=1}^{n} (P_{241,i} - P_{121,i})^2 / \sum_{i=1}^{n} P_{121,i}^2)} = 7.6 \times 10^{-3}$ where in the subscripts the number of elements used is given. On the other hand, Figure 5a shows the distribution of the damage parameter D at its different maximum values along the axis of the specimen at its middle section during the growth of the localized damage. It is also an equivalent way of modeling the development of macro-crack formation from the point of view of damage mechanics [30,49]. The D-distributions with a proportional 2-fold increase of all dimensions of the specimen model (Figure 5b) with the same values of parameters determining the accuracy of the solution were also calculated in a comparative way. A very similar variability of individual distributions was obtained, while the maximum width of the equivalent macro-crack zone reached a value approximately equal to $2l_c$. This confirms the correctness of the presented method of modeling the localized damage evolution in the case of tension in concrete. The presented considerations also justify the adoption of a specific resolution in tomographic imaging by means of wave velocity maps, i.e., dimensions δ_1 and δ_2, according to Figure 1. In an extreme case, they should not be greater than $2l_c$, however, in order to ensure adequate image sharpness and precise identification of the degree of damage, they should be adequately smaller. Taking into account the variability of the distribution D, it is reasonable that δ_1 and δ_2 were assumed in the interval of approximately $l_c/3 - l_c/5$. The presented result also shows that the knowledge of the value of l_c is crucial for the correct assessment of the distribution of damage around the crack in ultrasound tomography, and, on the other hand, it is the tomography that can be used for the direct assessment of this value.

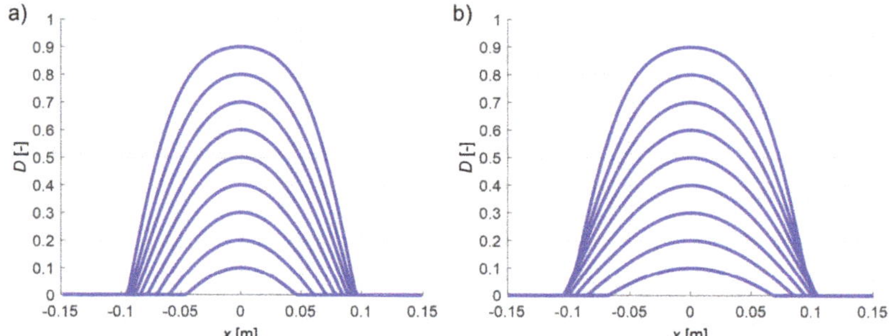

Figure 5. Calculated distributions of D during the development of the damaged zone in the middle of the specimen: (**a**) model of the specimen $d \times h = 20$ cm $\times$ 30 cm (**b**) specimen model with twice the total width and length increased $d \times h = 40$ cm $\times$ 60 cm (d, h according to Figure 3).

Given that the growth D according to relation (8) causes the decrease of Young's modulus of the material, i.e.,:

$$\frac{E_D}{E_0} = 1 - D, \tag{19}$$

where: E_D—tangential Young's modulus (Pa) of the damaged material during inactive growth of the damage, it is in its micro-cracked zones that the velocity of ultrasound waves decreases. Hence, based on the simplified isotropic damage evolution model for concrete developed in Reference [27–31] for three-dimensional stress state and assuming negligible change in material density during the evolution of brittle damage, the following estimated relationships can be given:

$$\frac{c_{LD}}{c_{L0}} = \sqrt{\frac{E_D}{E_0}} \text{ or } \frac{E_D}{E_0} = \left(\frac{c_{LD}}{c_{L0}}\right)^2, \tag{20}$$

where: c_{L0}, c_{LD}—the velocity of the longitudinal wave in the virgin and damaged material, respectively, (m/s). Based on relations (19), (20), variabilities of Young's modulus and longitudinal wave velocity are shown in Figure 6 which correspond to the damage parameter distributions from Figure 5b. They will be used in the computational examples presented in point 4 that illustrate the problem of tomographic detection of cracks in concrete members at various stages of their evolution.

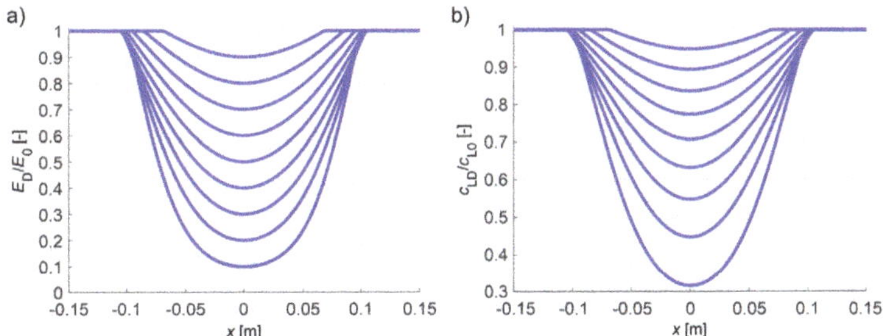

Figure 6. Relative changes of Young's modulus (**a**) and longitudinal wave velocities (**b**) for distributions of *D* from Figure 5b.

It needs to be highlighted that the numerical results presented in this point can be directly used only in the case of pure tension in concrete. However, they can be also generalized usefully for the case of bent reinforced concrete (RC) beams when estimating the damage level in concrete of such beams with the following simplifying assumptions: the planar cross-sections of RC beam remain planar after loading (as in Bernoulli–Euler theory), failure of the concrete is determined by the normal cross-sectional stresses, and the average normal strains along beam axis are equal in the bonded reinforcing longitudinal bars and surrounding concrete. These assumptions are also commonly used in the design of RC beams, for example, as described in the EN-1992-1-1:2004 standard. Under the mentioned conditions, the results shown in Figure 6 can be also applied to a damage estimation of individual fibers of RC beam. The authors used the numerical results in this way for damage level estimation of experimentally tested RC beams in point 5.

4. Determination of Times and Paths of the Fastest Sound Propagation Using Dijkstra's Algorithm

In this paper, Dijkstra's algorithm [40] was used to evaluate the shape of paths of the fastest propagation of ultrasound waves and the time needed for them to travel in concrete elements. The physical basis for this type of calculations is Fermat's principle [36]. Implementation of Dijkstra's algorithm in this type of problem in a compact material area $V \in R^2$ (plane case) may be described as follows (e.g., Reference [42]). Let a sonic pulse be given at point $S \in V$ which triggers the propagation of a specific wave type (e.g., longitudinal). Then, a graph from nodes with numbers i assigned to points $X_i \in V$ ($i = 1, 2, \ldots, m$) should be built. S is one of the points X_i, and edges of the graph will be created only between those nodes which the $X_i X_j \subset V$ ($i, j = 1, 2, \ldots, m$) condition is met for. Weight values $w(i, j)$ for the particular edges will be the pass-through times of sound between the points, assuming that it is moving in a straight line section $X_i X_j$. Hence:

$$w(i, j) = \int_{X_i}^{X_j} \frac{dl_{ij}}{c}, \tag{21}$$

where: c—considered wave speed (m/s), l_{ij}—variable describing the position on a straight line section joining points X_i and X_j (m). In general, it can be seen that the more nodes evenly covering the V area (e.g., assuming an orthogonal grid of points X_i—Figure 7), the more accurate the calculation will be carried out. To increase the time efficiency of the method, when we are only interested in the time of sound propagation between selected two points, the number of nodes can be reduced only to those that lie within a reasonable distance from the straight line connecting these points and/or do not place nodes in a sub-areas where it is known that $D \approx 0$. If the graph is constructed, it is possible to use Dijkstra's method in a standard way, i.e.,:

1. A set Q of all the nodes $i = 1, 2, \ldots, m$ is created and a vector **d** in which the times needed for a wave to travel from the node corresponding to the point $S = X_1$ to other nodes are recorded. First, it is assumed that $d_1 = 0$ and $d_i = \infty$ for $i \neq 1$.

2. In the set Q, a node (let us assign it a number j) is found for which the component **d** has a minimum value and is removed from this set. If the set Q is empty, the calculations are over.

3. In the case of other nodes $i \neq j$ belonging to the set Q, an inequality $d_i > d_j + w\,(i,j)$ is checked. If the inequality is satisfied, the value of $d_j + w\,(i,j)$ is assigned to the component d_i. The algorithm returns to point 2.

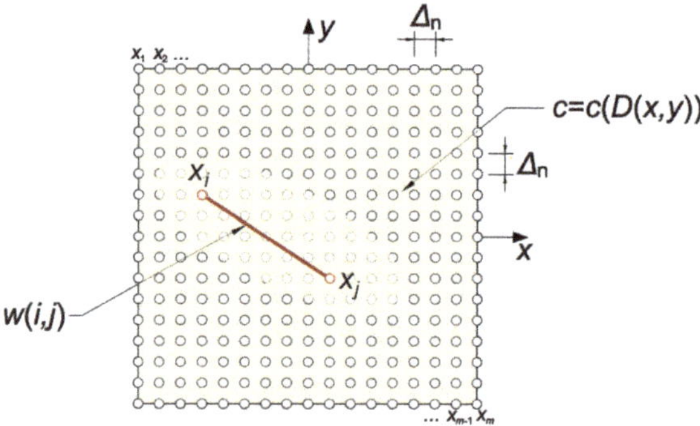

Figure 7. Illustrative orthogonal node arrangement for a graph in a plane problem (based on Reference [42]).

In the final effect of the algorithm operation, the vector components d_i are equal to the time after which, from the moment of excitation of the wave at point S, it reaches the point X_i assuming that the wave can only move on the edges of the graph. Remembering during the calculation also the next nodes indicated in point 3 of the algorithm, you can recreate the path of the fastest wave propagation in the graph.

4.1. Calculation Example 1

The primary purpose of the example is to show to what extent a change in stiffness in a localized, elastically degraded concrete zone can cause the paths of the fastest sound wave propagation to deflect. Another objective is to propose, on the basis of the performed analysis, a useful way to reduce the inaccuracies in tomographic imaging that may arise from this.

The example presents Dijkstra's method of calculating the paths and times of the fastest longitudinal ultrasound wave propagation between the assumed transmitting/receiving points in the longitudinal section of the concrete beam model. One damaged zone was assumed to be formed in the beam in its entire cross-section of 0.4 m × 0.4 m (these dimensions were arbitrarily selected as typical, but this does not change the general conclusions which were finally drawn on this basis). The damaged area is located in the middle of a beam section of 1.2 m in length. It was assumed that the brittle damage changes the distribution of wave propagation velocity along its length in the x function according to the results shown in Figure 6b. It means that a damage level is uniform in a given cross-section of the beam. Consideration was limited to cases where a minimum ratio of E_D/E_0 in the damaged area was 0.9, 0.8, 0.6, or 0.2 with the development of the defect, and, due to the geometry of the task, imaging was performed on the section of the beam along its central vertical plane of symmetry. In order to simplify the calculations, the impact of any reinforcement on the results is not taken into account in the example. It should be obviously noted that this influence can be important for practical research and must not be ignored in the case of higher reinforcement ratios, in particular for the sound wave paths along the reinforcing bars. It is recommended to avoid such measurement situations as much as possible (e.g., Reference [1]). For this reason, it should also be emphasized that the conclusions resulting from the presented calculation examples can be used for quantitative analyses directly only for beams with a low reinforcement ratio and arrangement of propagation paths between the primary reinforcing bars and not along them and arms of stirrups, as well. In other situations, the impact of reinforcement should be taken into account—e.g., using appropriate correction factors, reducing the measured wave velocities so that they correspond to propagation conditions in non-reinforced concrete [1].

The assumed scheme of the beam with transmitting/receiving points is shown in Figure 8. The transmitting/receiving points were selected in the system of opposite points distant from each other every $\Delta_p = 6.25$ mm. Using Dijkstra's algorithm, wave propagation times and paths were determined between the opposite points and with shifting of the receiving points relative to the transmitting points right or left so that the angle of inclination of the straight line between them to the axis x was 45° or 135°, respectively. In Figure 8, the tomographic rays corresponding to the wave propagation paths analyzed are also marked. Furthermore, in order to provide a convenient quantitative interpretation, the numerical results are presented in a dimensionless form, normalizing them to values that would have been obtained in the absence of the damage (i.e., in material characterized by longitudinal wave velocity equal to c_{L0}).

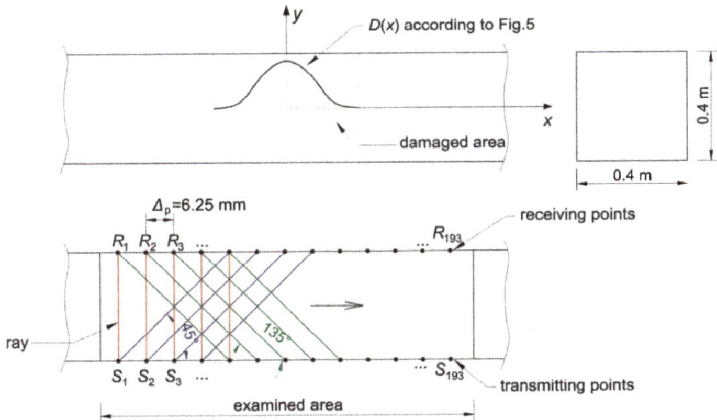

Figure 8. Scheme of the beam model with the assumed damage distribution and the system of transmitting/receiving points and rays.

First, it was checked how the distance between the nodes of the graph (Δ_n according to Figure 7) in Dijkstra's method influences the convergence of results. The graph nodes were placed in the transmitting/receiving points and inside the imaged longitudinal section in the intersection points of the orthogonal grid with the step of Δ_n vertically and horizontally. In order to reduce the calculation time, only the damaged area (in which $E_D/E_0 < 1$) was covered by the internal nodes. The weights of the graph edges were determined from relation (21) by integrating the speed distributions shown in Figure 6b with a step of 1 mm using the rectangular method. Figure 9 shows paths of the fastest propagation illustratively only between transmitting/receiving points distant by 5 cm from each other,

$\min(E_D/E_0) = 0.2$ and $\Delta_n \approx 100$ mm, 50 mm, 25 mm, 12.5 mm, 6.25 mm, or 3.125 mm. Adoption of Δ_n in this way caused that the number of graph nodes in the width of the damaged area changed in extreme cases from 3 to 70. The width of this area can be read from the diagrams in Figures 5 and 6. Table 1 shows the maximum relative changes of the determined wave propagation times on the fastest paths $t_{\text{path},i}$ (s) and their lengths $L_{\text{path},i}$ (m) as the step of the graph nodes decreases where i is the path index. When interpreting the shape of paths in Figure 9, it should be noted that if there is more than one path with the same wave propagation time due to the selection of points of graph nodes and symmetry of the problem geometry then the algorithm used shows the first one found among them. On the other hand, when analyzing the convergence of the solution (Table 1), it can be stated that, with the decrease in the step of the graph node grid, its relative changes become less and less significant and converge to zero. Due to the time needed to solve one such task and the owned computer hardware, further calculations were carried out using the solutions obtained at $\Delta_n = 3.125$ mm (PC with processor 3.06 GHz and RAM 8 GB allowing to determine data for one path on average in approximately 4 min at $\Delta_n = 3.125$ mm).

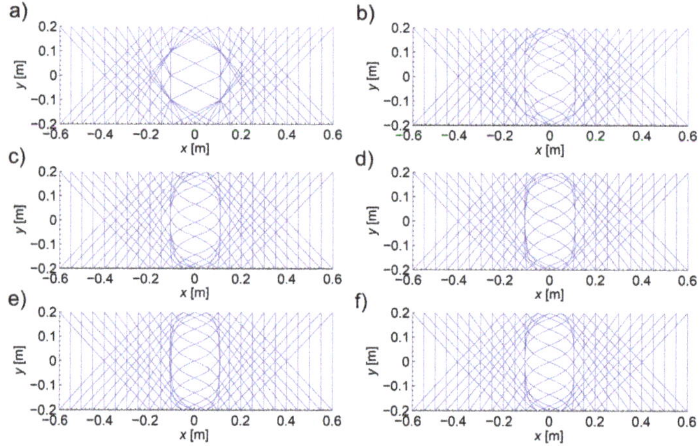

Figure 9. The shape of the paths of the fastest propagation between the selected transmitting/receiving points in the longitudinal section of the beam with the area damaged at $\min(E_D/E_0) = 0.2$ and $\Delta_n \approx$ (**a**) 100 mm, (**b**) 50 mm, (**c**) 25 mm, (**d**) 12.5 mm, (**e**) 6.25 mm, and (**f**) 3.125 mm.

Figure 10 collectively shows the shape of selected paths of the fastest wave propagation depending on the degree of damage. It is evident that as the minimum ratio E_D/E_0 decreases the wave undergoes increasing deflection in the damaged area. At the same time, it is obvious that the time needed to travel such a path by the wave is shorter and shorter than the one that would be calculated with a simplistic assumption of its propagation in a straight ray. The scale of this difference is illustrated in Figure 11 showing how the propagation times change along the fastest paths and rays connecting consecutive opposite transmitting/receiving points and points lying diagonally at an angle of 45° to each other in relation to the axis of the beam. Figure 11 does not show the variability of times along the paths between points lying diagonally to each other at an angle of 135° relative to the axis x because it is the same as for those lying at an angle of 45° taking into account the symmetry. In order to make the diagrams more readable, they are also presented in the form of continuous curves. In the case of a damaged zone with a course perpendicular to the axis of the beam analysed in the example, it can be observed that greater differences in wave propagation times along the real paths and calculated on the assumption of their straightness occur for those that connect the opposite points and are increasing as the damage increases. These differences are much smaller for paths connecting the points diagonally. This observation may be used to reduce the inaccuracy of tomographic calculations due to the fact that $t_{path,i}$ (in fact, the times determined from measurements) are inserted in real situations in place of $t_{ray,i}$ in the system of Equation (3) where it is assumed in a simplistic way that $t_{ray\,measured,i} = t_{path,i}$. Therefore, in the case of propagation times only for opposite points, an approximate relationship can be proposed between $t_{path,i}$ and $t_{ray,i}$, i.e.,:

$$t_{ray,i} \approx t_{ray\,approx,i} = t_{path,i} + \beta\left(t_{path,i} - \min\left(t_{path,i}\right)_{i=1,2,\ldots,I}\right), \tag{22}$$

where: $t_{ray\,approx,i}$—approximated propagation times assuming a wave passage over the ray (s), β—non-negative factor [-].

Table 1. Maximum relative changes in path length and wave propagation times when changing Δ_n in different defect evolution phases.

Δ_{n1}, Δ_{n2} [mm]	$\max\left\|\dfrac{t_{path,i}(\Delta_{n2})-t_{path,i}(\Delta_{n1})}{t_{path,i}(\Delta_{n1})}\right\|_{i=1,2,\ldots I}$ [-]			
	min (E_D/E_0) [-]			
	0.9	0.8	0.6	0.2
50, 100	0.0035	0.0085	0.0154	0.0150
25, 50	0.0030	0.0054	0.0104	0.0230
12.5, 25	0.0024	0.0020	0.0030	0.0057
6.25, 12.5	0.0011	0.0012	0.0021	0.0040
3.125, 6.25	0.0003	0.0006	0.0004	0.0011

Δ_{n1}, Δ_{n2} [mm]	$\max\left\|\dfrac{L_{path,i}(\Delta_{n2})-L_{path,i}(\Delta_{n1})}{L_{path,i}(\Delta_{n1})}\right\|_{i=1,2,\ldots I}$ [-]			
	min (E_D/E_0) [-]			
	0.9	0.8	0.6	0.2
50, 100	0.0199	0.0386	0.0794	0.0430
25, 50	0.0119	0.0245	0.0543	0.0281
12.5, 25	0.0071	0.0100	0.0262	0.0324
6.25, 12.5	0.0043	0.0072	0.0109	0.0180
3.125, 6.25	0.0016	0.0032	0.0067	0.0094

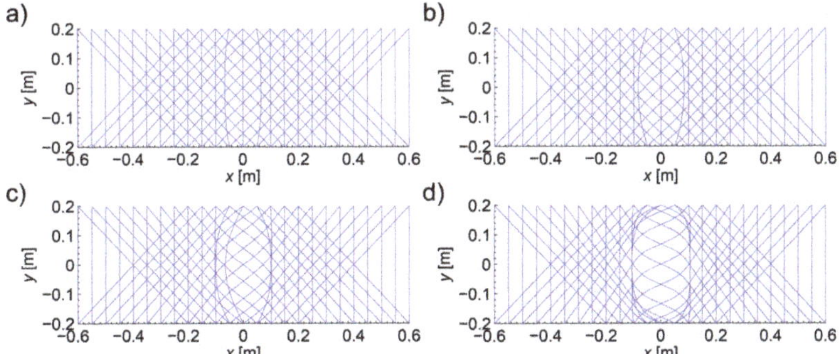

Figure 10. The shape of paths of the fastest propagation between the selected transmitting/receiving points in the longitudinal section of the beam depending on the degree of elastic degradation in the damaged zone $\min(E_D/E_0) =$ (**a**) 0.9, (**b**) 0.8, (**c**) 0.6, and (**d**) 0.2 (for $\Delta_n = 3.125$ mm).

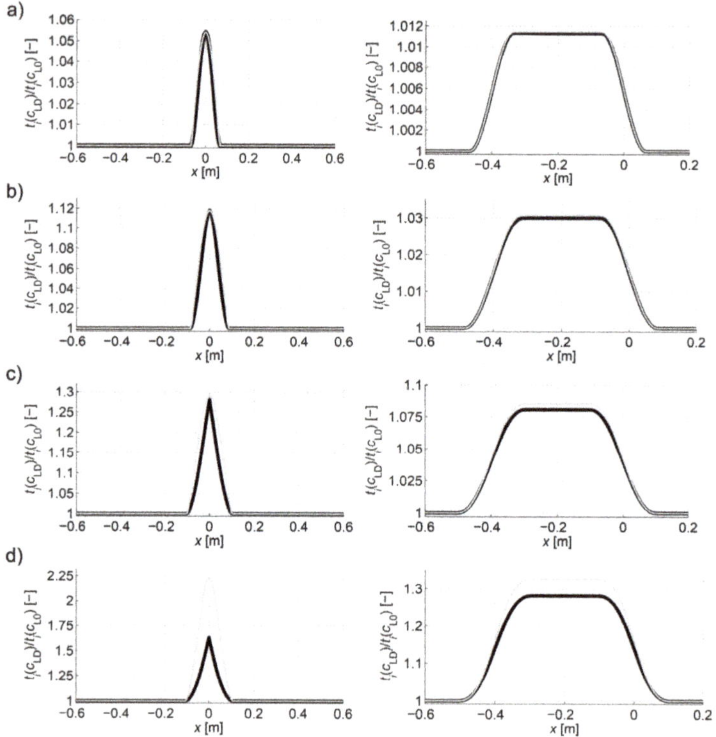

Figure 11. Times of wave propagation $t_{\text{path},i}$ on the fastest paths, and $t_{\text{ray},i}$ over the straight rays between the transmitting/receiving points in the different phases of the defect evolution: $\min(E_D/E_0) =$ (**a**) 0.9, (**b**) 0.8, (**c**) 0.6, and (**d**) 0.2. The results are presented as a function of the position of the transmitting points: a black line for the fastest paths determined by Dijkstra's algorithm (for $\Delta_n = 3.125$ mm); a grey line for the paths established as straight rays. The diagrams on the left refer to the paths connecting the opposite points and those on the right to the points lying diagonally at an angle of 45° to each other in relation to the axis x.

Thus, the factor β scales propagation times from real paths by elongating them in relation to the minimum value in the direction of positive values so as to bring their course as close as possible to the times that would correspond to the passage of a wave over the straight ray. As a result, this will clearly result in a reduction of calculation errors. Analyzing the graphs shown in Figure 11, this factor can be determined, e.g., from the least squares method. However, in a situation of actual measurement, due to the lack of such data, this is impossible. Therefore, the authors propose to apply the following heuristic approach to the determination of β. Approximate solution of the system of Equation (3) can be alternatively obtained using Tikhonov's method [51,52] minimizing the function of the mean square error of this solution (the so-called residual term F_{res}) with additional consideration of the regularisation term F_{reg} (e.g., Reference [24]). Using this approach in the measurement situation, when in Equation (3) we replace the appropriate components in the vector $\mathbf{b}$ $t_{\text{ray measured},i} = t_{\text{path},i}$ across $t_{\text{ray measured},i} = t_{\text{ray appox},i}(\beta)$, we will obtain:

$$F_{\text{res}} = \tfrac{1}{2}\|\mathbf{A}\mathbf{x}_{\alpha,\beta} - \mathbf{b}(\beta)\|^2, \ F_{\text{reg}} = \tfrac{1}{2}\alpha\|\mathbf{R}(\mathbf{x}_{\alpha,\beta} - \mathbf{x}_0)\|^2$$

$$F_{\text{res}}(\mathbf{x}_{\alpha,\beta}) + F_{\text{reg}}(\mathbf{x}_{\alpha,\beta}) = \min \rightarrow \mathbf{x}_{\alpha,\beta} = (\mathbf{A}^\mathrm{T}\mathbf{A} + \alpha\mathbf{R}^\mathrm{T}\mathbf{R})^{-1}(\mathbf{A}^\mathrm{T}\mathbf{b}(\beta) + \alpha\mathbf{R}^\mathrm{T}\mathbf{R}\mathbf{x}_0),$$

(23)

where: α—regularization parameter [-], $\mathbf{R}$—regularization matrix (m), $\mathbf{x}_{\alpha,\beta}$—solution of the problem for given values α and β (s/m), $\mathbf{x}_0$—the point in the vicinity of which a regularized solution is sought (s/m). Matrix $\mathbf{R}$ was assumed in further consideration as a unitary one multiplied by 1 m for the sake of the unit account, and $\mathbf{x}_0$ as the vector according to Equation (6). In the proposed method of determination of β, it is assumed that the less data contained in the vector $\mathbf{b}$ will be affected by measurement errors due to the selection of β, the smaller the values of the term F_{reg} in the case of an optimal solution to the problem. It is, therefore, proposed to determine the optimal value of the coefficient $\beta = \beta_{opt}$ from the relationships:

$$\beta_{opt} = \max \left(g(z) \right) \text{ for } z \geq 0, \tag{24}$$

where

$$g(z) = \arg \min \left(F_{reg}\left(\mathbf{x}_{\alpha=z,\beta>0} \right) \right) \text{ for } z \geq 0. \tag{25}$$

Selection of β_{opt} as the maximum value of function (25), in turn, has the following meaning: determine what is the maximum possible increase in the propagation time of the wave in the vector $\mathbf{b}$ using formula (22) with a minimum effect of regularization so that the original information contained in system of Equation (3) is lost as little as possible at the same time. Using formulas (22)–(25), the following values of β_{opt} were obtained as in Table 2 for the data from the graphs in Figure 11. Illustratively, the variability of the wave propagation time $t_{ray\ approx,i}$ modified according to relation (22) between opposite transmitting/receiving points at $\beta = \beta_{opt}$ and $\min(E_D/E_0) = 0.2$ is shown in Figure 12. At the same time, it was compared with $t_{ray,\ i}$ and $t_{path,i}$, where $t_{ray\ approx,i}$ has a much more similar course to that of $t_{ray,i}$ than $t_{path,i}$.

Table 2. β_{opt} calculated on the basis of data from Figure 11 in case of the opposite transmitting/receiving points in different phases of defect growth.

$\min (E_D/E_0)$ [-]	0.9	0.8	0.6	0.2
β_{opt} [-]	0.09	0.15	0.33	1.24

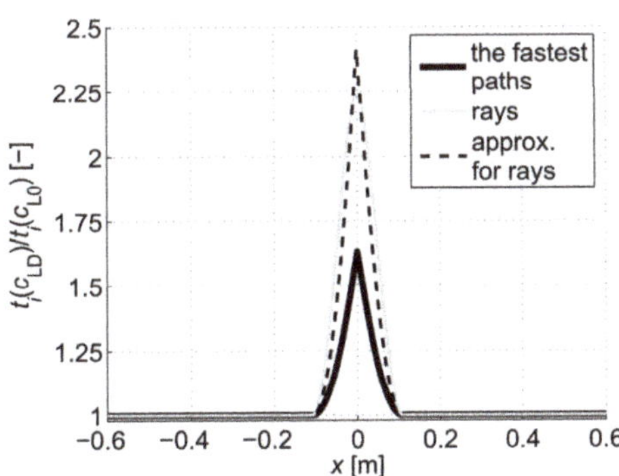

Figure 12. Propagation times $t_{path,i}$, $t_{ray,i}$ and $t_{ray\ approx,i}$ between the opposite transmitting/receiving points for a defect with $\min(E_D/E_0) = 0.2$. $t_{ray\ approx,i}$ were calculated according to relation (22) with $\beta = \beta_{opt}$.

In order to assess the correctness of the proposed approximation method, appropriate tomographic reconstructions of the longitudinal wave velocity maps in the beam model section from Figure 8 are presented in Figure 13. The reconstructions were determined by the randomized Kaczmarz method in accordance with the information presented in point 2, where c_{L0} was adopted as $c_{L\ ref}$. The results are shown only for the central area of the beam separated by a red dashed line through which all types of rays passed due to their inclination. Wave velocity distributions in the damaged area according to Figure 6b were assumed with minimal ratio E_D/E_0 equal to 0.9, 0.8, 0.6, or 0.2. The resolution of

$\delta_1 \times \delta_2 = 3.33$ cm $\times 3.33$ cm was applied, as well as the number and arrangement of rays, as shown in Figure 8. For comparison purposes, in addition to the reconstruction with the use of times $t_{\text{ray approx},i}$ according to relation (22) and $t_{\text{path},i}$, the original map is also shown on the basis of data from Figure 6b, but in the resolution used, i.e., in each cell, its average speed value is taken from the formula:

$$c_{\text{L mean},k} = \frac{\int_{A_k} c_{\text{L}} \, dx \, dy}{A_k},\tag{26}$$

where: A_k—rectangular area occupied by the k-th cell (m^2). Integral (26) was calculated in an approximate way using the rectangular method, taking the step of integration after x and y, respectively as $\delta_1/10$ and $\delta_2/10$. Therefore, relative errors were calculated for each reconstruction: global—mean square $e_g = \sqrt{(\sum_{k=1}^{K}(c_{\text{L mean},k} - c_{\text{L tom},k})^2 / \sum_{k=1}^{K} c_{\text{L mean},k}^2)}$ and local—maximal $e_{l \max} = \max((c_{\text{L mean},k} - c_{\text{L tom},k})/c_{\text{L mean},k})_{k=1,2,...,K}$, where $c_{\text{L tom},k}$ was determined from system of Equation (3). Because $c_{\text{L mean},k}$ is not known in a real measuring situation, it is also necessary to introduce a measure which will allow to estimate the reconstruction errors due to the assumption of straight-lined rays without this information.

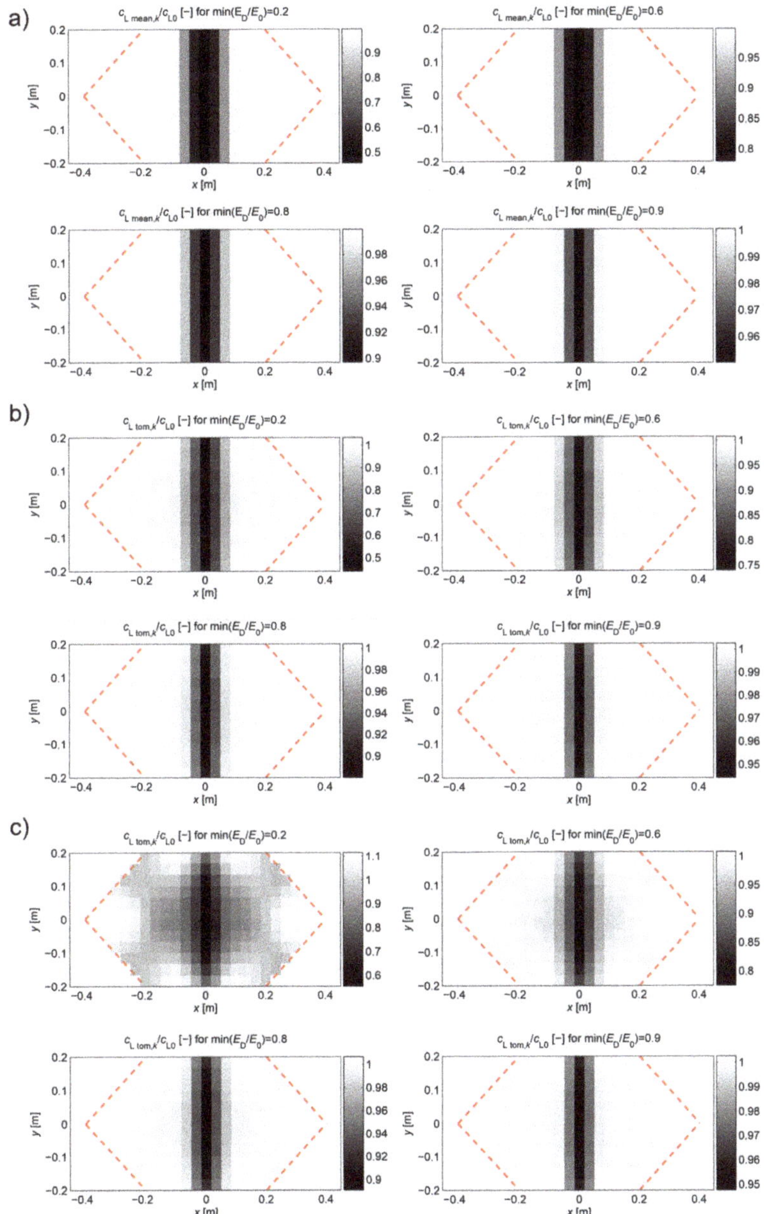

Figure 13. Originally assumed distributions of wave velocity in the longitudinal section of the beam in different stages of defect evolution (**a**) and their tomographic reconstructions according to Equation (3): (**b**) calculated using the propagation times $t_{\text{ray},i} = t_{\text{ray approx},i}$ for the paths connecting the opposite points and $t_{\text{ray},i} = t_{\text{path},i}$ for the paths connecting the points diagonally, (**c**) calculated using only the propagation times $t_{\text{ray},i} = t_{\text{path},i}$ (red dotted lines—explanation in the text).

In this paper, it is proposed to calculate index $d_{\text{cL ray max}}$: the maximal relative difference between $c_{\text{L ray measured},i}$ (wave velocities averaged over the i-th ray based in real directly on the measured propagation times) and $c_{\text{L ray approx},i}$ (wave velocities averaged over the i-th ray calculated, respectively, on the basis of approximated propagation times according to relation (22)). This leads to the formula:

$$d_{\text{cL ray max}} = \max\left(\frac{c_{\text{L ray measured},i} - c_{\text{L ray approx},i}}{c_{\text{L ray measured},i}}\right)_{i=1,2,\ldots,I} = \max\left(1 - \frac{t_{\text{path},i}}{t_{\text{ray approx},i}}\right)_{i=1,2,\ldots,I}. \tag{27}$$

A summary of e_g, $e_{\text{l max}}$, and $d_{\text{cL ray max}}$ is presented in Table 3, depending on the adopted calculation strategy, while for $d_{\text{cL ray max}}$, the exact value of it is given in a comparative manner (if calculated with the use of times of propagation $t_{\text{ray},i}$ instead of $t_{\text{ray approx},i}$).

Table 3. Global—mean square (e_g) and local—maximal ($e_{\text{l max}}$) relative tomographic reconstruction error using the propagation times for the opposite points in Equation (3) $t_{\text{ray},i} = t_{\text{ray approx},i}$ or $t_{\text{ray},i} = t_{\text{path},i}$. Maximal relative difference in average wave velocities over the rays $d_{\text{cL ray max}}$ due to the use of $t_{\text{ray},i} = t_{\text{ray approx},i}$ or $t_{\text{ray},i} = t_{\text{path},i}$.

	e_g [-]			
$t_{\text{ray},i}$	$\min(E_D/E_0)$ [-]			
	0.9	0.8	0.6	0.2
$t_{\text{ray approx},i}$	0.001	0.004	0.011	0.027
$t_{\text{path},i}$	0.002	0.005	0.018	0.087
	$e_{\text{l max}}$ [-]			
$t_{\text{ray},i}$	$\min(E_D/E_0)$ [-]			
	0.9	0.8	0.6	0.2
$t_{\text{ray approx},i}$	0.007	0.022	0.058	0.186
$t_{\text{path},i}$	0.010	0.025	0.099	0.700
	$d_{\text{cL ray max}}$ [-]			
$\min(E_D/E_0)$ [-]	0.9	0.8	0.6	0.2
Equation (27), Exact value	0.005, 0.009	0.016, 0.023	0.067, 0.068	0.325, 0.323

The analysis of the presented results shows that the reconstruction error increases with the degree of development of localized brittle damage. If propagation times $t_{\text{path},i}$ are used in vector **b** of Equation (3) (as would be obtained directly from the measurement in the real situation), then, in the analyzed case, the relative decrease of Young's modulus in the damaged zone from 10% to 80% is connected with the increase of e_g from approximately 0.002 to 0.09. In parallel, $e_{\text{l max}}$ changes from 0.01 to 0.70. Introduction to the vector **b** propagation times for the opposite transmitting/receiving points according to formula (22) with the optimum value of β allows reducing the relative reconstruction error of both e_g, as well as $e_{\text{l max}}$, for ranges from approximately 0.001 to 0.03 and from 0.007 to 0.19, respectively. The use of formula (22) also allows for a more accurate evaluation of the shape and spatial range of the defect (Figure 13), i.e., regardless of the degree of damage, its original shape is visualized correctly. Using only propagation times $t_{\text{path},i}$ in the calculations allows to strictly assess the shape of the analyzed defect only when Young's modulus drops to about 20%. On the other hand, the value of $d_{\text{cL ray max}}$ also allows a rough estimation of the differences in the reconstructed wave velocity maps that may occur due to the adoption of the fastest straight-line propagation paths in calculations (although, taking into account the limitations that result from its definition according to Equation (27)). Its advantage in turn is that it can be determined in research only on the basis of measurement data. It starts to increase noticeably when Young's modulus drops in the defected zone above 20%.

4.2. Calculation Example 2

The purpose of the second calculation example is to show that the desired resolution of the image of changes in ultrasound wave velocity around localized damage can be obtained by using a reduced number of transmitting/receiving points. To this end, so-called fictitious transmitting/receiving points should be introduced for which propagation times will be interpolated on the basis of data from the original set of points (e.g., Reference [33,47]). In a real measurement situation, this may often result in a significant reduction of transmitting/receiving points, costs and time consumption of tests. The simulation of such a measurement strategy for the data of the first calculation example is shown below. The same geometry of the beam model and the method of its damage as in the first example were adopted, but the number of points for which the original information about wave propagation times is available was reduced in such a way that the gaps between them amount to $\Delta_p = 10$ cm. There are fictitious points every 6.25 mm between them at the same intervals as the points in the first example. For each group of paths, depending on the location of their transmitting/receiving points relative to each other (the opposite points, the points lying diagonally at an angle of 45° and 135° in relation to the axis x) times for paths connecting fictitious points were interpolated. The interpolation in the function of transmitting points position was carried out with the use of a cubic Hermite spline assuming continuity to the first derivative (*pchip* function in the MATLAB program environment was used). The results are shown in Figure 14, where interpolation nodes were marked with circles. They were also compared with the propagation times calculated with Dijkstra's algorithm in the first example. Figure 14 does not show the variability of times for the paths between the points lying diagonally to each other at an angle of 135° relative to the axis x because it is the same as for those lying at an angle of 45° taking into account the symmetry of the problem. In order to make the graphs easier to read, the interpolated times and those of the first example are presented in the form of continuous curves. Relative global interpolation errors are summarized in Table 4, i.e., $e_g = \sqrt{ \left(\sum_{i=1}^{I} \left(t_{\text{path},i} - t_{\text{path int},i} \right)^2 / \sum_{i=1}^{I} t_{\text{path},i}^2 \right) }$ was calculated where $t_{\text{path int},i}$ (s) is the interpolated propagation time of the wave on the i-th path.

Table 4. Relative global interpolation errors of propagation times from the first example using a cubic Hermite spline in different phases of defect growth.

$\min(E_D/E_0)$ [-]	0.9	0.8	0.6	0.2
e_g [-]	0.002	0.003	0.007	0.020

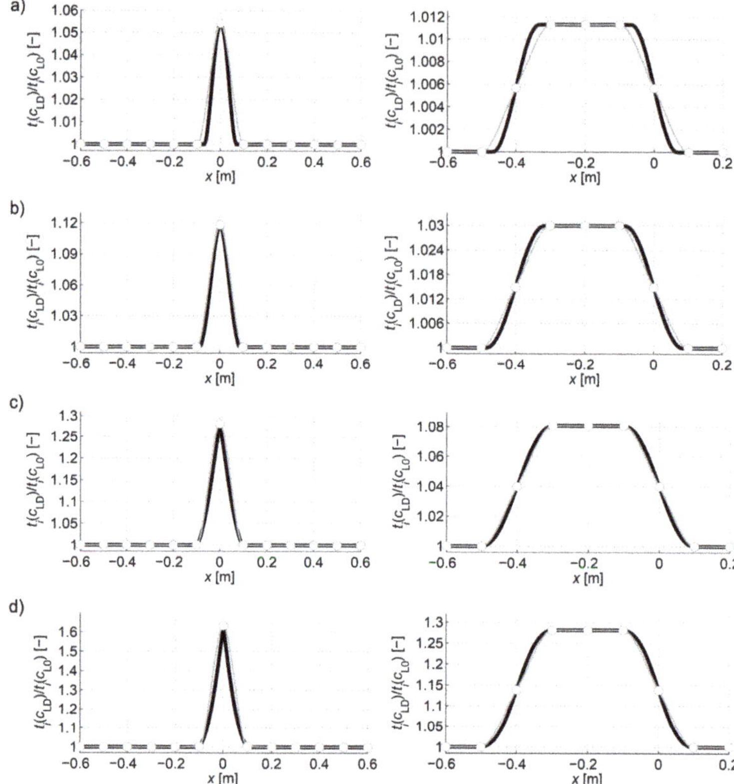

Figure 14. Times of wave propagation $t_{path,i}$ and $t_{path\ int,i}$ on the fastest paths between transmitting/receiving points in different phases of defect evolution: $\min(E_D/E_0)$ = (**a**) 0.9, (**b**) 0.8, (**c**) 0.6, and (**d**) 0.2. The results are presented as a function of the position of the transmitting points: a black line for the fastest paths determined by Dijkstra's algorithm (for Δ_n = 3.125 mm); a grey line in case of interpolation with nodes marked with circles. The diagrams on the left refer to the paths connecting the opposite points and those on the right to the points lying diagonally at an angle of 45° to each other in relation to the axis x.

In the same way as in the first example, the propagation times for the fastest ultrasound wave paths connecting opposite transmitting/receiving points were modified to bring them as close as possible to the propagation times of straight rays, but with the use of times $t_{path\ int,i}$, i.e.,:

$$t_{ray,i} \approx t_{ray\ approx,i} = t_{path\ int,i} + \beta\left(t_{path\ int,i} - \min\left(t_{path\ int,i}\right)_{i=1,2,...,I}\right). \tag{28}$$

Then, using formulas (23)–(25), the values of β_{opt} were obtained for data from Figure 14 as in Table 5. Illustratively, the variability of the wave propagation time $t_{ray\ approx,i}$ modified according to relation (28) between opposite transmitting/receiving points at $\beta = \beta_{opt}$ and $\min(E_D/E_0) = 0.2$ is shown in Figure 15. At the same time, it was compared with $t_{ray,i}$ and $t_{path,i}$, where $t_{ray\ approx,i}$ has also a much more similar course to that of $t_{ray,i}$ than $t_{path,i}$.

Table 5. β_{opt} calculated on the basis of data from Figure 14 in case of the opposite transmitting/receiving points in different phases of defect growth.

$\min(E_D/E_0)$ [-]	0.9	0.8	0.6	0.2
β_{opt} [-]	0	0	0.10	0.68

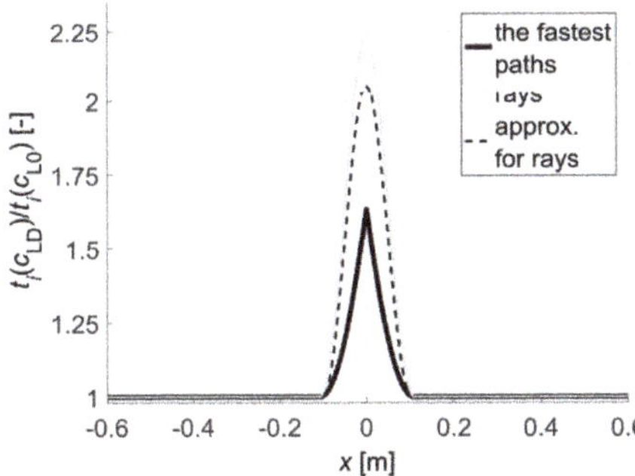

Figure 15. Propagation times $t_{\text{path},i}$, $t_{\text{ray},i}$ and $t_{\text{ray approx},i}$ between the opposite transmitting/receiving points for a defect with $\min(E_D/E_0) = 0.2$. $t_{\text{ray approx},i}$ were calculated according to relation (28) with $\beta = \beta_{\text{opt}}$.

In order to assess the correctness of the proposed propagation time interpolation method, tomographic reconstructions of the wave velocity maps in the beam model longitudinal section from Figure 8 are presented in Figure 16. The reconstructions were determined by the randomized Kaczmarz method in accordance with the information presented in point 2 where c_{L0} was adopted as $c_{L\,\text{ref}}$. The results are shown only for the central area of the beam section separated by a red dashed line through which all types of rays passed due to their inclination. For comparison purposes, reconstructions with the use of times $t_{\text{ray approx},i}$ according to relation (28) and $t_{\text{path int},i}$ were presented. For the individual reconstructions, their relative errors were also calculated: global—mean square e_g, local—maximal $e_{l\,\text{max}}$. The relative maximal difference in average speed over rays $d_{cL\,\text{ray max}}$ was also determined due to the use of $t_{\text{ray},i} = t_{\text{ray approx},i}$ or $t_{\text{ray},i} = t_{\text{path int},i}$, i.e., in this particular case:

$$d_{cL\,\text{ray max}} = \max\left(1 - \frac{t_{\text{path int},i}}{t_{\text{ray approx},i}}\right)_{i=1,2,\ldots,I}. \tag{29}$$

A summary of e_g, $e_{l\,\text{max}}$ and $d_{cL\,\text{ray max}}$ is presented in Table 6, depending on the calculation strategy adopted, where the exact value of $d_{cL\,\text{ray max}}$ is also shown for a comparison as in Table 3.

Table 6. Global—mean square (e_g) and local—maximal ($e_{l\,\text{max}}$) relative tomographic reconstruction error using the propagation times for the opposite points in Equation (3) $t_{\text{ray},i} = t_{\text{ray approx},i}$ or $t_{\text{ray},i} = t_{\text{path int},i}$. Maximal relative difference in average wave velocities over the rays $d_{cL\,\text{ray max}}$ due to the use of $t_{\text{ray},i} = t_{\text{ray approx},i}$ or $t_{\text{ray},i} = t_{\text{path int},i}$.

	e_g [-]			
$t_{\text{ray},i}$	$\min(E_D/E_0)$ [-]			
	0.9	0.8	0.6	0.2
$t_{\text{ray approx},i}$	0.0018	0.0024	0.0083	0.0191
$t_{\text{path int},i}$	0.0018	0.0024	0.0100	0.0632
	$e_{l\,\text{max}}$ [-]			
$t_{\text{ray},i}$	$\min(E_D/E_0)$ [-]			
	0.9	0.8	0.6	0.2
$t_{\text{ray approx},i}$	0.0108	0.0156	0.0467	0.1326
$t_{\text{path int},i}$	0.0108	0.0156	0.0598	0.5053
	$d_{cL\,\text{ray max}}$ [-]			
$\min(E_D/E_0)$ [-]	0.9	0.8	0.6	0.2
Equation (29) Exact value	0, 0.009	0, 0.023	0.021, 0.068	0.208, 0.323

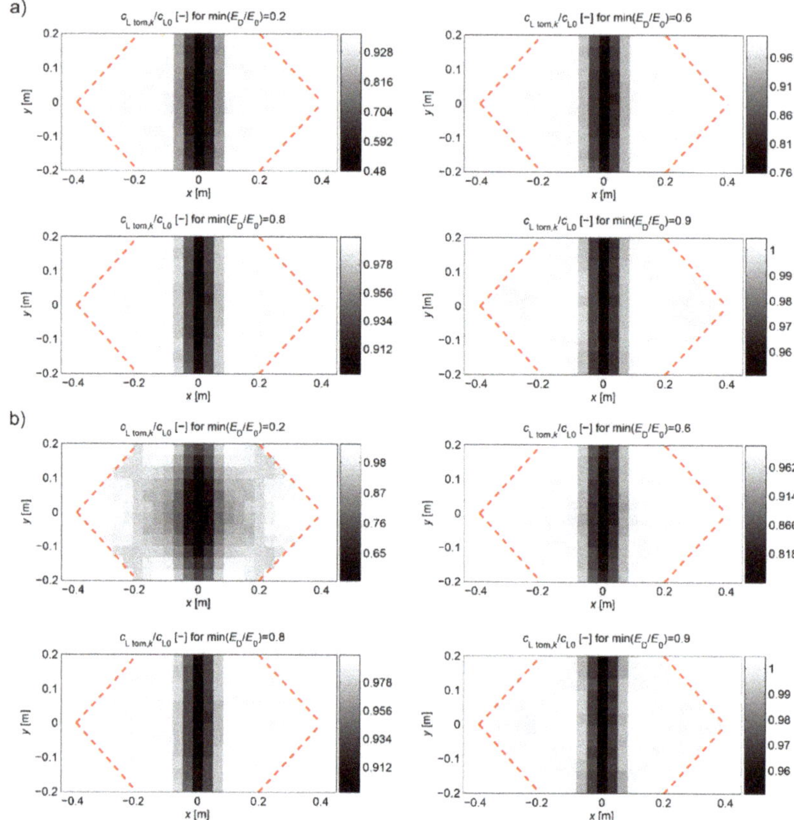

Figure 16. Tomographic reconstructions of the wave velocity distribution from Figure 13a according to Equation (3): (**a**) calculated with the propagation times $t_{\text{ray},i} = t_{\text{ray approx},i}$ for the paths connecting the opposite points and $t_{\text{ray},i} = t_{\text{path int},i}$ for the paths connecting the points diagonally, (**b**) calculated only with the propagation times $t_{\text{ray},i} = t_{\text{path int},i}$ (red dotted lines—explanation in the text).

Analyzing the presented results one can draw general conclusions similarly to the reconstructions from the first example. In addition, it can be noted that reconstruction errors, compared to those in Table 3 for damaged zones with Young's modulus drop of more than 10%, were decreased by up to 2 times at most. Again, it can also be stated that, in the case of a defect in a concrete member with a higher degree of elastic degradation, the introduction into the vector **b** in Equation (3) of wave propagation times after appropriate scaling (Equation (28) with the optimal value of β) allows for effective, even several fold reduction of calculation errors and more correct evaluation of the defect shape. The obtained results also confirm that the use of so-called fictitious points with interpolated propagation times allows to increase the resolution of tomographic reconstructions of elastically degraded concrete areas without the need to use "too dense" system of real transmitting/receiving points. In this case, the value of $d_{\text{cL ray max}}$ also allows rough estimation of differences in reconstructed wave velocity maps, which can occur due to the adoption of the fastest propagation paths as straight in calculations. However, the use of fictitious points causes that it deviates from the exact values much more than in the first example. Nevertheless, it is important at this point that it starts to increase noticeably when Young's modulus in the defect drops above 20%, and, for example, such an estimation can be rationally increased by 2–3 times for safety reasons.

5. Experimental Study

As part of the tomographic experiments, studies were carried out on RC elements after cracking initiation—three beams on a laboratory scale and one prefabricated beam on a natural scale. Since the calculation methodology presented in point 3 concerned the detection of elastically degraded zones with a course perpendicular to the beam axis, the experiments illustrating the possibilities of its practical application also focused on the evaluation of this type of defects. Hence, the laboratory beams were bent until the first perpendicular cracks were formed in the middle area. In the second case, a prefabricated industrially manufactured beam, that was damaged during transport to the construction site, was inspected. There were cracks perpendicular to the beam axis and visible to the naked eye. This beam was specially selected for an assessment to also test the presented calculation method in near-real conditions. An important aspect of this study was also the willingness to check whether ultrasound tomography in the presented approach could potentially be used for quality control of prefabricated RC elements in industrial conditions.

The cross-sections of beams for tomographic imaging were selected so that the measurements were disturbed as little as possible by their reinforcement (between longitudinal bars and vertical arms of stirrups) [1,53]. In the case of laboratory beams, it was also decided to show how changing the care method can affect tomographic detection of brittle damage. For this purpose, three samples were stored under water for 1 to 28 days from the time of forming. Two of them were tested after 28 days and the third one after 35 days (after its removal from water at the age of 28 days and storage at the room conditions for the next 7 days). The storage temperature of all the beams was about 20 °C. Taking into account also disturbances which may be caused by uneven distribution of humidity [53], the beams were examined in conditions in which the distribution of humidity in them would be as homogeneous as possible. The 28-day laboratory beams were tested for up to about 1 h after the removal from the water bath, and the 35-day beam, after the removal from the water, was protected by polyethylene film against moisture exchange with the ambient air until the test. In turn, the prefabricated beam was stored about 7 months before the test inside the laboratory at an average relative humidity of about 50% and a temperature of 20 °C. The age of the prefabricated beam at the time of testing was about 9 months.

On the basis of theoretical considerations discussed in previous points, a detailed course of tomographic measurements and method of processing data obtained this way was also established and used during own experiments. The general scheme of this procedure is shown in Figure 17.

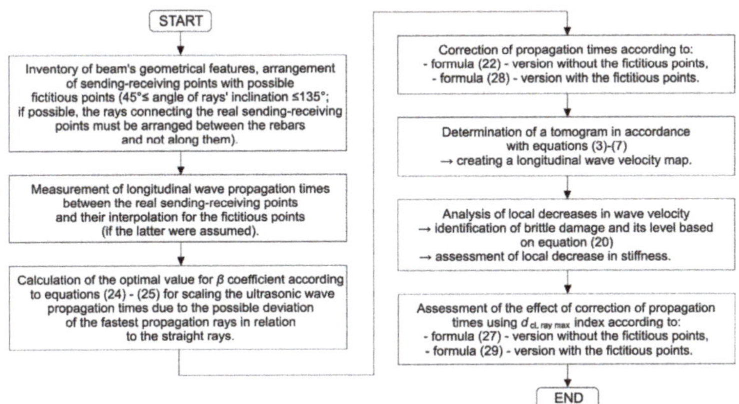

Figure 17. Flowchart of the calculation process.

5.1. Laboratory Beams

A diagram of the beams is shown in Figure 18. Their dimensions were 10 cm × 10 cm × 50 cm and they were made of ready market mixture of concrete with a mean compression strength of $f_{cm,cube} = 48$ MPa after 28 days. The maximum aggregate diameter was $d_{a\,max} = 8$ mm. The longitudinal lower reinforcement consisted of two bars of 8 mm in diameter, made of steel with characteristic yield point declared by the manufacturer of $f_{yk} = 400$ MPa. The studies consisted in the measurement of the time of propagation of the longitudinal ultrasound wave using a Pundit-Lab tester and transreceiver heads with a frequency of 250 kHz. The coupling of the heads and element was provided by special gel for ultrasonic tests. The adopted system of transmitting/receiving points distant from each other by $\Delta_p = 10$ cm is shown in Figure 18. Tomographic rays were assumed between the opposite points and lying diagonally at an angle of 45° and 135° in relation to the beam axis (Figure 18). The longitudinal vertical section of the beams for tomographic imaging was located in the middle between the reinforcement bars. For comparison, the ultrasound tests were carried out in two stages: before and after the first load-unload cycle in static three-point bending (Figure 19). The first load stage was finished at the moment when the first cracks appeared, controlling the registered load (P) and deflections at the centre of the span (u), i.e., until the slope change occurs in the function $P - u$. After the second stage of ultrasonic testing, the beams were bent until their load capacity was exhausted. The obtained values of the cracking (P_{cr}) and maximum (P_{max}) loads are summarized in Table 7.

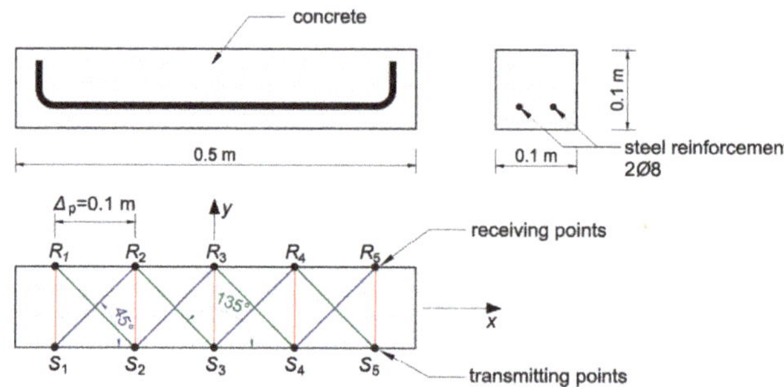

Figure 18. Scheme of the beams and the assumed system of transmitting/receiving points and rays.

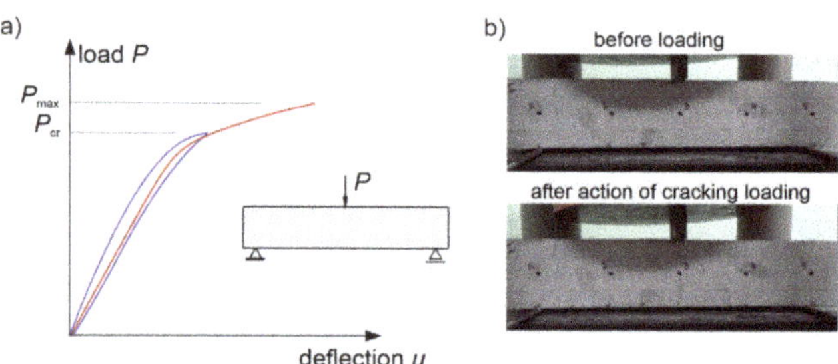

Figure 19. (**a**) Beam loading scheme. (**b**) Illustrative pictures of beam No. 3 before and after an action of cracking load.

Table 7. Cracking and maximal loads of the lab-beams.

Load	No. of the Beam			Mean
	1	2	3	
P_{cr} [kN]	27	30	28	28.3
P_{max} [kN]	43	45	46	44.7

Taking into account the measured longitudinal wave propagation times and basic frequency of ultrasonic pulses, the average wavelengths for beam No. 1, 2, and 3 were, respectively, ~1.7 cm, ~2.0 cm, and ~2.1 cm before the loading and ~1.7 cm, ~1.8 cm, and ~2.0 cm after the loading. At the same time, it allowed satisfying the basic requirement described in ASTM D2845-08 regarding the selection of frequency so that measurable longitudinal ultrasonic waves could be generated in the samples, i.e.,: dominant wavelength $\geq 3\times$ the average grain size equal to ~4 mm.

The first visible crack appeared in each case in the middle of the beam span as perpendicular to the beam axis (Figure 19). Measured and interpolated longitudinal wave propagation times $t_{\text{path int},i}$ are shown in Figure 20 where interpolated times are determined using a cubic Hermite spline. The diagrams also show $t_{\text{ray approx},i}$ determined according to relation (28) with $\beta = \beta_{\text{opt}}$ according to (24), (25). In Figure 20, by comparing wave propagation times before and after the loading, the evolution of cracks can be clearly observed and their location initially made in the sections where these times have increased the most. It can also be seen that not taking into account the increase in the propagation time of ultrasonic pulses along the straight rays compared to the times measured for the fastest paths would lead to their underestimation of approximately 5–8% in the most damaged areas of the beams. Figure 21 shows an example of a recorded signal by the receiving head together with a reading the time of longitudinal wave propagation. In Figure 21, the signal caused by longitudinal wave propagation is visible first, and, a moment later, the signal connected with the propagation of transverse and Rayleigh waves of much higher amplitude can be noticed, but without the possibility of precise distinction of the initial moment of their registration. This was also the main reason why the authors decided to use longitudinal waves in their studies, taking into account the capabilities of their research equipment.

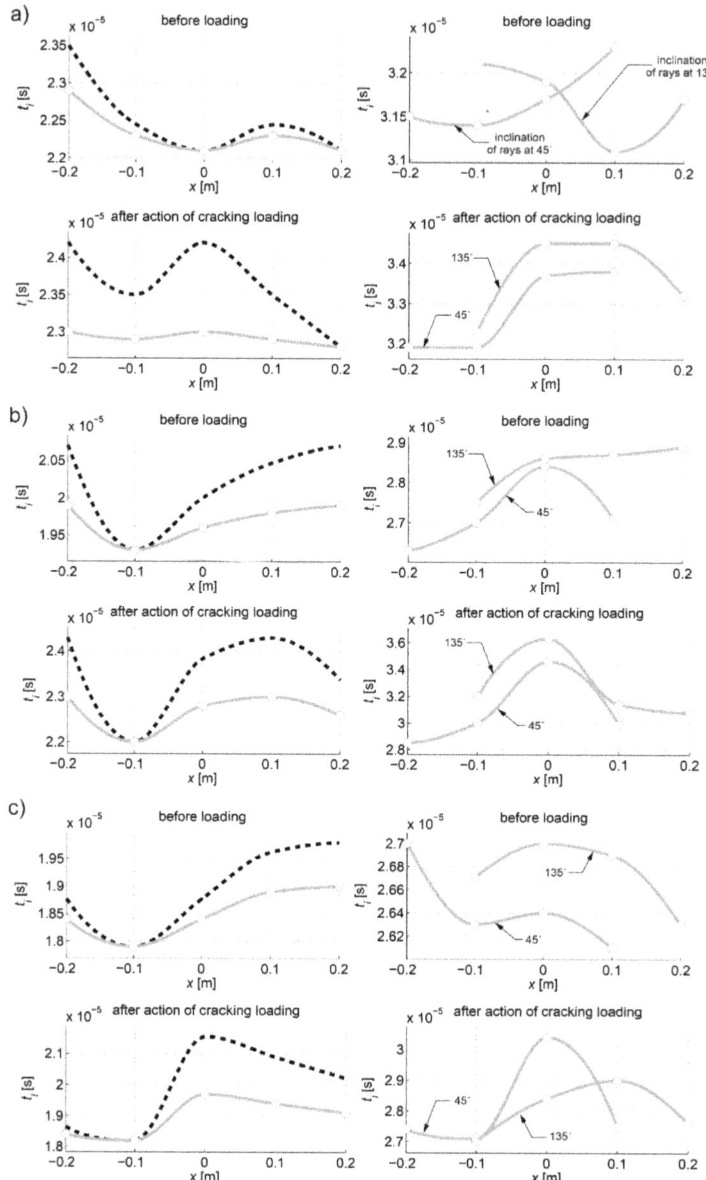

Figure 20. Propagation times $t_{\text{path int},i}$ and $t_{\text{ray approx},i}$ between the transmitting/receiving points in the beams before and after an action of cracking load: (**a**) No. 1, (**b**) No. 2, and (**c**) No. 3. The results are presented as a function of the position of the transmitting points: $t_{\text{path int},i}$ a grey line with nodes marked with circles, and $t_{\text{ray approx},i}$ a black, dashed line. The diagrams on the left refer to the paths connecting the opposite points and those on the right to the points lying diagonally at an angle of 45° and 135° to each other in relation to the axis x.

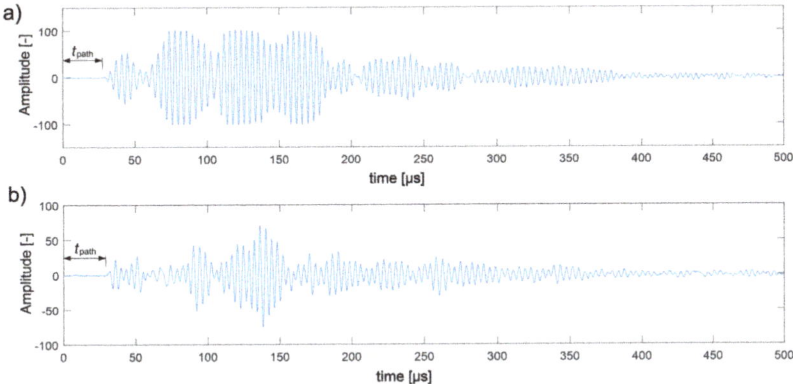

Figure 21. Example of signal recorded by the receiving head (transmitting point No. 4 and receiving point No. 3 in beam No. 3): (**a**) before load; (**b**) after an action of cracking load.

Figure 22 shows the tomographic reconstructions of longitudinal wave velocity maps in the vertical longitudinal section of the beams. Reconstructions were determined by a randomized Kaczmarz method in accordance with the information presented in point 2, where for $c_{L\ ref}$ the maximum measured mean speed of the longitudinal wave over all rays before loading was adopted (4547 m/s, 5377 m/s, and 5587 m/s for beams No. 1, No. 2, and No. 3, respectively). The resolution of $\delta_1 \times \delta_2 = 2\ cm \times 2\ cm$ was applied and the arrangement of rays as in Figure 18 with addition of rays between real rays connecting fictitious transmitting/receiving points at a distance of every 1 cm. The results are shown only in the middle area of the beam section, separated by a red dashed line through which all types of rays passed due to their inclination. The maps presented here are calculated on the basis of Equation (3) with propagation times $t_{ray,i} = t_{ray\ approx,i}$ in accordance with (28) for the paths connecting the opposite points and $t_{ray,i} = t_{path\ int,i}$ for the diagonal paths. For this purpose, the values of $t_{path\ int,i}$ are taken as shown in Figure 20. Optimal coefficients β necessary for the determination of $t_{ray\ approx,i}$ were calculated in accordance with (24) and (25) and their values are summarized in Table 8. The table also shows the values of $d_{cL\ ray\ max}$ determined from Equation (29).

Table 8. β_{opt} calculated on the basis of data from Figure 20 in case of the opposite transmitting/receiving points and $d_{cL\ ray\ max}$ according to Equation (29) before and after an action of cracking load.

Optimal Scaling Factor β and $d_{cL\ ray\ max}$ Index	No. of the Beam		
	1	2	3
β_{opt} [-] before loading	0.74	1.34	0.73
β_{opt} [-] after action of cracking load	6.02	1.29	1.24
$d_{cL\ ray\ max}$ [-] before loading	0.025	0.039	0.041
$d_{cL\ ray\ max}$ [-] after action of cracking load	0.050	0.053	0.086

Figure 22 shows clearly formed elastically degraded zones caused by load $P = P_{cr}$. Because of the static scheme of the bent beams, they were created in the middle of their span in the lower part of the cross-section. Based on Equation (20), the maximal tangential changes of Young's moduli, defined by the ratio of $\min(E_D/E_0)$ at the level 0.81, 0.50, and 0.68 in beam No. 1, No. 2, and No. 3, respectively, can be estimated for these zones. In turn, their widths in these places is within the range of 16–18 cm. They are very close to the width of the localized elastically degraded zones which were calculated theoretically in point 3. For the ratio $\min(E_D/E_0)$ in the 0.5–0.8 interval, this corresponds to the width of the damaged area from approximately 17 cm to 19 cm, which can be read from Figure 5a or Figure 6a. This result indirectly pre-confirms the validity of the identification method for the internal length

l_c of concrete proposed in point 3. However, these preliminary out-comes need necessarily further intense experimental verification. In addition, it can be seen in Figure 22 that, under real conditions, the heterogeneity of the concrete itself can have a non-negligible effect on the results, as can be seen in the reconstructions of wave velocity distributions before the loading the beams. This is also evidenced by the determined values of $d_{cL\ ray\ max}$ at this stage of the study (Table 8). In turn, after the bending moment load initiating the appearance of the first cracks, the values of $d_{cL\ ray\ max}$ from Table 8 show that, in the investigated case, not taking into account deflections of the fastest propagation paths may lead to overestimation of the velocity of longitudinal waves on average by approximately 5–9%.

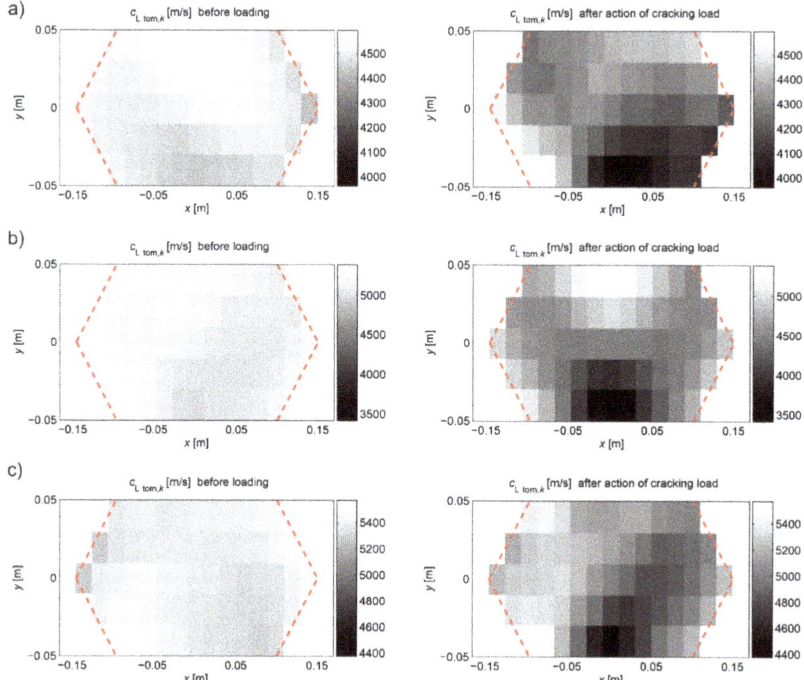

Figure 22. Tomographic reconstructions of the distribution of wave velocity according to Equation (3) in the longitudinal section of the beam before and after an action of the cracking load: (**a**) No. 1, (**b**) No. 2, and (**c**) No. 3 (red dotted lines—explanation in the text).

Another interesting issue that can be seen is that, despite the same period of care in the water of all beams, the speed of longitudinal waves in the intact configuration in beam No. 1 (stored an additional 1 week in the room conditions and isolation) was lower by about 20% if compared to the speed in beams No. 2 and 3 (which were tested right after removing from water). The explanation for this may be the fact of the phenomenon of self-drying of young concrete as a result of hydration processes [54]. On the other hand, as predicted by poromechanics [55], the initial tangent Young's modulus of the porous material and the Poisson's ratio in the state of full saturation is higher than in the dry state, which may result in a corresponding decrease in longitudinal wave velocity. In the case of cement matrix materials, such changes in Young's modulus and Poisson's ratio were measured by means of static tests among others in works [56–58]. For example, the Young's modulus of concrete at the age of 51 days with a mean compressive strength $f_{cm,cube} = 64.6$ MPa (in the state of water saturation) varied from 47.2 GPa to 45.1 GPa and the Poisson's ratio from 0.25 to 0.15 at the transition from the saturation with water to a moisture concentration reduced by approximately 2.2% by weight [58].

Assuming proportional changes for dynamic values of this parameters and taking into account the formula expressing the speed of longitudinal waves:

$$c_L = \sqrt{\frac{E_0(1-v)}{\rho(1-2v)(1+v)}}\,, \tag{30}$$

where: v—Poisson's ratio [-], ρ—density (kg/m^3), its relative decrease with the quoted range of changes in E_0, v and density would be about 8%. The effect can be further enhanced by micro-cracks in concrete arising as a result of its autogenic and/or moisture shrinkage [54] (the latter in the case of absence of drying protection). The comparison of preliminary results from beams No. 1, 2, and 3 presented in this paper obviously requires further testing on a larger number of samples. Nevertheless, it can be unequivocally stated that obtaining reliable, reference longitudinal wave speed, necessary to assess the scale of damage evolution in tomographic tests, must always be determined for concrete without damage of the same composition, and which is stored in the same conditions as the assessed concrete. In practice, this may be the maximal speed determined at the time of the test on the concrete structural member in a place where there are no defects, or on a sample without defects taken from the member. It should be emphasized in the light of outcomes of the tested beams for which only self-drying concrete and extending the period before the tomographic investigation by 7 days changed the reference speed by approximately 20%.

5.2. Prefabricated Beam

A scheme of a beam is shown in Figure 23. Its dimensions were 20 cm × 40 cm × 360 cm and it was made of concrete with a mean compression strength after 28 days of $f_{cm,cube} = 38$ MPa. The maximal aggregate diameter was $d_{a\ max} = 16$ mm. The reinforcement was made of steel with characteristic yield point declared by the manufacturer of $f_{yk} = 500$ MPa. The longitudinal lower reinforcement consisted of four bars with a diameter of 12 mm, top reinforcement of two bars with a diameter of 10 mm and the transverse reinforcement was made of bi-armed stirrups with a diameter of 8 mm with a spacing of 125 mm. As mentioned at the beginning of this point, the beam was damaged during transport and there were three cracks crosswise to its axis, two of which were in the central area of the beam selected for tomographic imaging. The shape and width of this defects are shown in Figure 24. The studies consisted in the measurement of the time of propagation of the longitudinal ultrasound wave using a Pundit-Lab tester and transreceiver heads with a frequency of 54 kHz. The coupling of the heads and beam was provided by special gel for ultrasonic testing. Taking into account the measured longitudinal wave propagation times and basic frequency of the ultrasonic pulses, the average wavelength was ∼ 7 cm and it allowed satisfying the basic requirements described in ASTM D2845-08 regarding the selection of frequency from the point of view of average grain size (as in point 5.1). The adopted system of transmitting/receiving points distant from each other by $\Delta_p = 10$ cm is shown in Figure 23. Rays were assumed between the opposite points and those lying diagonally at an angle of ∼ 26.6° and ∼ 116.6° in relation to the beam axis (Figure 23). These angles have been changed from those used in computational examples and experiments on the laboratory beams to shorten to a reasonable minimum the length of diagonal paths taking account of the attenuation of ultrasound signals and to ensure the most correct reading of the longitudinal wave propagation times. In turn, the section for tomographic examinations was the vertical longitudinal plane of symmetry of the system running simultaneously between the longitudinal reinforcement bars. Measured and interpolated longitudinal wave propagation times $t_{path\ int,i}$ are shown in Figure 25, where interpolated times are determined using a cubic Hermite spline. The diagrams also show $t_{ray\ approx,i}$ determined according to relation (28) with $\beta = \beta_{opt}$ according to formulas (24) and (25). On the other hand, Figure 26 shows an example of a recorded signal by the receiving head together with a reading of the time of longitudinal wave propagation. As in the case of the tests presented in point 5.1, Figure 26 shows first the signal caused by the propagation of the longitudinal wave and then the signal induced by the propagation of transverse and Rayleigh waves of much higher amplitude.

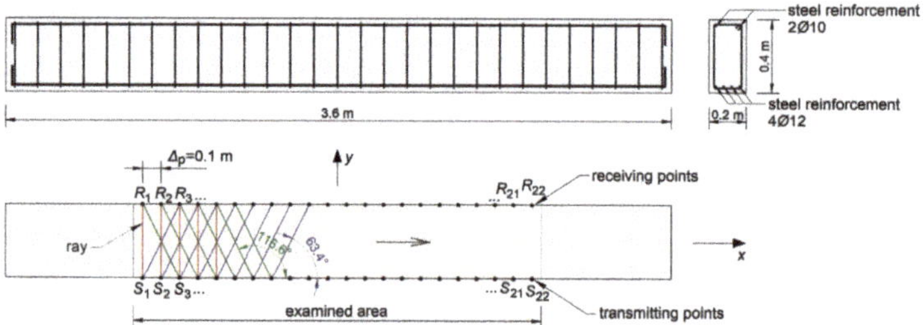

Figure 23. Scheme of the beam and the assumed system of transmitting/receiving points and rays.

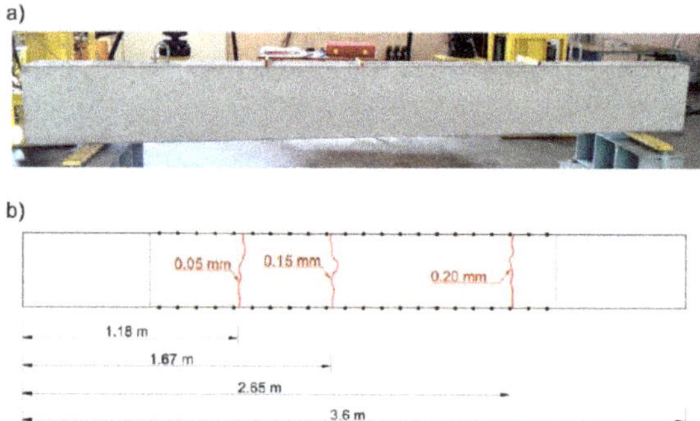

Figure 24. (**a**) Picture of the beam (the places of visible cracks are marked with the slats put on the upper surface of the beam). (**b**) Shape and width of the visible cracks.

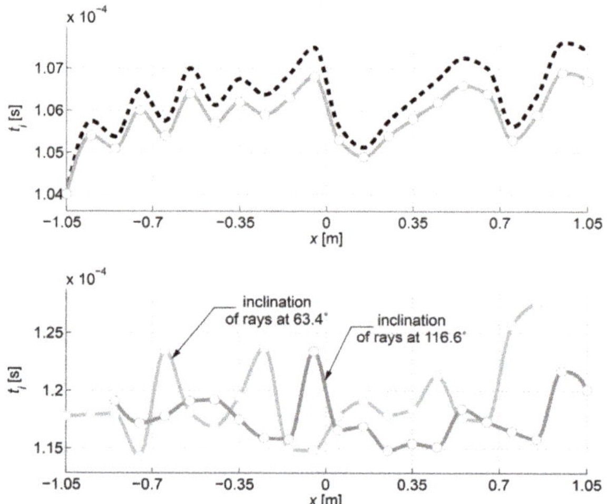

Figure 25. Propagation times $t_{\text{path int},i}$ and $t_{\text{ray approx},i}$ between the transmitting/receiving points in the beam. The results are presented as a function of the position of the transmitting points: $t_{\text{path int},i}$ grey lines with nodes marked with circles, and $t_{\text{ray approx},i}$ a black dashed line. The top diagram refer to the paths connecting the opposite points and the bottom one to the points lying diagonally at an angle of $63.4°$ and $116.6°$ to each other in relation to the axis x.

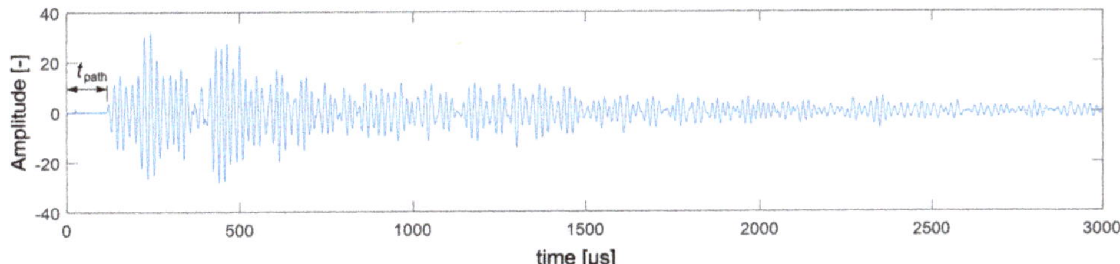

Figure 26. Example of a signal recorded by the receiving head (transmitting point No. 3 and receiving point No. 1).

Figure 27 shows a tomographic reconstruction of the longitudinal wave velocity map in the longitudinal section of the beam which was determined by a randomized Kaczmarz method according to the information presented in point 2. The maximum measured average velocity of a longitudinal wave along all rays, i.e., 3913 m/s, was assumed to be $c_{L\ ref}$. The resolution of $\delta_1 \times \delta_2 = 3.33$ cm $\times$ 3.33 cm was applied and the arrangement of rays as in Figure 23 with addition of rays between real ones connecting fictitious transmitting/receiving points at a distance of every 6.25 mm. The results are shown only for the central section of the beam separated by a red dashed line through which all types of rays passed due to their inclination. The maps presented here are calculated on the basis of Equation (3) with propagation times $t_{ray,i} = t_{ray\ approx,i}$ in accordance with (28) for the paths connecting opposite points and $t_{ray,i} = t_{path\ int,i}$ for the diagonal paths. For this purpose, the values of $t_{path\ int,i}$ are taken, as shown in Figure 25. The optimal coefficient β necessary for the determination of $t_{ray\ approx,i}$ was calculated in accordance with formulas (24)–(25) and amounted to 0.25. At the same time $d_{cL\ ray\ max}$ according to Equation (29) amounted to 0.007, which, in the considered case, proves the lack of significant influence of the generated cracks on the deflecting the fastest ultrasound wave propagation paths. This may also demonstrate the high degree of homogeneity of the concrete in the prefabrication plant.

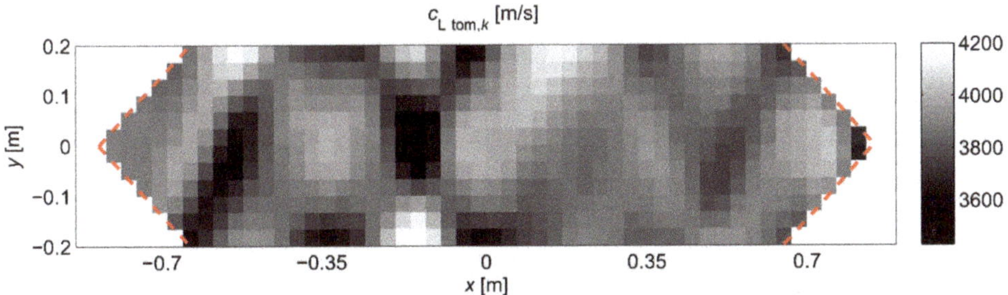

Figure 27. Tomographic reconstructions of the distribution of wave velocity according to Equation (3) in the longitudinal section of the beam (red dotted line—explanation in the text).

In Figure 27, on the left side, clearly formed 2 elastically degraded zones can be seen which were created around the visible cracks (at $x \approx -0.6$ m and $x \approx -0.15$ m). In addition, there are also two other zones of this type which can be tomographically observed and were not signaled by visible defects (at $x \approx 0.2$ m and $x \approx 0.5$ m) and a few smaller ones, reaching up to about $6 \div 9$ cm deep into the beam from its lower and upper surfaces. The latter may have been created before the beam was damaged as a result of shrinkage stresses occurring while the element was drying out after dismantling the beam formwork. Based on Equation (20), the maximal change of Young's tangent modulus defined by the $\min(E_D/E_0)$ ratio at the 0.77 level can be estimated, if one assumes in this case that $c_{L0} = c_{L\ ref}$ (in a zone that goes across the beam at $x \approx -0.15$ m). The width of the defect at this point is approximately 17 cm. It is very close to the width of the elastically degraded zone, which was calculated theoretically in point 3. For the $\min(E_D/E_0) = 0.77$ ratio, this corresponds to a damaged area width of approximately

18 cm, which can be read from Figure 5b or Figure 6b. This result also indirectly pre-confirms the validity of the identification method for the internal length l_c of concrete proposed in point 3.

6. Conclusions

As a summary of this work, the following general conclusions should be highlighted:

(1) The accuracy of transmission ultrasonic tomography for the detection of brittle damage in concrete can be effectively supported by the graph theory and, in particular, by Dijkstra's algorithm. What is important, it allows the determination of real paths of the fastest ultrasonic wave propagation in concrete containing localized elastically degraded zones at any stage of their evolution. Thanks to the analyses conducted on this basis, the authors developed the method of reducing errors in reconstructions of longitudinal wave speed maps. In this approach, the errors are decreased which are caused by using a simplification of straightness of the fastest wave propagation paths assumed in the typical mathematical apparatus for ultrasonic tomography. The method is based on the appropriate elongation of the measured propagation times of the wave travelling between opposite transmitting-receiving transducers if the actual propagation paths deviate from straight lines.

(2) Transmission ultrasonic tomography allows the estimation of the internal length of concrete defined in accordance with the methodology of damage mechanics. This problem is very important in the case of studies carried within this field of mechanics (e.g., Reference [29–31,49,50]) when predicting the extent and degree of brittle damage evolution in concrete structures. However, this conclusion may be addressed to testing RC beams with certain restrictions regarding issues of bonding between concrete and reinforcing bars and is appropriate for beams of lower reinforcement ratios with arrangement of tomographic rays not along reinforcing bars.

(3) Knowledge of the internal length of concrete allows rational determination of the appropriate resolution in ultrasonic tomography imaging and assessment of evolution of localized elastically degraded zones.

(4) The use of fictitious transmitting-receiving points in ultrasonic tomography, for which wave propagation times are calculated by interpolation of measured times, can contribute to the reduction of the required number of transducers and possible costs in the considered approach to concrete beams assessment while maintaining proper resolution of tomographic images. This outcome was well-grounded in the case of numerical analysis conducted in the work and usefulness of this approach was pre-confirmed in the own experiments. However, due to the limited number of these tests, it needs further justification and experimental studies using different arrangements of distances between sending-receiving points for ultrasonic pulses on the same RC members and confrontation with tests based on, e.g., acoustic emission or X-ray tomography.

Author Contributions: Conceptualization Z.P.; Data curation, K.T.; Formal analysis, Z.P.; Investigation, Z.P. and K.T.; Methodology, Z.P.; Software, Z.P.; Validation, Z.P.; Visualization, K.T.; Writing—original draft, Z.P.; Writing—review & editing, K.T. All authors have read and agreed to the published version of the manuscript.

Funding: This research received no external funding.

Acknowledgments: Special thanks are dedicated to Bronisław Jędraszak for bringing our attention to the possibilities related with testing prefabricated RC beams.

Conflicts of Interest: The authors declare no conflict of interest.

References

1. Drobiec, Ł.; Jasiński, R.; Piekarczyk, A. *Diagnostyka Konstrukcji Żelbetowych*; PWN: Warsaw, Poland, 2010. (In Polish)
2. Jacobs, F. Durability of concrete in structural members. *Beton-und Stahlbetonbau* **2019**, *114*, 383–391. [CrossRef]
3. Breysse, D.; Balayssac, J.P.; Biondi, S.; Corbett, D.; Goncalves, A.; Grantham, M.; Luprano, V.A.M.; Masi, A.; Monteiro, A.V.; Sbartai, Z.M. Recommendation of RILEM TC249-ISC on non destructive in situ strength assessment of concrete. *Mater. Struct.* **2019**, *52*, 71. [CrossRef]

4. Bai, Y.; Basheer, M.; Cleland, D.; Long, A. State-of-the-art applications of the pull-off test in civil engineering. *Int. J. Struct. Eng.* **2009**, *1*, 93–103. [CrossRef]

5. Dillon, R.; Rankin, G.I.B. Cube, cylinder, core and pull-off strength relationships. *Proc. Inst. Civ. Eng. Struct. Build.* **2013**, *166*, 521–536. [CrossRef]

6. Kowalski, R.; Wróblewska, J. Application of a Sclerometer to the Preliminary Assessment of Concrete Quality in Structures After Fire. *Arch. Civ. Eng.* **2018**, *64*, 171–186. [CrossRef]

7. Janků, M.; Cikrle, P.; Grošek, J.; Anton, O.; Stryk, J. Comparison of infrared thermography, ground-penetrating radar and ultrasonic pulse echo for detecting delaminations in concrete bridges. *Constr. Build. Mater.* **2019**, *225*, 1098–1111. [CrossRef]

8. Na, S.; Paik, I. Application of Thermal Image Data to Detect Rebar Corrosion in Concrete Structures. *Appl. Sci.* **2019**, *9*, 4700. [CrossRef]

9. Casas, J.R.; Aparicio, A.C. Structural Damage Identification from Dynamic-Test Data. *J. Struct. Eng.* **1994**, *120*, 2437–2450. [CrossRef]

10. Capozucca, R. Vibration analysis of damaged RC beams strengthened with GFRP. *Compos. Struct.* **2018**, *200*, 624–634. [CrossRef]

11. Szelag, M. Development of Cracking Patterns in Modified Cement Matrix with Microsilica. *Materials* **2018**, *11*, 1928. [CrossRef]

12. Suzuki, T.; Shiotani, T.; Ohtsu, M. Evaluation of cracking damage in freeze-thawed concrete using acoustic emission and X-ray CT image. *Constr. Build. Mater.* **2017**, *136*, 619–626. [CrossRef]

13. Xargay, H.; Folino, P.; Nuñez, N.; Gómez, M.; Caggiano, A.; Martinelli, E. Acoustic Emission behavior of thermally damaged Self-Compacting High Strength Fiber Reinforced Concrete. *Constr. Build. Mater.* **2018**, *187*, 519–530. [CrossRef]

14. Barnes, C.L.; Trottier, J.F.; Forgeron, D. Improved concrete bridge deck evaluation using GPR by accounting for signal depth–amplitude effects. *NDT E Int.* **2008**, *41*, 427–433. [CrossRef]

15. Muldoon, R.; Chalker, A.; Forde, M.; Ohtsu, M.; Kunisue, F. Identifying voids in plastic ducts in post-tensioning prestressed concrete members by resonant frequency of impact–echo, SIBIE and tomography. *Constr. Build. Mater.* **2007**, *21*, 527–537. [CrossRef]

16. Whitehurst, E.A. *Evaluation of Concrete Properties from Sonic Tests, Monograph 2*; American Concrete Institute: Detroit, MI, USA, 1966.

17. Dumoulin, C.; Karaiskos, G.; Deraemaeker, A. Monitoring of crack propagation in reinforced concrete beams using embedded piezoelectric transducers. In *Acoustic Emission and Related Non-Destructive Evaluation Techniques in the Fracture Mechanics of Concrete*; Woodhead Publishing: Oxford, UK, 2015; pp. 161–175.

18. Saint-Pierre, F.; Philibert, A.; Giroux, B.; Rivard, P. Concrete Quality Designation based on Ultrasonic Pulse Velocity. *Constr. Build. Mater.* **2016**, *125*, 1022–1027. [CrossRef]

19. Molero, M.; Aparicio, S.; Al-Assadi, G.; Casati, M.; Hernández, M.; Anaya, J. Evaluation of freeze–thaw damage in concrete by ultrasonic imaging. *NDT E Int.* **2012**, *52*, 86–94. [CrossRef]

20. Ferraro, C.C.; Boyd, A.J.; Consolazio, G.R. Evaluation of damage to bridge piers using pulse velocity tomography. *Constr. Build. Mater.* **2013**, *38*, 1303–1309. [CrossRef]

21. Schabowicz, K. Ultrasonic tomography—The latest nondestructive technique for testing concrete members—Description, test methodology, application example. *Arch. Civ. Mech. Eng.* **2014**, *14*, 295–303. [CrossRef]

22. Haach, V.G.; Ramirez, F.C. Qualitative assessment of concrete by ultrasound tomography. *Constr. Build. Mater.* **2016**, *119*, 61–70. [CrossRef]

23. Choi, H.; Ham, Y.; Popovics, J.S. Integrated visualization for reinforced concrete using ultrasonic tomography and image-based 3-D reconstruction. *Constr. Build. Mater.* **2016**, *123*, 384–393. [CrossRef]

24. Perkowski, Z.; Tatara, K.; Czabak, M. Imaging elastic degradation in reinforced concrete slab using methodology of ultrasonic tomography and Tikhonov regularization. In *Shell Structures: Theory and Applications, Proceedings of the 11th Int. Conference "Shell Structures: Theory and Applications—SSTA 2017", Gdańsk, Poland, 11–13 October 2017*; Pietraszkiewicz, W., Witkowski, W., Eds.; CRC Press: Boca Raton, FL, USA, 2017; pp. 445–448.

25. Perlin, L.P.; de Andrade Pinto, R.C. Use of network theory to improve the ultrasonic tomography in concrete. *Ultrasonics* **2019**, *96*, 185–195. [CrossRef] [PubMed]

26. Radon, J. Über die Bestimmung von Funktionen Durch ihre Integralwerte längs Gewisser Mannigfaltigkeiten. In *Berichte über die Verhandlungen der Königlich Sächsischen Akademie der Wissenschaften zu Leipzig, Mathematisch-Physikalische Klasse*; 69; B.G. Teubner: Leipzig, Germany, 1917; pp. 262–277. (In German)

27. Chaboche, J.L. Continuum Damage Mechanics: Part II—Damage Growth, Crack Initiation, and Crack Growth. *J. Appl. Mech.* **1988**, *55*, 65–72. [CrossRef]

28. Mazars, J. A description of micro- and macroscale damage of concrete structures. *Eng. Fract. Mech.* **1986**, *25*, 729–737. [CrossRef]

29. Pijaudier-Cabot, G. Non-local damage. In *Continuum Models for Materials with Microstructure*; Mühlhaus, H.B., Ed.; John Wiley & Sons: Chichester, UK, 1995; Volume 4, pp. 105–144.

30. Mazars, J.; Pijaudier-Cabot, G. From damage to fracture mechanics and conversely: A combined approach. *Int. J. Solids Struct.* **1996**, *33*, 3327–3342. [CrossRef]

31. Pijaudier-Cabot, G.; Mazars, J. Damage Models for Concrete. In *Handbook of Materials Behavior Models*; Elsevier: Amsterdam, The Netherlands, 2001; Volume 2, pp. 500–512.

32. Perkowski, Z.; Gozarska, K. Identyfikacja kruchych mikrouszkodzeń elementów betonowych za pomocą tomografii ultradźwiękowej. *Rocz. Inżynierii Bud.* **2014**, *14*, 57–62. (In Polish)

33. Tatara, K. Identyfikacja kruchych uszkodzeń w konstrukcjach betonowych z wykorzystaniem transmisyjnej tomografii ultradźwiękowej i algorytmu Dijkstry. Ph.D. Thesis, Opole University of Technology, Opole, Poland, 2019. (In Polish).

34. Van Vliet, M.R.; Van Mier, J.G. Experimental investigation of size effect in concrete and sandstone under uniaxial tension. *Eng. Fract. Mech.* **2000**, *65*, 165–188. [CrossRef]

35. Kak, A.C.; Slaney, M. *Principles of Computerized Tomographic Imaging*; The Institute of Electrical and Electronics Engineers, Inc.: New York, NY, USA, 1999.

36. Resnick, R.; Haliday, D. *Physics, Part II*; John Willey & Sons: New York, NY, USA, 1978.

37. Johnson, S.A.; Greenleaf, J.F.; Samayoa, W.A.; Duck, F.A.; Sjostrand, J. Reconstruction of three-dimensional velocity fields and other parameters by acoustic ray tracing. In Proceedings of the Ultrasonics Symposium, Los Angeles, CA, USA, 22–24 September 1975; IEEE: New York, NY, USA, 1975; pp. 46–51.

38. Kobayashi, Y.; Shiotani, T.; Aggelis, D.G.; Shiojiri, H. Three-dimensional seismic tomography for existing concrete structures. In Proceedings of the 2nd International Operational Modal Analysis Conference, Copenhagen, Denmark, 30 April–2 May 2007; pp. 595–600.

39. Yanli, C. The shortest path ray tracing algorithm in concrete computerized tomography. In Proceedings of the 2010 International Forum on Information Technology and Applications, Kunming, China, 16–18 July 2010; pp. 392–396.

40. Dijkstra, E.W. A note on two problems in connexion with graphs. *Numer. Math.* **1959**, *1*, 269–271. [CrossRef]

41. Moser, T.J. Shortest path calculation of seismic rays. *Geophysics* **1991**, *56*, 59–67. [CrossRef]

42. Perkowski, Z.; Gozarska, K. Szacowanie czasu propagacji fal dźwiękowych w ośrodkach materialnych za pomocą algorytmu Dijkstry. *Rocz. Inżynierii Bud.* **2012**, *12*, 43–48. (In Polish)

43. Gordon, R.; Bender, R.; Herman, G.T. Algebraic Reconstruction Techniques (ART) for three-dimensional electron microscopy and X-ray photography. *J. Theor. Biol.* **1970**, *29*, 471–481. [CrossRef]

44. Hounsfield, G.N. Computerized transverse axial scanning (tomography): Part Description of system. *Br. J. Radiol.* **1973**, *46*, 1016–1022. [CrossRef] [PubMed]

45. Rokita, E. Podstawy matematyczne tomografii komputerowych. In *Fizyczne Metody Diagnostyki Medycznej i Terapii*; Hrynkiewicz, A.Z., Rokita, E., Eds.; PWN: Warsaw, Poland, 2013; pp. 36–63. (In Polish)

46. Hansen, P.C.; Saxild-Hansen, M. AIR tools—A MATLAB package of algebraic iterative reconstruction methods. *J. Comput. Appl. Math.* **2012**, *236*, 2167–2178. [CrossRef]

47. Polakowski, K.; Sikora, J.; Filipowicz, S.F.; Wójtowicz, S.; Biernat, K. Zastosowanie tomografii ultradźwiękowej do obrazowania stanu zawilgocenia ścian. *Prace Inst. Elektrotechniki* **2007**, *233*, 109–123. (In Polish)

48. EN 12504-4. *Testing Concrete—Determination of Ultrasonic Pulse Velocity*; CEN: Brussels, Belgium, 2004.

49. Bazant, Z.P.; Jirásek, M. Nonlocal Integral Formulations of Plasticity and Damage: Survey of Progress. *J. Eng. Mech.* **2002**, *128*, 1119–1149. [CrossRef]

50. Marzec, I. Zastosowanie modelu sprężysto-plastycznego betonu z degradacją sztywności i nielokalnym osłabieniem do modelowania elementów betonowych cyklicznie obciążonych. Ph.D. Thesis, Gdańsk University of Technology, Gdańsk, Poland, 2009. (In Polish).

51. Phillips, D.L. A Technique for the Numerical Solution of Certain Integral Equations of the First Kind. *J. Assoc. Comput. Mach.* **1962**, *9*, 84–97. [CrossRef]
52. Tikhonov, A.N. Solution of incorrectly formulated problems and the regularization method. *Soviet Math. Dokl.* **1963**, *4*, 1035–1038.
53. Popovics, J.S. NDE techniques for concrete and masonry structures. *Prog. Struct. Eng. Mater.* **2003**, *5*, 49–59. [CrossRef]
54. Neville, A.M. *Properties of Concrete*, 4th ed.; Longman: London, UK, 1995.
55. Coussy, O. *Poromechanics*; John Wiley & Sons: New York, NY, USA, 2004.
56. Yurtdas, I.; Burlion, N.; Skoczylas, F. Triaxial mechanical behaviour of mortar: Effects of drying. *Cem. Concr. Res.* **2004**, *34*, 1131–1143. [CrossRef]
57. Yurtdas, I.; Peng, H.; Burlion, N.; Skoczylas, F. Influences of water by cement ratio on mechanical properties of mortars submitted to drying. *Cem. Concr. Res.* **2006**, *36*, 1286–1293. [CrossRef]
58. Perkowski, Z. *Modelowanie Mikrouszkodzeń w Kruchych Materiałach Budowlanych z Uwzględnieniem Zjawisk Powierzchniowych*; PAN KILiW IPPT: Warsaw, Poland, 2009. (In Polish)

Article

Influence of Crack Size on Stress Evaluation of Ferromagnetic Low Alloy Steel with Metal Magnetic Memory Technology

Bin Liu [1,*], Peng Fu [1], Ruifeng Li [1], Peng He [2] and Shiyun Dong [3]

[1]   Material Science and Engineering, Jiangsu University of Science and Technology, Zhenjiang 212003, China; baix2415674055@163.com (P.F.); li_ruifeng@just.edu.cn (R.L.)

[2]   State Key Laboratory of Advanced Welding and Joining, Harbin Institute of Technology, Harbin 150001, China; hepeng@hit.edu.cn

[3]   National Key Laboratory for Remanufacturing, Academy of Armored Forces Engineering, Beijing 100072, China; liubindely@163.com

*   Correspondence: Binliudely@163.com; Tel.: +86-511-8440-1181

Received: 9 October 2019; Accepted: 2 December 2019; Published: 4 December 2019

Abstract: Based on the magneto-mechanical effect, the influence of crack size on stress evaluated with metal magnetic memory (MMM) technology was discussed in this paper. Based on equivalent theory, the regular rectangular grooves, with different widths and depths, were precut in the surface of an experimental sample for simulating surface crack, and a three dimensional electrically controlled displacement system was used to collect the $H_p(y)$ signal of the sample under different stresses, and the fracture morphology was observed by using scanning electron microscopy (SEM). The results show that the influence of detection line on $H_p(y)$ signal can be ignored; as stress increases, the $H_p(y)$ signal turns counterclockwise around zero-crossing point and its mutation, corresponding to the location of groove, becomes distinct gradually. When groove depth is constant, the magnetic intensity gradient changes in the form of quadratic polynomial as groove width increases, and when the groove width is the same, the magnetic intensity gradient is a linear function of groove depth. When stress reaches the yield strength of the material, the magnetic intensity gradient decreases gradually as stress increases further, and the orientation of magnetic domain is seen as the main reason for that result. At last, the experimental results are discussed based on the piezomagnetic effect and leakage magnetic field theory of finite depth slit model, and the change of magnetic domain orientation is considered to be the main reason.

Keywords: non-destructive evaluation; crack effect; metal magnetic memory; magnetic intensity gradient; stress

1. Introduction

The safety and reliability of equipment in the working environment has attracted the attention of scholars. However, there is no absolute method for evaluating reliability, so finding a feasible method appears to be particularly urgent. To solve that problem, lots of research has been designed and carried out, and it indicates that stress could be seen as a parameter to characterize reliability, so the method of stress evaluation was studied. In general, the method of stress evaluation can be divided into two categories, including non-destructive and destructive methods [1–7]. Because of non-destruction, low cost, high efficiency and other merits, the metal magnetic memory (MMM) technology, which is one kind of magnetic flux leakage method, becomes a commonly used method for stress evaluation of ferromagnetic material.

Based on the magnetic memory effect of ferromagnetic material in the geomagnetic field, the MMM technology was firstly proposed by a Russian scholar [8–11], and it stated clearly that the normal component of MMM signal—$H_p(y)$ signal, which was defined in the geomagnetic field, could be employed to evaluate stress. Up to now, lots of research results have been found, however lots of influential factors are found in the application of stress evaluation with this method, and its mechanism is still not elaborated clearly. Thus, lots of experimental and theoretical studies were carried out. For example, the change in the orientation of magnetic domain was explained as the reason for evaluating stress with this technology [12–14]. The results of stress effect on $H_p(y)$ signal and its influence mechanism were both clarified qualitatively [15]. The grain size effect on $H_p(y)$ signal for stress evaluation of low carbon steel was analyzed, and the relation of magnetic intensity gradient and grain size was also obtained [16]. As fatigue cycle numbers increased, the change in the maximal value of surface magnetic flux density of 304 stainless steel was discussed, and the phase transformation was seen as the main reason [17]. Besides that, the fatigue stress of ferromagnetic material was also evaluated, and the relation of fatigue cycle and $H_p(y)$ signal was discussed [18–22]. While it should be noted that defects exist inevitably in material, its effect on stress evaluation is discussed rarely, so designing the method for discussing its influence on stress evaluation and explain its results clearly are very important. As a typical defect, the influence of crack on stress evaluated with MMM technology was studied, and its influence degree affected by the crack size was explained in this study. To solve the problem, the equivalent theory was employed, and the regular rectangular groove is used to simulate cracks. After that, the $H_p(y)$ signals were collected with the uniaxial static tensile experiment, and then the parameter of $H_p(y)$ signal for characterizing stress was determined. At last, the relations between the characteristic parameter of $H_p(y)$ signal and crack size was obtained.

2. Experimental Material and Methods

2.1. Experimental Material

In this study, the experimental material is 12CrMoV, and its mechanical properties are tested and shown in Table 1.

Table 1. Mechanical property of the experimental material.

Mechanical Property	σ_s/MPa	σ_b/MPa	δ_5/%	φ/%
12CrMoV	337.5	547	22	50

In order to discuss and determine the crack effect on stress evaluation, the regular rectangular grooves with different depths and widths were precut in the surface of 12CrMoV by using wire cutting technology, and its sketch map was shown in Figure 1. For experimental requirements, a sample with the size of 280 mm × 40 mm × 7 mm was prepared. In the same sample, the groove depth, the symbol of which is *d* as shown in Figure 1, is the same, and the widths are 0.5 mm, 1.0 mm, 1.5 mm, 2.0 mm, 2.5 mm and 3.0 mm, respectively. For different samples, the groove depths are 0.5 mm, 1.0 mm, 1.5 mm, 2.0 mm, 2.5 mm and 3.0 mm, respectively. To avoid the influence of surface roughness on $H_p(y)$ signal, the surfaces of all samples were rubbed, and its surface roughness was Ra1.2. To avoid the deformation influence of material, the spacing between adjacent grooves is determined as 40 mm, and seven detection lines, the spacing between which is 5.0 mm, are drawn on the surface of sample as shown in Figure 1. To eliminate the initial $H_p(y)$ signal effect caused by machining, all samples were annealed at 860 °C in the vacuum heat treatment furnace, the vacuity of which was 8×10^{-4} Pa. Afterwards, the temperature was reduced to 200 °C and the samples were taken out from the vacuum heat treatment furnace.

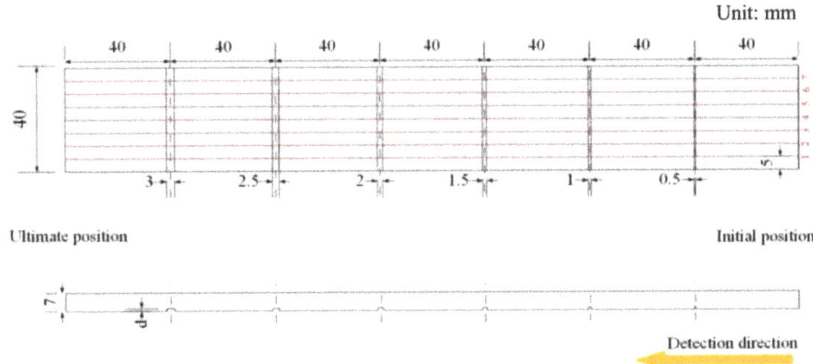

Figure 1. The sketch map of the experimental sample with regular rectangular groove.

2.2. Experimental System

The MMM detection system, shown in Figure 2, consists of three parts. The metal magnetic memory detector—EMS-2003, shown in Figure 2b, is used to store $H_p(y)$ signal. The three-dimensional electrically controlled displacement system, shown in Figure 2c, is used to move the two-channel pen sensor probe with a constant scanning speed and lift-off. The displacement system is made of aluminum alloy, so the influence of material magnetism on $H_p(y)$ signal can be ignored. The two-channel pen sensor probe, shown in Figure 2d, is used to collect $H_p(y)$ signal. For matching the sampling rate of EMS-2003 and the displacement system, the scanning speed of the sensor probe is optimized, and 40 mm/s is optimal. Before collecting $H_p(y)$ signal, the samples are firstly placed horizontally on the platform of the displacement system, and then the $H_p(y)$ signal is collected along detection lines from initial position to ultimate position, as shown in Figure 1. The static tensile test is done by using a CMT2502 testing machine, shown in Figure 2a, the dynamic load error of which is ±1.0%, meeting the requirements of the experiment.

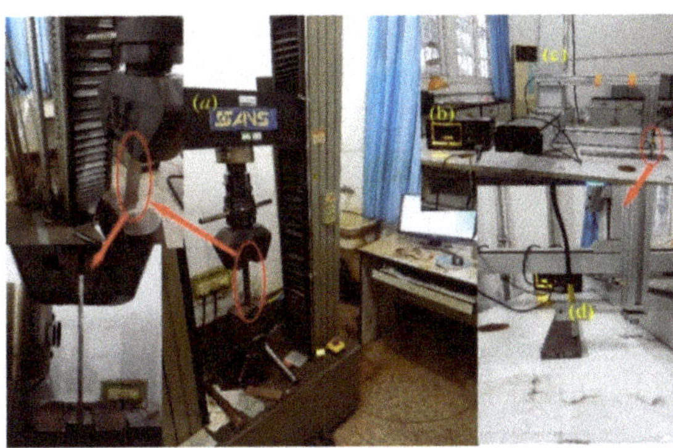

Figure 2. Metal magnetic memory detection system for stress evaluation. (**a**) CMT2502 testing machine. (**b**) EMS-2003. (**c**) Three-dimensional electrically controlled displacement system. (**d**) Two-channel pen sensor probe.

Lots of experiments pointed out that lift-off of the sensor probe is an important factor for the amplitude of the $H_p(y)$ signal, so the best optimal lift-off is discussed in this study. For the experiment, the sample with the same width and different depths of crack was prepared by using the same technology, which was employed in Figure 1. Increasing the lift-off gradually, the $H_p(y)$ signals were collected and shown in Figure 3. It can be seen that when lift-off is lower than 5.0 mm, the mutation of $H_p(y)$ signal, at the location of the groove, is clear, and the amplitude of the $H_p(y)$ signal is high. As lift-off increased, that mutation becomes smaller and smaller, and the amplitude of the $H_p(y)$ signal also becomes lower. After lift-off reaches 15.0 mm, and as lift-off increases further, that mutation disappears and that amplitude becomes lower. With comprehensive consideration, 1.0 mm is seen as the best optimal lift-off of sensor probe in this study.

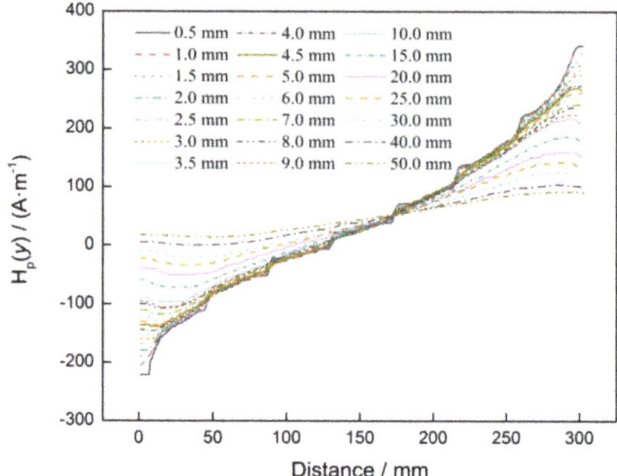

Figure 3. $H_p(y)$ signal corresponding to different lift-offs of sensor probe.

3. Results and Discussion

3.1. Experimental Results

To discuss the crack size effect on stress evaluation, the $H_p(y)$ signal, corresponding to different stresses, should be collected in advance with the tensile test, so the preload was determined based on the mechanical properties of material, as shown in Table 1. In this study, the load interval is 6 kN, and the loading rate is 0.5 kN/s. When the preload was reached, the load was held about 120 s, and then the sample was taken down from the CMT2502 testing machine. After that, the sample was placed on the platform, and the $H_p(y)$ signals were sequentially collected along detection lines. Comparing the $H_p(y)$ signals, it can be seen that although the sizes of cracks are different, the change regulation of the $H_p(y)$ signal is very similar as stress changes. Therefore, when the groove depth is 3.0 mm, the $H_p(y)$ signals of the sample along seven detection lines, corresponding to different stresses, are shown in Figure 4.

From Figure 4a, it can be seen that when the stress is 0 MPa, the distribution of $H_p(y)$ signal is irregular, and its amplitude is in the range of $-2.6{\sim}67.8$ (A·m^{-1}). For the result, it can be known that the amplitude is in the range of intensity of the geomagnetic field, thus the initial $H_p(y)$ signal effect on stress evaluation can be ignored. From Figure 4b–e, it can be seen that as stress increases, the amplitude of the whole $H_p(y)$ signal increases gradually, and the $H_p(y)$ signal mutation, corresponding to groove, also becomes more obvious gradually. Compared with the amplitude of $H_p(y)$ signal in Figure 4e, it can be seen that the amplitude in Figure 4f decreases as stress increases. Comparing $H_p(y)$ signals in different detection lines, it can be seen that when the stress is the same, the $H_p(y)$ signals and its changes are basically the same. Based on that result, the $H_p(y)$ signals along the middle line are analyzed and shown in Figure 5.

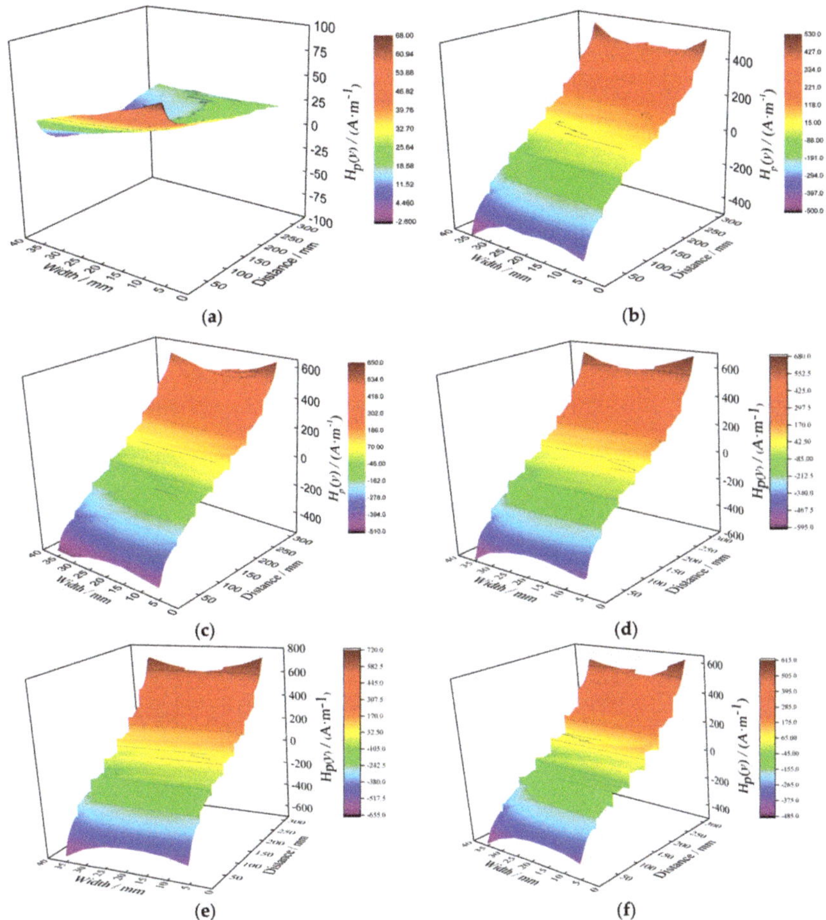

Figure 4. $H_p(y)$ signals under different tensile loads. (**a**) 0 MPa, (**b**) 75 MPa, (**c**) 150 MPa, (**d**) 225 MPa, (**e**) 337.5 MPa, (**f**) 375 MPa.

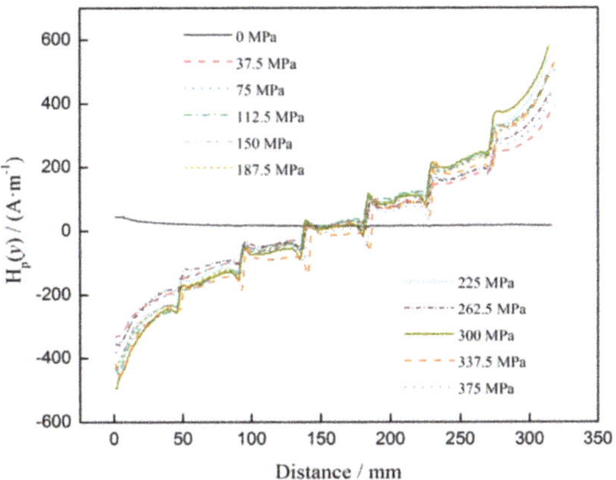

Figure 5. $H_p(y)$ signals corresponding to different stresses.

Figure 5 shows that when stress is 0 MPa, the $H_\mathrm{p}(y)$ signal appears nearly as a horizontal line. As stress increases, the $H_\mathrm{p}(y)$ signals present linear distribution, then turn anticlockwise and approximately intersect at one point, which is defined as zero crossing point in this paper. For the mutation of $H_\mathrm{p}(y)$ signal, corresponding to groove, it can be seen that when the groove size is different, the mutation degree also changes, so it can be used to evaluate the stress. Therefore, the mutation of $H_\mathrm{p}(y)$ signal, corresponding to groove, was extracted and defined as magnetic intensity gradient K, which was calculated with Equation (1).

$$K = \frac{H_p(y)_\mathrm{max} - H_p(y)_\mathrm{min}}{\Delta l} \tag{1}$$

where $H_\mathrm{p}(y)_\mathrm{max}$ and $H_\mathrm{p}(y)_\mathrm{min}$ are the maximal and minimal values of $H_\mathrm{p}(y)$ signal, Δl is the spacing between the location of $H_\mathrm{p}(y)_\mathrm{max}$ and $H_\mathrm{p}(y)_\mathrm{min}$.

For different crack sizes, the K of $H_\mathrm{p}(y)$ signal was calculated based on Equation (1). It can be seen that when the depth of groove is the same, the relationship between K and groove width is very similar. For that reason, not all the experimental results were shown in this paper, and only when the groove depth is 3.0 mm, the relation of K and groove width was shown in this paper.

When the stresses are different, the curves of K and groove width were shown in Figure 6. It can be seen that as groove width increases, the change regulation of K is very similar, and it presents in the form of parabola. When stress is 337.5 MPa, which is the yield strength of material, the K reaches the maximal value. Based on that, the value of K and groove width is fitted with quadratic polynomial function.

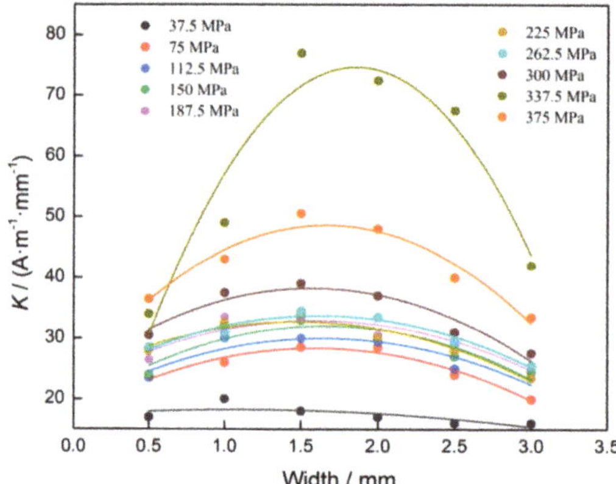

Figure 6. Relation of K and width of groove.

Similarly, for discussing the depth effect on stress evaluation, the relation of groove depth and K was analyzed. For that purpose, the $H_\mathrm{p}(y)$ signals, corresponding to the same widths and different stresses, were collected, and its K was also calculated. Comparing experimental results, it can be seen that although groove depths are different, the change regulation of K is very similar as stress increases gradually. In view of that, when groove widths were 2.0 mm and 3.0 mm, which were the common widths, the results of K and groove depth were shown in Figure 7.

As shown in Figure 7, it can be seen that the relationship of K and groove depth is nearly linear, and the slope between K and groove depth increases gradually as stress increases. When stress reaches 337.5 MPa, the slope reaches the maximal value, and then becomes smaller as stress increases further. To discuss the influence of stress on the value of K, the result of K and stress was fitted with linear function. In this study, the fitting coefficient was defined as K_F, so the relationship between K_F and stress was determined and shown in Figure 8.

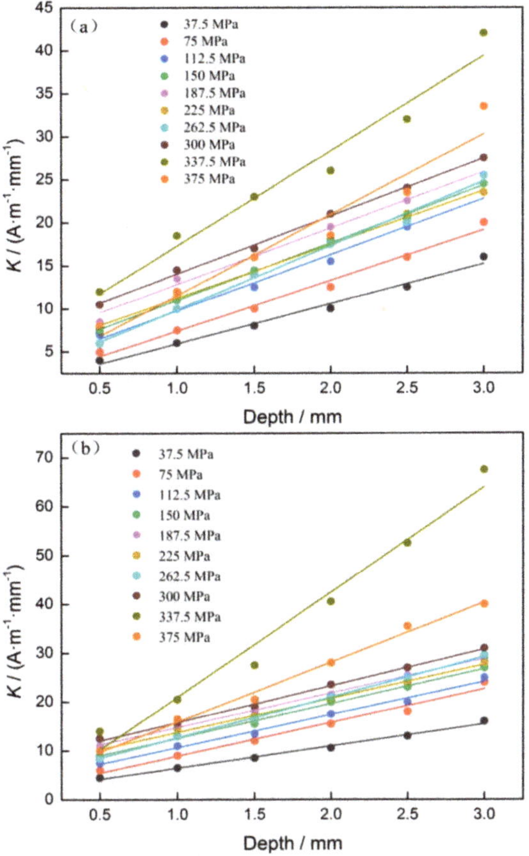

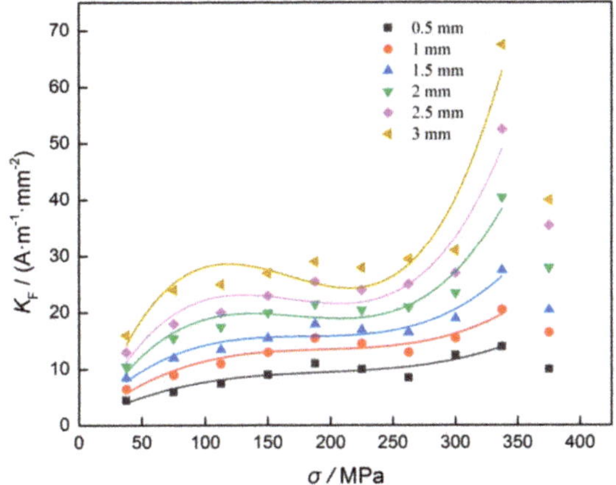

Figure 7. Relationship of K and groove depth. (**a**) 2.0 mm, (**b**) 3.0 mm.

Figure 8. Result of K_F and tensile stress.

As shown in Figure 8, it can be seen that although the groove widths are different, the change regulation of K_F is basically the same as stress increases. In detail, when the stress is 337.5 MPa, the K_F corresponds to the maximal value, and it decreases obviously as stress increases further. For the same stress, the K_F increases gradually as the groove width increases.

3.2. Discussion and Analysis

In order to explain the experimental results, the magneto-mechanical effect and the finite depth slit leakage magnetic field theory are employed. It is generally known that after demagnetization of magnetic material, the orientation of magnetic domain is random, so when stress is 0 MPa, the amplitude of $H_p(y)$ signal is very low in experimental results. When the stress is lower than the yield strength of material, as stress increases, the orientation of the magnetic domain becomes orderly gradually. For that process, the change in orientation of the magnetic domain is described as three stages, including magnetic domain movement, magnetic domain merger and magnetic domain rotation; after that, the orientation is parallel to the stress direction. It indicates that as stress increases, the magnetic characteristic of material becomes stronger, so the amplitude of $H_p(y)$ signal becomes higher gradually as shown in Figure 4. When the stress reaches yield strength, as the stress increases further, lots of dislocations are generated and gathered together in material, and an internal stress field is formed simultaneously. Because there is a pinning effect of dislocation on the magnetic domain, which has been reported by many studies, the orientation change in the magnetic domain is restricted as stress increases. As we all know, the magnetization degree of material largely depends on the order degree of magnetic domain orientation, thus the magnetization degree is decreased as stress increases in the plastic deformation stage. Thist means that the amplitude of $H_p(y)$ signal decreases, as shown in Figure 4f. However, there were some studies showing that when the material was in the state of plastic deformation, the magnetization degree still increased gradually as stress increased, because that pinning effect could be broken. However, some other scholars had different opinions, and they thought that for the material with good plastic deformation capability, that pinning effect was hardly broken [23]. For the material in this study, its plastic deformation capability is very good, so a larger number of dislocations are generated in the plastic deformation stage, and the pinning effect of dislocation cannot be broken. To verify the theoretical analysis, the fracture morphology of experimental material was observed and shown in Figure 9.

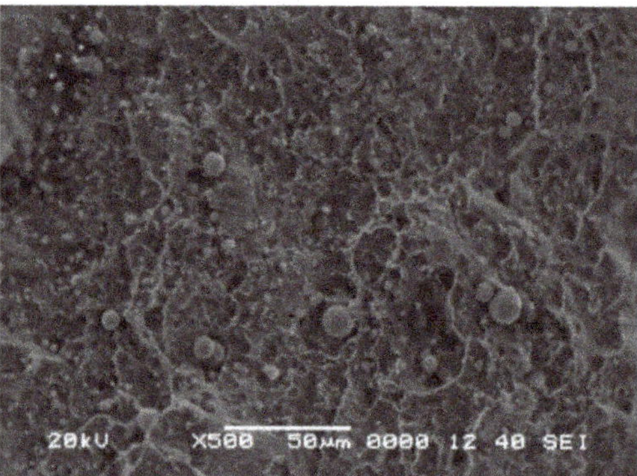

Figure 9. The fracture morphology of the experimental sample.

From Figure 9, it can be seen that many dimples are generated, which can be seen in the material fracture. The fracture theory of metal states clearly that more dimples represent better deformation performance, thus the experimental result agrees well with the theoretical analysis.

To explain the change in $H_p(y)$ signal, corresponding to grooves, the leakage magnetic field model of finite depth slit, which was built by Förster [24], was employed, and it was shown in Figure 10.

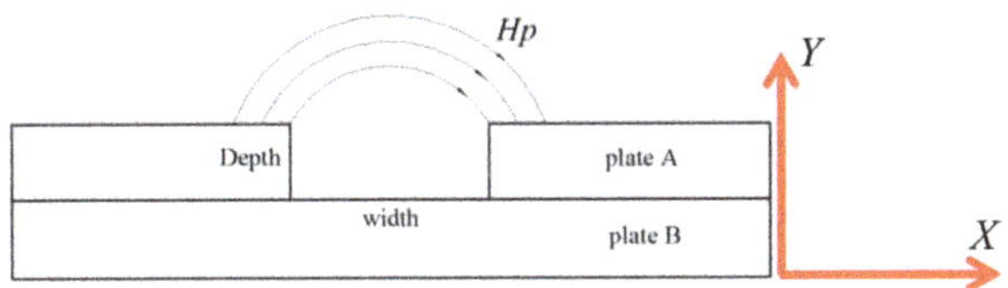

Figure 10. The leakage magnetic field model of finite depth slit.

As the definition of the coordinate (X,Y), shown in Figure 10, the leakage magnetic field intensity can be written as:

$$H_y = \frac{(l+W)\mu H_0 W}{\pi(l+\mu W)}\left[\frac{x}{x^2+y^2} - \frac{x}{x^2+(y+D)^2}\right] \tag{2}$$

where H_y is the normal component of leakage magnetic field, x is the horizontal distance between testing point and the center of slit, y is the vertical distance between testing point and the center of slit, W and D are the width and depth of the slit, H_0 is internal magnetic field intensity of material, l is the length of sample, and μ is the magnetic permeability.

Generally speaking, the H_0, which is determined by the earth's magnetic field and the magnetic field induced by stress, can be written as:

$$H_0 = H_E + \frac{H_E M_S(\alpha\mu_0 + 3\gamma\sigma)}{3\alpha\mu_0 - M_S(\alpha\mu_0 + 3\gamma\sigma)} \tag{3}$$

where H_E is the earth's magnetic field intensity, M_S is the saturation magnetization of sample, α is the parameter of molecular field determined by material, μ_0 is the vacuum permeability, and γ is determined by the magnetostrictive coefficient.

Taking Equation (3) into Equation (2), the H_y can be written as:

$$H_y = \left[H_E + \frac{H_E M_S(\alpha\mu_0 + 3\gamma\sigma)}{3\alpha\mu_0 - M_S(\alpha\mu_0 + 3\gamma\sigma)}\right]\frac{(l+W)\mu W}{\pi(l+\mu W)}\left[\frac{x}{x^2+y^2} - \frac{x}{x^2+(y+D)^2}\right] \tag{4}$$

Then, taking Equation (4) into Equation (1), the magnetic intensity gradient K, which is defined in Equation (1), can be expressed as:

$$K = \frac{H_{y1}-H_{y2}}{x_1-x_2}$$

$$= \left[H_E + \frac{H_E M_S(\alpha\mu_0+3\gamma\sigma)}{3\alpha\mu_0-M_S(\alpha\mu_0+3\gamma\sigma)}\right]\cdot\frac{(l+W)\mu W}{\pi(l+\mu W)}\cdot\left[\frac{x_1(y+D)}{(x_1^2+y^2)(x_1^2+(y+D)^2)} - \frac{x_2(y+D)}{(x_2^2+y^2)(x_2^2+(y+D)^2)}\right]\cdot\frac{1}{x_1-x_2} \tag{5}$$

To discuss the relationship of K and groove width, the partial derivation of K respect to W is calculated and reduced as:

$$K' = \left[H_E + \frac{H_E M_S(\alpha\mu_0+3\gamma\sigma)}{3\alpha\mu_0 - M_S(\alpha\mu_0+3\gamma\sigma)}\right]\cdot\frac{1}{x_1-x_2}\cdot\frac{\mu W^2+2WL+L^2}{(L+\mu W)^2}\cdot\left[\frac{x_1(y+D)}{(x_1^2+y^2)[x_1^2+(y+D)^2]} - \frac{x_2(y+D)}{(x_2^2+y^2)[x_2^2+(y+D)^2]}\right] \tag{6}$$

where K' is defined as partial derivation of K respect to W.

Comparing the value of L and μ, it can be known that the value of $L >> \mu$ can be accepted, so Equation (6) is simplified to:

$$K' = \left[H_E + \frac{H_E M_S(\alpha\mu_0+3\gamma\sigma)}{3\alpha\mu_0 - M_S(\alpha\mu_0+3\gamma\sigma)}\right]\cdot\frac{1}{x_1-x_2}\cdot\frac{2W+L}{L}\cdot\left[\frac{x_1(y+D)}{(x_1^2+y^2)[x_1^2+(y+D)^2]} - \frac{x_2(y+D)}{(x_2^2+y^2)[x_2^2+(y+D)^2]}\right] \tag{7}$$

In this study, the groove depth is constant in the same one sample, meaning that the value of D is a constant, so the relation of K' and W is linear in Equation (7). Based on the derivative theory, it can be known that K is a quadratic polynomial function of D, so it agrees well with the experimental result, as shown in Figure 6.

Similarly, the relation of K and D is discussed. The partial derivation of K respect to D is calculated, so the Equation (5) is reduced to:

$$K'' = \left[H_E + \frac{H_E M_S(\alpha\mu_0+3\gamma\sigma)}{3\alpha\mu_0 - M_S(\alpha\mu_0+3\gamma\sigma)}\right] \cdot \frac{1}{x_1-x_2} \cdot \frac{\mu(L+W)W}{(L+\mu W)\pi} \cdot$$
$$\left[\frac{1}{(x_1^2+y^2)[x_1^2+(y+D)^2]} \cdot \left[\frac{x_1^3}{x_1^2+(y+D)^2} - x_1\right] - \frac{1}{(x_2^2+y^2)[x_2^2+(y+D)^2]} \cdot \left[\frac{x_2^3}{x_2^2+(y+D)^2} - x_2\right]\right] \tag{8}$$

where K'' is defined as the partial derivation of K respect to D.

Simplifying Equation (8), it can be written as:

$$K'' = \left[H_E + \frac{H_E M_S(\alpha\mu_0+3\gamma\sigma)}{3\alpha\mu_0 - M_S(\alpha\mu_0+3\gamma\sigma)}\right] \cdot \frac{1}{x_1-x_2} \cdot \frac{\mu(L+W)W}{(L+\mu W)\pi} \cdot$$
$$\left[\frac{x_1^3}{(x_1^2+y^2)[x_1^2+(y+D)^2]^2} - \frac{x_2^3}{(x_1^2+y^2)[x_2^2+(y+D)^2]^2} - \frac{x_1}{(x_1^2+y^2)[x_1^2+(y+D)^2]} + \frac{x_2}{(x_2^2+y^2)[x_2^2+(y+D)^2]}\right] \tag{9}$$

Because the value of x is far bigger than the value of D in this study, the influence of change in D on K'' can be ignored. It means that the relationship of K and D can be seen as linear, so it agrees well with results as shown in Figure 7. In a word, the experimental result agrees well with theoretical discussion.

4. Conclusions

The influence of crack size on stress evaluation of low alloy steel with MMM technology was discussed in this study. It can be concluded that:

1. The K of $H_p(y)$ signal is a quadratic polynomial function of groove width, and the value of K increases gradually as stress increases. When the stress reaches yield strength of the material, the value of K also reaches maximum.
2. For different tensile stresses, the K is a linear function of groove depth, and its linear slope increases as stress increases gradually. When the stress reaches yield strength of the material, the linear slope reaches the maximal value.
3. For different groove widths, the relationship of K_F and stress is very similar, and it appears nonlinearly.

Author Contributions: Conceptualization and methodology, B.L.; formal analysis, P.F.; validation, R.L.; writing—review and editing, P.H.; supervision, S.D.

Funding: This research was funded by the National Natural Science Foundation of China, grant number 51775254.

Conflicts of Interest: The authors declare no conflict of interest.

References

1. Liu, C.; Yang, J.; Shi, Y.; Zhao, Y. Investigation on narrow-gap welding residual stresses in ultra-thick ring-type mockups. *Mater. Res. Express* **2018**, *5*, 016526. [CrossRef]
2. Liu, C.; Yan, Y.; Cheng, X.; Wang, C.; Zhao, Y. Residual stress in a restrained specimen processed by post-weld ultrasonic impact treatment. *Sci. Technol. Weld. Join.* **2019**, *24*, 193–199. [CrossRef]
3. Liu, B.; Dong, S.Y. Anisotropic structure affects thickness measurement of laser cladding coating with surface ultrasonic wave based on cross-correlation function. *Mater. Res. Innov.* **2015**, *19*, S5-194–S5-198. [CrossRef]
4. Liu, B.; Miao, W.; Dong, S.; He, P. Grain size correction of welding residual stress measurement in a carbon steel plate using the critical refraction of longitudinal waves. *Res. Nondestruct. Eval.* **2019**, *30*, 112–126. [CrossRef]
5. Liu, C.; Luo, Y.; Yang, M.; Fu, Q. Three-dimensional finite element simulation of welding residual stress in RPV with two J-groove welds. *Weld. World* **2017**, *61*, 151–160. [CrossRef]
6. Liu, C.; Chen, D.; Hill, M.R.; Tran, M.N.; Zou, J. Effects of ultrasonic impact treatment on weld microstructure, hardness, and residual stress. *Mater. Sci. Technol.* **2017**, *33*, 1601–1609. [CrossRef]
7. Liu, B.; Miao, W.; Dong, S.; He, P. Grain size effect on Lcr elastic wave for surface stress measurement of carbon steel. *Nondestruct. Test. Eval.* **2018**, *33*, 139–153. [CrossRef]

8. Doubov, A.A. A study of metal properties using the method of magnetic memory. *Met. Sci. Heat Treat.* **1997**, *39*, 401–402. [CrossRef]

9. Doubov, A.A. Screening of weld quality using the magnetic metal memory effect. *Weld. World* **1998**, *41*, 196–198.

10. Doubov, A.A. A Technique for monitoring the bends of boiler and steam-line tubes using the magnetic memory of metal. *Therm. Eng.* **2001**, *48*, 289–295.

11. Doubov, A.A. Diagnostics of steam turbine disks using the metal magnetic memory method. *Therm. Eng.* **2010**, *57*, 16–21. [CrossRef]

12. Huang, H.; Yang, C.; Qian, Z.; Han, G.; Liu, Z. Magnetic memory signals variation induced by applied magnetic field and static tensile stress in ferromagnetic steel. *J. Magn. Magn. Mater.* **2016**, *416*, 213–219. [CrossRef]

13. Yin, D.W.; Xu, B.S.; Dong, S.Y.; Yang, S.L.; Dong, L.H. Characteristics of magnetic memory signals for medium carbon steel under static tensile conditions. *J. Cent. South Univ.* **2005**, *12*, 106–111. [CrossRef]

14. Shi, C.L.; Dong, S.Y.; Xu, B.S.; He, P. Metal magnetic memory effect caused by static tension load in a case-hardened steel. *J. Magn. Magn. Mater.* **2010**, *322*, 413–416. [CrossRef]

15. Leng, J.; Liu, Y.; Zhou, G.; Gao, Y. Metal magnetic memory signal response to plastic deformation of low carbon steel. *NDT E Int.* **2013**, *55*, 42–46. [CrossRef]

16. Liu, B.; Xue, X.; Li, J.; Li, R.; Dong, S.; Fang, J. Grain size effect on metal magnetic memory signal for stress damage evaluation of low carbon steel. *Nondestruct. Test. Eval.* **2019**, *34*, 267–282. [CrossRef]

17. Takaya, S.; Nagae, Y. Magnetic property change of type 304 stainless steel due to accumulation of fatigue damage at elevated temperature. *Int. J. Appl. Electromagn. Mech.* **2007**, *25*, 211–217. [CrossRef]

18. Venkatachalapathi, N.; Raju, G.J.; Raghavulu, P. Characterization of fatigued steel states with metal magnetic memory method. *Mater. Today Process.* **2018**, *5*, 8645–8654. [CrossRef]

19. Leng, J.; Xu, M.; Xu, M.; Zhang, J. Magnetic field variation induced by cyclic bending stress. *NDT E Int.* **2009**, *42*, 410–414. [CrossRef]

20. Ni, C.; Hua, L.; Wang, X.K. Crack propagation analysis and fatigue life prediction for structural alloy steel based on metal magnetic memory testing. *J. Magn. Magn. Mater.* **2018**, *462*, 144–152. [CrossRef]

21. Dong, L.H.; Xu, B.S.; Dong, S.Y.; Chen, Q.Z.; Wang, Y.Y.; Zhang, L.; Wang, D.; Yin, D.W. Metal magnetic memory testing for early damage assessment in ferromagnetic materials. *J. Cent. South Univ.* **2005**, *12*, 102–106. [CrossRef]

22. Xu, M.X.; Chen, Z.H.; Xu, M.Q. Micro-mechanism of metal magnetic memory signal variation during fatigue. *Int. J. Min. Metall. Mater.* **2014**, *21*, 259–265. [CrossRef]

23. Lihong, D.; Binshi, X.; Shiyun, D.; Qunzhi, C.; Dan, W. Variation of stress induced magnetic signals during testing of ferromagnetic steels. *NDT E Int.* **2008**, *41*, 184–189. [CrossRef]

24. Udpa, S.S.; Moore, P.O. *Nondestructive Testing Handbook: Electromagnetic Testing*; The American Society for Nondestructive Testing: Columbus, OH, USA, 2004.

Article

Detection of Delamination with Various Width-to-depth Ratios in Concrete Bridge Deck Using Passive IRT: Limits and Applicability

Van Ha Mac [1,†], Quang Huy Tran [2,†], Jungwon Huh [1,*], Nhu Son Doan [1], Choonghyun Kang [3] and Dongyeob Han [1]

[1] Department of Ocean Civil Engineering, Chonnam National University, Yeosu 59626, Korea; 188456@jnu.ac.kr (V.H.M.); 188444@jnu.ac.kr (N.S.D.); hozilla@chonnam.ac.kr (D.H.)
[2] Department of Civil Engineering, Nha Trang University, Khanh Hoa 57000, Vietnam; huytq@ntu.edu.vn
[3] Department of Ocean Civil Engineering, Gyeongsang National University, Tongyeong 53064, Korea; chkang@gnu.ac.kr
*   Correspondence: jwonhuh@chonnam.ac.kr; Tel.: +82-61-659-7247
†   These authors contributed equally to this work.

Received: 28 October 2019; Accepted: 29 November 2019; Published: 2 December 2019

Abstract: In bridge structures, concrete decks have a higher risk of damage than other components owing to the direct impact of traffic. This study aims to develop a comprehensive system for bridge inspection using passive infrared thermography (IRT). Experiments were conducted on a concrete specimen (assumed as the surface of the bridge deck) embedded artificial delaminations with different width-to-depth ratios (WTDRs). Both professional handheld IR camera (H-IRC) and a UAV mounted with an IR camera (UAV-IRC) were employed simultaneously to capture the surface temperature of the structure. The present work indicates that the passive IRT technique with an H-IRC can be used to detect delaminations located at depths of 4 cm or less from the structure surface if the WTDRs are not lesser than 1.9 for daytime and 2.5 for nighttime when testing on a sunny day. In addition, the larger the WTDR, the higher the temperature difference can be produced, thus delaminations could be observed more clearly. Furthermore, our study suggests that the concrete bridge deck inspection using passive IRT can produce appropriate results if the inspection is performed from 10:00 to 15:00 or from 19:30 to approximately 2:00 on a sunny day. Good agreement between the results obtained from tests using H-IRC and UAV-IRC was observed, which validates the application of UAV-IRC in real structure inspection.

Keywords: delamination; width-to-depth ratio (WTDR); concrete bridge deck; handheld IR camera (H-IRC); UAV IR camera (UAV-IRC); passive IRT; non-destructive technique; concrete structure

1. Introduction

In bridge structures, compared to other components, defects occur more frequently in the concrete deck because it is directly subjected to traffic loads [1,2]. Among the various defects, delamination often develops in the concrete cover due to the corrosion effect of steel bars in reinforced concrete [2,3]. Delamination is one of the most dangerous deteriorations because it is usually invisible under visual inspection and can lead to potential spalls (Figure 1). Therefore, delaminations need to be detected as accurately as possible to maintain normal working conditions of the bridge structure.

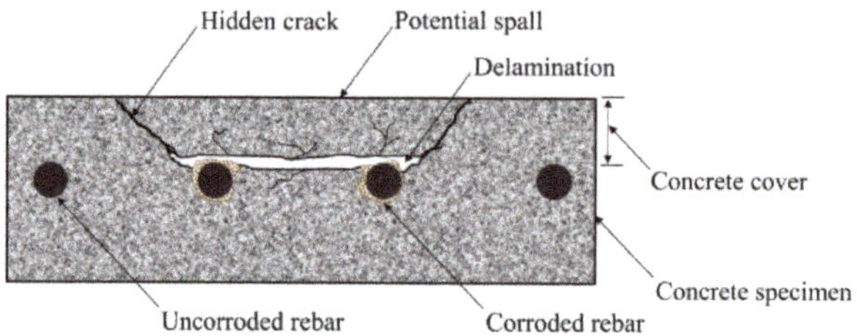

Figure 1. Locations of delaminations in concrete structure.

To investigate the health of a concrete bridge deck in terms of delamination, conventional methods such as chain dragging and hammer sounding are widely used, coupled with modern non-destructive techniques (NDTs) to enhance the accuracy and quality of the inspection. The NDTs that are currently used in concrete bridge inspections include infrared thermography (IRT), ground penetrating radar, ultrasonic surface waves, impact echo, and electrical resistivity [3–13]. In the IRT method, there are two different approaches: passive and active [14]. In active IRT, several minutes to a few hours should be spent to inspect just small regions of structures that are not exposed directly to the sun because the structure surface must be heated by an artificial heat source [15–27]. On the other hand, passive IRT has been proven to be the preferred method to inspect structure components exposed directly to the sunlight [4,12,13,28–31]. The information related to potential delaminations can be extracted accurately without using any extra artificial heat source. Therefore, the concrete bridge deck is scanned more rapidly and simply using passive IRT compared to the active approach.

According to the American Association of State Highway and Transportation Officials (AASHTO)-Manual for Bridge Element Inspection (2015), the condition of the concrete bridge deck is divided into four states from one to four based on the level of the severity of the damages as shown in Figure 2 [32]. If delaminations/spalls occur inside the concrete bridge deck, it means that it does not belong to the condition state 1 (good status). The condition states 2 (fair status) and 3 (poor status) correspond to the areas of moderate and severe delaminations/spalls, respectively. In the condition state 2, the delamination depth (d) is smaller than 2.54 cm or diameter (w) is less than 15.24 cm, while d is greater than 2.54 cm or w is larger than 15.24 cm in the condition state 3. When a structure is in the condition state 4 (severe status), load reduction, bridge closure, or replacement with a new structure should be performed. It is found that both the size and depth of delamination are critical indicators used to classify the health condition of concrete bridge deck.

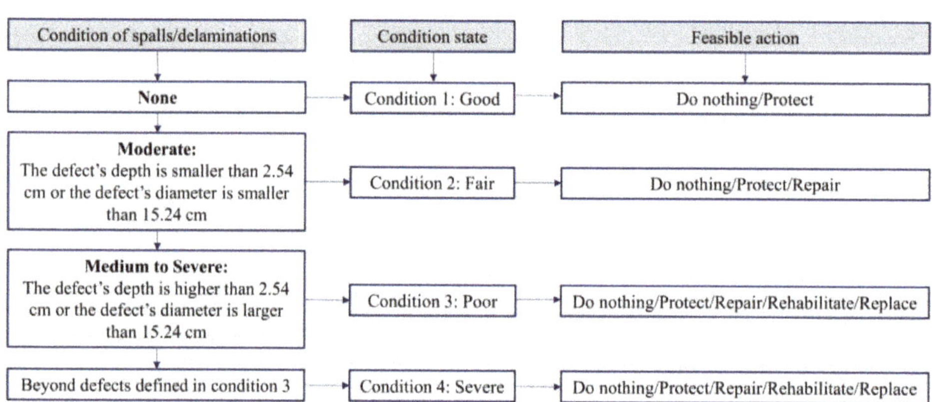

Figure 2. Classification of concrete bridge deck conditions required by American Association of State Highway and Transportation Officials (AASHTO) 2015.

Moreover, although the effect of the size and depth of the delamination on the detectability of delaminations has been studied in several works, it still remains to be a controversial issue. A shallower delamination with a small size sometimes may not be detected, but a deeper delamination with a larger size may be identified. Thus, in consideration of the practical application for concrete bridge deck inspection, the employment of the passive IRT technique with various width-to-depth ratios (WTDRs, the ratios between the size and depth of delaminations) is investigated in detail in this work. This implies that the effect of the size and depth of delamination is not considered independently, like previous works, but is studied using the WTDR [1,29,30].

In addition, as the environment strongly affects the detectability of defects in passive IRT, suggesting a suitable inspection time is mentioned in previous researches [12,13,29,30,33,34]. However, this topic might not be studied fully, especially for defects with small size or WTDR, it therefore is one of the main tasks in the present work. In addition, the applicability of unmanned aerial vehicles (UAVs) is studied for both quantitative as well as qualitative assessment of the concrete bridge deck conditions. The UAV mounting an IR camera (UAV-IRC) is used alongside the handheld IR camera (H-IRC). The UAV-IRC is more suitable for places where the H-IRC might not be used owing to heavy and high-speed traffic. The results of the UAV-IRC would be verified by those from the H-IRC. This study determines the limits and applicability of passive IRT in detecting delamination with different WTRDs in concrete bridge deck. Hence, a comprehensive system for bridge inspection combining H-IRC and UAV is expected to be developed.

2. Literature Review

Regarding the detectability of subsurface defects in concrete structures using passive IRT techniques, the main concerns growing in recent years are the effect of delamination parameters like the size, depth, and thickness, effect of the type of the infrared detector, impact of weather conditions, and effective inspection time, as well as the minimum threshold of the surface temperature difference used to define a defect as detectable or undetectable [1,29,30,35,36].

The detected depth of delaminations has been studied by researchers like Yehia et al. in 2007, Farrag et al. in 2016, Sultan et al. in 2017, and Hiasa et al. in 2018 [1,13,36,37]. Detected and undetected depths of delaminations with various sizes are depicted in Figure 3 as summarized from previous works [1,2,13,29,30,34–39]. It is noted that the results shown in Figure 3 were based on normal concrete specimens, using an uncooled IR camera, and monitored via the passive IRT approach. In Figure 3, the blue filled-in circle represents undetectable delaminations while the red unfilled-in diamond stands for detectable delaminations.

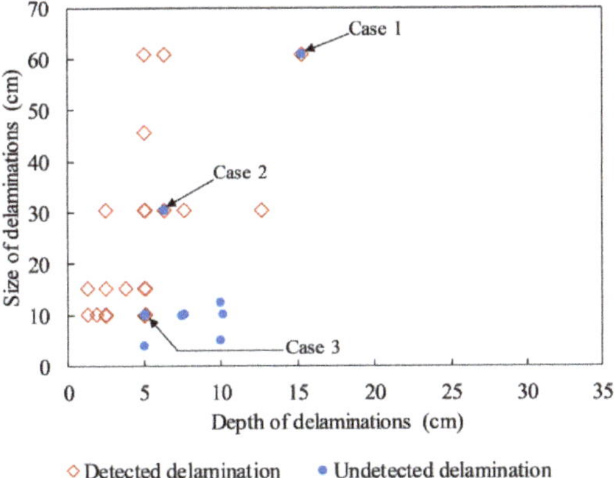

Figure 3. Detected and undetected delaminations summarized from the literature.

It can be observed in Figure 3 that despite having the same WTDR, delaminations could be detected or undetected. The reason is that the experiments were performed in different environmental conditions, and the parameters of the artificial delaminations, the dimensions of concrete specimens and the specification of IR detector are not the same. For example, in the case of delaminations with the same WTDR of 4.0 shown as "Case 1" in Figure 3, one delamination could be detected while other one was not identified. The same phenomenon can be observed as pointed out in "Case 2" (WTDR = 4.8) and "Case 3" (WTDR = 2.0). Thus, the conclusions given in the previous works regarding the detected depth and size of delaminations still remain a controversial topic. Hence, further studies need to be conducted, especially in the case of heterogeneous material, such as concrete. In this work, WTDR is considered instead of the depth or size separately unlike in previous studies.

The most effective time to conduct the concrete inspection (i.e., either daytime or nighttime) proposed in the above-mentioned studies is also inconsistent. Washer et al. in 2009 and 2010, conducted tests on a concrete specimen for approximately six months on the field [29,30]. During the test, every 10 min, the surface temperature of the specimen was monitored by an IR camera. Washer et al. recommended that the optimum times of the day for identifying delaminations at depths of 25, 51, 76, and 127 mm are after sunrise 5 h and 40 min, 6 h, 7 h and 9 h, respectively [29,30]. However, in 2018, Hiasa et al. proposed different times to perform the delamination inspection in concrete structures based on the results of the field test on four concrete specimens in which an IR camera was utilized to capture the surface temperature of the specimen from 7:00 am to 12:00 am [1]. They concluded that delaminations at depths of 5.1 and 7.6 cm could not be detected. It was suggested that delaminations at depths of 1.3 and 2.5 cm can be detected at any time except during the interchange periods of 09:00–9:30 and 15:00–16:00. The authors stated that the inspections should be conducted during nighttime because the available time during nighttime is longer than that during daytime, and the probability of the misdetection of the delaminations may be reduced.

Furthermore, Yehia et al. (2007), Kee et al. (2012), and Gucunski et al. (2013) gave their own recommendations about the available inspection time [12,13,34]. Yehia mentioned that during the period from 10:00 to noon, delaminations with depths equal to or smaller than 5.1 cm were shown more clearly than from noon until 3 pm [13]. Focusing on deep delaminations of 6.35 and 15.24 cm, Kee et al. stated that delaminations could be detected clearly at 45 min after sunrise during the cooling cycle or after sunrise 7 h and 45 min while no delaminations were shown at 3 h and 45 min after sunrise [34]. Gucunski et al. recommended that 40 min after sunrise is a better time than around noon for the delamination inspection [12]. As reviewed above, the proposed times are completely different and this can be explained by the effects of different environmental conditions, delamination characteristics, concrete specimen dimensions, infrared detectors, and experiment setup. Hence, the determination of feasible inspection time should be studied further, especially in the case of small WTDR delamination in order to reduce the misdetection problem in the concrete bridge inspection.

In passive IRT, the IR camera is normally kept in hand or mounted on a car to scan the concrete bridge deck. In these cases, although the lane closure is not required, the vehicle speed on the bridge is reduced and it is inconvenient in certain aspects. Recently, some researchers have utilized UAV-IRC to capture thermal images in order to conduct bridge inspection [40–44]. In 2017, Omar et al. used UAV-IRC to detect defects in a concrete bridge [40]. In their study, the severity of the delaminated area in the concrete bridge deck was indicated by a condition map based on their proposed objective thresholds. Khan et al. in 2015 used the UAV-IRC to collect thermal images of a mock-up bridge. As a result, potential defects in the concrete bridge deck were indicated [41]. In addition, the effect of technical data of IR camera, UAV and the overlap of individual images were considered by Vasterling and Meyer in 2013 and Gillins et al. in 2016 [43,44]. They stated that the uncooled sensor IR camera with 8–14 μm of bandwidth is a good choice for the UAV-IRC because of its lightweight. The flight altitude should be selected based on the camera resolution and separated images should be taken with a minimum overlap of 50%. In all works mentioned above, the effects of WTDRs as well as the ranges of the WTDRs in which delaminations may be detected by using UAV-IRC have not been presented. These matters will be clarified in this study and the applicability of UAV-IRC is examined by comparing its results with those of experiments from H-IRC.

In the present work, the experiments were carried out on a concrete slab embedded artificial delaminations with various WTDRs using passive IRT. Two IR detectors with different specifications are used including an H-IRC and a UAV-IRC. To develop a comprehensive system in inspecting the bridge using both H-IRC and UAV, the aims of this study is to: (1) Examine the effect of WTDRs on the detectability of delaminations in a concrete specimen; (2) Determine the minimum WTDR of a delamination that can be detected during the daytime and nighttime; (3) Propose a suitable time for concrete bridge deck inspection using passive IRT technique; (4) Evaluate the applicability of UAV-IRC on delamination detection in concrete bridge decks.

3. Fundamentals of Passive IRT

In the passive IRT technique, there are two configuration modes including transmission and reflection [28]. In transmission mode, heat sources and IR cameras are placed on two opposite sides of the specimen whereas they are placed on the same side of the specimen in the reflection mode [28]. In the present work, the reflection mode is chosen (during daytime) to detect delaminations in a concrete slab because this mode is more popular and applicable for concrete bridge deck inspection, especially in cases of bridges with great heights. The mechanism of the detectability of delaminations during both daytime and nighttime is briefly explained as shown in Figure 4.

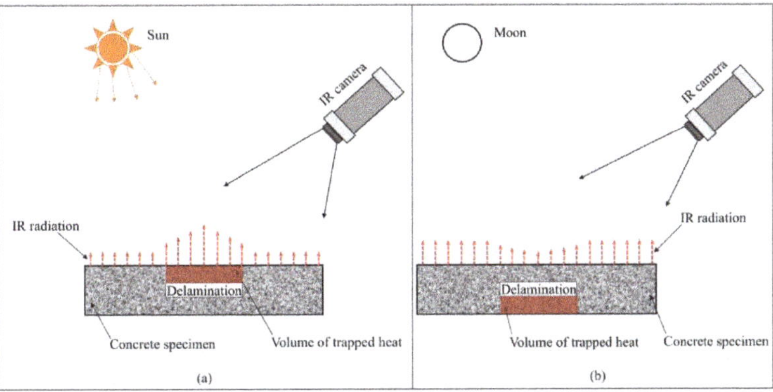

Figure 4. Fundamentals of defect detection using passive IRT: (**a**) during daytime; (**b**) during nighttime.

During the daytime, energy from the sun heats up the concrete surface. If a delamination exists inside the concrete structure, the volume of the trapped heat develops above the delamination because the thermal conductivity of air (approximately 0is much lower than that of the concrete (between 0.4 and 1.8 W/m°C), then the energy from the trap.024 W/m°C) ped heat volume turns back to the surface [1,2,45,46]. Therefore, the concrete surface above the delaminations becomes hotter than the neighborhoods during the daytime. In contrast, the heat is radiated back to the sky, thus the volume of the trapped heat is under the delamination during nighttime [1]. As a result, the concrete surface above the delamination is cooler than its surroundings during nighttime. When utilizing an IR camera to capture the thermal images at a suitable time, defects can be observed based on the difference of the surface temperature between the delaminated and surrounding areas.

The total radiation (W_{tot}) is captured by an IR camera not only from the object but also from ambient sources and the atmosphere, as shown in Equation (1) and Figure 5 [47]. The surface temperature of the object can be computed automatically by using Equation (2).

$$W_{tot} = \varepsilon \times \tau \times \sigma \times T_{obj}^4 + (1 - \varepsilon) \times \tau \times \sigma \times T_{amb}^4 + (1 - \tau) \times \sigma \times T_{atm}^4, \tag{1}$$

$$T_{obj} = \left\{ \frac{W_{tot} - (1 - \varepsilon) \times \tau \times \sigma \times T^4_{amb} - (1 - \tau) \times \sigma \times T^4_{atm}}{\varepsilon \times \tau \times \sigma} \right\}^{\frac{1}{4}}, \tag{2}$$

where $\varepsilon \times \tau \times \sigma \times T^4_{obj}$ is the emission from the object, $(1 - \varepsilon) \times \tau \times \sigma \times T^4_{amb}$ is the reflected emission from ambient sources, $(1 - \tau) \times \sigma \times T^4_{atm}$ is the emission from the atmosphere, T_{obj} is the temperature of the object, T_{amb} is the reflected temperature from ambient sources, T_{atm} is the atmospheric temperature, ε is the emissivity of the object, τ is the transmittance of the atmosphere, σ is the Stefan-Boltzmann constant ($\sigma = 5.67 \times 10 - 8$ W/m^{2}K^4), thereby $(1 - \varepsilon)$ is the reflectance of the object, and $(1 - \tau)$ is the emissivity of the atmosphere. The values of ε, T_{amb}, and T_{atm} must be set in the camera as input data.

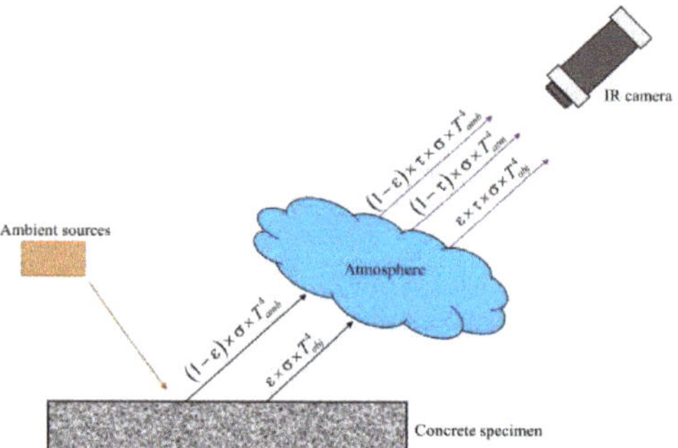

Figure 5. Radiation received by an infrared (IR) camera.

4. Experimental Work

4.1. Concrete Specimen Design

In this study, a concrete slab was made with dimensions 125 cm × 105 cm × 10 cm and design compression strength of 30 MPa. The coarse aggregate with maximum size of 25 mm and cement PC40 were employed to make the specimen. The proportion of water, fine aggregate and coarse aggregate to cement was 0.49, 1.48 and 1.90, respectively. The thermal diffusivity of concrete specimen was estimated as $\alpha = 1.076$ cm^2/min in our previous study [20]. Twelve squared artificial delaminations with the same thickness of 1 cm but different sizes ranging from 5.0 cm × 5.0 cm to 15.75 cm × 15.75 cm, and depths of 2, 3, and 4 cm from the back face and 5, 6, and 7 cm from front face were embedded inside the specimen that creates WTDRs from 1.0 to 7.9. Artificial delaminations were made by square pieces of polystyrene with the thermal conductivity of approximately k = 0.027 W/m°C that is expected to have very similar thermal behavior compared to the air k = 0.024 W/m°C [26,48]. In addition, artificial delaminations were glued on the stone pieces (made by concrete having same mixture ratio with the specimen) that were attached on the bottom wooden plate of formwork to create defects with exact depths as expected. Therefore, the specimen is anticipated to have similar response under heat transfer in comparison with real concrete bridge deck structure. The arrangement of delaminations and their parameters are shown in Figure 6 and Table 1.

It should be noted that if subsurface defects such as delaminations or voids develop in the concrete bridge, the air fills up inside them. These defects may achieve a similar behavior when receiving a heat flux. Therefore, the results of this study can be applied to not only delamination but also void in concrete bridge deck.

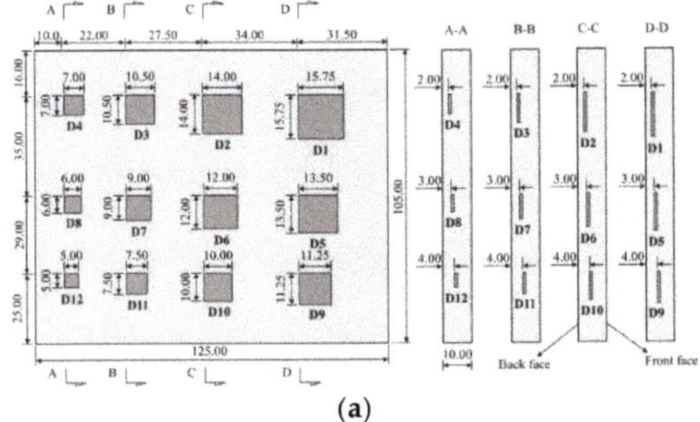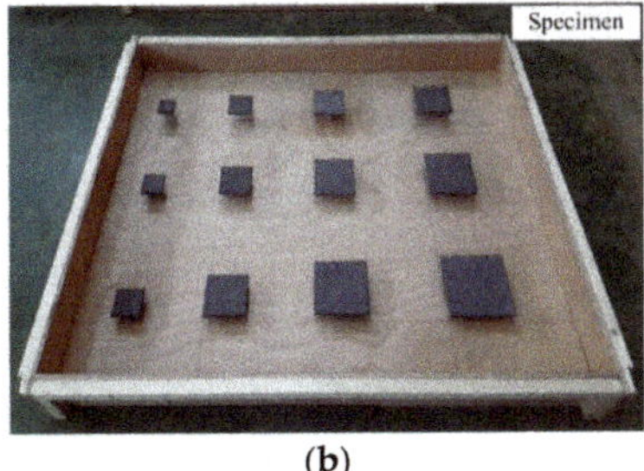

(a)   (b)

Figure 6. Concrete specimen with embedded imitating delaminations: (**a**) arrangement of delaminations; (**b**) formwork to cast the concrete specimen.

Table 1. Characteristics of artificial delaminations inside the concrete specimen.

Delamination	Sizes (cm × cm × cm)	From Back Face			From Front Face		
		Depth (cm)	WTDR	Denotation	Depth (cm)	WTDR	Denotation
D1	15.75 × 15.75 × 1.0	2	7.9	B-D1	7	2.25	F-D1
D2	14.0 × 14.0 × 1.0	2	7.0	B-D2	7	2.0	F-D2
D3	10.5 × 10.5 × 1.0	2	5.25	B-D3	7	1.5	F-D3
D4	7.0 × 7.0 × 1.0	2	3.5	B-D4	7	1.0	F-D4
D5	13.5 × 13.5 × 1.0	3	4.5	B-D5	6	2.25	F-D5
D6	12.0 × 12.0 × 1.0	3	4.0	B-D6	6	2.0	F-D6
D7	9.0 × 9.0 × 1.0	3	3.0	B-D7	6	1.5	F-D7
D8	6.0 × 6.0 × 1.0	3	2.0	B-D8	6	1.0	F-D8
D9	11.25 × 11.25 × 1.0	4	2.8	B-D9	5	2.25	F-D9
D10	10.0 × 10.0 × 1.0	4	2.5	B-D10	5	2.0	F-D10
D11	7.5 × 7.5 × 1.0	4	1.9	B-D11	5	1.5	F-D11
D12	5.0 × 5.0 × 1.0	4	1.25	B-D12	5	1.0	F-D12

4.2. Experimental Setup

The passive IRT test was conducted at Chonnam National University, Republic of Korea (Figures 7 and 8a). Two IR detectors were employed in this study, i.e., an H-IRC (FLIR, Wilsonville, OR, USA) and a UAV-IRC (FLIR, Wilsonville, OR, USA) (Figure 8b,c and Figure 9). The H-IRC is a long-wavelength handheld IR camera (FLIR SC660) in the 7.5–13.5 μm bandwidth [49]. This camera has a focal plane array of 640 × 480 pixels, thermal sensitivity ≤ 0.03 °C (at + 30 °C), spatial resolution (IFOV) of 0.65 mrad, field of view (FOV) of (24° × 18°)/0.3 m, and an accuracy of ±1 °C. It should be noted that long-wavelength cameras in the 7.5–13.5 μm range work well when atmospheric attenuation is involved because the atmosphere tends to act as a high-pass filter above 7.5 μm. The H-IRC is located at a height of 195 cm and the distance from the closest edge of the specimen to the camera lens is 360 cm that allows capturing the temperature of the whole specimen surface in only one thermal image (Figure 8b). Overall, the surface temperature was captured at intervals of 30 min.

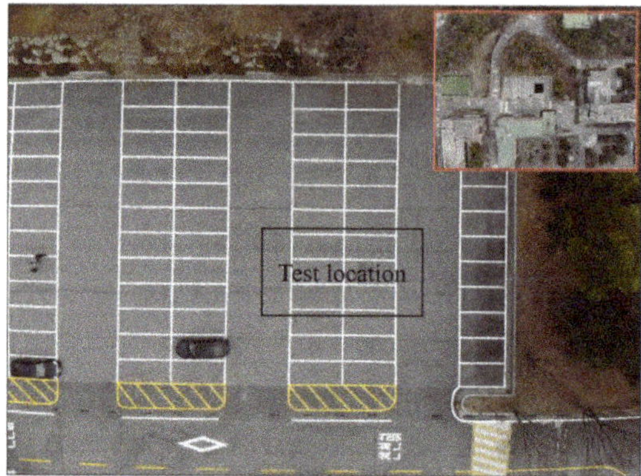

Figure 7. Location of the test.

Another long-wavelength IR camera (FLIR Zenmuse XT2 - FLIR, Wilsonville, OR, USA) with the same bandwidth as that of the H-IRC is mounted on the UAV-IRC DJI M200 version [50]. The thermal sensitivity of the UAV-IRC is equal to or less than 0.05 °C, and the FOV is 57.12° × 42.44°. The technical data of the UAV is shown in Table 2 [50].

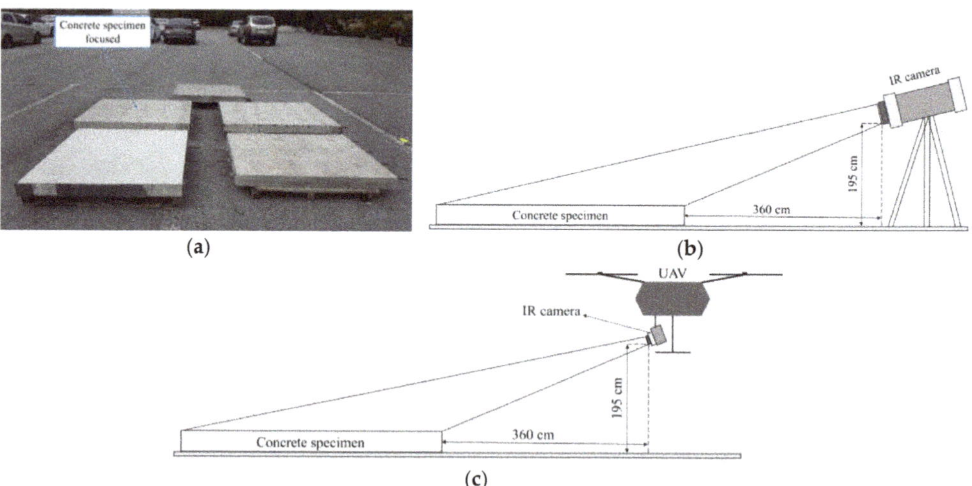

Figure 8. Arrangement of the experiment: (**a**) location of specimen; (**b**) with H-IRC; (**c**) with UAV-IRC.

Table 2. Main technical data of the DJI M200 version UAV.

Items	Parameters
Dimensions (unfolded)	887 mm × 880 mm × 378 mm
Self-weight (with two TB50 batteries)	~3.80 kg
Max payload (with two TB50 batteries)	~2.34 kg
Max angular velocity	Pitch: 300 °/s; Yaw: 150 °/s
Max speed	60.8 km/h
Max wind resistance	12 m/s
Max flight time	27 min

All experiments except the case of using UAV-IRC were conducted on both faces of the specimen to consider depths from 2 to 7 cm. The passive IRT test was conducted for five days. The collection time was from 6:00 to 6:00 (24 h), 8:30 to 24:00 (15 h and 30 min), 9:00 am to 21:00 pm (12 h), 8:00 to 8:00 (24 h), and 5:30 to 13:30 (8 h) on the 1st, 2nd, 3rd, 4th, and 5th days respectively.

The scope of each experiment cycle is indicated in Table 3. The thermal images were captured only by the H-IRC during the first four days whereas both H-IRC and UAV-IRC were used on day 5. The capability of the UAV-IRC in detecting delaminations in the concrete deck is analyzed by comparing it with respect to the obtained data from the experiments using H-IRC.

The relative humidity and ambient temperature are important parameters that the camera must be fed with before capturing thermal images. Then, the surface temperature is automatically calculated by the IR camera (FLIR, Wilsonville, OR, USA) using Equation (2) in Section 3. Therefore, the ambient temperature, relative humidity as well as wind speed were measured by the Kestrel 3000 (Nielsen-Kellerman, Chester, PA, USA) before taking thermal images. The equipment used in this study is shown in Figure 9.

Table 3. Time cycle for the experiment.

Day	Duration (h)	Face Exposed Directly to the Sun	Camera Used
Day 1	6:00–24:00 0:00–6:00	Back face Back face	H-IRC
Day 2	8:30–24:00	Back face	
Day 3	9:00–21:00	Front face	
Day 4	8:00–24:00 0:00–8:00	Front face Front face	
Day 5	5:30–13:30	Back face	H-IRC and UAV-IRC

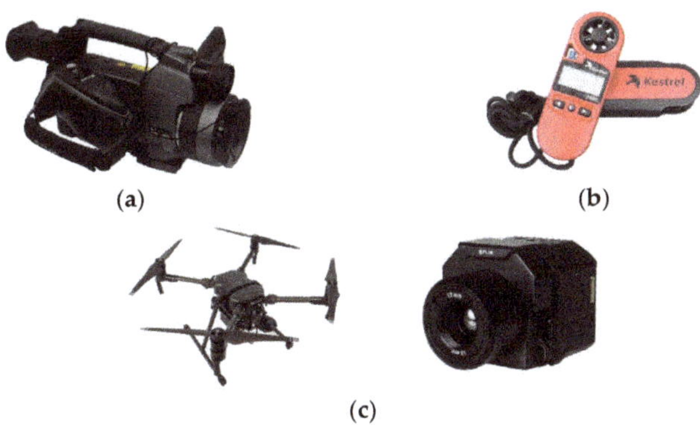

Figure 9. Equipment used in the experiment: (a) H-IRC; (b) Kestrel 3000; (c) UAV-IRC.

4.3. Experimental Condidion

The environmental conditions on the experimental days are summarized in Table 4. The ambient temperature, relative humidity, and wind speed are calculated as the average of the values for every three hours. Generally, the weather condition is sunny on day 1, day 3, and day 5 while day 2 is rainy and day 4 is cloudy. According to the ASTM Standards, the inspection should not be conducted when the wind speed is higher than 24 km/h [51]. The maximum wind speed on experimental days was 3.2 km/h, so all experiments were carried out under appropriate weather conditions in terms of wind speed.

As recommended in the ASTM D4788-03 Standards, the sunshine should be included because the temperature difference of 0.5 °C within a delaminated area and its neighborhood can be obtained after at least 3 h of direct sunshine [51]. Based on the data listed in Table 4, three experiment cycles including days 1, 3, and 5 are in the range while days 3 and 4 are out of the scope recommended by the ASTM

Standards [51]. The non-sunny days selected in this study are for developing a comprehensive system for concrete bridge deck inspection with the combination of H-IRC and UAV-IRC under different weather conditions.

In 2009, Washer et al. stated that the ambient temperature change is an important factor affecting the inspection outcome in passive IRT [30]. The positive ambient temperature change leads to the positive surface temperature difference above a delamination and its surrounding while it demonstrates a negative trend under the effect of negative ambient change. In addition, relative humidity change is another important factor that significantly impacts the surface temperature difference [30]. Thus, the ambient temperature and relative humidity during the test are presented in detail in Table 4 as well as Figure 10. This data is necessary for the discussion of the results as well as for the application of the study's propositions to real structures.

The changes in relative humidity and ambient temperature during a sunny day (day 1) and rainy day (day 2) are shown in Figure 10a,b, respectively. In these figures, the red line represents relative humidity while the blue line is for ambient temperature.

Table 4. Weather conditions during the experiments.

Day	Time (h)	Average of Ambient Temperature (°C)	Average of Relative Humidity (%)	Average of Wind Speed (km/h)	Weather Condition	Overall Condition
Day 1	6:00–9:00	17.2	79.5	0.8	Sunny, partly cloudy	Sunny day
	9:00–12:00	23.0	58.8	0.6	Sunny, clear sky	
	12:00–15:00	27.0	48.3	2.2	Sunny, partly cloudy	
	15:00–18:00	26.1	52.1	1.2	Sunny, clear sky	
	18:00–21:00	22.2	63.5	0.6	Clear sky	
	21:00–24:00	19.2	71.1	0.4	Clear sky	
Day 2	0:00–3:00	18.3	74.0	0.1	Clear sky	Rainy day
	3:00–6:00	16.7	84.6	0.0	Clear sky	
	8:30–9:00	19.8	84.9	0.4	Moderate rain	
	9:00–12:00	19.0	88.2	1.2	Moderate rain	
	12:00–15:00	19.3	89.2	1.0	Light rain	
	15:00–18:00	20.6	88.2	0.3	Cloudy	
	18:00–21:00	20.4	89.0	0.4	Heavy rain	
	21:00–24:00	20.4	83.9	0.8	Light rain	
Day 3	9:00–12:00	28.3	66.6	0.4	Sunny, clear sky	Sunny day
	12:00–15:00	30.1	59.8	0.9	Sunny, clear sky	
	15:00–18:00	28.1	61.6	0.9	Sunny, partly cloudy	
	18:00–21:00	22.8	74.8	0.6	Partly cloudy	
Day 4	8:00–9:00	20.6	83.5	0.8	Cloudy	Cloudy day
	9:00–12:00	19.0	89.9	1.2	Cloudy	
	12:00–15:00	20.3	83.5	1.5	Cloudy	
	15:00–18:00	23.4	73.8	1.0	Mostly cloudy	
	18:00–21:00	20.5	78.2	0.4	Mostly cloudy	
	21:00–24:00	18.5	90.5	0.0	Cloudy	
	0:00–3:00	18.9	89.5	0.0	Cloudy	
	3:00–6:00	19.4	90.0	0.0	Cloudy	
	6:00–8:00	21.9	76.3	0.4	Cloudy	
Day 5	5:30–8:00	13.8	67.3	3.2	Sunny, partly cloudy	Sunny day
	8:00–11:00	16.7	56.2	2.8	Sunny, clear sky	
	11:00–13:30	21.9	41.8	2.9	Sunny, clear sky	

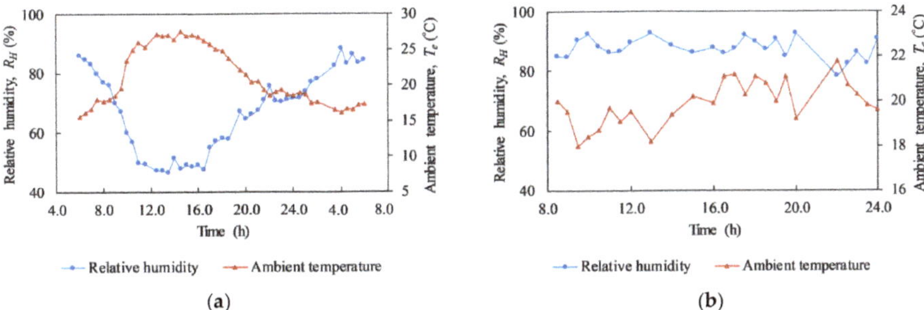

Figure 10. Ambient temperature and relative humidity: (**a**) day 1 (sunny day); (**b**) day 2 (rainy day).

On the sunny day, it can be observed that relative humidity (R_H) is inversely related to the air temperature (T_e); if temperature increases, the relative humidity decreases and vice versa as shown in Figure 10a. Overall, the temperature tends to go up from 6:00 and reaches the maximum value around 14:00 before falling down and towards the smallest value at 4:00 in the next day. After that, there is a small rise in the temperature from 4:00 to 6:00. In contrast, during the time between 6:00 and 14:00, the relative humidity descends dramatically. Thereafter, the relative humidity increases rapidly from 14:00 to nighttime (until 4:00). During the rainy day (Figure 10b), the tendency of both ambient temperature and relative humidity is not observed clearly. The temperature and humidity lines fluctuate significantly and there is no correlation between these two variables.

5. Detectability of Delaminations Using H-IRC

5.1. Surface Temperature

The temperature of the concrete specimen surface is analyzed focusing on the back face with delaminations at depths equal to or less than 4 cm during both sunny day (day 1) and rainy day (day 2). It should be noted that in the experiment, sunrise appeared at 5:30 and sunset was at 18:40 on day 1. However, the sun started heating the specimen from 6:45 and stopped heating at 17:35 owing to the shadow of trees and buildings near the experiment site as shown in Figure 7.

The thermal images of the back face at 11:00 and 20:30 during the sunny day (day 1) are shown in Figure 11a,b, respectively. There are five images shown in each of the Figure 11a,b. This is because in the experiment one thermal image was captured for the entire specimen surface, and then four individual thermal images were recorded for four areas of specimen as shown by the red dashed rectangular. The reason in capturing four individual images is to reduce the temperature range of the thermal image for avoiding the misdetection of delamination that may occur if only one thermal image was recorded. There is a gap between the surface temperature above a delaminated area (T_{de}) and its surrounding (T_{so}). Particularly, T_{de} is higher than T_{so} owing to the effect of daytime heating (Figure 11a) while T_{de} becomes cooler than T_{so} during the nighttime (Figure 11b). Based on the temperature difference between T_{de} and T_{so}, delaminations can be observed on the thermal images during both the daytime and nighttime. In this section, the surface temperatures above a defect (T_{de}) and its surrounding (T_{so}) calculated as the mean value of surface temperatures within "delaminated area" and "sound area" respectively as presented in Section 5.2. The time when the temperature on the surface above a delamination changes from hotter to cooler in relation to its neighborhood or vice versa is called "interchange period". The interchange period can be observed more clearly in consideration of the absolute contrast that is discussed in detail in Section 5.2.

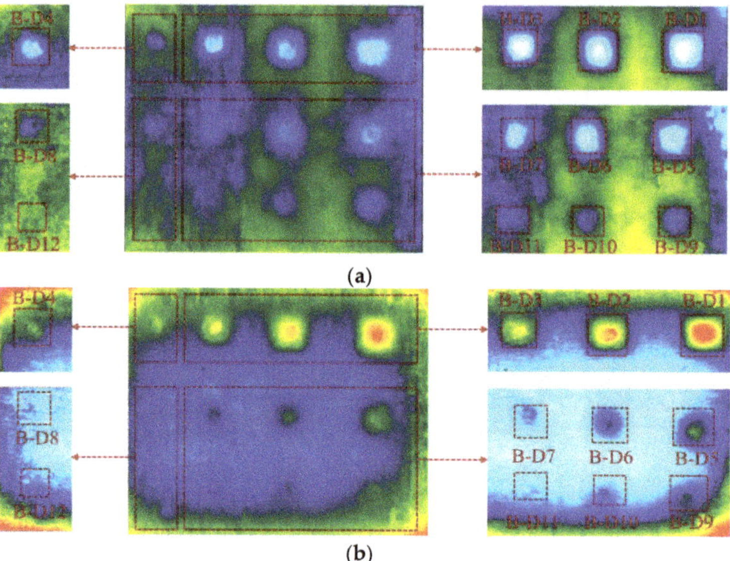

Figure 11. The thermal image: (**a**) at 11:00; (**b**) at 20:30.

Figure 12 shows the surface temperature (T_{de}) within delaminated areas on day 1 and day 2. Figure 12a–c plot T_{de} values during day 1 (the sunny day) for delaminations with depths of 2, 3, and 4 cm, respectively. It should be noted that each graph depicts delaminations that have the same depth but different sizes. Certain key phenomena can be indicated as follows:

First, all T_{de} value lines have the same tendency. There is a small increase of T_{de} in the first 2 h from 6:00 before it increases considerably and reaches the peak area around noon. Around 13:30, the sky became cloudy, which causes a significant decrease in the surface temperature, as mentioned by Hiasa et al. [1]. After noon, the surface temperature rapidly drops until 22:00; however, from 22:00 on the previous day to 5:30 the next day the T_{de} lines go down with a smaller slope. Then, the surface temperature increases slightly from 5:30 to 6:00 the next day.

Secondly, the effect of the delamination size on the surface temperature of the specimen during daytime and nighttime can be seen clearly. Small figures are added on the graphs in Figure 12 to zoom out the difference between the T_{de} value lines during daytime and nighttime. Overall, with the same depth, a larger delamination has a higher surface temperature than a smaller one during daytime while it becomes smaller during nighttime. For example, at 14:00, the surface temperatures are 51.66, 50.83, 50.72, and 48.81 °C corresponding to delaminations B-D1, B-D2, B-D3, and B-D4, as indicated in Figure 12. For the same set of delaminations, the observed surface temperatures are 27.00, 27.64, 28.37, and 28.90 °C at 20:30. The reason for this phenomenon is the effect of the trapped heat volume. During daytime, the trapped heat volume is developed above delamination under the heat energy from the sun. However, at the same depth, in comparison to a larger delamination, the trapped heat volume of a smaller delamination has a smaller intensity and diffuses more significantly in all directions leading to the decrease in the surface temperature [45]. During nighttime, the heat energy is radiated back into the air from the ground; thus, the volume of trapped heat is located under the delamination. Therefore, a higher surface temperature above a smaller delamination compared to a larger one is observed during nighttime.

The surface temperature above the delaminations with different sizes at a depth of 2 cm during the rainy day is presented in Figure 12d. There is no tendency in the surface temperature change during daytime and nighttime compared to the sunny day. During the rainy day, a slightly higher surface temperature can be produced in the case of a larger delamination, but it is quite difficult to observe this as the difference is small. For instance, at 12:00 the surface temperature above delamination B-D1 (21.66 °C) has a slightly higher value (0.01 °C) than delamination B-D2 (21.65 °C).

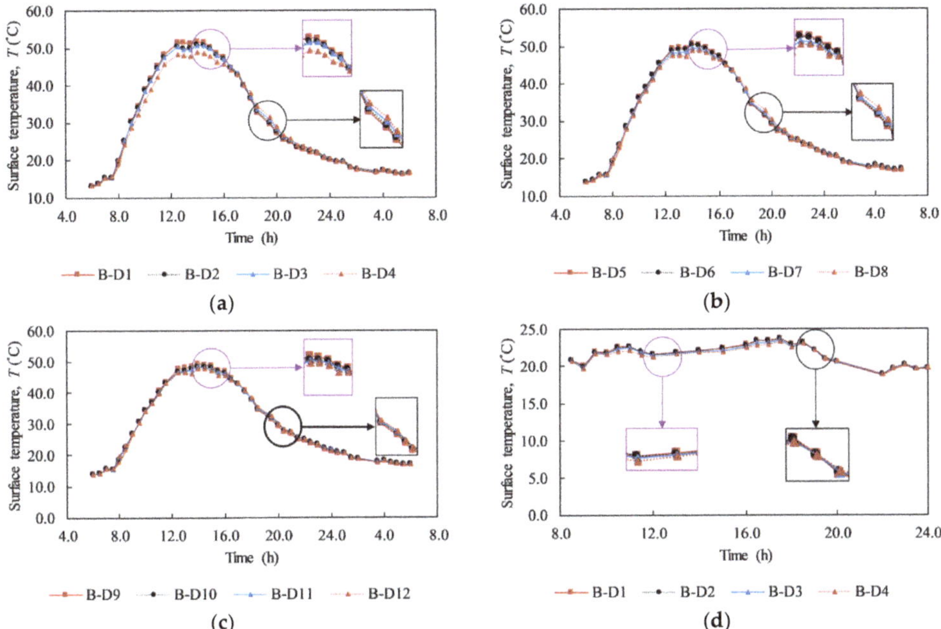

Figure 12. Surface temperature above delaminations: (**a**) at 2 cm depth on day 1; (**b**) at 3 cm depth on day 1; (**c**) at 4 cm depth on day 1; (**d**) at 2 cm depth on day 2.

5.2. Absolute Contrast

Contrast-based methods are the simplest image processing techniques that can be applied to enhance the thermograms quality. Even though a requirement of selection of sound area is the disadvantage of these methods, they are still the most common techniques utilized to preprocess image sequences [18,28]. In this study, in order to evaluate the detectability of delaminations with different WTDRs using passive IRT, the absolute contrast ($\Delta T(t)$), which is defined as the temperature difference above a defect ($T_{de}(t)$) and its neighborhood ($T_{so}(t)$) at the same time, as shown in Equation (3), is employed [28].

$$\Delta T(t) = T_{de}(t) - T_{so}(t),\tag{3}$$

The selection of areas of region of interest (ROI) used to compute $T_{de}(t)$ and $T_{so}(t)$ is depicted in Figure 13, where $T_{de}(t)$ is the average surface temperature of a ROI above a defect, called as the "delaminated area" whereas $T_{so}(t)$ is determined from of a ROI nearby the delamination denoted as the "sound area". The average surface temperature is employed as mentioned by Vaghefi in 2013 that is more effective compared to only one pixel or a group of three pixels [26,52,53]. The size of delaminated areas is similar with the dimension of respective delamination while the size of all sound areas is same (around of 8 cm × 8 cm). Nine sound areas at the middle position of defects are selected for twelve delaminations. The level of upper edges of delaminated and sound area for each delamination are similar. There are three couples of delaminations at the middle region of specimen (D2 and D3, D6 and D7, D10 and D11) in which each pair refers to the same sound area to avoid the misdetection phenomenon. As per the ASTM Standards, a delamination is considered as detectable if the amplitude of the absolute contrast ($\Delta T(t)$) known as "temperature difference" is 0.5 °C or higher [51]. Thus, this threshold is used in our present study. The higher the absolute contrast, the greater the certainty that delamination may appear.

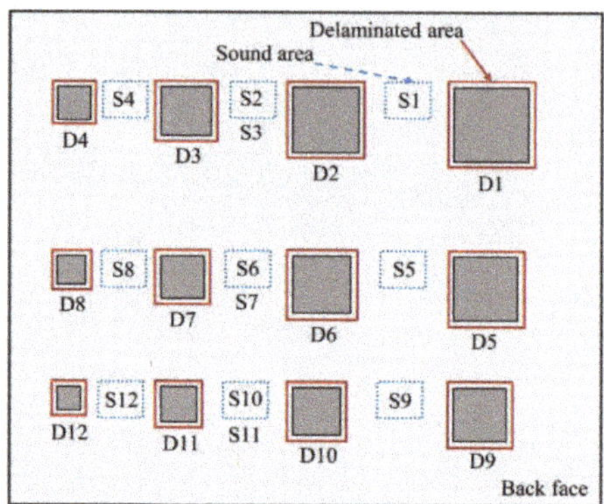

Figure 13. Selection of delaminated and sound areas on the structure surface.

In the experiment, delaminations with depth of 5, 6, and 7 cm (F-D1 to F-D12) were not detected at any given time during the test (day 3 and day 4). Then, the absolute contrast is analyzed focusing on the back face that delaminations (B-D1 to B-D12) are located at depths equal to or less than 4 cm (day 1 and day 2). Figure 14 shows the absolute contrast of all delaminations during the sunny days. In each graph, delaminations with the same depth but different sizes are depicted. The absolute contrast profiles have a similar tendency in comparison with the surface temperature and ambient temperature. In general, the absolute contrast increases quickly and reaches a positive peak value around noon. After that, it reduces and attains the maximum negative value before increasing again during early morning in the next day.

Owing to the formation of the volume of trapped heat under the daytime heating effect, the surface above a delamination becomes warmer than its surrounding while it is cooler at night because of the nighttime cooling effect. It is shown in the graphs in Figure 14 that there are two interchange periods between positive values (daytime heating effect) and negative values (nighttime cooling effect) of the absolute contrast. The first interchange period occurred due to the shift from cooling effect during the nighttime to heating effect during the daytime, while the second period is caused by the change from the daytime heating effect to the nighttime cooling effect. At the interchange period, delaminations cannot be observed by the IR camera because the temperature difference between the delaminated area and sound area is small (≤ 0.5 °C). The interchange period lasts for approximately 2 h during both the morning time (from 6:00 to 8:00) and afternoon time (from 16:30 to 18:30) under the given experimental conditions.

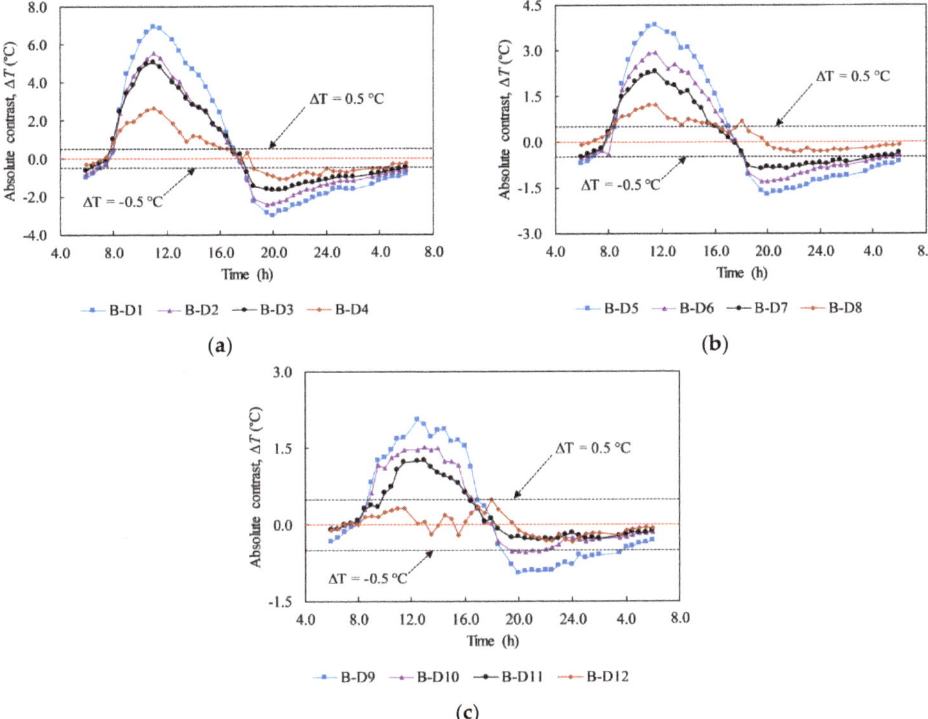

Figure 14. Absolute contrast of delaminations with different depths and sizes on day 1: (**a**) at the depth of 2 cm; (**b**) at the depth of 3 cm; (**c**) at the depth of 4 cm.

The effect of the size of delaminations on the temperature difference is studied as well. Under the heating effect during daytime and cooling effect during nighttime, a larger delamination produces a higher temperature difference than a smaller delamination. This phenomenon is caused by the intensity of the trapped heat volume and effect of the diffusion heat. For example, at 11:30, the temperature difference of delaminations B-D1, B-D2, B-D3, and B-D4, which have the same depth of 3 cm and different sizes of 13.5, 12.0, 9.0, and 6.0 cm, respectively, corresponds to 6.86, 5.32, 4.80, and 2.48 °C. This implies that delamination B-D1 can be observed more clearly than delaminations B-D2, B-D3, and B-D4 on the thermal image, as illustrated in Figure 15.

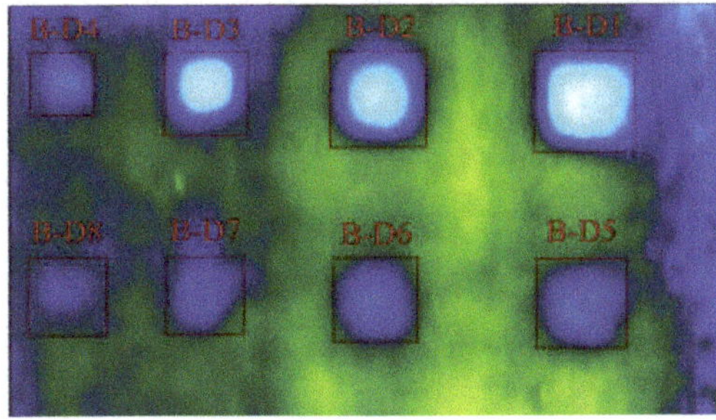

Figure 15. Delaminations (B-D1 to B-D8) on thermal image at 11:30 on day 1.

In terms of WTDR, it is indicated that a larger WTDR delamination can produce a higher temperature difference than a smaller WTDR one at the same depth. In addition, the conclusion can be given from the present study that a delamination, which has a larger WTDR, can produce a higher temperature difference even though it is located at greater depth than a smaller delamination as shown in Figure 16. For example, in Figure 16, delamination B-D5 and B-D6 obtained temperature differences of 3.84 °C and 2.94 °C whereas delamination B-D4 attained a temperature difference of 2.48 °C at 11:30. Moreover, the temperature differences are 1.73 °C, 1.28 °C, and 0.82 °C corresponding to delaminations B-D5, B-D6, and B-D4 at 20:00. Therefore, delamination B-D5 and B-D6 can appear more certainly than B-D4 during both the daytime and nighttime (Figures 11 and 15). Furthermore, a delamination placed at a relatively greater depth experiences a time delay to obtain the maximum temperature difference than a delamination located at a relatively shallower depth from the surface as demonstrated in Figure 16. In detail, delaminations B-D5 and B-D6 achieve the maximum temperature difference at 11:30 whereas delamination B-D4 obtains the maximum temperature difference at 11:00.

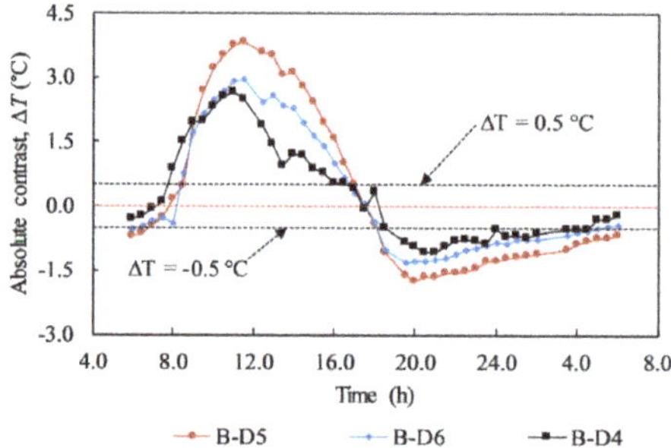

Figure 16. Absolute contrasts of delaminations with different WTDRs on day 1.

The maximum absolute contrasts of all delaminations composed in this study are graphically depicted in Figure 17. The red solid and red dashed line represent the maximum absolute contrast during daytime and nighttime on a sunny day, respectively. It can be concluded that a higher WTDR delamination obtains a greater maximum temperature difference than a smaller WTDR delamination on a sunny day. The maximum temperature difference rises from 0.49 °C to 6.95 °C (positive value) and from 0.32 °C to 3.00 °C (negative value) corresponding to the daytime and nighttime when the WTDR increases from 1.25 to 7.9. However, under the effect of rain, the above-mentioned trend is not observed clearly although the maximum temperature difference may tend go up with the rise of WTDR, as indicated in Figure 17. In addition, almost all delaminations obtain the maximum temperature difference smaller than 0.5 °C. Thus, it is recommended that the concrete bridge deck inspection must not be conducted while it is raining or after rain.

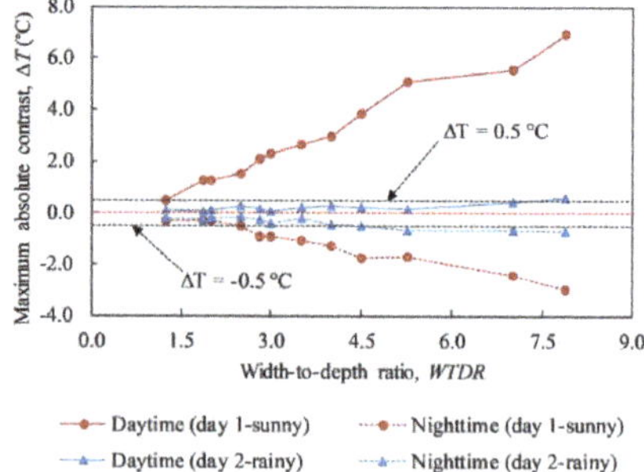

Figure 17. Maximum absolute contrast of delaminations during daytime and nighttime on both days 1 and 2.

5.3. Detection of Delamination

The detectability of delaminations during both daytime and nighttime on the sunny day are pointed out in Figure 18. The black filled-in squares and red unfilled-in circles represent detected and undetected delaminations respectively. It should be noted that delaminations with depths equal to or higher than 5.0 cm and the given sizes in this study were not detected. Thus, the red dashed line is used to divide the graph into two regions: detectable and undetectable regions, considering delaminations with depths equal to or smaller than 4.0 cm.

In Figure 18, the horizontal axis shows the delamination depth while the vertical axis is for the delamination size; hence the detectability of delaminations can be indicated in terms of WTDR. Certain conclusions can be given based on Figure 18 and information presented under Section 5.2 as follows.

For delaminations with depths higher than 4.0 cm (F-D1 to F-D12) on both sunny and cloudy days (day 3 and day 4): All delaminations could not be detected at any given time of the experiment period. This implies that it is impossible to identify delaminations with depths deeper than 4.0 cm and WTDRs equal to or smaller than 2.25 under the conditions of this study.

For delaminations with depths equal to or smaller than 4.0 cm (B-D1 to B-D12) on the sunny day (day 1): In the daytime, the dashed line has a slope of 1.9 that implies a delamination whose size is not smaller than 1.9 times its depth (WTDR ≥ 1.9) may be detected (Figure 18a). However, a delamination with a WTDR of 1.9 could not be identified during the nighttime (Figure 18b). The WTDR during the nighttime should be at least 2.5 for a delamination to become detectable (the dashed line with a slope of 2.5).

For delaminations with depths equal to or smaller than 4.0 cm (B-D1 to B-D12) on the rainy day (day 2), delamination could hardly be identified.

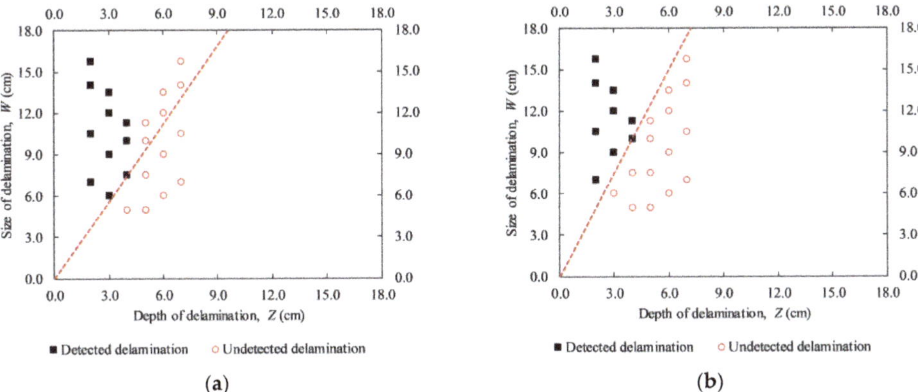

Figure 18. Detected and undetected delaminations on the sunny day (day 1 and day 3): (**a**) daytime; (**b**) nighttime.

5.4. Effective Time for Delamination Detection

In passive IRT, the favorable time in capturing thermal images is a very important factor in order to be applied effectively to real structure inspections. On the rainy day (day 2), almost all delaminations could not be detected owing to the small temperature differences. Therefore, the concrete bridge deck inspection should be carried out on a sunny day.

The detected time of delaminations from the back face on days 1 and 3 is graphically depicted in Figure 19. The blue filled-in square, blue filled-in circle, red unfilled-in square, and red unfilled-in circle represent the first time during daytime, first time during nighttime, last time during daytime, and last time during nighttime when the delaminations can be detected respectively. Under the effect of the heating cycle during the daytime and cooling cycle during the nighttime, there are two ranges of detectable time when the defects with depths equal to or smaller than 4.0 cm may be detected.

Under the effect of the heating cycle (daytime): The structure inspection should be conducted from 10:00 until 16:00 (6 h) to detect delaminations with WTDRs equal to or higher than 1.9 as shown by the green rectangular in Figure 19. If a delamination with a WTDR of 1.9 is eliminated, the feasible time is from 9:00 to 16:00 (7 h) to identify delamination with the WTDRs equal to or higher than 2.0. However, the inspection of the structure should be carried out from 10:00 to 15:00 (5 h) such that significantly high temperature difference can be obtained, and a clearer observation of delamination can be achieved as shown in Figures 11 and 15.

Under the effect of the cooling cycle (nighttime): If the structure is inspected from 19:30 to 3:30 (8 h), delaminations with the WTDR equal to or higher than 2.8 can be identified. However, in the case of delaminations with WTDRs from 2.5 to 2.8, the suitable collection time is shortened as the period from 19:30 to 21:30 (2 h). In addition, the temperature difference of delaminations achieves a significantly high value from 19:30 to around 2:00 (6.5 h), the structure inspection therefore should be conducted during this time to achieve good results.

In fact, the detected time of the delaminations lasts significantly during both daytime and nighttime. Therefore, the concrete bridge deck inspection can be conducted effectively throughout the day. However, during the daytime, the WTDRs of delaminations that can be detected using the passive IRT technique are smaller in comparison with the nighttime. Thus, daytime is a better choice for concrete bridge inspection than nighttime even if the delaminations with small WTDRs (≤ 2.5) are considered in the inspection. In this case, the limitation of WTDR is approximately 1.9. Conversely, if the inspection focuses only on delamination with WTDRs of 2.8 or higher, the nighttime offers a more efficient selection because of its longer detected time compared to daytime.

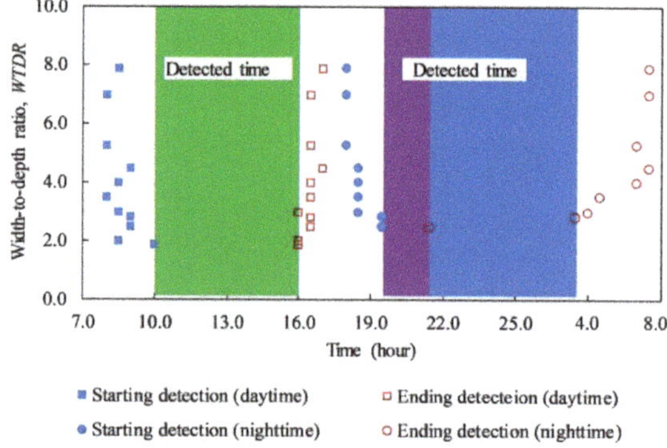

Figure 19. Feasible time on a sunny day to detect delamination in concrete bridge deck.

6. Detectability of Delaminations Using UAV-IRC

In the case of using the UAV-IRC, the detectability of delamination is also evaluated by using the absolute contrast measurement that is calculated from Equation (3). The temperature difference is the amplitude of the absolute contrast.

Figure 20 shows the thermal image at 11:30 by using UAV-IRC. The temperature difference of delaminations (B-D1 to B-D11) ranges from 5.73 to 0.54 °C that is higher than 0.5 °C, while it is 0.4 °C for the delamination B-D12. The largest WTDR delamination obtains the highest temperature difference (5.73 °C) whereas the smallest temperature difference is achieved in the case of the least WTDR delamination (0.4 °C). Therefore, all the given delaminations can be seen on the thermal images captured by using UAV-IRC except for delamination B-D12. Hence, it can be concluded that the UAV-IRC might detect delaminations at the depth equal to or lower than 4 cm and with WTDRs not smaller than 1.9 during daytime on the sunny day. However, it should be noted that delamination B-D11 (WTDR = 1.9) is still quite difficult to be recognized as shown in Figure 20.

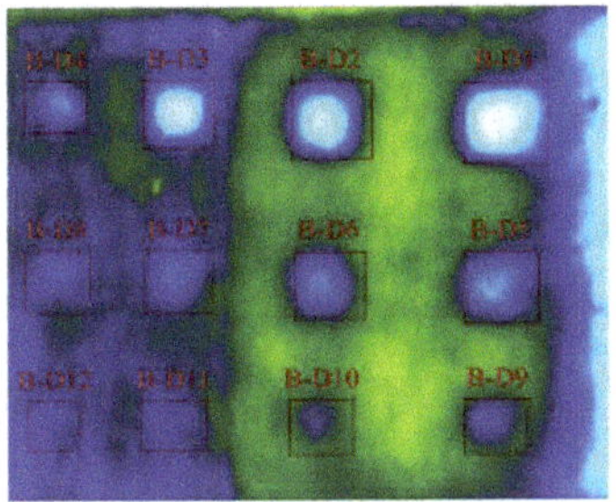

Figure 20. Thermal image of specimen surface using UAV-IRC at 11:30.

Figure 21 shows the absolute contrast of delaminations having different WTDRs on day 5. Particularly, Figure 21a shows the case of defects at the same depth of 2 cm (B-D1, B-D2, B-D3, and B-D4) while Figure 21b depicts defects at depths of 2 (B-D4) and 3 cm (B-D5 and B-D6). The WTDRs of delaminations B-D1, B-D2, B-D3, and B-D4 are 7.9, 7.0, 5.25, and 3.5, respectively, whereas the WTDR is 4.5 and 4.0 corresponding to delaminations B-D5 and B-D6. The same tendency can be observed in comparison with the results obtained from the H-IRC that a larger delamination WTDR achieves higher temperature difference than a smaller one, although it is located at the same depth or further from the surface of the structure. For example, in the case of different WTDRs at the same depth of 2 cm, at 10:30, the temperature differences are 5.22, 3.78, 3.30, and 1.55 °C corresponding to delaminations B-D1, B-D2, B-D3, and B-D4. On considering delaminations with different WTDRs and depths, delamination B-D5, B-D6, and B-D4 obtained the temperature difference of 2.70, 2.40, and 1.75 °C at 10:30, respectively. Consequently, delamination B-D5 and B-D6 can appear more clearly than B-D4.

The interchange period occurred in the period from 06:30 to 08:30 on day 5 that is very similar to the results from experiments using H-IRC on day 1. It should be noted that the experiments on both days 1 and 5 focus on the back face and the overall weather conditions are sunny during these days. There is only 30 min of the delay time of the interchange period between day 5 and day 1. It might be caused by the difference between the interval in capturing thermal images that are 30 min and 1 h in the case of day 1 and day 5, respectively.

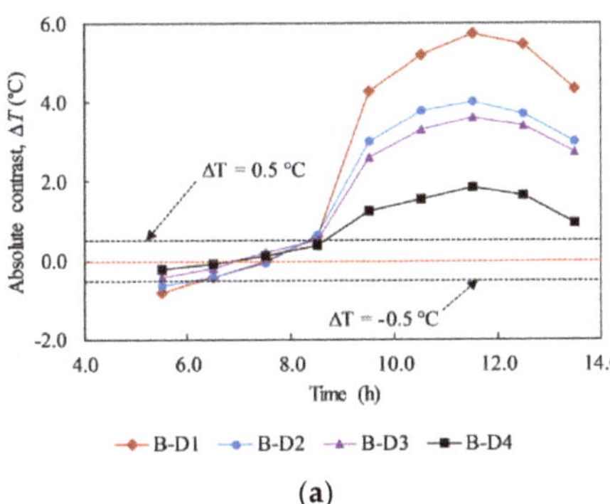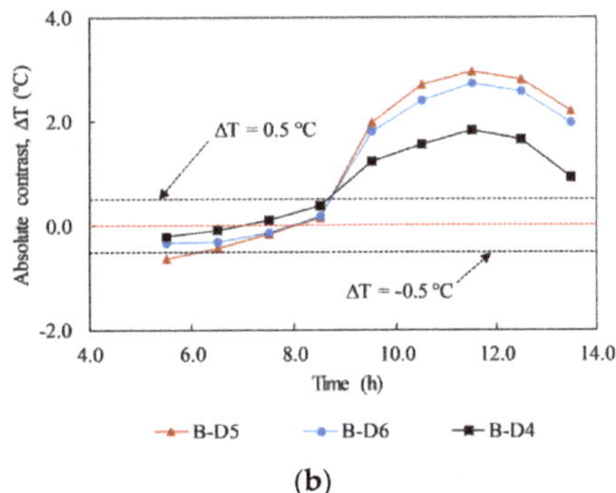

Figure 21. Absolute contrast of delaminations with different WTDRs on day 5 from UAV-IRC: (**a**) at same depth of 2 cm; (**b**) at different depths of 2 and 3 cm.

Figure 22 shows the absolute contrast of delamination B-D2 (depth of 2 cm) and B-D5 (depth of 3 cm) based on the results of experiments using UAV-IRC and H-IRC on day 5. In Figure 22a,b, the blue line represents H-IRC while the red line stands for UAV-IRC. At early hours in the morning, the variance of temperature difference between UAV-IRC and H-IRC, known as DIF is much smaller than at other times because the surface temperature in the early morning is more stable than at noon as well as the afternoon, and that time is nearby the interchange period. For example, the DIF is 0.05 °C at 5:30 while it is 0.29 °C at 12:30 in the case of delamination B-D2. Furthermore, at 11:30, both H-IRC and UAV-IRC obtained the maximum temperature differences. These maximum values are 3.48 °C and 2.95 °C for B-D5 corresponding to the outcome from H-IRC and UAV-IRC. In addition, it is shown that larger the delamination depth, higher the DIF obtained except for the observation at 9:30 am. For instance, DIFs are 0.29 °C and 0.44 °C at 12:30 corresponding to delamination B-D2 and B-D5.

Moreover, as shown in Figure 22a,b, both lines show the same tendency, but the red line is located more closely to zero axis compared to the blue line. This can be explained by the effect of the wind from the fans of UAV and the smaller thermal sensitivity of IR camera mounted on the UAV in comparison with H-IRC leading to the decrease in absolute contrast in case of UAV-IRC. However, the gap between the two lines is not significant, which implies that there is a good agreement between the results obtained from experiments using UAV-IRC and H-IRC. Consequently, it can be stated that under the given environmental conditions, using UAV-IRC can give appropriate results. The effective application of UAV-IRC in the defect detection of concrete bridge decks is confirmed.

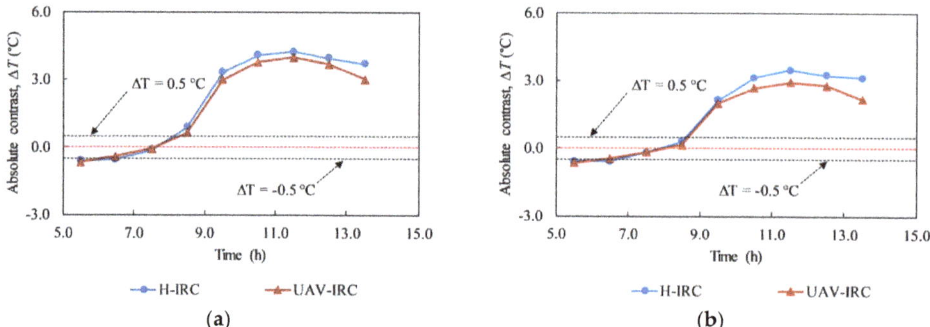

Figure 22. Comparison between the absolute contrast captured by H-IRC and UAV-IRC on a sunny day (day 5): (**a**) delamination B-D2; (**b**) delamination B-D5.

7. Conclusions

In this study, an effort was made to consider the possibility of simultaneously using the results from the handheld IR camera (H-IRC) and the IR camera hanged on a UAV (UAV-IRC). From the results presented in this work under given experimental conditions and processing method, the following conclusions are obtained:

For delaminations at the depths of 4.0 cm or lesser, the larger the WTDRs of the delaminations, the higher is the temperature difference produced on the thermal image, which implies that there is a strong correlation between the size and detected depth of delamination.

The minimum WTDR of a delamination that can be detected by using H-IRC was determined in this study. Delaminations with WTDRs of 1.9 or larger located at depths of 4.0 cm or lesser from the structure surface may be identified during the daytime on a sunny day, whereas during the nighttime of a sunny day, only delaminations with WTDRs of 2.5 or larger can be discovered. However, if the delamination depth is 5.0 cm or higher and its WTDR is under 2.25, it may not be noticeable on the thermal images.

Passive IRT must be applied to scan the concrete bridge deck on a sunny day because adequate heat energy is provided that helps to extract accurately the delamination occurrence. Generally, the optimal time to inspect the concrete bridge deck is from 10:00 to 15:00 and from 19:30 to around 2:00, respectively, during the daytime and nighttime on a sunny day.

Although the absolute contrast produced in the case of UAV-IRC is slightly smaller than H-IRC, UAV-IRC also can be employed to detect delaminations with the depths equal to or less than 4.0 cm and WTDRs of 1.9 or larger during the daytime of a sunny day. Thus, it is proven that UAV-IRC in the passive IRT technique is a feasible solution that can be utilized separately or simultaneously with H-IRC to discover the delamination with a shorter inspection time compared to H-IRC in the concrete bridge deck.

Author Contributions: Conceptualization, V.H.M., Q.H.T. and J.H.; methodology, V.H.M., Q.H.T. and J.H.; software, V.H.M., Q.H.T. and N.S.D.; validation, V.H.M., Q.H.T., J.H. and D.H.; formal analysis, V.H.M.; resources, V.H.M., Q.H.T., N.S.D. and C.K.; data curation, V.H.M., N.S.D. and J.H.; writing-original draft preparation, V.H.M., N.S.D. and C.K.; writing-review & editing, J.H., V.H.M. and Q.H.T.; supervision, J.H. and Q.H.T.; project administration, D.H.; funding acquisition, D.H.

Funding: This research was funded by [Ministry of Land, Infrastructure and Transport, Korea] grant number [19RDRP-B076564-06] and [Ministry of Oceans and Fisheries, Korea] grant number [20180323].

Acknowledgments: This research was supported by a grant (19RDRP-B076564-06) from Regional Development Research Program funded by Ministry of Land, Infrastructure and Transport of Korean government and is a part of the project (20180323) titled 'Development of Design Technology for Safe Harbor from Disasters', funded by the Ministry of Oceans and Fisheries, Korea.

Conflicts of Interest: The authors declare no conflict of interest.

References

1. Hiasa, S.; Birgul, R.; Matsumoto, M.; Necati Catbas, F. Experimental and numerical studies for suitable infrared thermography implementation on concrete bridge decks. *Meas. J. Int. Meas. Confed.* **2018**, *121*, 144–159. [CrossRef]

2. Vaghefi, K. Infrared Thermography Enhancements for Concrete Bridge Evaluation. Ph.D. Thesis, Michigan Technological Univesity, Houghton, MI, USA, 2013.

3. Rodríguez, J.; Andrade, C. *CONTECVET A Validated Users Manual for Assessing the Residual Service Life of Concrete Structures—Manual for Assessing Corrosion-Affected Concrete Structures*; EC Innovation Programme IN30902I; GEOCISA and Torroja Institute: Madrid, Spain, 2002.

4. Milovanović, B.; Banjad Pečur, I. Review of Active IR Thermography for Detection and Characterization of Defects in Reinforced Concrete. *J. Imaging* **2016**, *2*, 11. [CrossRef]

5. Sun, H.; Pashoutani, S.; Zhu, J. Nondestructive evaluation of concrete bridge decks with automated acoustic scanning system and ground penetrating radar. *Sensors* **2018**, *18*, 1955. [CrossRef] [PubMed]

6. Sultan, A.A.; Washer, G.A. Comparison of Two Nondestructive Evaluation Technologies for the Condition Assessment of Bridge Decks. *Transp. Res. Rec.* **2018**, *2672*, 113–122. [CrossRef]

7. Sirieix, C.; Lataste, J.F.; Breysse, D.; Naar, S.; Dérobert, X. Comparison of nondestructive testing: Infrared thermography, electrical resisitivity and capacity methods for assessing a reinforced concrete structure. *J. Build. Apprais.* **2007**, *3*, 77–88. [CrossRef]

8. Oh, T.; Kee, S.H.; Arndt, R.W.; Popovics, J.S.; Zhu, J. Comparison of NDT methods for assessment of a concrete bridge deck. *J. Eng. Mech.* **2013**, *139*, 305–314. [CrossRef]

9. Li, M.; Anderson, N.; Sneed, L.; Maerz, N. Application of ultrasonic surface wave techniques for concrete bridge deck condition assessment. *J. Appl. Geophys.* **2016**, *126*, 148–157. [CrossRef]

10. Fan, G.; Zhang, H.; Zhu, W.; Zhang, H.; Chai, X. Numerical and experimental research on identifying a delamination in ballastless slab track. *Materials* **2019**, *12*, 1788. [CrossRef]

11. Diamanti, N.; Annan, A.P.; Redman, J.D. Concrete Bridge Deck Deterioration Assessment Using Ground Penetrating Radar (GPR). *J. Environ. Eng. Geophys.* **2017**, *22*, 121–132. [CrossRef]

12. Gucunski, N.; Imani, A.; Romero, F.; Nazarian, S.; Yuan, D.; Wiggenhauser, H.; Shokouhi, P.; Taffe, A.; Kutrubes, D. *Nondestructive Testing to Identify Concrete Bridge Deck Deterioration*; SHRP 2 Report, S2-R06A-RR-1; Transportation Research Board: Washington, DC, USA, 2013; ISBN 978-0-309-12933-6.

13. Yehia, S.; Abudayyeh, O.; Nabulsi, S.; Abdelqader, I. Detection of Common Defects in Concrete Bridge Decks Using Nondestructive Evaluation Techniques. *J. Bridge Eng.* **2007**, *12*, 215–225. [CrossRef]

14. Usamentiaga, R.; Venegas, P.; Guerediaga, J.; Vega, L.; Molleda, J.; Bulnes, F. Infrared Thermography for Temperature Measurement and Non-Destructive Testing. *Sensors* **2014**, *14*, 12305–12348. [CrossRef] [PubMed]

15. Abdel-Qader, I.; Yohali, S.; Abudayyeh, O.; Yehia, S. Segmentation of thermal images for non-destructive evaluation of bridge decks. *NDT E Int.* **2008**, *41*, 395–405. [CrossRef]

16. Tran, Q.H.; Han, D.; Kang, C.; Haldar, A.; Huh, J. Effects of Ambient Temperature and Relative Humidity on Subsurface Defect Detection in Concrete Structures by Active Thermal Imaging. *Sensors* **2017**, *17*, 1718. [CrossRef]

17. Maldague, X.P.V. *Nondestructive Evaluation of Materials by Infrared Thermography*; Springer: London, UK, 1993.

18. Tran, Q.H.; Huh, J.; Mac, V.H.; Kang, C.; Han, D. Effects of rebars on the detectability of subsurface defects in concrete bridges using square pulse thermography. *NDT E Int.* **2018**, *100*, 92–100. [CrossRef]

19. Cotič, P.; Kolarič, D.; Bosiljkov, V.B.; Bosiljkov, V.; Jagličić, Z. Determination of the applicability and limits of void and delamination detection in concrete structures using infrared thermography. *NDT E Int.* **2015**, *74*, 87–93. [CrossRef]

20. Tran, Q.H.; Huh, J.; Kang, C.; Lee, B.Y.; Kim, I.T.; Ahn, J.H. Detectability of Subsurface Defects with Different Width-to-Depth Ratios in Concrete Structures Using Pulsed Thermography. *J. Nondestruct. Eval.* **2018**, *37*, 32. [CrossRef]

21. Maierhofer, C.; Brink, A.; Röllig, M.; Wiggenhauser, H. Transient thermography for structural investigation of concrete and composites in the near surface region. *Infrared Phys. Technol.* **2002**, *43*, 271–278. [CrossRef]

22. Maierhofer, C.; Arndt, R.; Röllig, M. Influence of concrete properties on the detection of voids with impulse-thermography. *Infrared Phys. Technol.* **2007**, *49*, 213–217. [CrossRef]

23. Maierhofer, C.; Arndt, R.; Röllig, M.; Rieck, C.; Walther, A.; Scheel, H.; Hillemeier, B. Application of impulse-thermography for non-destructive assessment of concrete structures. *Cem. Concr. Compos.* **2006**, *28*, 393–401. [CrossRef]

24. Huh, J.; Mac, V.; Tran, Q.; Lee, K.-Y.; Lee, J.-I.; Kang, C. Detectability of Delamination in Concrete Structure Using Active Infrared Thermography in Terms of Signal-to-Noise Ratio. *Appl. Sci.* **2018**, *8*, 1986. [CrossRef]

25. Weritz, F.; Arndt, R.; Röllig, M.; Maierhofer, C.; Wiggenhauser, H. Investigation of concrete structures with pulse phase thermography. *Mater. Struct.* **2005**, *38*, 843–849. [CrossRef]

26. Huh, J.; Tran, Q.H.; Lee, J.-H.; Han, D.; Ahn, J.-H.; Yim, S. Experimental Study on Detection of Deterioration in Concrete Using Infrared Thermography Technique. *Adv. Mater. Sci. Eng.* **2016**, *2016*, 53856. [CrossRef]

27. Cheng, C.C.; Cheng, T.M.; Chiang, C.H. Defect detection of concrete structures using both infrared thermography and elastic waves. *Autom. Constr.* **2008**, *18*, 87–92. [CrossRef]

28. Maldague, X. *Theory and Practice of Infrared Thermography for Nondestructive Testing*; John Wiley & Sons, Inc.: New York, NY, USA, 2001; p. 684.

29. Washer, G.; Fenwick, R.; Bolleni, N. Effects of Solar Loading on Infrared Imaging of Subsurface Features in Concrete. *J. Bridge Eng.* **2010**, *15*, 384–390. [CrossRef]

30. Washer, G.A.; Fenwick, R.G.; Bolleni, N.K. *Development of Hand-Held Thermographic Inspection Technologies*; MoDOT Project Number RI06-038; University of Missouri: Columbia, MO, USA, 2009; p. 120.

31. Kashif Ur Rehman, S.; Ibrahim, Z.; Memon, S.A.; Jameel, M. Nondestructive test methods for concrete bridges: A review. *Constr. Build. Mater.* **2016**, *107*, 58–86. [CrossRef]

32. American Association of State Highway and Transportation Officials (AASHTO). *Manual for Bridge Element Inspection*; AASHTO: Washington, DC, USA, 2015.

33. Washer, G.; Fenwick, R.; Bolleni, N.; Harper, J. Effects of environmental variables on infrared imaging of subsurface features of concrete bridges. *Transp. Res. Rec.* **2009**, *2018*, 107–114. [CrossRef]

34. Kee, S.-H.; Oh, T.; Popovics, J.S.; Arndt, R.W.; Zhu, J. Nondestructive Bridge Deck Testing with Air-Coupled Impact-Echo and Infrared Thermography. *J. Bridge Eng.* **2012**, *17*, 928–939. [CrossRef]

35. Hiasa, S.; Birgul, R.; Catbas, F.N. Infrared thermography for civil structural assessment: Demonstrations with laboratory and field studies. *J. Civ. Struct. Health Monit.* **2016**, *6*, 619–636. [CrossRef]

36. Sultan, A.A.; Washer, G. A pixel-by-pixel reliability analysis of infrared thermography (IRT) for the detection of subsurface delamination. *NDT E Int.* **2017**, *92*, 177–186. [CrossRef]

37. Farrag, S.; Yehia, S.; Qaddoumi, N. Investigation of Mix-Variation Effect on Defect-Detection Ability Using Infrared Thermography as a Nondestructive Evaluation Technique. *J. Bridge Eng.* **2016**, *21*, 04015055. [CrossRef]

38. Vaghefi, K.; de Melo e Silva, H.A.; Harris, D.K.; Ahlborn, T.M. Application of thermal IR imagery for concrete bridge inspection. In Proceedings of the PCI National Bridge Conference, Salt Lake City, UT, USA, 22–26 October 2011.

39. Rocha, J.H.A.; Póvoas, Y.V.; Santos, C.F. Detection of Delaminations in Sunlight-Unexposed Concrete Elements of Bridges Using Infrared Thermography. *J. Nondestruct. Eval.* **2019**, *38*, 1–12. [CrossRef]

40. Omar, T.; Nehdi, M.L. Remote sensing of concrete bridge decks using unmanned aerial vehicle infrared thermography. *Autom. Constr.* **2017**, *83*, 360–371. [CrossRef]

41. Khan, F.; Ellenberg, A.; Mazzotti, M.; Kontsos, A.; Moon, F.; Pradhan, A.; Bartoli, I. Investigation on bridge assessment using unmanned aerial systems. In Proceedings of the Structures Congress 2015, Portland, OR, USA, 23–25 April 2015; pp. 404–413. [CrossRef]

42. Brooks, C.N.; Dobson, R.J.; Dean, D.B.; Banach, D.; Oommen, T.; Havens, T.C.; Ahlborn, T.M.; Cook, S.J.; Clover, A. Evaluating the use of unmanned aerial vehicles for transportation purposes: A michigan demonstration. In Proceedings of the 21st World Congress on Intelligent Transport Systems, ITSWC 2014: Reinventing Transportation in Our Connected World, Detroit, MI, USA, 7–11 September 2014.

43. Gillins, M.N.; Gillins, D.T.; Parrish, C. Cost-Effective Bridge Safety Inspections Using Unmanned Aircraft Systems (UAS). In Proceedings of the Geotechnical and Structural Engineering Congress 2016, Phoenix, AZ, USA, 14–17 February 2016; pp. 1931–1940.

44. Vasterling, M.; Meyer, U. Challenges and opportunities for UAV-borne thermal imaging. *Remote Sens. Digit. Image Process.* **2013**, *17*, 69–92. [CrossRef]

45. Kretzmann, J. Evaluating the Industrial Application of Non-Destructive Inspection of Composites Using Transient Thermography. Master's Thesis, Stellenbosch University, Stellenbosc, South Africa, 2016.

46. Lamond, J.; Pielert, J. *Significance of Tests and Properties of Concrete and Concrete-Making Materials*; ASTM International: West Conshohocken, PA, USA, 2006. [CrossRef]

47. FLIR Systems Inc. *The Ultimate Infrared Handbook for R & D Professionals*; Shatin, N.T., Ed.; FLIR Systems Inc.: Wilsonville, OR, USA, 2012.

48. Pollock, D.G.; Dupuis, K.J.; Lacour, B.; Olsen, K.R. *Detection of Voids in Prestressed Concrete Bridges Using Thermal Imaging and Ground-Penetrating Radar*; Project DTFH61-05-C-00008; Washington State University: Pullman, WA, USA, 2008.

49. FLIR System Inc. *SC660 Catalog, Technical Data of FLIR SC660 Infrared Camera*; FLIR System Inc.: Hong Kong, China, 2014.

50. FLIR Systems Inc. *Featuring DJI Matrice 200 Series V2 and Zenmuse XT2 Thermal Imagers*; FLIR System Inc.: Wilsonville, OR, USA, 2018.

51. ASTM. *ASTM D4788-03, Standard Test Method for Detecting Delaminations in Bridge Decks Using Infrared Thermography*; ASTM International: West Conshohocken, PA, USA, 2013. [CrossRef]

52. Ahlborn, T.M.; Brooks, C.N. *Evaluation of Bridge Decks using Non-Destructive Evaluation (NDE) at Near Highway Speeds for Effective Asset Management*; RC-1617; Michigan Technological University: Houghton, MI, USA, 2015.

53. Brown, J.R.; Hamilton, H.R. Heating methods and detection limits for infrared thermography inspection of fiber-reinforced polymer composites. *ACI Mater. J.* **2007**, *104*, 481–490. [CrossRef]

Article

Determination of Thermophysical Parameters Involved in The Numerical Model to Predict the Temperature Field of Cast-In-Place Concrete Bridge Deck

Aleksandra Kuryłowicz-Cudowska

Department of Mechanics of Materials and Structures, Faculty of Civil and Environmental Engineering, Gdansk University of Technology, Narutowicza 11/12, 80-233 Gdansk, Poland; aleksandra.kurylowicz-cudowska@pg.edu.pl; Tel.: +48-58-348-6149

Received: 14 August 2019; Accepted: 11 September 2019; Published: 22 September 2019

Abstract: The paper deals with a concept of a practical computation method to simulate the temperature distribution in an extradosed bridge deck. The main goal of the study is to develop a feasible model of hardening of concrete consistent with in-situ measurement capabilities. The presented investigations include laboratory tests of high performance concrete, measurements of temperature evolution in the bridge deck and above all, numerical simulations of temperature field in a concrete box bridge girder. A thermal conductivity equation in the author's program, using finite difference method has been solved. New approach for identification of the model parameters and boundary conditions (heat transfer coefficients) has been proposed. The numerical results are verified by means of a wide set of experimental tests carried out on three stages of the extradosed bridge studies. A high agreement between the concrete temperature distribution in the time and space domain was obtained. The temperature history of concrete hardening, supplemented with maturity method equations, made it possible to estimate an early-age compressive strength of the cast-in-place concrete. The proposed solution could be applied in a Structural Health Monitoring system for concrete objects.

Keywords: cast-in-place concrete; temperature; boundary conditions; on-line monitoring; numerical simulations; strength; extradosed bridge; SHM system

---

1. Introduction

Attention to the quality, durability and rapid construction process is one of the major challenges for the concrete industry. Prediction of the temperature field in concrete objects allows us to use the proper care of young concrete as well as to estimate development of compressive strength. In the case of massive structures, it is particularly important to not exceed the temperature difference between the interior and the concrete surface. Too high temperature gradient generates an increase in thermal-shrinkage stresses, what could be the reason of exceeding the tensile strength and as a consequence cracking occurrence [1]. Knowledge about the temperature distribution is important not only for the type of construction discussed above, but also for medium-weight and thin-walled concrete elements, for which time is a major factor to achieve the required strength to start prestressing or formwork removal. The temperature field in concrete combined with maturity method equations gives the possibility to estimate concrete strength changes over time and space [2,3].

In recent years, there has been growing interest in rationalization of the building schedule. Supporting the process of designing, curing and monitoring concrete features with the use of computer technologies, provide a lot of benefits, especially for contractors [3,4]. ElSafty et al. [5] developed the complex tool (The Deck Cracking Spreadsheet) to predict the early-age cracking of concrete bridge decks. The thermal aspect of concrete hardening is described there using time-dependent parameters. The concrete heat of hydration is calculated based on the concrete mixture proportions and the constituent material properties. The detailed input data in this tool can improve temperature forecast. However, the developed spreadsheet is limited to the decks with thickness about 20 cm and there are no analysis comparing the measured and predicted temperature evolution of concrete bridge deck [5].

The key aspect in the prediction of temperature distribution of cast-in-place (CIP) concrete is the choice of the numerical model, which should be consistent with the measurement capabilities. The multi-field mathematical models of young concrete [6,7] are well documented in a literature, but very often an application of the theoretical formulations are adopted only at the laboratory level. The major problem is that there are many constants that cannot be determined under field conditions. Therefore, the main purpose of this work is the determination of thermophysical model parameters of CIP concrete for numerical modeling of concrete temperature distribution in the bridge deck, as a part of the service offered by the Structural Health Monitoring (SHM) system.

The well-defined model parameters and initial-boundary conditions corresponding to the real environmental conditions at the construction site are the basis of reliable predictions of the in-place concrete temperature distribution. In this paper, two thermo-chemical models of hardening of concrete [8,9] were used. The heat flow equations were solved with the finite difference method in Matlab environment. The preliminary calculations of temperature evolution in bridge slab were published in the works [10,11], wherein the selection of model parameters based on the literature. A novel element of the paper is a complex proposition to identify model parameters and boundary conditions i.e., the heat transfer coefficients of free and protected concrete surfaces. Numerical simulations were confirmed by concrete temperature data measured on the construction site of extradosed bridge located in Poland. An own monitoring system was applied to register the advancement of hydration reaction of high performance concrete, C 60/75 class. Based on the temperature field and augmented maturity method, the dates of prestressing of individual sections of bridge deck, was also determined.

2. Multifield Modeling of Early-Age Concrete

The plurality and unrecognizability of phenomena occurring during concrete curing, leads to extensive experimental research and the development of a model, in which concrete is defined as a material with non-stationary properties. Mathematical models, available in the literature, takes into account four aspects of concrete hardening. It can be said that in the early age concrete there is a strong coupling of thermal, chemical, moisture and mechanical phenomena. In this study complex models presented i.e., in Azenha's [6,12] and Di Luzio-Cusatis's [7] papers were considered. However, having in mind the application of model to on-line monitoring of the structure under the construction process and limited number of measurement parameters, the thermo-chemical models described below, in Sections 2.1 and 2.2, have been chosen.

2.1. Cervera's et al. Model

The process of cement hydration is related to the change of the phase composition of the medium. Phase transitions are accompanied by thermal phenomena associated with the heat release, as a result of hydration reactions of cement components. As a reason of heat conduction, the temperature field appears in every area of the body. In the macroscopic approach, concrete can be treated as an isotropic and homogeneous material with constant conductivity ($\lambda = const.$)

The model proposed by Cervera et al. [8] consists of two coupled equations: Thermal (1) and chemical kinetic equation, expressed as a function of evolution of hydration degree (2):

$$\nabla \cdot \lambda \nabla T + Q_\xi \dot{\xi} = \rho c \dot{T}, \tag{1}$$

$$\dot{\xi} = \widetilde{A}(\xi) \exp\left(-\frac{E_a}{R \cdot T}\right), \tag{2}$$

where the following parameters are responsible for heat transfer: Concrete density ρ, specific heat of concrete c, thermal conductivity λ, material constant Q_ξ (heat of cement hydration in concrete), activation energy E_a and gas constant R. In this model, a normalized internal variable, i.e., a degree of hydration ξ, which evolution allows us to predict the advancement of the hardening process has been introduced. Due to the thermally activated nature of cement hydration, a strong dependence ξ on temperature T is observed, and the Arrhenius type law is responsible for the kinetics of these changes. The normalized chemical affinity $\widetilde{A}(\xi)$ is expressed through the chemical affinity $\overline{A}$ and permeability η [8]. Hence, the rate of hydration $\dot{\xi}$ can be finally obtained:

$$\dot{\xi} = \underbrace{\frac{\kappa}{n_0}\left(\frac{A_0}{\kappa}\frac{1}{\xi_{\max}} + \xi\right)(\xi_{\max} - \xi)\exp\left(-\bar{n}\frac{\xi}{\xi_{\max}}\right)}_{\widetilde{A}(\xi)=\overline{A}\eta\,(normalised\,chemical\,affinity)}\ \underbrace{\exp\left(-\frac{E_a}{R \cdot T}\right)}_{(Arrhenius\,typ\,law)}, \tag{3}$$

where κ/n_0, A_0/κ and $\bar{n}$ are material constants. The value of the final degree of hydration $\xi_{\max}$ depends on the water-cement ratio w/c and can be determined in an approximate way, using the formula suggested by Mills [13]:

$$\xi_{\max} = (1.031 \cdot w/c)/(0.194 + w/c), \tag{4}$$

or Waller [14]:

$$\xi_{\max} = 1 - 1/\exp(3.38(w/c - \delta)), \tag{5}$$

where without fly ash and silica fume additives $\delta = 0$. A comparison of both approaches is presented in Figure 1. Two propositions give similar results for the ratio $w/c < 0.27$. However, the above formulas do not include the type and fineness of cement.

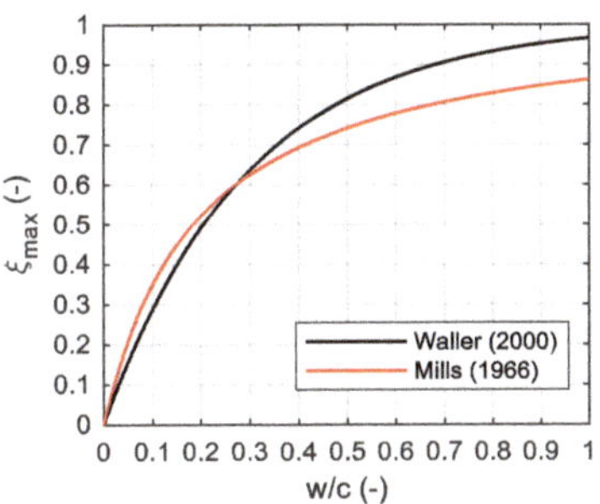

Figure 1. Final hydration degree $\xi_{\max}$ versus w/c ratio.

The normalized chemical affinity is directly measurable during an adiabatic test in which heat is not exchanged with the environment, thus the Equation (1) is simplified to:

$$Q_\xi\dot{\xi} = \rho c\dot{T}, \tag{6}$$

At the end of an adiabatic test $T = T_{\max}$ and $\xi = \xi_{\max}$. Marking the initial temperature of concrete by T_0, it is possible to express a Q_ξ constant as:

$$Q_\xi = \rho c\,(T_{\max} - T_0)/\xi_{\max}, \tag{7}$$

By substituting the Equation (2) and (7) into (6) we get:

$$\widetilde{A}_{test} = \frac{\xi_{\max}\dot{T}}{(T_{\max} - T_0)\exp(-E_a/(R \cdot T))},$$ (8)

Formula No. 8 allows us to calculate the normalized affinity by measuring the temperature rate $\dot{T}$ during the adiabatic calorimetric test. Parameters κ/n_0, A_0/κ and $\bar{n}$ can be determined from the regression analysis by fitting the function $\widetilde{A}(\xi)$ to the results of the experiment $(\widetilde{A}_{test})$ [8].

The described model [8] is suitable for both, ordinary and high performance concrete, and it's capabilities are presented by a wide set of experimental studies (application on the viaduct bridge, located between Denmark and Sweden).

2.2. Martinelli's et al. Model

The model presented in the work of Martinelli et al. [9,15] is also a thermo-chemical model, however, it differs from the previous original proposition to simulate the temperature evolution of concrete in adiabatic and non-adiabatic conditions.

The cement hydration is the reason of concrete hardening, so the concrete maturity can be expressed by the degree of cement hydration, which in general, is defined as the ratio between the amount of hydrated cement at time t to the total amount of cement contained in the mixture. However, this is not evident due to the fact that there is no experiment that shows directly the degree of hydration. Byfors [16] presents five different definitions of the degree of hydration, adapted to specific research possibilities. According to one of them and to Martinelli's work [9], the degree of hydration $\xi(t)$ is expressed by the formula:

$$\xi(t) = Q(t)/Q_{\max},$$ (9)

where $Q(t)$ is the heat of hydration produced at time t, and $Q_{\max}$ means the total amount of heat. The generated heat depends on the temperature, which, in turn, is determined by the size of the sample and the boundary conditions. The analytical relationship between the degree of hydration $\xi(t)$ and the corresponding temperature increase, in the ideal adiabatic conditions $\Delta T_a(t)$ is expressed by the following relationship:

$$\Delta T_a(t) = \frac{C}{\rho c}Q_a(t) = \frac{C}{\rho c}\xi(t)Q_{\max},$$ (10)

where C is the cement content per 1 m^3 volume. Based on the results of adiabatic experiments on concrete samples, the evolution of the hydration heat can be approximated using equation [9,17]:

$$Q_a(t) = Q_{\max}\xi_{\max}e^{-(\frac{a}{t})^b},$$ (11)

where a and b control the shape of the function.

In real structures, setting and hardening of concrete in non-adiabatic conditions take place, which generates non-stationary temperature field. In general, heat transfer through a solid can be described by the Fourier equation 1. The rate of heat source of cement hydration in concrete $(Q_\xi\dot{\xi})$ expressed in equation 1 is described by the formula [9]:

$$q_c(x,t) = C\frac{dQ_c(t)}{dt}.$$ (12)

The current value of the concrete temperature has a significant effect on the rate of heat source q_c, while the temperature depends on the produced heat. Therefore, the feedback effect between q_c and T occurs. The temperature influence on the rate of chemical reactions is given by the relation:

$$k(T) = A_k\exp\left(-\frac{E_a}{R \cdot T}\right).$$ (13)

Arrhenius constant A_k and apparent activation energy E_a can be determined by measuring the reaction rate $k(T)$ versus curing temperature. American standard ASTM C1074 [2] provides procedures for the calculation of mentioned values. The Arrhenius equation 13 is useful to represent the relationship between the actual heat source $q_c(T)$ and the corresponding one $q_a(T_a)$, measured in adiabatic conditions at the same stage of reaction (Figure 2a). If concrete hardening in non-adiabatic conditions has reached the degree of hydration $\xi(t)$ at time t, it is possible to define an equivalent time t_{eq}, in which the same degree of hydration will be achieved under adiabatic conditions ($\xi_a(t_{eq}) = \xi(t)$) [9]. Thus, it will be assumed that $Q_a(t_{eq}) = \xi(t) \cdot Q_{\max}$. An equivalent time t_{eq} may be computed from Equation (11):

$$t_{eq} = a \bigg/ \left(-\ln\left(\frac{\xi(t)}{\xi_{\max}}\right) \right)^{\frac{1}{b}}. \tag{14}$$

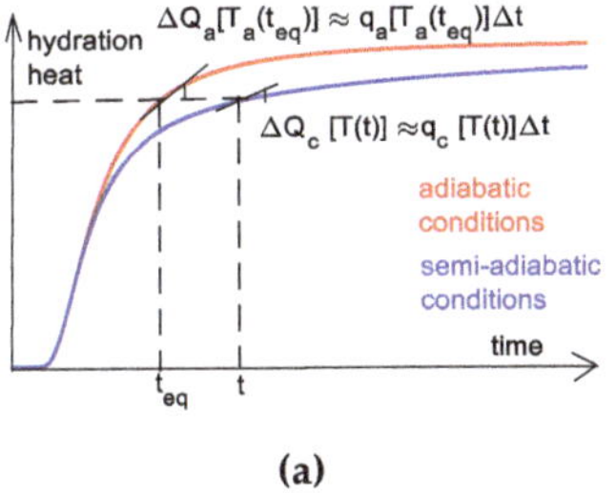

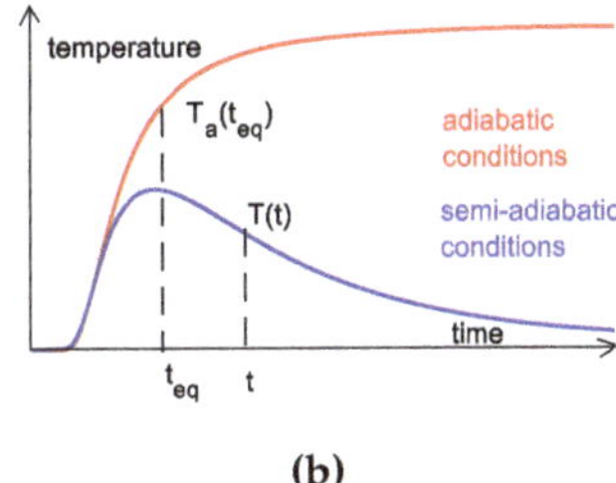

(a)   (b)

Figure 2. Adiabatic and semi-adiabatic conditions: (**a**) Hydration heat and (**b**) temperature development [9].

Although the heat released in non-adiabatic conditions $Q_c(T(t))$ at time t is equal to the heat generated in the adiabatic process $Q_a(T_a(t_{eq}))$ at time t_{eq}, the temperature values $T(t)$ and $T_a(t_{eq})$ are not equal (Figure 2b). Considering the above dependence, the ratio of heat source of cement hydration in concrete, in both indicated conditions, can be expressed using Equation (13):

$$\frac{q_c(T(t))}{q_a(T_a(t_{eq}))} = \frac{e^{-\frac{E_a}{R \cdot T(t)}}}{e^{-\frac{E_a}{R \cdot T_a(t_{eq})}}} = e^{-\frac{E_a}{R}\frac{T_a(t_{eq}) - T(t)}{T_a(t_{eq}) \cdot T(t)}}. \tag{15}$$

Thus, describing the heat source $q_c(T(t))$ using the formula 15, the partial differential equation representing the thermal energy balance in one-dimensional space is equal to:

$$\lambda \cdot \frac{\partial^2 T}{\partial x^2} + q_a(T_a(t_{eq})) \cdot e^{-\frac{E_a}{R}\frac{T_a(t_{eq}) - T(t)}{T_a(t_{eq}) \cdot T(t)}} = \rho c \frac{\partial T}{\partial t}. \tag{16}$$

Equation (16) with the defined initial-boundary conditions allows us to determine the temperature distribution of the concrete in non-adiabatic conditions.

In analyzed paper [9], the validation of the mathematical model was carried out using 150 mm cubic samples and prismatic specimens. The experimental and numerical results showed a high compatibility and the possibility to estimate the compressive strength of concrete.

3. Initial-Boundary Conditions

Assuming that the model describes the temperature development immediately after concrete mixing and placing, the initial condition can be written in the following form:

$$T(x, y, z; t) = T_0(x, y, z), \tag{17}$$

where T_0 is the initial temperature of concrete.

When considering boundary conditions, two mechanisms of heat transfer between a solid and the environment should be distinguished: Convection (natural and forced) and radiation (Figure 3). In natural convection, the heat flow from the concrete surface takes place as a result of the difference between the concrete and ambient temperature. In the process of forced convection, an additional factor accelerates the heat exchange, e.g., the action of wind. The second mechanism of heat transport is radiation. Longwave radiation applies to any solid substance that emits and receives radiation from the environment. Shortwave radiation is a type of energy transmission usually considered in relation to the energy emitted by the sun. It needs to be marked, that when the steel formwork is used, it can absorb large energy from the sun. As a result, the temperature of concrete can be significantly affected by the conduction heat between the steel formwork and concrete.

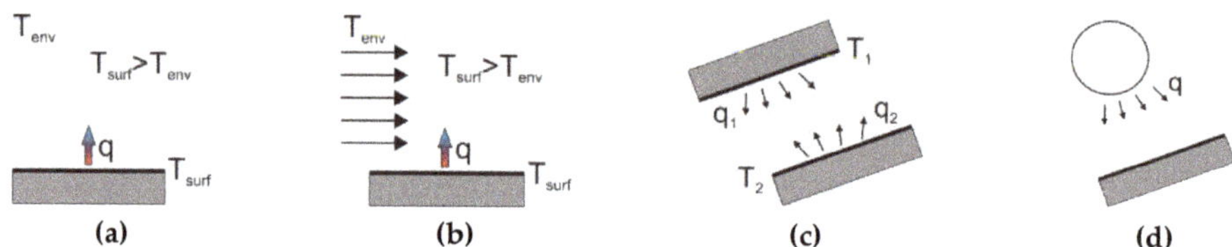

Figure 3. Heat transfer between a solid and the environment: (**a**) Natural convection; (**b**) forced convection; (**c**) longwave radiation and (**d**) shortwave radiation [12].

The equation of the energy balance is supplemented by the Newton's [18] or Stefan-Boltzmann's condition. From the physical point of view, heat exchange between the surface and the environment takes place, both, through convection and radiation. Formally, the heat transfer coefficient is a convective type factor [18]. The heat flow coming to the concrete surface must be absorbed by the surrounding air. Irrespective of the driving force of the air movement, the convective heat transfer can be expressed by Newton's law:

$$q_0 = \alpha \left(T_{surf} - T_{env} \right), \tag{18}$$

where α is the convective heat transfer coefficient and T_{env} and T_{surf} means the ambient and concrete surface temperature. Although, many theoretical attempts have been made to establish prediction equations for the heat transfer coefficient, accurate predictions are available only for very simple geometries and controlled environmental conditions. In irregular cases, such as variable external conditions and the complex geometry of engineering structures, the forecasts were limited to empirical correlations [12]. According to the Institute of Physics of the Cracow University of Technology [19], the heat transfer coefficient on the concrete surface for natural convection α_{nc} is 7 W/(m²·K). Klemczak gives the value of 6 W/(m²·K) [20] and also proposes a formula [21]:

$$\alpha_{nc} = 2.62 \left(T_{surf} - T_{env} \right)^{0.25}. \tag{19}$$

For forced convection, the heat transfer coefficient on the concrete surface α_{fc} depends on the wind speed v_w and can be calculated according to the McAdams [22], Jonasson [23], Branco [24], Ruitz [25], Silveira [26] and Jayamaha [27] (Figure 4). It should be emphasized that suggestions shown in Figure 4 apply to the perfect reproduction of previous tests, in other situations they can only provide approximate solutions [12].

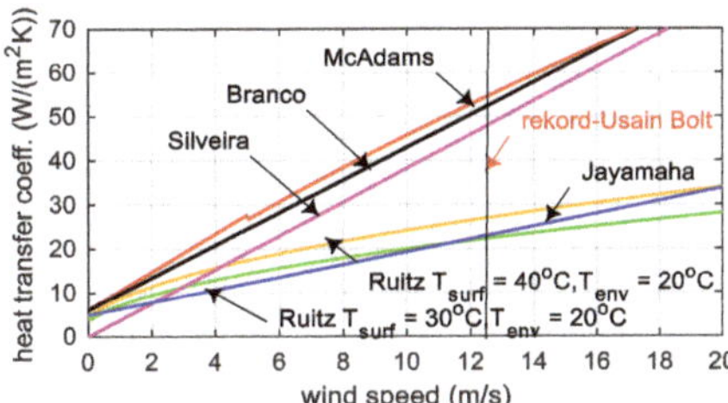

Figure 4. The convective heat transfer coefficient α_{fc} versus wind speed v_w.

On the construction site, an additional concrete surface insulation in the form of the styrofoam layer or foil is often used. For thermally fixed conditions, where the heat flux is perpendicular to the surface, a substitute heat transfer coefficient α_s can be determined taking into account the thickness l_i and conduction coefficient of a particular layer λ_i:

$$\alpha_s = \left(\frac{1}{\alpha} + \sum_{i=1}^{i=n} \frac{l_i}{\lambda_i} \right)^{-1}. \tag{20}$$

Jonasson states that for young concrete the heat transfer coefficient is theoretically correct only for free surface, but the above formula can be applied as some approximation, when the insulation has a lower heat capacity than concrete and the volume of other materials in the formwork (plywood and styrofoam) it is small, i.e., the heat accumulated in the formwork can be omitted compared with the thermal energy in the concrete [18]. In case where there is insulation between the formwork girders, the heat flux is not perpendicular to the surface and equation 20 cannot be used [18].

4. Concept of a Numerical Model Consistent with the In-Situ Measurement Capabilities

The discussed models [8,9] provide very good insight into the concrete hardening process, especially for engineering applications (e.g., SHM system), where very limited data is available. The temperature of concrete and heat released during hardening are the main measurement parameters, around which hypotheses regarding the course of phenomena, mathematical theory and numerical model can be built [28,29]. Relative humidity measurements can be assessed in the surface, however, the registration of changes in concrete moisture content at any depth of the element is complicated, for example due to the construction of this type of sensor. The detection of concrete strains and microcracks could be implemented using ultrasonic methods, more and more often used successfully, but requiring extremely precise evaluation. The implementation of such an extensive system that monitors thermal, moisture and mechanical parameters at the construction site conditions, combined with numerical analysis, becomes an extremely laborious task to rival with traditional methods of thermal and mechanical properties evaluation. However, this is not an obstacle to integrate the possibility of thermal changes registration in a structure with numerical verification and prediction of temperature distribution in the time and space domain.

The thermal aspect of early age concrete should focus on the integral parameters, used by engineers, during controlling the hardening process. These parameters should have a clear technical sense, be easily measurable in the monitoring system and have their equivalents in the mathematical model. The main thermal parameters are the:

- Initial temperature of the mixture (°C),
- Ambient temperature (°C),
- Temperature at the concrete surface (°C),
- Maximum temperature of concrete and time of its occurrence (°C), (h),

- Maximum increase of temperature, as a difference between maximum temperature and initial temperature (°C),
- Maximum rate of temperature increase determined as the quotient of the temperature difference in the period of its increase up to time of this period (°C /h),
- Spatial temperature gradient defined as the quotient of the maximum temperature difference measured at the same time in two selected points,
- Released heat of cement hydration calculated based on the approximate formulas.

In addition, it is necessary to control the wind speed (m/s), the thickness of the element (m), type of formwork and insulation.

In the function of the above parameters, simulations of the temperature field in the concrete bridge deck were performed. For this purpose, two thermo-chemical models presented in Cervera's [8] and Martinelli's [9] papers were used. Although the mentioned models take into account only thermal and chemical aspects, its implementation for real structures is a difficult task. The discussed models simplify the problem of heat transfer omitting the heat movement by migrating moisture and assuming that the heat transport only by conduction is carried out. According to many authors, the influence of moisture diffusion on heat movement in concrete is not significant [30,31].

Prediction of temperature field of CIP concrete requires knowledge of the temperature development of concrete in adiabatic conditions. Due to the difficulties in the availability of such commercial devices, a novel method for determination of adiabatic hydration curve was proposed. To identify this curve, the concrete temperature measurements inside 150 mm cubic samples should be taken. These specimens are curing under isothermal conditions, i.e., in a water bath at constant temperature and under semi-adiabatic conditions in styrofoam containers, especially prepared for this purpose. The suggested approach and equations described in Martinelli's paper [9] allow us to plot the adiabatic hydration curve and then to identify three material parameters: κ/n_0, $\bar{n}$ and A_0/κ, necessary in the mathematical model described by Cervera et al. [8]. Parameters obtained in laboratory conditions are only a reference point for the considered concrete, because they strongly depend on the geometry and atmospheric conditions prevailing at the construction site. Therefore, in the present work, for the analyzed concrete deck of the extradosed bridge, the own functions for κ/n_0 and $\bar{n}$ were defined., and

The initial condition of heat transfer phenomenon does not cause many difficulties, because the initial temperature of the mixture delivered on the building site is usually known. Definitely, more complications generate the determination of boundary conditions, which constantly change under field tests (ambient temperature, wind speed, curing activities and insolation). Mainly two possibilities of heat exchange between the concrete surface and the environment deal with analyzed box bridge girder. The surface of the plate can be free or protected by a shuttering layer. In the second variant, the heat transfer coefficient on the concrete surface protected by the formwork layer ($\alpha_s{}^{form}$) depends on the difference between the initial temperature of the mixture (T_0) and the ambient temperature (T_{env}). In turn, for the convective heat transfer coefficient for the free surface (α_{fc}), the own pattern as a function of wind speed was proposed. The presented concept for estimation thermal changes of the concrete elements of the extradosed bridge was applied.

5. The Description of the Extradosed Bridge

The research works on the longest span in the European extradosed bridge located in Poland were carried out (Figure 5). The aim of the studies was to determine the compressive strength of young concrete embedded in the bridge's span in real time. Laboratory tests included compression tests of mortar and concrete samples and allowed it to develop a maturity–strength relationship for the considered concrete. Field works were focused on temperature measurements of concrete bridge deck and also on compression tests of validation specimens.

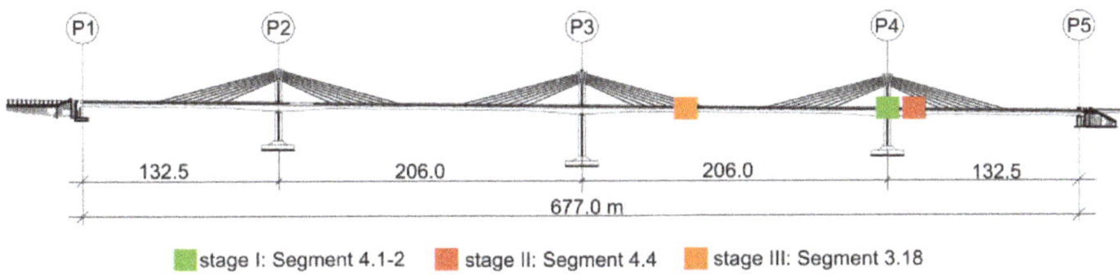

Figure 5. View of the object and monitored zone.

The bridge object (Figure 5) was monitored three times. For every considered section, the formwork of the system was made of 3-SO plywood and H20N timber formwork beams spaced at a distance of about 30 cm. The first stage involved monitoring of the bottom slab of the starting section (No. 4.1–2, Figure 6a). The temperature sensors were mounted in two cross-sections (A–A and B–B) of the bottom plate (points p1–p6). The ambient temperature was also monitored (o1, o2). In two cubes and cylinders, the temperature measurements of the concrete (p7–p10) and the measurements of the water temperature (o3) were carried out. Figure 6b illustrates the location of the measurement points.

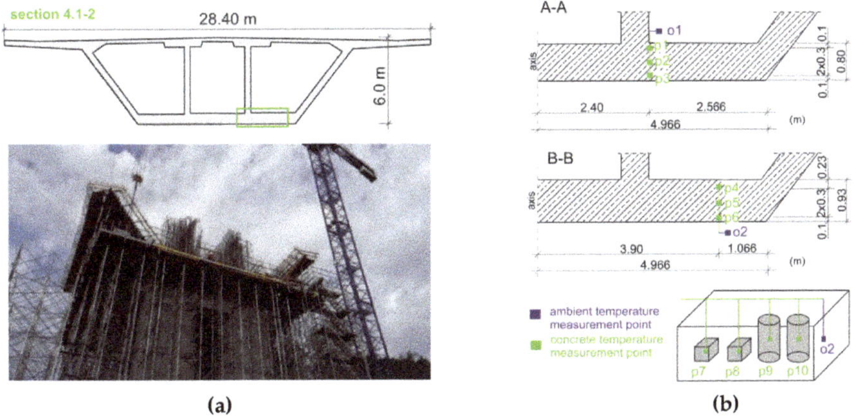

Figure 6. The monitored section No. 4.1–2: (**a**) View and cross section and (**b**) localization of temperature measurement points.

The second stage concerned the overhanging section (No. 4.4) with variable height (Figure 7a). In this case, the concrete temperature in the top slab and the web within the prestressing cables zone was monitored. This segment was concreted in two stages: The bottom plate and webs at first, and a day later, the top plate. The temperature sensors were installed in the inner web (points p1–p7) and in the top slab (p8–p14; Figure 7b). The measurements, with dedicated prefabricated modules consisting of nine sensors (seven for concrete temperature and two for air temperature) were realized. The p7 and p8 sensors were destroyed during the constructions works.

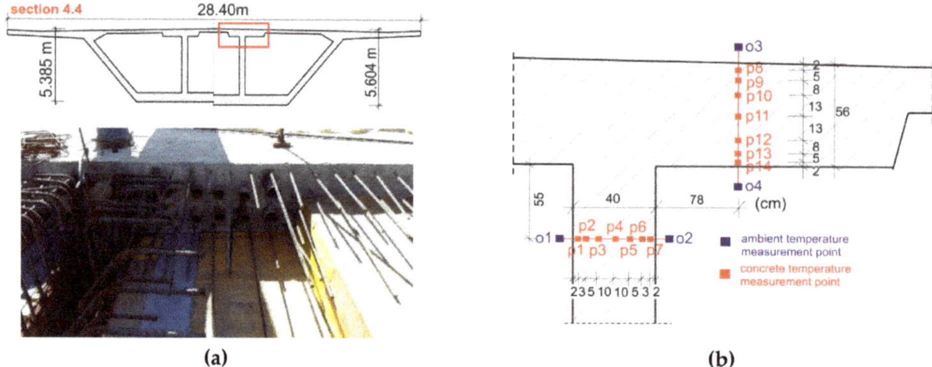

(a)   **(b)**

Figure 7. The monitored section No. 4.4: (**a**) View and cross section and (**b**) localization of temperature measurement points.

The last, third part of the in-situ research, was dedicated to the overhanging section (No. 3.18) with a fixed height (Figure 8). The entire segment was concreted in one cycle. The temperature sensors were installed in the top plate (p1–p5), the inner web (p6–p8) and the bottom slab (p9–p13). Especially for this monitoring a prefabricated measurement module consisted of 13 sensors was designed (Figure 9b). The temperature in the cylindrical and cubic samples (p14, p15) and air temperature (o16–o19) was also registered.

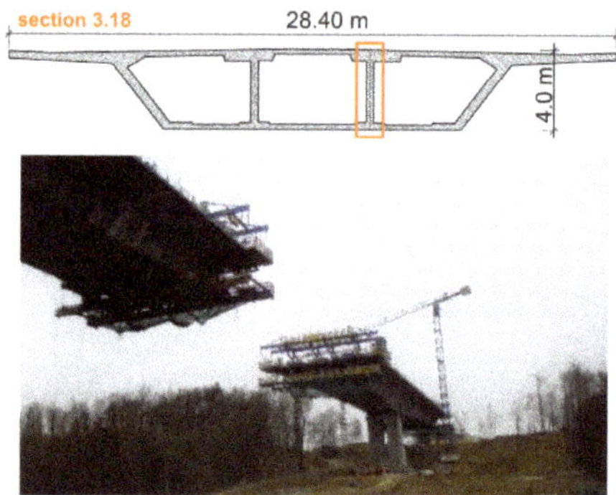

Figure 8. The monitored section No. 3.18.

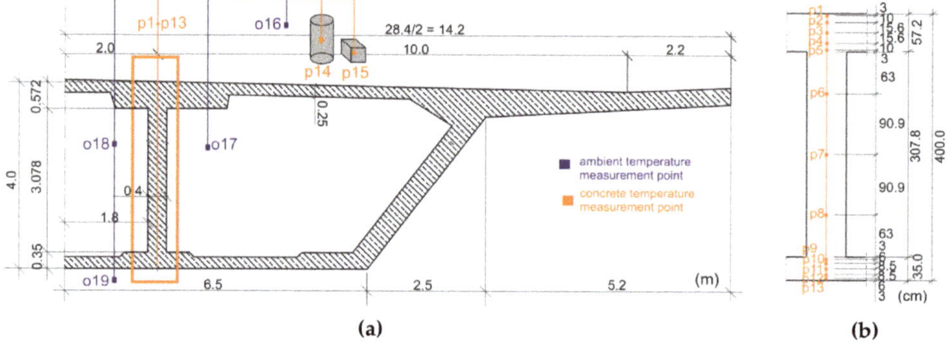

(a)   **(b)**

Figure 9. Localization of measurement points—section No. 3.18: (**a**) Global view and (**b**) detailed view.

Especially for field tests, the own system was designed. The recorder makes it possible to measure concrete temperature at 20 points simultaneously and works on water-resistant, digital, 1-wire sensors type ds18b20, which do not need to be calibrated. 1-wire interface requires a single digital pin, and it is possible to connect multiple ones to the same pin, each one has a unique 64-bit ID burned in at the factory to differentiate them. The ds18b20 sensor provides a temperature measurement in the range from −55 to +125 °C with an accuracy of ± 0.5 °C. The sensor is fairly precise and can give up to 12 bits of precision from the onboard digital-to-analog converter. It is powered from 3.0 V to 5.5 V. Since the sensor is digital, it does not get any signal degradation even over long distances. Measurement points can be assembled using individual sensors or by using a prefabricated measurement strip with a series of sensors. The device can be powered with a battery for about 30 days. Temperature data is transmitted via a GSM modem.

6. Identification of the Model Parameters of High Performance Concrete

6.1. Thermophysical Properties of C 60/75 Concrete

The tests were carried out with Portland cement CEM I 52,5N SR3/NA (Table 1), [32]. The composition of the concrete for 1 m^3 is as follows: Cement, 440 kg; water, 143 kg; sand 0/2, 632 kg; basalt aggregate 2/8, 498 kg; basalt aggregate 8/16, 785 kg and plasticizers, 8.49 kg.

Table 1. Chemical and physical properties of the cement.

Component	CEM I 52.5 N
SO_3	2.6%
Cl^-	0.07%
Na_2O_{eq}	0.5%
Al_2O_3	3.9%
C_3A	1.8%
$C_4AF + 2C_3A$	18.9%
Bulk Density	3.2 g/cm^3
Blain area	3789 g/cm^2
Initial setting time	204 min.
Final setting time	257 min.

The used cement characterizes by a normal (N) initial strength gain, low content of alkali (NA: Na_2O_{eq} < 0.6%) and very high corrosion resistance, especially sulphate (SR3: C_3A < 3%) and chloride [33]. The considered cement is indicated for applications where a high initial and final strength is required, primarily in communication engineering and for the rapid stripping of prefabricated elements. The composition of the aggregate affects on the final strength of concrete and on the modulus of elasticity, which for concrete with basalt aggregate is about 15%–20% greater, than with granite or pebble aggregate. In the analyzed mixture, three types of admixtures: Liquefying, plasticizing and aerating were also used. The composition of the mixture was a complex process, which included selecting ingredients, performing a series of laboratory experiments and technological trials. The designed concrete had to meet the strength requirements of class C 60/75 and reach a minimum 47 MPa after 48 h. Other mix requirements were as follows: Water–cement ratio w/c < 0.34, exposure class XC4 (carbonation of concrete), XD1 (chloride other than sea water) and XF2 (freeze), consistency S4 (slump range 160–210 mm) according to [34], air content 4.0%–6.0% in place, good pumpability and workability.

In order to determine the activation energy divided by the gas constant (E_a/R), tests on 50 mm mortar samples (acc. to ASTM C1074 [2]) were carried out. Three sets of 18 specimens were prepared and placed in a water bath at constant temperatures: $T_{w1} = 5\,°C$, $T_{w2} = 24\,°C$ and $T_{w3} = 35\,°C$. A detailed scope of testing mortar samples is described in [3]. The rate constants k for strength development and curing temperatures were determined on the basis of approximation of strength data (Figure 10a) using non-linear regression according to the equation proposed by Freiesleben Hansen et al. [35]:

$$S = S_u \cdot \exp(-(\tau/t)^\beta), \tag{21}$$

where S means compressive strength at time t, S_u is the ultimate strength, τ is the time constant and β is the shape constant. The values of τ and β were determined using approximation by the least squares method. For temperature T_{w1} : $\tau = 3.658$ day, $\beta = 1.338$ (−), for T_{w2} : $\tau = 1.335$ day, $\beta = 1.549$ (−) and for T_{w3} : $\tau = 0.716$ day, $\beta = 0.715$ (−). The k-values for temperature 5, 24 and 35 °C equaled respectively 0.273, 0.749 and 1.397 (1/day). The results of the presented studies were $E_a/R = 4620$ K (Figure 10b).

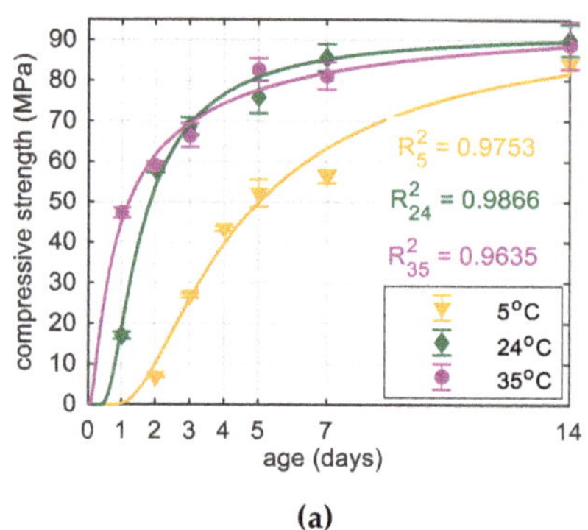

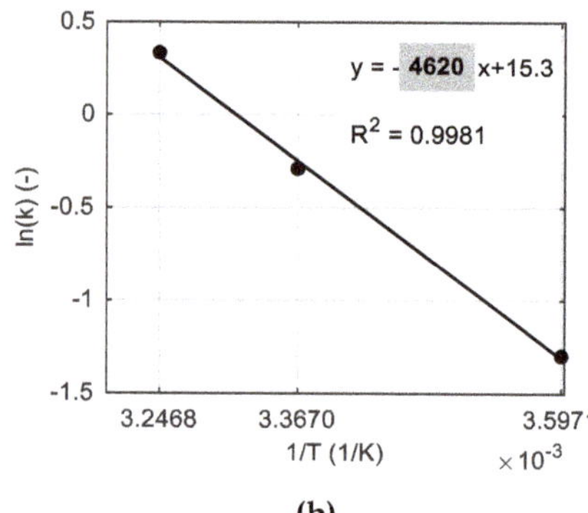

Figure 10. (a) The average compressive strength for mortar cubes and the (b) determination of E_a/R [3].

The cement content per cubic meter of concrete C, density ρ and water–cement ratio w/c based on the composition of the mixture, was assumed. The parameters such as: ξ_{max}, c and λ using literature formulas were calculated.

The final hydration degree ξ_{max} for $w/c = 0.325$ can be obtained from the Mills' (4):

$$\xi_{max} = (1.031 \cdot 0.325)/(0.194 + 0.325) = 0.65, \tag{22}$$

and Waller's (5) formula, where the parameter δ is equal to zero due to the absence of fly ash and silica fume:

$$\xi_{max} = 1 - 1/\exp(3.38(0.325 - 0)) = 0.67. \tag{23}$$

Similar values of ξ_{max} were obtained, however, the calculations according to Mills' proposal [13] were adopted.

The hydration heat of CEM I 52,5 N SR3/NA marked in the semi-adiabatic calorimetry test given by the cement producer after 48 h was about 302 kJ/kg. In Figure 11 the extrapolated value of the total heat $Q_{max} = 330$ kJ/kg was highlighted. For one cubic meter of concrete, the Q_{max} value had to multiply by the cement content. It is also worth remarking that the considered cement was sulphate resistant (SR) and the amount of tricalcium aluminate (C_3A) was restricted to lower than 3%, which affected the relatively low heat of cement hydration.

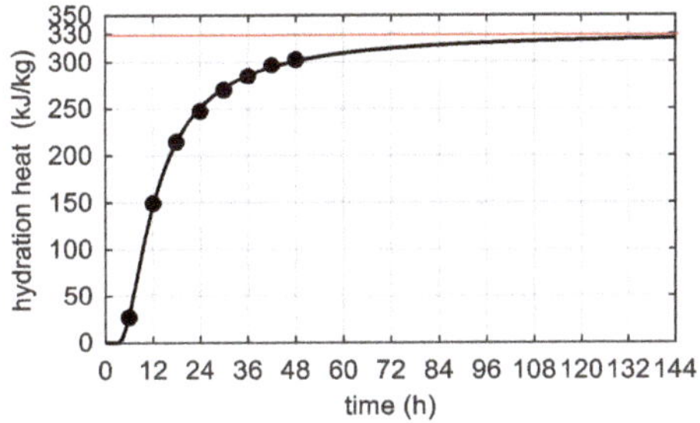

Figure 11. Hydration heat of CEM I 52, 5 N SR3/NA.

Prediction formulas also exist for specific heat of concrete, for instance, in the work of Lura and Breugel [36] the value of c is based on the mass component per cubic meter (W) and specific heat of each fraction of the mix (cement c_{cem}, basalt aggregate c_{bas}, quartz aggregate c_{quar} and water c_w). For the degree of hydration $\xi = 0$, the specific heat expressed in J/(kg·K) is given by relation:

$$c = \frac{W_{cem} \cdot c_{cem} + W_{bas} \cdot c_{bas} + W_{quar} \cdot c_{quar} + W_w \cdot c_w - 0.2 \cdot W_{cem} \cdot \xi \cdot c_w}{W_{cem} + W_{bas} + W_{quar} + W_w} = \frac{440 \cdot 456 + 1283 \cdot 766 + 632 \cdot 699 + 143 \cdot 4187}{440 + 1283 + 632 + 143} = 890.23 \tag{24}$$

and for $\xi = \xi_{max} = 0.65$ the specific heat is equal to 794.41 J/(kg·K). The high value of c reduces the extremum temperature of concrete and for a low value of specific heat, the temperature peak increases. Generally, the specific heat of concrete depends not only on the composition of the concrete mixture, but also on the temperature and humidity, which change during the hardening process. However, the models described in the literature often omit this relation and the constant specific heat estimated on the basis of the mixture composition is applied. Therefore, based on the above calculations, the model adopted the average specific heat value equals 840 J/(kg·K) = 0.84 kJ/(kg·K).

The heat conduction coefficient of concrete, expressed in W/(m·K), was estimated based on the thermal conductivities of the mix components using the following equation [36]:

$$\lambda = \frac{W_{cem} \cdot \lambda_{cem} + W_{bas} \cdot \lambda_{bas} + W_{quar} \cdot \lambda_{quar} + W_w \cdot \lambda_w}{W_{cem} + W_{bas} + W_{quar} + W_w} = \frac{440 \cdot 1.23 + 1283 \cdot 1.91 + 632 \cdot 3.09 + 143 \cdot 0.6}{440 + 1283 + 632 + 143} = 2.01 \tag{25}$$

According to Neville [37], for concrete with basalt aggregates, this coefficient equals 2.0 W/(m·K). A similar approach has been reported by Breugel [38], which says, that the value of λ ranges between 1.9 and 2.2 W/(m·K). Thus, the computed value of the thermal conductivity is consistent with the literature proposals and for further calculations $\lambda = 2.0$ W/(m·K) was assumed.

Table 2 summarizes all thermophysical parameters, fixed for the C 60/75 high-performance concrete analyzed in this paper.

Table 2. Thermophysical properties of concrete C 60/75 class.

C (kg/m³)	ρ (kg/m³)	E_a/R (K)	w/c (–)	ξ_{max} (–)	Q_{max} (kJ/kg)	c (kJ/(kg·K))	λ (W/(m·K))
440	2570	4620	0.325	0.65	330	0.84	2.0

6.2. Three Selected Parameters: κ/n_0, $\overline{n}$ and A_0/κ

The model parameters (κ/n_0, $\overline{n}$ and A_0/κ) required to perform numerical simulations were identified experimentally. The laboratory tests involved the concrete temperature measurements in the cubic samples curing under isothermal and semi-adiabatic conditions. For this purpose, four 150 mm cubic specimens made according to the recipe of the analyzed concrete were prepared. Immediately after molding, two cubes were placed in a water bath at a constant temperature of 24 °C and the concrete temperature changes under isothermal conditions were recorded. Two more cubes were inserted in specially prepared styrofoam containers with 10 cm thick insulation layer. The insulation around the specimens slows down the rate of heat loss and limits the heat exchange with the environment, thus cubes can be cured in semi-adiabatic conditions. The measurement boxes allow free samples placement in the mold and to insert the temperature sensors (Figure 12a). The experimental set-up illustrated in Figure 12b consists of a computer, 1-wire temperature sensors, semi-adiabatic containers and a water bath.

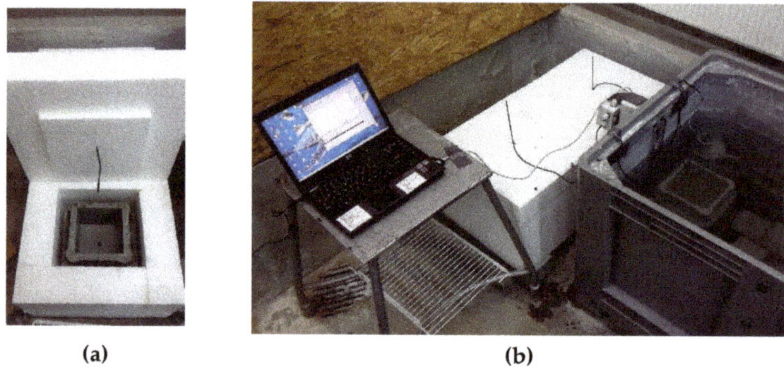

(a) **(b)**

Figure 12. (**a**) Styrofoam container to semi-adiabatic concrete curing and (**b**) experimental set-up.

The maximum temperature in concrete cubes reached under semi-adiabatic conditions was 10.4 °C higher than in isothermal conditions and the rate of temperature increase was more than twice bigger (Figure 13). The tests have proved how large the influence of boundary conditions on the temperature is, even with such a small volume of concrete. The average characteristic values from two concrete cubes are summarized in Table 3.

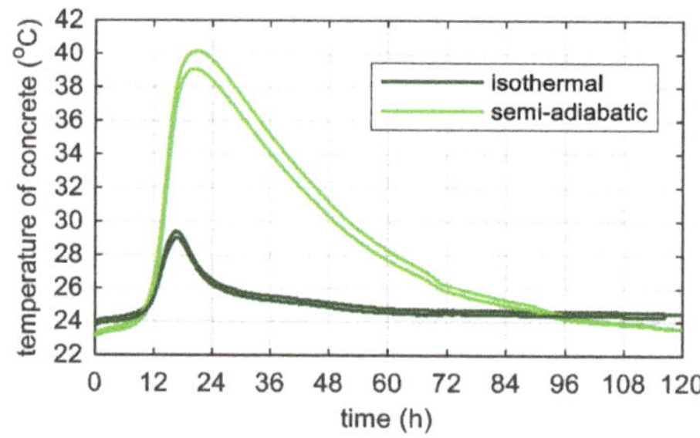

Figure 13. Temperature evolution in the centre of a concrete cubes cured in isothermal and semi-adiabatic conditions.

Table 3. Characteristic temperature values.

Conditions	T_0 (°C)	T_{env} (°C)	T_{max} (°C)	t_{max} (h)	ΔT (°C)	$\Delta T/t_{max}$ (°C /h)
Isothermal	24.0	24.0	29.2	16.5	5.2	0.32
Semi-adiabatic	23.2	24.0	39.6	20.0	16.4	0.82

6.2.1. Determination of Model Parameters Using the Martinelli's Approach

The partial differential equation of heat flow, reported by Cervera [8] and Martinelli [9], was solved using the finite difference method in the author's programs written in the MATLAB environment. For the models described in paragraph No. 2, the validation process was successfully performed based on literature data [8,9] and then used to analyze the own cases.

Experimental results, presented in above paragraph (Figure 13) were used to determine the adiabatic heat curve. Regression coefficients (a, b) for the function (11) and the heat transfer coefficients of the concrete hardening under isothermal (α_{nc}^{w}) and semi-adiabatic conditions (α_s) were calculated by the least squares regression. Table 4 reports the input values adopted in the model.

Table 4. Model parameters according to Martinelli et al. [9].

a	b	α_s	α_{nc}^{w}
(–)	(–)	(W/(m^2·K))	(W/(m^2·K))
15.5	2.3	3.0	20.0

The adiabatic temperature evolution represents the response of the simulations of the semi-adiabatic and isothermal temperature development (Figure 14). In both cases, the maximum temperature of self-heating of concrete hardening in adiabatic conditions was mutually convergent and the temperature value equals:

$$T_{\max} = T_0 + \frac{C \cdot \xi_{\max} \cdot Q_{\max}}{c \cdot \rho} = 23.2 + \frac{440 \cdot 0.65 \cdot 330}{0.84 \cdot 2570} = 23.3 + 43.7 = 66.9 \ (^\circ\text{C}). \tag{26}$$

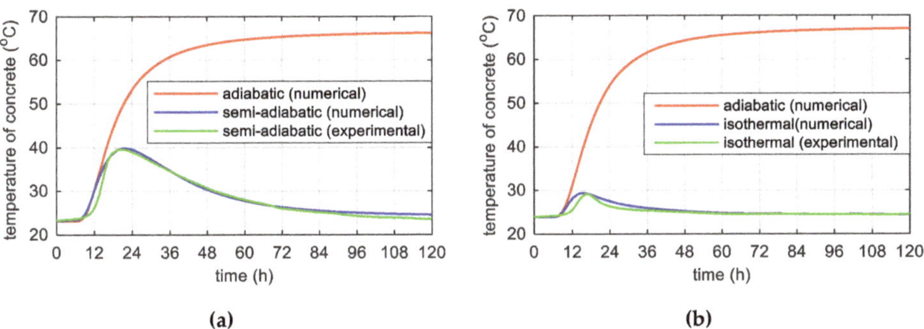

Figure 14. Adiabatic hydration curve for two various conditions: (a) Semi-adiabatic and (b) isothermal.

In addition, this value was consistent with the maximum concrete temperature (67.8 °C) measured in the middle of the 93 cm thick bottom slab (point p5, Figure 6b), for which conditions close to adiabatic could be assumed. Very good agreement between real measured value of $T_{\max}$ and numerical response was confirmed.

The results of calculations using the finite difference method (FDM) for the considered specimens allowed us to plot the development of hydration degree ξ versus time and chemical affinity $\widetilde{A}_{test}$ as a function of hydration degree according to equation 8, (Figure 15a). Founded on the nonlinear approximation of affinity $\widetilde{A}(\xi)$ to experimental data (Figure 15b), three sought parameters of the model were obtained: $\kappa/n_0 = 6.6 \cdot 10^6\,\text{h}^{-1}$, $\bar{n} = 5.2$ and $A_0/\kappa = 1 \cdot 10^{-4}$.

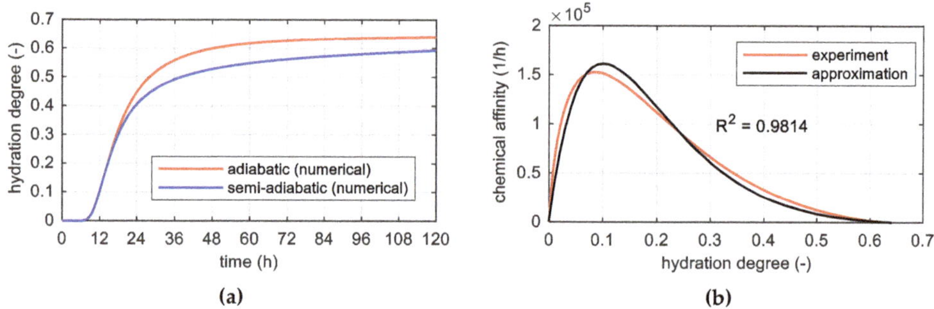

Figure 15. (**a**) Time evolution of the hydration degree and (**b**) chemical affinity vs. hydration degree.

6.2.2. Determination of Model Parameters Using the Cervera's Approach

The next step included verification of the determined constants in reliance on temperature simulations of the concrete specimens according to the Cervera's approach and also involved a possible modification of the three parameters. For this purpose two simulations were carried out: "v1" for designated constants: $\kappa/n_0 = 6.6 \cdot 10^6\,\mathrm{h^{-1}}$, $\bar{n} = 5.2$, $A_0/\kappa = 1 \cdot 10^{-4}$ and "v2" for coefficient $A_0/\kappa = 1 \cdot 10^{-5}$. The actualization of A_0/κ parameter improved the solution, which agreed with the measurement data (Figure 16).

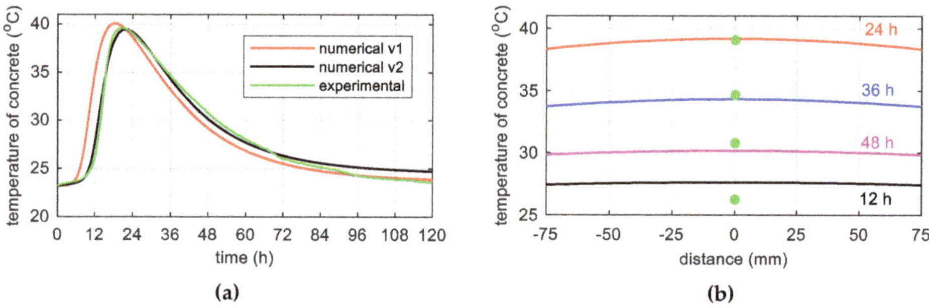

Figure 16. Temperature of concrete specimens cured in semi-adiabatic conditions: (**a**) In the time domain and (**b**) space variation of the temperature field in a 150-mm cube at different curing times (simulations + four measured values).

The predicted adiabatic temperature evolution was consistent with Martinelli's proposal (Figure 17). The results received in this part of the work constituted a check of the models themselves as well as determined parameters and indicated the correctness of the adopted assumptions.

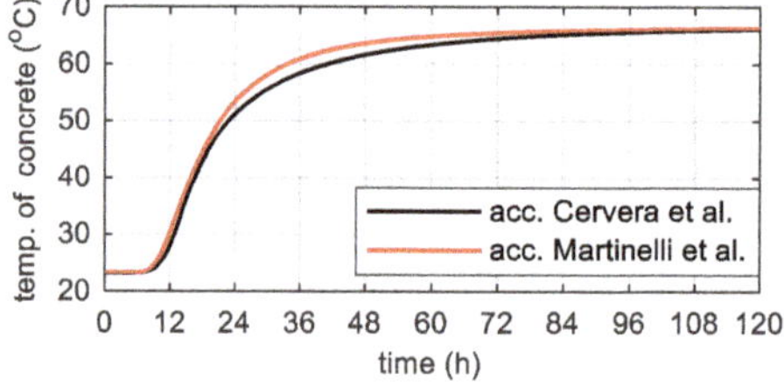

Figure 17. The concrete temperature in adiabatic conditions for two approaches.

Additionally, in order to recognize the influence of $\kappa/n_0, \bar{n}$ and A_0/κ parameters on the temperature distribution of concrete, three cases were analyzed. In each case, one of the mentioned parameters was changed, and two subsequent remained constant:

- Case 1: Variable coefficient κ/n_0, fixed $\bar{n} = 5.2$ and $A_0/\kappa = 1 \cdot 10^{-5}$,
- Case 2: Variable coefficient $\bar{n}$, fixed $\kappa/n_0 = 6.6 \cdot 10^6 \, \text{h}^{-1}$ and $A_0/\kappa = 1 \cdot 10^{-5}$,
- Case 3: Variable coefficient A_0/κ, fixed $\bar{n} = 5.2$ and $\kappa/n_0 = 6.6 \cdot 10^6 \, \text{h}^{-1}$.

Figure 18 shows the results of calculations for the temperature development in a concrete cubic sample. The temperature measured experimentally with a dashed, black line was marked. On the basis of Figure 18 it was found, that the coefficient κ/n_0 influenced the reaction rate, $\bar{n}$ was responsible for the extreme temperature value and A_0/κ for the time of its occurrence. If the parameter κ/n_0 increased, the reaction rate also rose. The maximum peak value of concrete temperature increased every ~1 °C and time occurrence of peak was reduced twice from $\kappa/n_0 = 4 \cdot 10^6 \, \text{h}^{-1}$ to $\kappa/n_0 = 12 \cdot 10^6 \, \text{h}^{-1}$. When $\bar{n}$ rose, the extreme concrete temperature dropped gradually from 2 to 1 °C for every half from $\bar{n} = 7$ to $\bar{n} = 4$. If we consider coefficient A_0/κ the time occurrence of peak was shorter every ~3 h form $A_0/\kappa = 1 \cdot 10^{-7}$ to $A_0/\kappa = 1 \cdot 10^{-2}$ with step equals 10^{-1}.

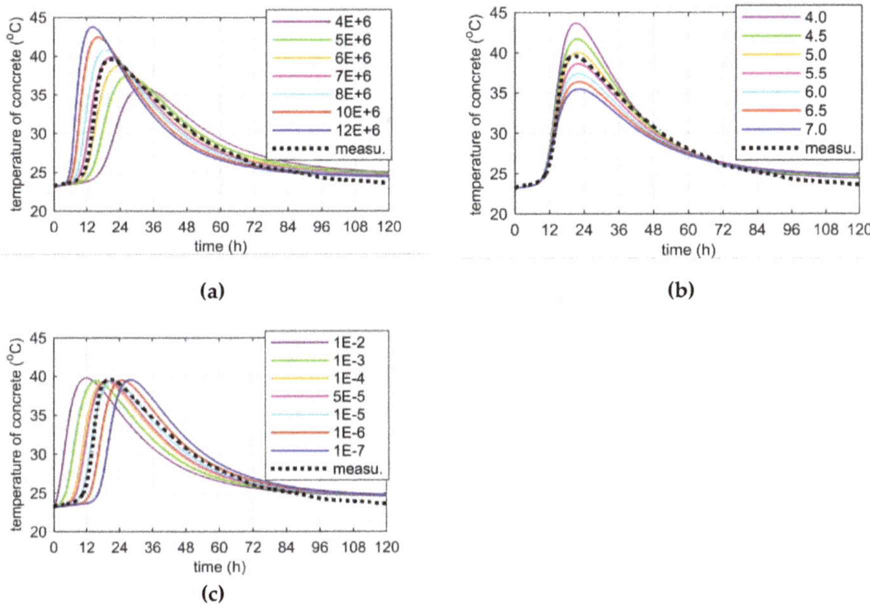

Figure 18. Three considered model parameters: (a) κ/n_0; (b) $\bar{n}$ and (c) A_0/κ.

7. Concrete Temperature Evolution of the Bridge Deck

The one-dimensional model (1D) was used to predict the temperature development of regular concrete slabs. In case of any irregularities such as the proximity of the protruding reinforcement or the influence of the temperature of earlier concreted parts of element, previously determined model parameters are no longer valid. Therefore, the calculations were performed for elements with clearly defined boundary conditions. This applies to the B-B section of the bottom plate of stage I, the web and the top plate (stage II) and the web monitored during III stage of research on the bridge.

Numerical simulations were carried out relying on experimentally determined constants for each research stage. The selected model parameters and heat transfer coefficients were adopted from the own propositions dedicated to the analyzed high-performance concrete and element thickness up to

100 cm. In this kind of structure, semi-adiabatic conditions of concrete hardening might be assumed. The prediction of concrete temperature for compressive strength assessment is the most useful for the medium-weight structure, which constitute the building contractor's interest. The proposed relationships are not intended for massive or thin-walled concrete elements. On the basis of laboratory and field tests, the coefficient κ/n_0 as a function of ambient temperature (Figure 19a), and the parameter $\bar{n}$ versus the thickness of the element (Figure 19b) was expressed. In both cases a linear function was used, and the accuracy of the approximated data (black dots) describes the determination coefficient R^2. As shown in Figure 18c, the parameter A_0/κ is mainly responsible for the time of extreme temperature occurrence, so for elements with thicknesses higher than 60 cm (80 and 93 cm) the value $1\cdot10^{-5}$ was assumed, and for elements thinner than 60 cm (35, 40 and 56 cm) A_0/κ equals $1\cdot10^{-4}$.

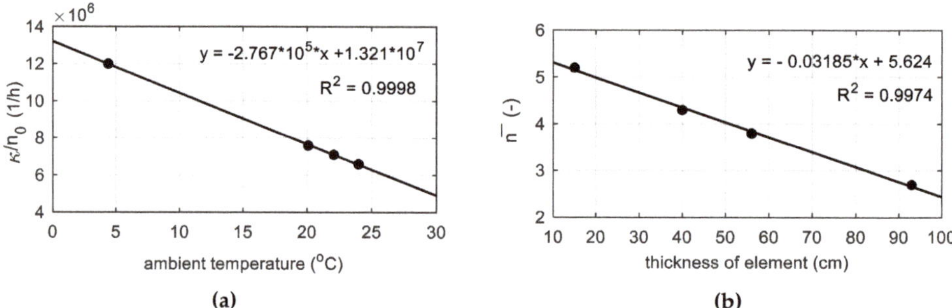

Figure 19. The proposition of determination model parameters: (**a**) κ/n_0 and (**b**) $\bar{n}$.

Literature propositions for heat transfer coefficients for forced convection or for insulation layer are often the result of specific experimental investigations and are only relevant in this particular case. In this study, the convective heat transfer coefficient on the concrete surface protected by the formwork layer α_s^{form} depended on the difference between the initial temperature of the mixture T_0 and the ambient one T_{env} (Figure 20a). In Figure 20 dots corresponded to the experimental data, which came from the three stages of field tests and the existing atmospheric conditions. For the heat transfer coefficient for the free surface of the slab α_{fc}, the formula as a function of wind speed was proposed. The comparison with the suggestion of other researchers is illustrated in Figure 20b.

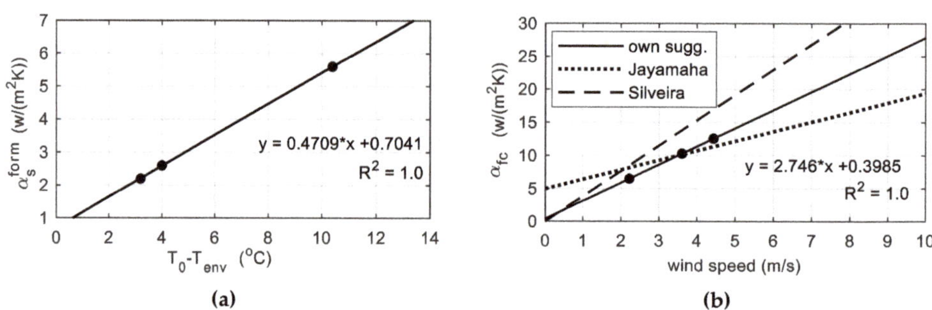

Figure 20. The own suggestion of connective heat transfer coefficient: (**a**) α_s^{form} and (**b**) α_{fc}.

In the future, to determine reliable nomograms (such as Figures 19 and 20), a series of concrete tests should be performed for different storage conditions and different concrete volumes. This approach would be justified in the case of repetitive elements, manufactured e.g., in a prefabrication plant. However, for individual constructions, at least measurements of the temperature of concrete cubes hardening in an adiabatic calorimeter or isothermal and semi-adiabatic conditions are required.

Additionally, the measurement system should be installed in one regular section on the object, e.g., the start segment.

7.1. Bottom Slab (Stage I)

The slab was concreted in June, when the average ambient temperature of 10 days was 22.1 °C. It was assumed that throughout the analyzed period, on the bottom surface of the slab the formwork (plywood and girders) was mounted. The top surface of the plate was exposed to the weather conditions, but between 23 and 94 h it was covered with a styrofoam layer, 5 cm thick. The average wind speed in this period was approximately 16 km/h = 4.44 m/s [39], thus according to the proposition presented in Figure 20b, coefficient $\alpha_{fc} = 12.6\,\text{W}/(\text{m}^2 \cdot \text{K})$. The initial temperature of the mixture was 26.7 °C.

Using the finite difference method, the space domain was subdivided in $m_s = 94$ nodes whose distance Δx was 1 cm (Figure 21). Moreover, for the explicit numerical integration a time increment dt met stability criterion. The calculations were carried out for a constant (Figure 22a) and a daily time variation of ambient temperature measured over and above the concrete slab, interpolated to the adopted step dt (Figure 22b). In Figure 22, as well as in the forthcoming ones, dashed lines indicate the results from the numerical simulations, whereas solid lines correspond to the concrete temperature measured at the appropriate depths. Thermo-physical parameters were adopted according to Tables 2 and 5. Figure 23 shows the temperature distribution, over time and plate thickness, in two variants. In Figure 23b the influence of daily temperature fluctuations can be seen, especially on the upper surface of the slab. The numerical results of temperature evolution show good agreement with the experimental data for both cases. According to Figure 22b, the relative error between measured and numerical value of maximum concrete temperature was equal to 1.6%, 0.9% and 4.5% for points p4, p5 and p6. The bigger differences concerned the time occurrence of the peak because the relative error reached 12.7%, 20.5% and 29.6%, respectively.

The maximum concrete temperature of the bottom plate was noted after 25 h of hardening, reaching a temperature of 67.8 °C at point p5 (Figure 22). The temperature of 70 °C was not exceeded, which the standard [40] gives for the limit value. In the core of the monitored slab, the conditions close to adiabatic existed. The temperature of self-heating of concrete caused by hydration reaction was 41.1 °C.

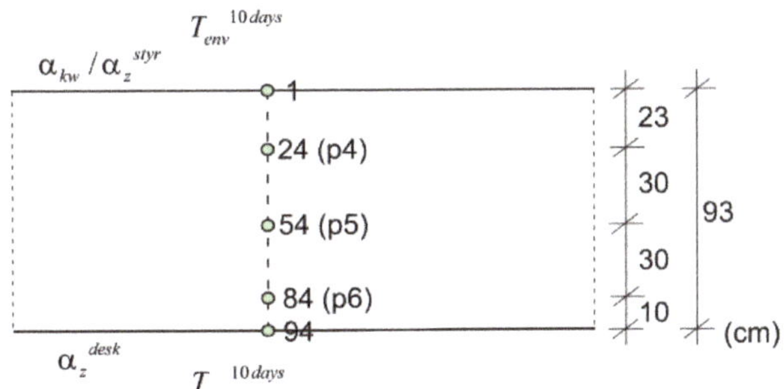

Figure 21. Space discretization.

Table 5. Thermophysical parameters—bottom slab, 93 cm thick, stage I.

κ/n_0	$\bar{n}$	A_0/κ	T_0	$T_{env}^{10\,days}$	t_{styr}	α_s^{styr}	α_{fc}	α_s^{form}
(1/h)	(–)	(–)	(°C)	(°C)	(h)	(W/(m²·K))	(W/(m²·K))	(W/(m²·K))
$7.1 \cdot 10^6$	2.7	$1 \cdot 10^{-5}$	26.7	22.1	23 – 94	0.40	12.6	2.2

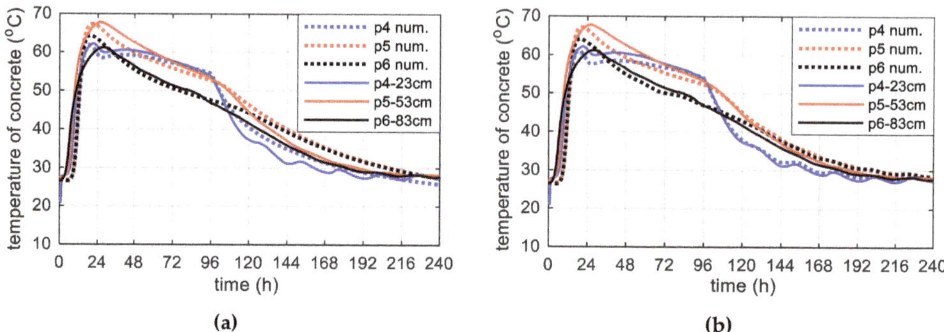

Figure 22. The concrete temperature of bottom slab, 93 cm thick (stage I): (**a**) Constant ambient temperature and (**b**) variable, measured ambient temperature.

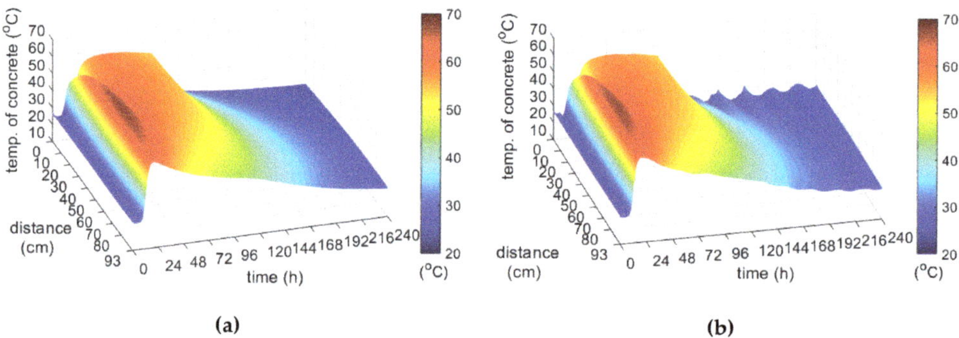

Figure 23. The temperature distribution of the concrete bottom slab (stage I): (**a**) Constant ambient temperature and (**b**) variable, measured ambient temperature.

7.2. Web and Top Slab (Stage II)

Another regular case is the web (40 cm) and the top slab (56 cm), i.e., medium-weight elements, monitored during the second stage of bridge investigations [11]. The recorded temperature history in the concrete web and the top plate are presented in Figure 24. The initial temperature of the mix, which was used to concrete the web was 29.1 °C, and in the case of the top plate it was 28.5 °C. The maximum concrete temperature of both the web and the top plate was recorded after 17.5 h. At point p4 and p12 its value was equal to 57.8 and 59.6 °C. Thus, the maximum temperature of self-heating caused by the cement hydration recorded in the web and the top plate was 28.7 and 31.1 °C.

In both monitored elements, boundary conditions had a strong influence on the temperature distribution. The concrete temperature of the web, 40 cm thick, surrounded on both sides with formwork, was very similar at all measurement points (Figure 24a). The top slab was on the one hand uncovered, and on the other protected with shuttering, which was the reason of temperature difference (e.g., Δp9-p12 = 5.2 °C; Figure 24b). Due to the high air temperature, these differences were not significant.

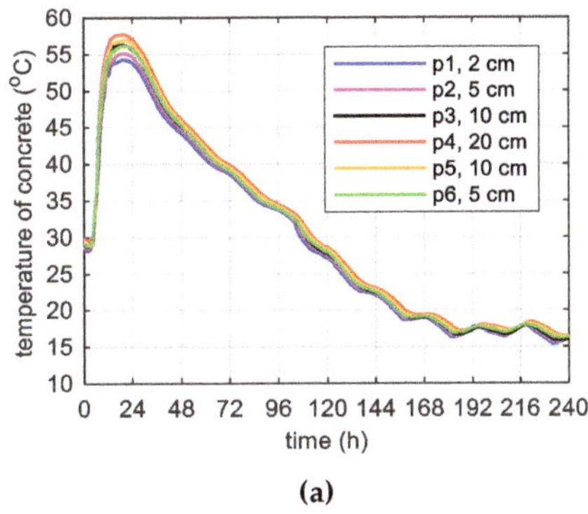
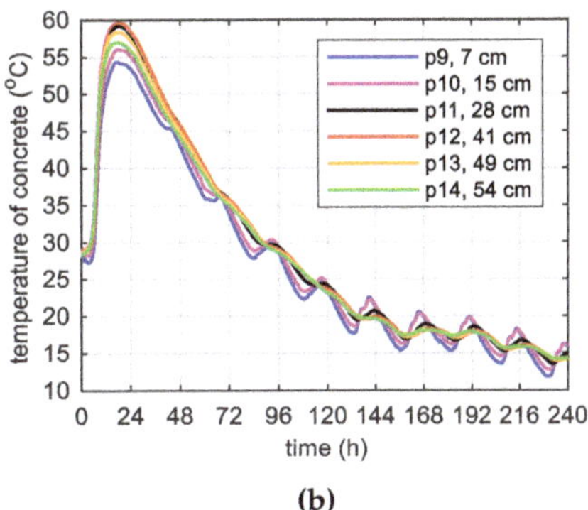

(a)　　　　　　　　　　　　　　(b)

Figure 24. Temperature development of the concrete during 240 h: (**a**) Web and (**b**) top slab.

During concreting of overhanging section of the bridge span, the average ambient temperature of 10 days was 20.1 °C. The average wind speed in this period reached 8 km/h = 2.22 m/s [39]. Calculations, as before, for constant and variable air temperature were made. Thermophysical parameters were adopted in accordance with Tables 2, 6 and 7. The temperature measurements in the web and the top plate in six points were carried out, while the graph presents results from three characteristic points (due to the readability of the graph). Better convergence of numerical and experimental data was obtained in the case of using a variable ambient temperature (Figure 25, Figure 26). According to Figure 25b the relative error for points p1, p3 and p4 was equal to 0.8%, 0.3% and 1.1% (max. temperature) and 5.8%, 0.9% and 2.7% (time occurrence of max. temperature). Based on Figure 26b the relative error for points p9, p10 and p12 was equal to 3.6%, 0.5% and 0.7% (max. temperature) and 1.9%, 1.0% and 0.8% (time occurrence of max. temperature). Good results have also been achieved for the temperature distribution in the cross-section of the web and the top plate, where the measured values were marked with dots (Figure 27, Figure 28). For the web the symmetry condition of the temperature distribution was met. Conducted research confirmed correctness of model assumptions and emphasized the importance of ambient temperature.

Table 6. Thermophysical parameters—web, 40 cm thick, stage II.

κ/n_0	$\bar{n}$	A_0/κ	T_0	$T_{env}^{10\,days}$	α_s^{form}
(1/h)	(–)	(–)	(°C)	(°C)	(W/(m²·K))
$7.6\cdot10^6$	4.3	$1\cdot10^{-4}$	29.1	20.1	2.6

Table 7. Thermophysical parameters—top slab, 56 cm thick, stage II.

κ/n_0	$\bar{n}$	A_0/κ	T_0	$T_{env}^{10\,days}$	α_{fc}	α_s^{form}
(1/h)	(–)	(–)	(°C)	(°C)	(W/(m²·K))	(W/(m²·K))
$7.6\cdot10^6$	3.8	$1\cdot10^{-4}$	28.5	20.1	6.5	2.6

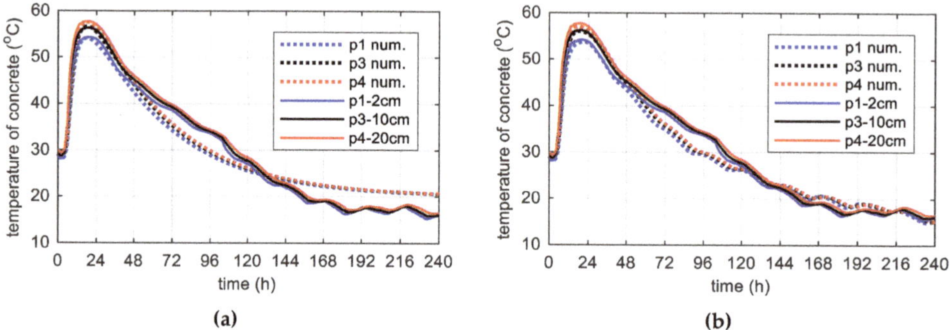

Figure 25. The concrete temperature of web, 40 cm thick (stage II): (**a**) Constant ambient temperature and (**b**) variable, measured ambient temperature.

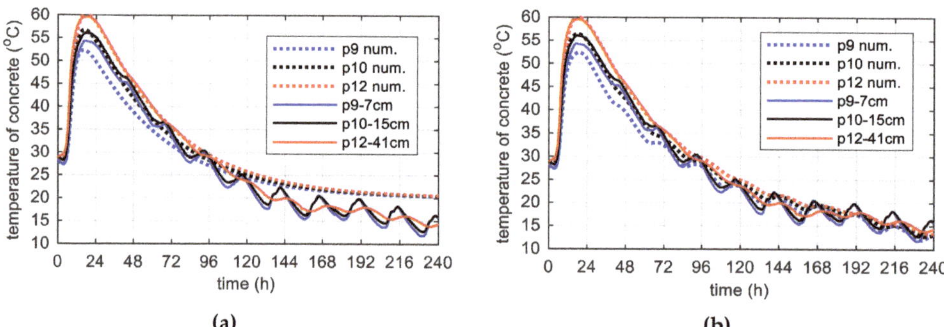

Figure 26. The concrete temperature of top slab, 56 cm thick (stage II): (**a**) Constant ambient temperature and (**b**) variable, measured ambient temperature.

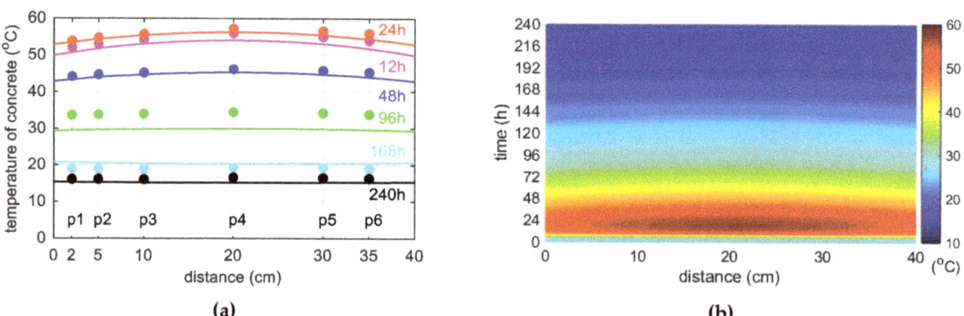

Figure 27. (**a**) Space variation of the web temperature field at different curing time and (**b**) concrete temperature distribution map (in both cases variable, measured ambient temperature).

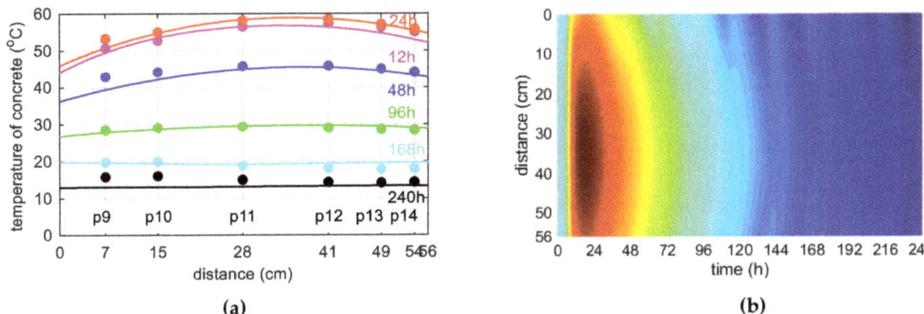

(a) **(b)**

Figure 28. (**a**) Space variation of the top slab temperature field at different curing time and (**b**) concrete temperature distribution map (in both cases variable, measured ambient temperature).

7.3. Web (Stage III)

The third stage of research on the bridge object was carried out in March, when the average ambient temperature was the lowest of the considered cases and it was around 4 °C. The average wind speed was equal in this period 13 km/h = 3.61 m/s [39]. The initial temperature of the concrete mix delivered to the construction site reached the value of 14.8 °C. The maximum temperature of the concrete web was equal 34.3 °C (in point p6, p8) and it was 23.5 °C lower than the maximum temperature recorded for the same element during the second stage of testing. The lowest temperature (32.5 °C) occurred in the middle of the web height (point p7).

As before, it was assumed, that throughout the analyzed time, the web was insulated by the formwork layer. Thermophysical parameters were adopted according to Tables 2 and 8. The temperature measurements in the web was made along its height, therefore the least-exposed point, located in the middle of the web height (p7) was chosen. The results of numerical simulations correspond very well to the measured values (Figure 29). Based on Figure 29b, the relative error of extreme concrete temperature was equal to 1.0% and the time occurrence of peak 6.3%.

Table 8. Thermophysical parameters—web, 40 cm thick, stage III.

κ/n_0	$\bar{n}$	A_0/κ	T_0	$T_{env}^{10\,days}$	α_s^{form}
(1/h)	(–)	(–)	(°C)	(°C)	(W/(m²·K))
$12.0 \cdot 10^6$	4.3	$1 \cdot 10^{-4}$	14.8	4.4	5.6

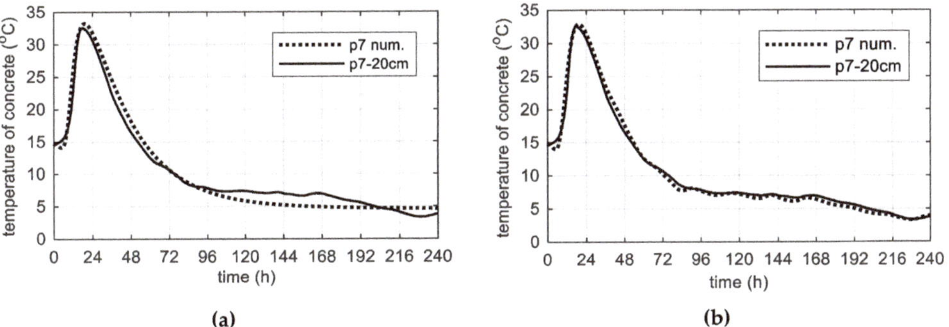

(a) **(b)**

Figure 29. The concrete temperature of web, 40 cm thick (stage III): (**a**) Constant ambient temperature and (**b**) variable, measured ambient temperature.

7.4. Limitations of One-Dimensional Approach

The proposed one-dimensional model was able to predict well both the heating and cooling phase, which is important in relation to the occurrence of undesirable tensile stresses. However, the one-dimensional model was not recommended to describe the temperature evolution of concrete plates with undefined strictly boundary conditions, as shown in the example discussed below. It concerns the bottom slab of stage I, for which the A-A measurement cross-section was located just next to the protruding web reinforcement and could not be insulated well with the styrofoam layer. The second example with irregular boundary conditions was the top and bottom plate monitored in stage No. III, where some disorder was the temperature coming from the hardening web. For these elements, the selected model parameters were modified. The obtained solutions were satisfactory, but for special cases, the calculations should be made with a two-dimensional model.

For the bottom slab, 80 cm thick (stage I) simulations in two variants were carried out. The first one for thermophysical coefficients specified for the 93 cm slab (Table 5) with modification of $\bar{n}$ parameter according to Figure 19b. The second approach with the change of heat transfer coefficients α_s^{styr} and α_{fc} (Table 9). The results of the calculations are shown in Figure 30. Considering the temperature distribution at the points located near the upper surface of the plate (p1, p2—Figure 30a) it could be concluded that there was no full insulation with styrofoam and the influence of protruding nearby reinforcement was substantial, hence the adopted coefficients of heat transfer α_s^{styr} and α_{fc} as in the case of a regular slab (93 cm, thick) was a wrong idea. Modification of these parameters improved the solution (Figure 30b).

For the A-A cross-section (80 cm thick), after 24 h, the temperature difference between points p1 and p2 was 19.3 °C (Figure 30a). Such information prompted the contractor to insulate the upper surface of the slab, which in 25–94 h was covered with a styrofoam layer, 5 cm thick. This was particularly visible in the thermal history of points located in the upper zone of the plate, i.e., p1. At this time, there was no cooling of the element, and a constant temperature was maintained.

Table 9. Thermophysical parameters—bottom slab, 80 cm thick, stage I.

κ/n_0	$\bar{n}$	A_0/κ	T_0	$T_{env}^{10\,days}$	t_{styr}	α_s^{styr}	α_{fc}	α_s^{form}
(1/h)	(–)	(–)	(°C)	(°C)	(h)	(W/(m2·K))	(W/(m2·K))	(W/(m2·K))
7.1·106	3.1	1·10^{-5}	26.7	22.1	23 - 94	5.4	18.6	2.2

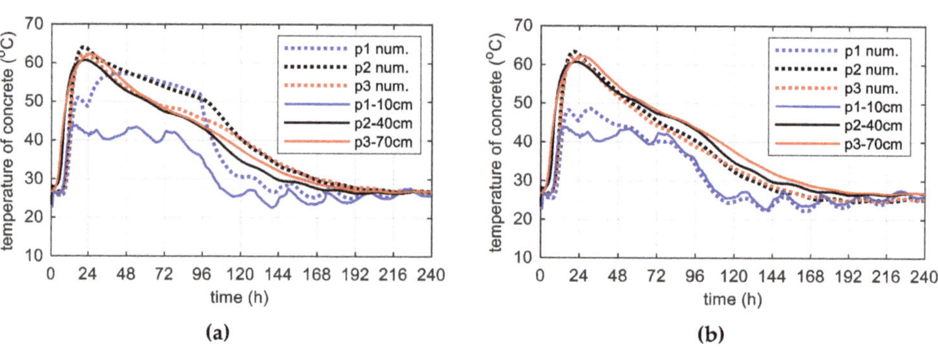

Figure 30. The concrete temperature of bottom slab, 80 cm thick (stage I): (**a**) Variant No. 1 and (**b**) variant No. 2.

In stage III, the box section of the bridge was concreted in one step. Thus, for the upper surface of the bottom slab and the bottom surface of the top slab, the influence of the temperature from the embedded web was significant. Therefore, a replacement heat transfer coefficient was adopted, indicating it with a symbol α_s^{conc} (Table 10, Table 11). The others parameters, except A_0/κ for the top plate and $\bar{n}$ for the bottom plate (modification $\bar{n}$ according to Figure 19b) remained unchanged relatively to the regular element for stage III, i.e., web (Table 10). For the bottom surface of the top plate the 10 days web temperature from point p6 ($T_{env\_t}^{10\,days}$) and for the upper surface of the bottom plate an average temperature from point p8 ($T_{env\_b}^{10\,days}$) was considered. The results (Figures 31 and 32) were satisfying, although they required additional assumptions.

Table 10. Thermophysical parameters—bottom slab, 35 cm thick, stage III.

κ/n_0	$\bar{n}$	A_0/κ	T_0	$T_{env\_t}^{10\,days}$	$T_{env\_b}^{10\,days}$	α_s^{form}	α_s^{conc}
(1/h)	(–)	(–)	(°C)	(°C)	(°C)	(W/(m²·K))	(W/(m²·K))
$12.0\cdot10^6$	4.5	$1\cdot10^{-4}$	14.8	11.3	4.4	5.6	4.0

Table 11. Thermophysical parameters—top slab, 57.2 cm thick, stage III.

κ/n_0	$\bar{n}$	A_0/κ	T_0	$T_{env\_t}^{10\,days}$	$T_{env\_b}^{10\,days}$	α_{fc}	α_s^{conc}
(1/h)	(–)	(–)	(°C)	(°C)	(°C)	(W/(m²·K))	(W/(m²·K))
$12.0\cdot10^6$	3.8	$1\cdot10^{-5}$	14.8	4.4	11.9	10.3	4.0

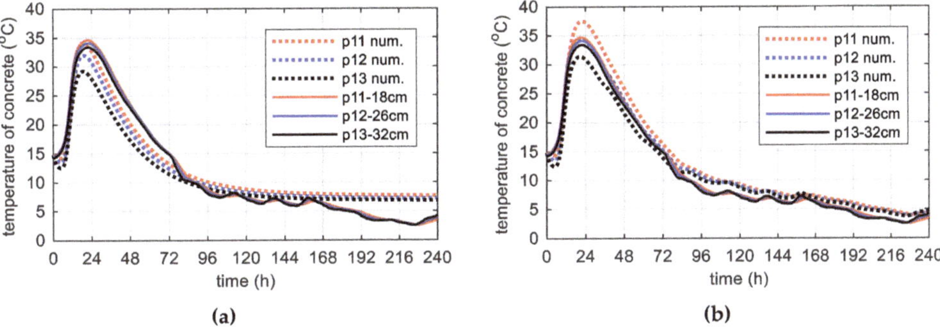

Figure 31. The concrete temperature of bottom slab 35 cm thick (stage III): (**a**) Constant ambient temperature and (**b**) variable, measured ambient temperature.

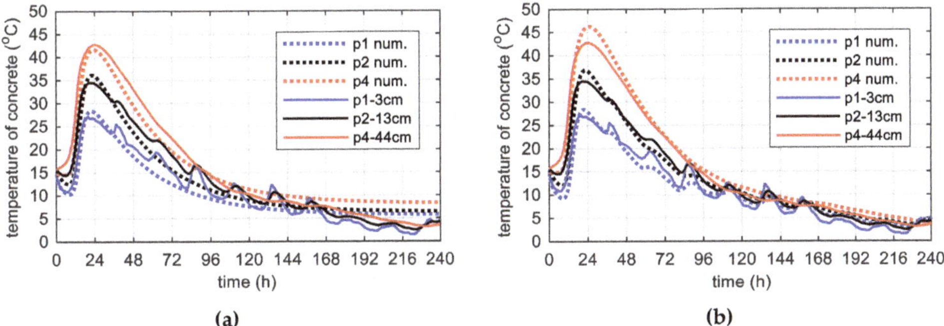

Figure 32. The concrete temperature of top slab 57.2 cm thick (stage III): (**a**) Constant ambient temperature and (**b**) variable, measured ambient temperature.

The thermal changes in the top plate differed the most (Figure 32). The max. concrete temperature occurred at point p4 was 42.8 °C. This temperature was lower by 16.8 °C than the maximum temperature recorded for the same element during the second stage of investigations. The various boundary conditions (formwork and free surface) and low ambient temperature generate difference (15.7 °C) between the center (p4) and the surface (p1) of the element. In general, guidelines state that the temperature difference must remain smaller than 15–20 °C depending on the element thickness.

The one-dimensional approach has several limitations. Early-age concrete is difficult for modeling because it is a complex material, which is additionally subjected to transformations as a result of cement hydration. The contractors use different types of formwork, insulation and the weather is changing all the time during construction process. Due to this fact, the 1–D way is not universal. Nevertheless, this fundamental method gives a possibility to control the early age behavior of concrete with quite good agreement. In the case of 1–D problem we could solve the partial differential equation using the own code of the finite difference method. There are no hardware restrictions and calculation time is short. A numerical, one-dimensional model is always a good complement to the measuring system. Many researchers [8,9,12] use this method to predict temperature evolution of concrete.

8. Distribution of Concrete Compressive Strength for Selected Structural Elements of the Bridge

As mentioned at the beginning of the paper, the in-place concrete temperature measurements were performed with a focus on estimation the real concrete compressive strength. The contractor planned to prestress every typical segment after reaching 60% of characteristic strength for class C 60/75, so it was exactly 45 MPa for cubic strength. Due to that fact, the maturity method was implemented [2,41]. The curve plotted after the third stage of the study is shown in Figure 33.

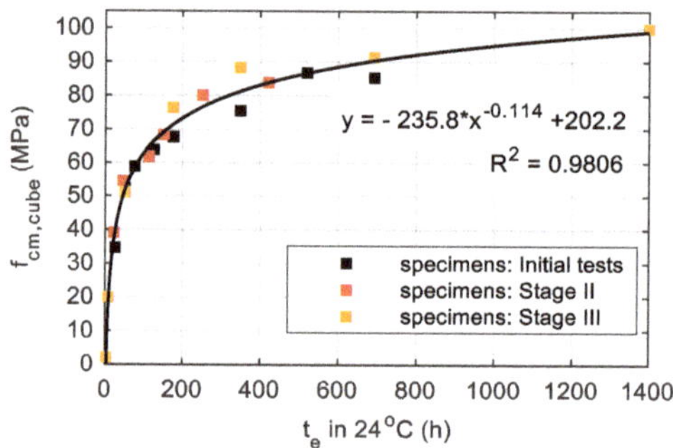

Figure 33. Maturity curve related to cubic strength after three stages of research.

Based on the numerically obtained maps of the temperature field and the maturity–strength relationship expressed in the equivalent age t_e domain, the distribution of early concrete compressive strength was presented. In each of the considered cases, the strength was calculated according to the concrete maturity curve for cubic samples, developed after the first, second and third stage of the specimen's validation tests performed during bridge operations. Figures 34–36 illustrates the compressive strength changes of young concrete for the bottom plate (stage I) and top slabs (stage II and III) versus the first three days of hardening. It can be seen that the age needed to achieve the required strength to prestress, and differed significantly in each monitored element. For plates with the same thickness, concrete reached 45 MPa in 15–48 h (Figure 35, Figure 36). In the case of the top plate monitored during third stage of research, the desirable concrete strength in the surface zone was achieved definitely later than for the bottom and middle part of the slab. The presented studies confirmed the impact of boundary conditions on early age compressive strength of cast-in-place concrete. Based on conducted investigations, the actual strength of in-place concrete and the possible dates of prestressing were determined.

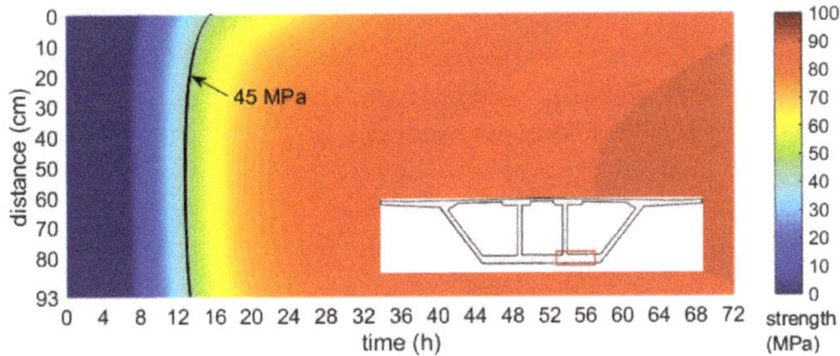

Figure 34. Map of the early age compressive strength distribution of the bottom slab (stage I, 93 cm thick, $T_{env}^{10\,days} = 22.1\,°C$).

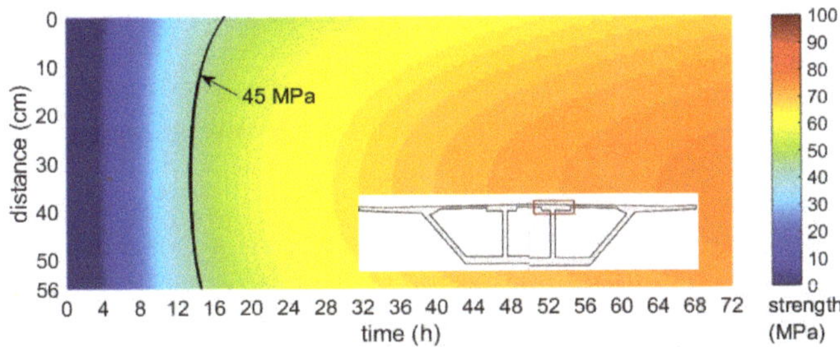

Figure 35. Map of the early age compressive strength distribution of the top slab (stage II, 56 cm thick, $T_{env}^{10\,days} = 20.1\,°C$).

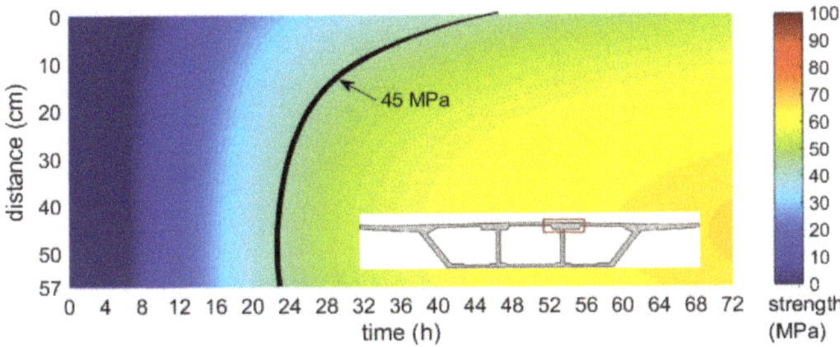

Figure 36. Map of the early age compressive strength distribution of the top slab (stage III, 57,2 cm thick, $T_{env}^{10\,days} = 4.4\,°C$).

9. Conclusions

This paper provided the guidelines to predict the temperature field in a concrete structure based on limited input data. The proposed solution, taking into account the thermal and chemical aspect of the hardening concrete, allowed us to simulate successfully the temperature changes in cast-in-place concrete of box bridge girder. The main results of the experimental and numerical studies could be summarized as follows:

- For on-line monitoring of concrete structures, the numerical model of the hardening concrete must be relatively simple, so that it can be an integral part of the SHM system.
- To register temperature of cast-in-place concrete it was suggested to use 1-wire digital sensors and install them on one measuring strip.
- The performed tests of cubic specimens curing under isothermal and semi-adiabatic conditions demonstrated the potential of experimental identification of the adiabatic hydration curve and thus

model parameters: κ/n_0, $\bar{n}$ and A_0/κ. In order to increase the quality of determined parameters, it was recommended to perform both experiments, for two mentioned curing conditions.

- The generated temperature of young concrete depends on the element dimensions and the weather conditions. Therefore, two novel nomograms for determination κ/n_0 as a function of an ambient temperature and $\bar{n}$ as a function of an element thickness were developed. An own proposition was dedicated for elements made of high-performance concrete with the thickness range of 10–100 cm.
- The boundary conditions were responsible for the heat exchange with the environment and had a huge impact on the rate gain of concrete temperature, the temperature drop and on the extreme temperature value. Therefore, the suggestions of connective heat transfer coefficient for the concrete surface with a formwork layer α_s^{form} and for forced convection α_{fc} were proposed.
- To reliably forecast thermal changes in concrete in a multi-segment structure, it was necessary to carry out preliminary tests on the starting section. The temperature should be measured inside the concrete element and on the formwork surface. Additionally, the ambient temperature as well as the wind speed needs to be registered.
- The numerical evolutions of temperature of high-performance concrete were positively verified in experimental measurements. The daily fluctuations of ambient temperature improved the accuracy of estimated temperature, however the assumption of constant air temperature provided very good results.
- The considered bridge deck proved that the FDM method was appropriate for structural elements with clearly defined boundary conditions as well as quite regular and simple geometry (e.g., concrete slabs and walls). In the case of more complex geometry, the calculations of the temperature evolution should be performed using the finite element method.
- The maturity method gave possibilities to estimate compressive strength distribution with an indication of time to achieve a required strength. The reduced air temperature resulted in a lower initial temperature of the mixture and a lower self-heating temperature, which delayed the development of the concrete strength, which was clearly visible for section No. 3.18.
- The knowledge of the temperature and strength development of concrete before the starting construction process allowed us to select temperature measurement points, to modify the designed concrete mixture and to take proper care of early-age concrete.

Funding: This research received no external funding.

Acknowledgments: The author thanks for support received from Krzysztof Wilde and Wilde Engineering Company.

Conflicts of Interest: The author declares no conflict of interest.

References

1. Klemczak, B.; Knoppik-Wróbel, A. Early age thermal and shrinkage cracks in concrete structures—Description of the problem. *Archit. Civ. Eng. Environ.* **2011**, *4*, 35–48.
2. ASTM C1074. *Standard Practice for Estimating Concrete Strength by the Maturity Method*; ASTM International: West Conshohocken, PA, USA.
3. Mariak, A.; Kurpińska, M.; Wilde, K. Maturity curve for estimating the in-place strength of high performance concrete. In Proceedings of the 64 Scientific Conference of the Committee for Civil Engineering of the Polish Academy of Sciences and the Science Committee of the Polish Association of Civil Engineers, Krynica, Poland, 16–20 September 2018; Volume 262, p. 06007.
4. Kurpinska, M.; Kułak, L. Predicting Performance of Lightweight Concrete with Granulated Expanded Glass and Ash Aggregate by Means of Using Artificial Neural Networks. *Materials* **2019**, *12*, 2002. [CrossRef] [PubMed]
5. ElSafty, A.; Graeff, M.K.; El-Gharib, G.; Abdel-Mohti, A.; Jackson, N.M. Analysis, prediction, and case studies of early-age cracking in bridge decks. *Int. J. Adv. Struct. Eng.* **2016**, *8*, 193–212. [CrossRef]

6. Azenha, M.; Sousa, C.; Faria, R.; Neves, A. Thermo–hygro–mechanical modelling of self-induced stresses during the service life of RC structures. *Eng. Struct.* **2011**, *33*, 3442–3453. [CrossRef]

7. Di Luzio, G.; Cusatis, G. Solidification–microprestress–microplane (SMM) theory for concrete at early age. Theory, validation and application. *Int. J. Solids Struct.* **2013**, *50*, 957–975. [CrossRef]

8. Cervera, M.; Faria, R.; Oliver, J.; Prato, T. Numerical modelling of concrete curing, regarding hydration and temperature phenomena. *Comput. Struct.* **2002**, *80*, 1511–1521. [CrossRef]

9. Martinelli, E.; Koenders, E.A.B.; Caggiano, A. A numerical recipe for modelling hydration and heat flow in hardening concrete. *Cem. Concr. Compos.* **2013**, *40*, 48–58. [CrossRef]

10. Chróścielewski, J.; Mariak, A.; Sabik, A.; Meronk, B.; Wilde, K. Monitoring of concrete curing in extradosed bridge supported by numerical simulation. *Adv. Sci. Technol. Res. J.* **2016**, *10*, 254–262.

11. Mariak, A.; Chróścielewski, J.; Wilde, K. Numerical simulation of hardening of concrete plate. In *Shell Structures: Theory and Applications, Proceedings of the 11th International Conference, Gdansk, Poland, 11–13 October 2017*; Taylor and Francis Group: London, UK, 2017.

12. Azenha, M. Numerical Simulation of the Structural Behaviour of Concrete since Its Early Ages. Ph.D. Thesis, University of Porto, Porto, Portugal, 2009.

13. Mills, R.H. Factors influencing cessation of hydration in water cured cement pastes. Special Report No. 90. In Proceedings of the Symposium on the Structure of Portland Cement Paste and Concrete, Highway Research Board, Washington, DC, USA; 1966; pp. 406–424.

14. Waller, V. Relations entre compositions des betons, exothermie en cours de prise et resistance en compression. Ph.D. Thesis, École des Ponts ParisTech, Champs-sur-Marne, 2000; p. 35.

15. Koenders, E.A.B.; Pepe, M.; Martinelli, E. Compressive strength and hydration processes of concrete with recycled aggregates. *Cem. Concr. Res.* **2014**, *56*, 203–212. [CrossRef]

16. Byfors, J. *Plain Concrete at Early Ages*; Technical Rep. No. 3:80; Swedish Cement and Concrete Institute: Stockholm, Sweden, 1980.

17. Van Breugel, K. Simulation of Hydration and Formation of Structure in Hardening Cement-based Materials. Ph.D. Thesis, Delft University of Technology, Delft, The Netherlands, 1991.

18. Jonasson, J.E.; Groth, P.; Hedlund, H. Modelling of temperature and moisture field in concrete to study early age movements as a basis for stress analysis. In *International Symposium Thermal Cracking in Concrete at Early Ages*; RILEM publications: Munich, Germany, 1994; pp. 45–52.

19. Faculty of Material Engineering and Physics. Cracow University of Technology. Available online: http://www.fizyka.pk.edu.pl/ (accessed on 20 July 2019).

20. Klemczak, B. *Modelowanie efektów termiczno-wilgotnościowych i mechanicznych w betonowych konstrukcjach masywnych*; Wydawnictwo Politechniki Śląskiej: Gliwice, Poland, 2008.

21. Andreasik, M. Naprężenia Termiczno-skurczowe w Masywach Betonowych. Ph.D. Thesis, Cracow University of Technology, Cracow, Poland, 1982.

22. McAdams, W. *Heat Transmission*; McGraw-Hill: New York, NY, USA, 1954.

23. Jonasson, J.E. Modelling of Temperature, Moisture and Stresses in Young Concrete. Ph.D. Thesis, Lulea University of Technology, Lulea, Sweden, 1994.

24. Branco, F.; Mendes, P.; Mirambell, E. Heat of hydration effects in concrete structures. *ACI Mater. J.* **1992**, *89*, 139–145.

25. Ruiz, J.; Schindler, A.; Rasmussen, R.; Kim, P.; Chang, G. Concrete temperature modeling and strength prediction using maturity concepts in the FHWA HIPERPAV software. In Proceedings of the 7th International Conference on Concrete Pavements, Orlando, FL, USA, 9–13 September 2001.

26. Silveira, A. The Influence of Thermal Actions in the Behavior of Reinforced Concrete Bridges. Ph.D. Thesis, LNEC, Lisbon, Portugal, 1996.

27. Davies, M. *Building Heat Transfer*; Wiley: Hoboken, NJ, USA, 2004.

28. De Freitas, J.T.; Cuong, P.T.; Faria, R.; Azenha, M. Modelling of cement hydration in concrete structures with hybrid finite elements. *Finite Elem. Anal. Des.* **2013**, *77*, 16–30. [CrossRef]

29. Cervera, M.; Oliver, J.; Prato, T. Thermo-chemo-mechanical model for concrete. I: hydration and aging. *J. Eng. Mech.* **1999**, *125*, 1018–1027. [CrossRef]

30. Bažant, Z.P.; Thonguthai, W. Pore pressure and drying of concrete at high temperature. *J. Eng. Mech. Div.* **1978**, *104*, 1059–1079.

31. Majorana, C.E.; Salomoni, V.; Schrefler, B.A. Hygrothermal and mechanical model of concrete at high temperature. *Mater. Struct.* **1998**, *31*, 378–386. [CrossRef]
32. EN 197-1. *Cement—Part 1: Compositions, Specifications ad Conformity Criteria for Common Cements*; European Standards, European Committee for Standardization: Belgium, 2011.
33. PN-B-19707. *Compositions, Specifications ad Conformity Criteria*; Cement; Special cement; Polish Committee for Standardization: Warszawa, Poland, 2013.
34. EN 206-1. *Concrete Part 1: Specification, Performance, Production and Conformity*; European Standards, European Committee for Standardization: Belgium, 2003.
35. Hansen, P.F.; Pedersen, E.J. Curing of concrete structures. In *Draft DEB—Guide to Durable Cocnrete Structures*; Appendix 1; Comite Euro-International du Beton: Switzerland, 1985.
36. Lura, P.; Breugel, K. Thermal Properties of Concrete: Sensitivity Studies. IPACS Document, Subtask 2.5, 2001.
37. Neville, A.M. *Properties of Concrete*; Polski Cement: Kraków, Poland, 2000.
38. Breugel, K. Prediction of temperature development in hardening concrete. In *Prevention of Thermal Cracking in Concrete at Early Ages*; Report, 15; Springenschmid, R., Ed.; E & FN SPON: London, UK, 1998.
39. Weather. Available online: http://www.ekologia.pl/pogoda/polska/warminsko-mazurskie/ostroda/ (accessed on 22 July 2019).
40. EN 13670. *Execution of Concrete Structures*; European Standards, European Committee for Standardization: Belgium, 2010.
41. Carino, N.J.; Lew, H.S. The Maturity Method: From Theory to Application. In Proceedings of the Structures Congress & Exposition, Washington, DC, USA, 21–23 May 2001; American Society of Civil Engineers: Reston, VA, USA, 2001.

Article

On the Influence of Capillary-Based Structural Health Monitoring on Fatigue Crack Initiation and Propagation in Straight Lugs

Marc Moonens [1,*]**, Eric Wyart** [2]**, Dieter De Baere** [1]**, Michaël Hinderdael** [1]**, Julien Ertveldt** [1]**, Zoé Jardon** [1]**, Galid Arroud** [1] **and Patrick Guillaume** [1]

[1]   Mechanical Engineering Department, Vrije Universiteit Brussel, Pleinlaan 2, 1050 Brussels, Belgium; Dieter.De.Baere@vub.be (D.D.B.); Michael.Hinderdael@vub.be (M.H.); Julien.Ertveldt@vub.be (J.E.); Zoe.Jardon@vub.be (Z.J.); Galid.Arroud@vub.be (G.A.); Patrick.Guillaume@vub.be (P.G.)

[2]   Cenaero ASBL, Rue des frères Wright 29, 6041 Gosselies, Belgium; eric.wyart@cenaero.be

*   Correspondence: marc.moonens@vub.be

Received: 5 August 2019; Accepted: 9 September 2019; Published: 12 September 2019

Abstract: This paper addresses the influence on the fatigue life induced by the implementation of a capillary-based structural health monitoring methodology, patented under the name eSHM. It consists in integrating structurally small and pressurized capillaries into the component, so that when a fatigue crack breaches the capillary network, it results in a leak flow to the open atmosphere and loss of pressure in the galleries which is detected by a pressure sensor. The novelty of the proposed system resides in the opportunity to locate the capillary according to the designer's need, as one resorts to additive manufacturing for the part production. However, the presence of these galleries in highly stressed regions raises concerns about crack initiation at the capillary itself and accelerated fatigue crack growth. This paper aims at the quantification of the influence the eSHM has on the fatigue behavior of the component and the determination whether this influence is significant or not. To that purpose, numerical simulations on a straight lug component, using the finite elements and eXtended Finite Elements Methods (XFEM), are performed. Various capillary sizes and shapes are assessed, so as to enable a general conclusion on the impact of the eSHM methodology in straight lugs.

Keywords: capillary-based structural health monitoring; fatigue crack growth; fatigue crack initiation; XFEM; straight lug

1. Introduction

The history of aviation is sadly enough quite abundant with examples of propagating fatigue cracks that lead to a structural failure. Actually, the firsts phenomenological understandings of the fatigue phenomenon in metals came after the series of crashes experienced by the first jetliner to enter commercial service, namely the Comet I aircraft [1–3]. In the time when it was designed, the deterioration of mechanical properties due to the presence of cracks was largely underestimated. Nowadays, even if our knowledge concerning the fatigue of materials has dramatically progressed, fatigue failures continue to happen and in almost all sectors of the industry. An insight of this can be given through the review of failure causes in aircraft components done by Findlay and Harrisson in 2002 [4,5], which reveals that fatigue is the dominant failure mode in aeronautics (55% of the failures), well ahead of corrosion (16% of failures) and overload (14%; the rest being wear and stress/temperature corrosion). This illustrates very well that due to the complexity of the fatigue phenomenon in metals, design offices are still struggling with issues related to fatigue. Therefore, in the author's view, this calls for a new paradigm in the approach followed to deal with fatigue cracks. One of the innovative approaches that could be followed in that context and which is thought to be very promising is known under the generic terminology of "Structural Health Monitoring".

Formally, Structural Health Monitoring (SHM) can be defined as the process of acquiring and analyzing the data from on-board sensors to evaluate the health of a structure [6]. Up to now, multiple SHM systems have been developed, with different target applications. Most of these systems used to be vibrations based [7,8]. However, structural health monitoring systems have been a thriving research domain for several years now, implying the emergence of innovative approaches for performing SHM. It would be impossible to be complete here, but technologies such as carbon nanotubes based structural health monitoring or comparative vacuum monitoring sensors illustrate very well this diversity [9–11]. In all cases, robustness remains a crucial attention point in the design of structural health monitoring systems [7,12].

Aside from this, the advent of additive manufacturing opens the door to the development of innovative concepts, such as smart metals. Therefore, the authors of this work have been active on an alternative implementation of the SHM principle where the geometric freedom provided by additive manufacturing could be exploited. The previously conducted research efforts resulted in a technology that is now patented (patent number EP2801809A1, [13]) under the name "effective Structural Health Monitoring (eSHM)". The system consists in integrating over- or under-pressurized capillaries (the capillaries are thus internal channels) into the to-be-monitored component, so that when a fatigue crack breaches the capillary network, a leak flow is created, and the pressure equilibration between the capillary and the open atmosphere is detected by a pressure sensor. Figure 1 illustrates the working principle on a four point bending test sample equipped with an under-pressurized capillary. The pressure is initially set to 0.5 atm in the capillary before the experiment starts, but as soon as the propagating fatigue crack reaches it (clearly seen on the micro-XCT image), air from the open atmosphere enters the capillary, and internal capillary pressure rises sharply. This pressure rise is detected by a sensor, which triggers an alarm when a pre-defined value has been reached, thus revealing the presence of the crack [14]. In that sense, a parallel can be drawn between the proposed structural health monitoring approach and the previously developed Helicopter Blade Crack Detection System (patent number US5205710A, [15]). However, the proposed methodology is on purpose mainly targeted at additively manufactured components [6]. The geometric freedom offered by additive manufacturing implies that the capillaries can be placed according to the designer's needs. In other words, the proposed SHM methodology enables close and robust monitoring of the regions which were identified by design studies as prone to crack initiation. For illustration, it can be mentioned that it would be impossible to monitor the entire lug hole region of the lug studied below should it be manufactured with conventional methods. Moreover, the proposed technology requires no significant signal processing and electronic hardware, as only the capillary pressure should be measured and monitored (refer to the experimental investigations in [16]). Detailed information regarding the working principle of this novel SHM methodology can be found in the patent [13].

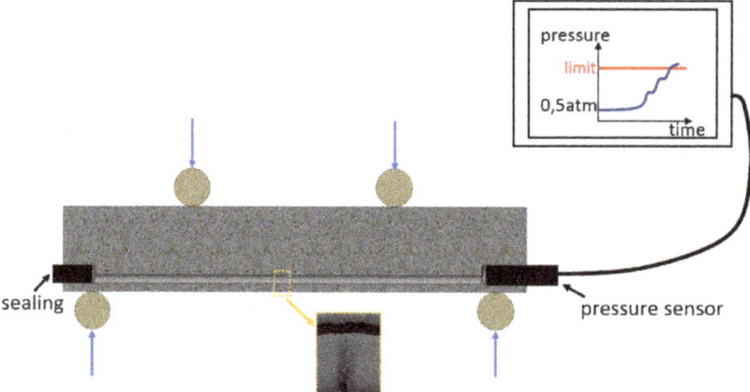

Figure 1. Illustration of the eSHM working principle on a four point bending test sample.

While additive manufacturing has shown significant potential for widespread usage in production of complex and/or customized parts, this potential has been dampened by, notably, the variation of mechanical properties, which includes the fatigue behavior of these parts [17]. Indeed, it is well-known in the literature that printing parameters such as technology used (powder bed fusion, laser metal deposition, etc.), scanning speed, powder flow rate (non-exhaustive list) have notably a strong influence on the microstructure and residual stresses present in the component, and hence, on the mechanical properties. Moreover, post-treatments applied (if any) on the part also result in a mechanical behavior different compared to the as-built conditions. As a matter of illustration, it has been reported that the Hot Isostatic Pressing process (hipping) contributes to a significant improvement of fatigue properties, and possibly to properties that are similar to their conventional counterparts [18–21]. The dispersal of mechanical properties of AM parts inevitably reduces their reliability and thus impinges the adoption of additive manufacturing in engineering applications [17]. In that context, it is thought that besides offering high level of robustness, the proposed SHM philosophy could also contribute in revealing the full potential of additive manufacturing.

However, the introduction of capillaries in regions which are potentially subject to high stresses might raise concerns on the influence the eSHM could have on the fatigue life of the structure. Günther et al. [22] have shown that test samples with internal channels have reduced fatigue lives compared to samples without internal channels, the reduction being dependent on the topology of the channels. However, still according to Günther et al., this is primarily due to rapid crack initiation at the primary roughness of the internal channels, and secondly to the modified topology of the specimens. In the present research, the effect of capillary roughness on the reduction of fatigue life has not been taken into account. Indeed, technologies such as hybrid manufacturing (combination of additive and subtractive manufacturing) or chemical etching could potentially enable a drastic reduction in capillary roughness levels in the near future [23–26]. Moreover, in the context of the eSHM system, the roughness influence has already been experimentally investigated (see [27]). Therefore, one focuses here on the consequences of the presence and topology of virtually smooth capillaries on fatigue life. This is particularly important for the further development of the system, as the eSHM system has to offer a quick detection of growing fatigue cracks (by being placed "as close as possible" to the most probable initiation site), while not affecting the component's function (the capillary should not trigger initiation nor reduce fatigue life). As a matter of fact, this study can be seen as the completion of our initial work [14], as both initiation and propagation are addressed, and as additional capillary topologies and initial flaws are studied. This should enable to draw a more general conclusion on the eSHM influence on fatigue behavior of equipped lugs.

2. Crack Initiation Considerations

2.1. Numerical Framework

The idea behind this research is to study the impact of the implementation of the eSHM technology in a potential practical application. Since the eSHM is originally targeted at aerospace applications, it has been decided to work on straight lug components. Indeed, this type of connecting element is frequently used in aeronautics, and is also very prone to fatigue failures (fretting and/or corrosion are very likely to initiate a crack) [28]. Amongst the abundant literature on lugs, it was decided to work on the configuration studied by Schijve in [29], where fatigue crack growth experiments have been performed on various aluminum straight lugs (in-plane geometry depicted in Figure 2, thickness of 5 mm) subject to a cyclically varying axial loading. The maximal pin force is $F_{max} = 21$ kN and the load ratio is $R = 0.33$. It must be noted that this force is modeled through this entire work as an equivalent pressure having a cosinusoidal distribution on the pin hole surface. Moreover, even though a lug with integrated capillary would be barely impossible to manufacture without resorting to additive manufacturing, the material properties that are used in the present work are similar to the properties of the Al2024-T3 material used in the experimental work done in [29]. These are: a Young's modulus of 73,000 MPa, a Poisson's ratio of 0.33, and a yield stress of 368 MPa. It is thus assumed that the material is isotropic, and that the variations of the mechanical properties due to the printing process do not influence the results of the comparison provided the same material properties are enforced in all the models.

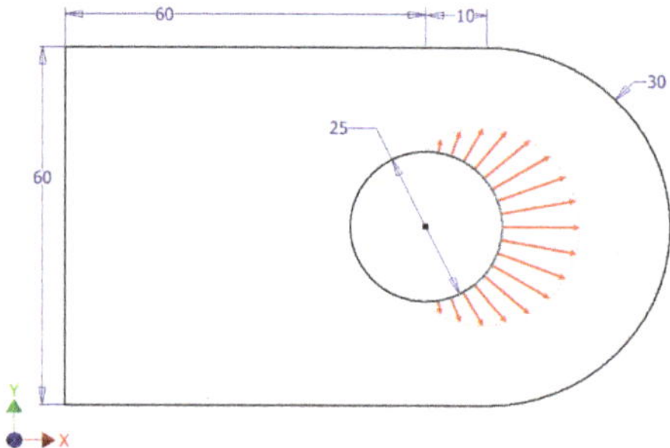

Figure 2. Dimensions (in mm) of the lug modeled, and pin loading modeled as an equivalent pressure with cosine distribution around the hole.

Figure 3 shows the axial stress field σ_{xx} (the x-axis is along the longitudinal direction, as shown on Figure 2), further noted as S_{11}, on the lug when subject to the maximal axial pin loading. The highest tensile stresses are found on the top and bottom regions of the pin hole surface and reach 336 MPa (effectively corresponding to a stress concentration factor of 2.8 with respect to the maximum net tension stress, see [29]). Excluding manufacturing defects, the highest probability of crack initiation is thus also found there. These two regions of high tensile stresses will hence further be referred as the "initiation regions" and are the focus in this part of the study. It must be noted that, considering this stress distribution, capillaries should be placed and limited to the neighborhood of these initiation regions. However, in an actual practical application, the loading could have a non-axial component. Moreover, in testings, notches are introduced in these area's in order to trigger initiation at a known place, which is of course not done in an operational component. As a consequence, the initiation region is never as clearly defined as what has been assumed here. Therefore, capillaries are integrated all around the hole of the studied lug, so as to enable detection of any crack that initiated at the hole surface (see Figure 4).

Figure 3. S_{11} field on the lug, highlighting the most probable region for initiation under cyclic and purely axial loading.

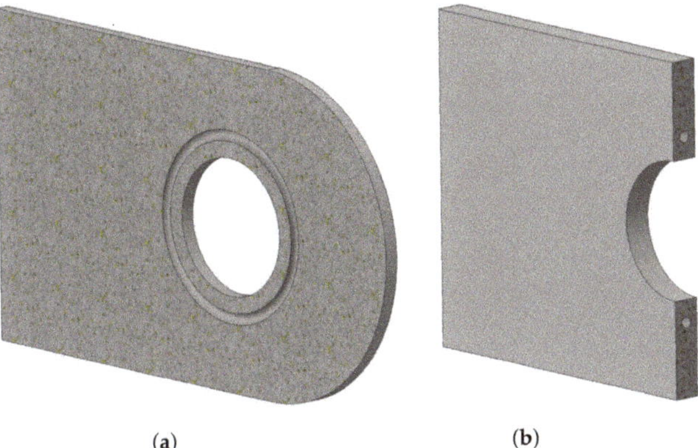

(a)   (b)

Figure 4. Cut views of the lug equipped with eSHM on a XY plane at mid-thickness (**a**) and on a YZ plane centered at the lug hole (**b**). The capillary around the hole is clearly visible (in the shown topology, the capillary is circular with a diameter of 2 mm).

While the eSHM system has the advantage of a very simple principle of operation, conferring a significant robustness to the system, and since the parts equipped with eSHM have to be manufactured by additive manufacturing, it still presents a considerable design freedom in terms of engineering. Indeed, the capillary shape is in principle arbitrary and can be "chosen" by the designer, as well as its dimensions. However, keeping in mind that one should avoid salient edges and convex shapes, this design freedom is restrained and one will limit oneself to capillaries of circular and elliptical cross section. Nevertheless, the diameter/aspect ratio of the capillary can be tuned, and this part of the study aims at quantifying in which proportion the capillary diameter/aspect ratio influences the crack initiation behavior. Besides, the "edge-to-edge" distance between the bottom of the capillary and the lug hole surface, referred here as distance "a" (in accordance with the terminology used in the works [6,26,27]), is also a parameter in the engineering of the crack detection system. A smaller edge-to-edge distance would imply a quicker detection of the propagating crack, and again, one aims at quantifying the influence of this design parameter on the crack initiation. It must be noted that the pressure inside the capillary also is another design parameter, but it has been shown in a previous study that, at the pressurization levels used, no effect was observable on the stress field in the component [14].

The parametric study led in this work therefore concentrates on the influence of capillary shape (circular or elliptical), dimension (diameter and aspect ratio), and edge-to-edge distance with respect to the monitored region (lug hole surface). Figure 5 illustrates the different configurations under consideration. Static simulations of the different configurations subject to the maximum pin loading ($F_{max} = 21,000$ N) have been run, and the resulting maximum tensile stresses on the lug hole surface and capillary surface are compared to the situation of a "standard" or "reference" lug where no capillary has been integrated. For the sake of accuracy of the numerical results, one has taken advantage of the symmetry of the problem and only a quarter of the lug has been modeled. The mesh, consisting of about 700k (exact number depending on the eSHM configuration) linear hexahedron elements, has been obtained after several refinement steps to ensure the use of a converged mesh.

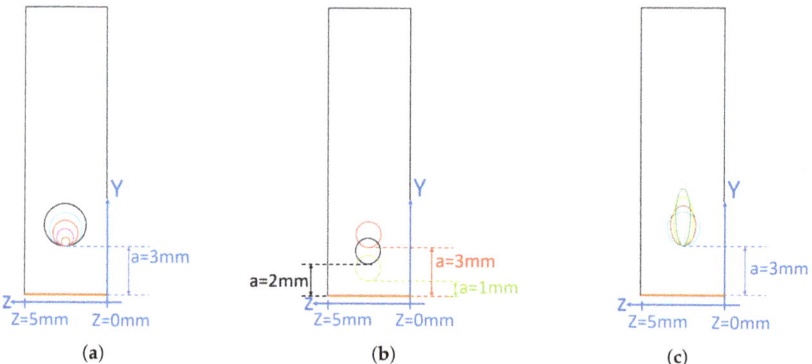

Figure 5. Schematics of the upper cross section of the lug when looking in a YZ plane centered at the lug hole (one can refer to Figure 4b), illustrating the different implementations of the eSHM (capillaries) that have been assessed. In particular, different capillary diameters (**a**), edge-to-edge distance (**b**) and aspect ratio's (**c**).

2.2. Circular Capillary—Influence of Diameter

In this section, capillaries of circular cross section are considered. The influence of different diameter sizes on the stress field in the component is studied. In particular, as here above mentioned, one focuses on the axial stress in the initiation region ($S_{11,ini}$) and on the capillary surface, as these are considered to be the best indicators of the influence the eSHM has on the initiation behavior. These results have been retrieved for five different diameters, namely 0.5 mm, 1 mm, 1.5 mm, 2 mm and 2.5 mm (respectively shown in brown, mauve, red, light blue and black on Figure 5a and on Figure 6), all being located so that the edge-to-edge distance "a" remains equal to 3 mm. These diameters corresponds to realistic values of what can be or will be achieved in a near future with the current or forthcoming additive manufacturing technologies. Indeed, technologies based on Powder Bed Fusion (PBF) have proven to be able to print capillaries down to 1 mm in diameter size, while technologies such as Direct Energy Deposition (DED) are currently limited to 2 mm. However, considering the evolution of the metal printing capabilities, and the advent of hybrid manufacturing, it is credible to think that smaller capillaries will be soon achievable. Therefore, it felt relevant to model capillaries as small as 0.5 mm in diameter, even if those are not yet printable.

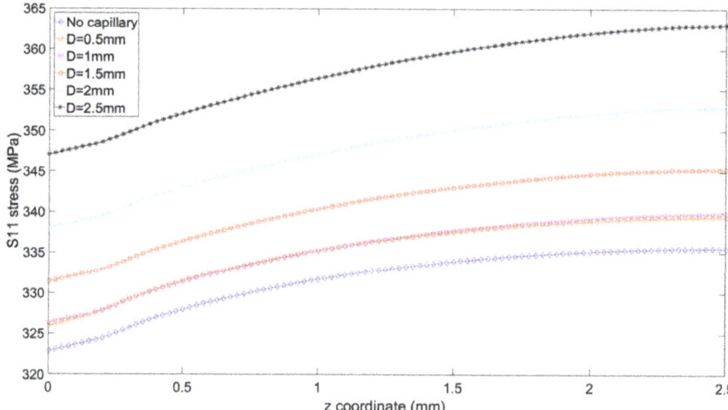

Figure 6. Axial stress S_{11} on the most critical line of the pin hole surface (bottom line in Figure 5, highlighted in orange), across the lug thickness, and as a function of the capillary diameter.

The axial stress values in the initiation region (cross-section bottom line in Figure 5) can be seen in Figure 6. As symmetry is used in the model, the results are presented for $z = 0$ mm to $z = 2.5$ mm, while the lug thickness is actually 5 mm. The reference lug experiences a maximum tensile axial stress of 336 MPa in the initiation region (also seen on Figure 3). On the graph of Figure 6, it can be seen that the presence of the capillary inside the lug raises the tensile axial stress level in that region, and therefore reduces the crack initiation life. However, it can be noticed that capillaries of very small dimensions (0.5 and 1 mm) have a very limited impact on the maximum stress level (339 MPa instead of 336 MPa). Table 1 summarizes this data, and also presents the maximum axial tensile stress on the capillary surface. It is noteworthy to realize that in all cases, the maximum stress on the capillary surface remains significantly inferior to the stress in the initiation region. This implies that in the case of a smooth capillary, a crack would normally not initiate at the capillary itself, as the lug hole surface remains the region where the amplitude of tensile stresses variations is the highest.

Table 1. Maximum axial tensile stress values in the initiation region ($S_{11,ini}$) and on the capillary surface ($S_{max,cap}$) in function of capillary diameter (D). The results are compared with the reference lug, in which no capillary is integrated.

D (mm)	$S_{11,ini}$ (MPa)	$\frac{S_{11,ini}}{ref}$	$S_{max,cap}$ (MPa)
Reference	336	-	-
0.5	339	+0.9%	198
1	340	+1.2%	204
1.5	345	+2.6%	214
2	352	+4.8%	226
2.5	363	+8.0%	245

2.3. Circular Capillary—Influence of Edge-to-Edge Distance

Bringing the capillary closer to the initiation region (thus, reducing the distance "a") is desirable for earlier detection of fatigue cracks that would develop at the lug hole surface. However, the influence it has on stress raise in the initiation region and at the capillary surface has to be studied. In the experimental investigations around the eSHM, the distance "a" is traditionally set to 2 mm [26,27]. Therefore, for the purpose of this parametric study, the diameter of the capillary was kept constant at 1.5 mm, and edge-to-edge distances equal to 3 mm, 2 mm and 1 mm have been considered (respectively shown in red, black and green on Figure 5b).

The results are presented in Table 2. These demonstrate that earlier crack detection (smaller "a" distance) inherently comes with a raised stress level in the initiation region, but this raising remains very limited (increase of 1.9% between $a = 3$ mm and $a = 1$ mm). Moreover, even for an edge-to-edge distance of 1 mm, the tensile stress level on the capillary surface remains inferior to the tensile stress level in the initiation region. Again, this implies that the lug hole surface would remain the initiation site. It is however noteworthy to realize that in practice, due to the current limitations in additive manufacturing technologies, the roughness level in the capillary acts as a significant stress raiser, and implies that when brought too close to the surface, cracks actually initiate at the capillary instead of at the surface. This has been thoroughly studied in the work by Günther et al. [22]. This is also the reason why experimental testing around the eSHM has traditionally used an edge-to-edge distance of 2 mm [26,27]. From this, it can be inferred that capillary roughness is the main challenge to the practical implementation of the eSHM, as the presence of the capillary itself should not affect the crack initiation site nor significantly the crack initiation life, provided the capillaries are small with respect to the lug dimensions.

Table 2. Maximum axial tensile stress values in the initiation region ($S_{11,ini}$) and on the capillary surface ($S_{max,cap}$) in function of edge-to-edge distance (a).

a (mm)	$S_{11,ini}$ (MPa)	$\frac{S_{11,ini}}{ref}$	$S_{max,cap}$ (MPa)
Reference	336	-	-
1	351	+4.5%	260
2	347	+3.3%	226
3	345	+2.7%	214

2.4. Elliptical Capillary—Influence of Ellipse Aspect Ratio

Finally, it has to be assessed whether there would be an added-value to using capillaries of elliptical cross section instead of circular cross section. Indeed, these would definitely be more complex to manufacture, but it is deemed that this could have a beneficial impact on the tensile stress level in the initiation region and more significantly on the capillary surface. To that purpose, capillaries of various aspect ratios, ranging from 1 (circular capillary) to 4, but of identical cross sectional area (1 mm^2) and constant edge-to-edge distance are considered. These configurations are shown in Figure 5c.

The results of the simulations performed on these lugs is shown in Table 3. It can be seen that the impact on the stress level in the initiation region is negligible. However, as it is expected, the use of an elliptical cross section affects quite significantly the stress level at the capillary surface, as the maximum value of tensile stress attained on the capillary surface drops from 226 MPa (circular capillary) to 174 MPa (ellipse of aspect ratio 4), thus resulting in a 23% decrease on the surface. As a consequence, it indicates that the use of an elliptical cross section could enable to position the capillary closer to the lug hole surface (hence achieving faster detection) while minimizing the risk of initiation at the capillary itself.

Table 3. Maximum axial tensile stress values in function of capillary cross section aspect ratio (AR).

AR	$S_{11,ini}$ (MPa)	$\frac{S_{11,ini}}{ref}$	$S_{max,cap}$ (MPa)
Reference	336	-	-
1 (circular)	352	+4.8%	226
1.56	350	+4.2%	199
2.25	349	+3.9%	183
3.06	348	+3.6%	176
4.0	347	+3.3%	174

3. Crack Propagation Considerations

3.1. Numerical Framework

When a crack grows from the initiation region, due to the reduction of cross-sectional area in the plane where the crack propagates, it can be expected that the propagation speeds will be higher than in the reference lug, at least when the crack propagates through the region where the capillary is located. Hence, the objective of this section is to quantify the repercussions of the acceleration on the crack growth life. This quantification is done in function of capillary shape and size. To that purpose, fatigue crack growth simulations have been run on lugs equipped with different configurations of the eSHM, namely those of Figure 5a,c. The results are then compared to computations run on the reference lug (in which there is no capillary). Two types of initial defect are considered: a part-through crack of initial length $a_0 = 1$ mm (to be consistent with the study of Schijve [29]), and a quarter-elliptical crack of initial lengths $a_0 = 1$ mm and $b_0 = 1$ mm, both located at the initiation region (top of the lug hole surface). Part-trough cracks and quarter-elliptical cracks are actually the most fore-coming type of defects in aeronautical lugs [30].

The objective here is to make a comparison between a reference lug and lugs equipped with the eSHM, so that to be able to infer up to what extend the capillaries do affect the crack growth life.

The crack propagation model used may thus remain simple, and therefore, the Paris law has been referred to [31,32], considering a Paris exponent n = 3.58 and a Paris coefficient of $C = 1.3651e - 13\frac{mm}{cycle} \cdot (MPa\sqrt{mm})^{-n}$. The fracture toughness of aluminum 2024-T3 is $K_c =$1010 MPa$\sqrt{mm}$, so that the lugs are considered to have failed when the cracks reach a length of $a_f =$12 mm (the maximum stress intensity factors (mode I) on the crack front reach K_c for that crack length).

All the fatigue crack growth simulations have been performed with Morfeo, developed by Cenaero [33,34]. In the software, the propagation is driven by a user defined crack propagation step Δa. The software then computes the corresponding ΔN, number of cycles required to propagate the crack by Δa. One should note that Δa has to be set to a value ensuring that the propagation path is properly computed, and that the time integration yielding the ΔN is correctly evaluated. Since the software is based on the XFEM method, the crack position in the mesh is spotted by level sets, which are at each step updated based on propagation length and direction [33–35]. All the meshes used in this section have been generated using the open source pre- and post-processor "Gmsh", developed by Geuzaine and Remacle [36]. They were obtained after several refinement steps, ensuring the use of a converged mesh. The mesh was highly refined in the crack region, and particularly in the initial defect region (see Figure 7), to ensure for converged stress intensity factors on the complete crack path. Depending on the configuration, the mesh totalizes between 850 k and 930 k linear tetrahedron elements.

(a)   (b)

Figure 7. Mesh used in the fatigue crack propagation simulations. (**a**): overview. (**b**): highly refined region around the crack.

3.2. Influence of Capillary Diameter on Crack Growth Life

In what follows, the "crack length" refers to the distance between the lug hole surface and the intersection between the crack front and the side surface of the lug. This choice has almost no influence on the crack growth curve of the part-through crack since the crack front curvature is very limited. However, concerning the quarter-elliptical crack, this definition implies that the number of cycles to detection (corresponding to the moment where the crack breaches the capillary) does not correspond to a crack length equal to 3 mm (see Figure 8).

The crack growth curves are presented in Figures 9 and 10. The number of cycles to failure and number of cycles to detection are given respectively in Tables 4 and 5. It is interesting to note that even though a simple crack propagation model has been used, the fatigue life obtained for the reference lug and for the part-through crack, namely 11,470 cycles, agrees fairly with the experimental works of Schijve, where the fatigue life of the specimens oscillated between 13,500 and 15,500 cycles [29]. For the quarter-elliptical initial defect, the numerical results cannot really be compared to the experimental works of Schijve, as the initial defect introduced in the experiment (a corner crack) is different than the one modeled in the present research.

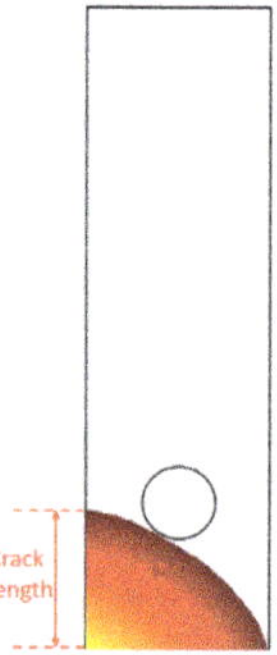

Figure 8. Crack front just before reaching the capillary (2 mm diameter) after propagation of the initial quarter-elliptical defect.

Table 4. Fatigue crack growth lives (N_f) and cycles to detection (N_d) as a function of capillary diameter, in the case of part-through crack. The variation of fatigue life compared to the reference lug (Δ) is also given, as well as the Remaining Useful Lifetime after detection (RUL_d).

D (mm)	N_f	Δ	N_d	RUL_d (% of N_f)
Reference	11,470	-	-	-
0.5	11,400	−0.6%	4500	61
1	11,160	−2.7%	4470	60
1.5	10,630	−7.3%	4070	62
2	10,050	−12.4%	3605	64
2.5	9290	−19%	3380	63

Table 5. Fatigue crack growth lives (N_f), variation with respect to the reference lug (Δ), cycles to detection (N_d) and post-detection Remaining Useful Lifetime (RUL_d) as a function of capillary diameter and in the case of quarter-elliptical crack.

D (mm)	N_f	Δ	N_d	RUL_d (% of N_f)
Reference	27,490	-	-	-
0.5	27,220	−1%	19,970	26
1	27,080	−1.5%	19,100	29
1.5	26,180	−4.8%	18,100	31
2	24,130	−12.2%	16,870	30
2.5	22,520	−18%	15,070	33

Several interesting observations can be made from these crack propagation computations. First, for both types of defects, small capillaries have a limited negative impact in terms of the fatigue crack growth lifes. The impact that would be tolerated by a component manufacturer would inevitably be application dependent. However, to set ideas, one considers here that a reduction by 5% of the crack growth life compared to a reference lug remains acceptable. With this assumption, capillaries of 0.5 mm and 1 mm are acceptable designs. Conversely, for large capillaries, the severe stress state on the propagation plane implies that the stress intensity factor field on the crack front is also more severe. In the first propagation step computed, the maximum mode I stress intensity factor on the reference lug for a part-through crack is 586 MPa$\sqrt{mm}$, while it already reaches 644 MPa$\sqrt{mm}$ for the lug equipped with a 2.5 mm diameter capillary. The propagation speeds are thus already larger, even before the crack reaches the capillary. This is clearly seen on Figure 9. Second, it is worth noting that this trend is consistent with the effect capillaries have in terms of crack initiation. This illustrates the interest of working with relatively small capillaries (compared to the dimensions of the component), which is also the direction aimed for in the research around the eSHM methodology. Third, for the part-through crack, the detection occurs early in the crack propagation life, and can be further improved, should it be needed, by bringing the capillary closer to the initiation region. Indeed, as shown in Table 4 and in Figure 9, the remaining post-detection crack growth life represents 60% of the total crack propagation life (N_f). However, due to the lower propagation rates in the early stage (clearly seen in Figure 10), this is not true anymore for the quarter-elliptical crack. Indeed, the Remaining Useful Lifetime after detection is reduced in this case to barely 30% of N_f. Therefore, alternative designs, such as integrating two ex-centered capillaries, could be envisaged to remedy this when needed. Finally, it must be noted that the presence of the capillary has no influence on the crack path, which is in all cases straight in the y-direction.

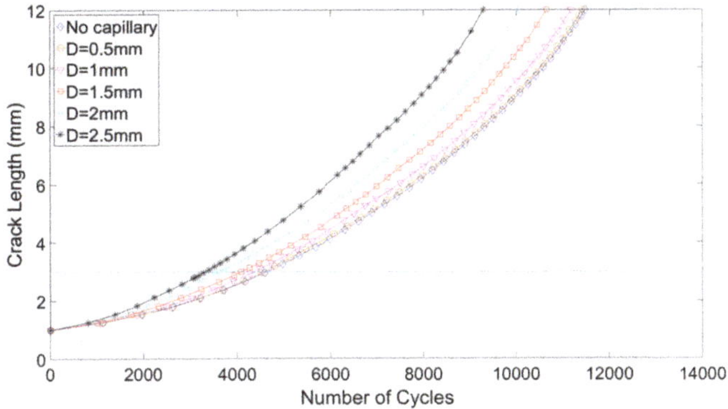

Figure 9. Fatigue crack growth curves for various capillary diameters, for a part-through crack.

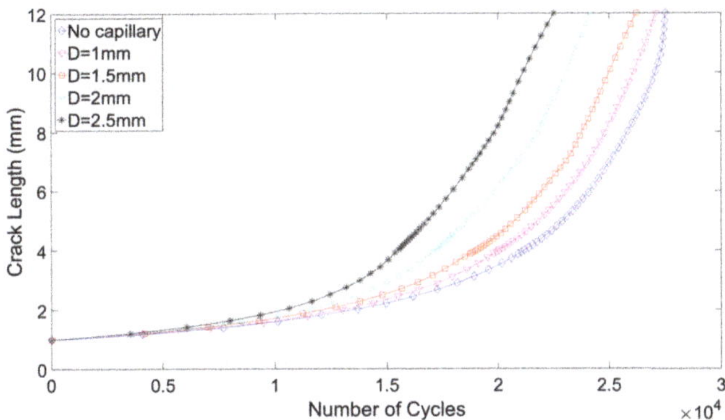

Figure 10. Fatigue crack growth curves for various capillary diameters, for a for a quarter-elliptical crack.

3.3. Influence of Capillary Shape on the Crack Growth Life

The crack propagation curves obtained for a part-through crack are shown in Figure 11. It can be clearly seen that varying the aspect ratio of the capillary actually has a negligible effect, as the impact on the crack propagation life remains below 2.5% variation (see Table 6). Increasing the aspect ratio of the capillary implies that the local reduction in area felt by the crack is less important than for circular capillaries. Crack propagation speeds in the neighborhood of the capillary are hence lower for elliptical capillaries. However, due to the vertical extension (along the y-axis) of the elliptical capillary, the time frame during which the crack propagates in a reduced available sectional area is also longer. This counterbalances the positive effect of lower propagation speeds, hence explaining why small difference in crack propagation life is obtained.

Table 6. Fatigue crack growth lives (N_f), variations with respect to the circular capillary case ($\Delta_{circular}$) and cycles to detection (N_d) as a function of capillary aspect ratio, in the case of part-through crack.

AR (-)	N_f	$\Delta_{circular}$	N_d
1	10,050	-	3605
1.56	9920	−1.3%	3805
2.25	9860	−1.9%	3900
3.06	9860	−1.9%	3990
4	9820	−2.3%%	4030

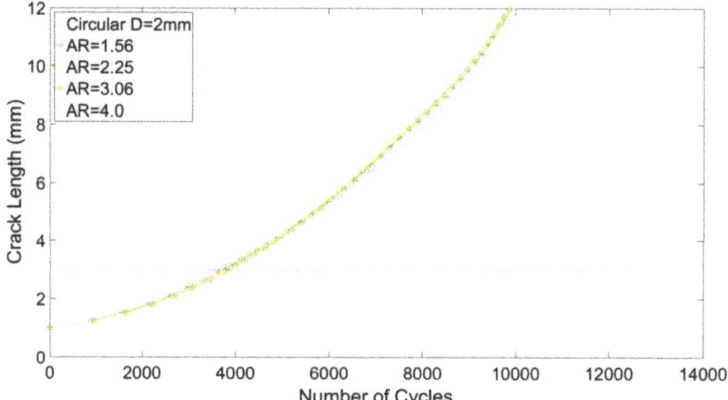

Figure 11. Fatigue crack growth curves for various capillary aspect ratio and in the case of part-through crack.

4. Conclusions

In the present work, the influence on fatigue life of the implementation of the effective Structural Health Monitoring (eSHM) methodology has been dealt within the context of straight lugs.

It has been shown that the inclusion of small capillaries (0.5 or 1 mm diameter) has a negligible influence on both crack initiation behavior and on fatigue crack propagation life. Conversely, large capillaries have a significant negative impact on crack growth life and crack initiation. This brings to the conclusion that proper application of the effective Structural Health Monitoring methodology requires extra engineering work at the design stage of the component, as a bad implementation can result in degraded fatigue performances. Moreover, this also sheds light on the importance of further research on hybrid manufacturing in the frame of the effective Structural Health Monitoring methodology, since that technology is deemed to enable the inclusion of capillaries that remain smooth while being very small in size.

It was also shown that the different configurations studied here present the drawback of late detection for quarter-elliptical cracks. However, thanks to the additive manufacturing process, many other configurations can be thought of to remedy this, should it be a problem for a particular application. One could think of integrating two side capillaries instead of one centered capillary, or use the weight spared with the inclusion of the capillary to increase the thickness of the lug when tolerances permit. In other words, the optimization space available in the frame of the configuration and implementation of the eSHM is very wide, and only a portion of that space has been explored here. This is especially true as additional features can be foreseen in the scope of the eSHM, such as crack localization and load monitoring. The scope of possibilities offered by the eSHM is very wide, and further research efforts should be devoted to the development of the eSHM methodology.

In this work, one important challenge to the possible practical utilization of the eSHM has been overcome. Several other hurdles exist before certification could be envisaged, but this work has demonstrated the interest and potential of equipping engineering components with the effective Structural Health Monitoring methodology. In particular, the authors intend to build a demonstrator lug equipped with this novel structural health monitoring technology, following the design recommendations of the present work.

Author Contributions: M.M. and E.W. performed the crack growth simulations. M.M. performed the other simulations and wrote the paper. E.W., D.D.B., M.H., J.E., Z.J. and G.A. contributed to the work through discussions and improvements to the text. P.G. supervised the works.

Funding: This research received no external funding.

Acknowledgments: The help provided and the sound advices given by the all the researchers at Cenaero is highly acknowledged.

Conflicts of Interest: The authors declare no conflict of interest.

References

1. Suresh, S. *Fatigue of Materials*; Cambridge University Press: Cambridge, UK, 1998.
2. Noels, L. Fracture Mechanics Online Class. Available online: http://www.ltas-cm3.ulg.ac.be/FractureMechanics/?p=welcome (accessed on 4 August 2019).
3. Federal Aviation Administration. Lessons Learned from Civil Aviations Accidents: De Havilland DH-106 Comet 1. Available online: https://lessonslearned.faa.gov/ll_main.cfm?TabID=1&LLID=28&LLTypeID=0 (accessed on 4 August 2019)
4. Findlay, S.J.; Harrrison, N.D. Why Aircraft Fail. *Mater. Today* **2002**, *5*, 18–25. [CrossRef]
5. Gorelik, M. Additive Manufacturing in the Context of Structural Integrity. *Int. J. Fatigue* **2017**, *94*, 168–177. [CrossRef]
6. De Baere, D.; Strantza, M.; Hinderdael, M.; Devesse, W.; Guillaume, P. Effective Structural Health Monitoring with Additive Manufacturing. In Proceedings of the 7th European Workshop on Structural Health Monitoring, Nantes, France, 8–11 July 2014; pp. 2314–2321.
7. De Baere, D.; Hinderdael, M.; Moonens, M.; Jardon, Z.; Lison, M.; Strantza, M.; Guillaume, P. Additive Manufactured Metallic Smart Structures to Monitor the Mechanical Behavior In Situ. *Proceeding* **2018**, *2*, 500. [CrossRef]
8. Deraemaeker, A.; Worden, K. *New Trends in Vibration Based Structural Health Monitoring*; CISM Lecture Notes Volume 520; Springer: Vienna, Austria, 2010.
9. Ahmed, S.; Thostenson, E.T.; Schumacher, T.; Doshi, S.M.; McConnell, J.R. Integration of Carbon Nanotube Sensing Skins and Carbonfiber Composites for Monitoring and Structural Repair of Fatigue Cracked Metal Structures. *Compos. Struct.* **2018**, *203*, 182–192. [CrossRef]
10. Abot, J.L. Structural Health Monitoring Using Carbon Nanotube Yarns: Sensing Concept and Applications in Composites. In Proceedings of the 5th International Symposium on Sensor Science (I3S 2017), Barcelona, Spain, 27–29 September 2017; Volume 1, p. 852.
11. Roach, D. Real Time Crack Detection Using Mountable Comparative Vacuum Monitoring Sensors. *Smart Struct. Syst.* **2009**, *5*, 317–328. [CrossRef]
12. Farrar, C.R.; Worden, K. An Introduction to Structural Health Monitoring. *Philos. Trans. R. A Math. Phys. Eng. Sci.* **2007**, *365*, 303–315. [CrossRef] [PubMed]
13. De Baere, D. Structural Health Monitoring Using Hollow Cavity Structure. European Patent EP2801809A1, 12 November 2014.
14. Moonens, M.; Wyart, E.; Hinderdael, M.; De Baere, D.; Guillaume, P. Numerical Simulation of Fatigue Crack Growth in Straight Lugs Equipped with Efficient Structural Health Monitoring. *Procedia Struct. Integr.* **2018**, *13*. [CrossRef]
15. Engels, G.P.; Thomas, M.C. Helicopter Blade Crack Detection System. U.S. Patent US5205710A, 27 April 1993.
16. Strantza, M.; De Baere, D.; Rombouts, M.; Maes, G.; Guillaume, P.; Van Hemelrijck, D. Feasibility Study on Integrated Structural Health Monitoring Produced by Metal Three-Dimensional Printing. *Struct. Health-Monit. Int. J.* **2015**, *14*, 622–632. [CrossRef]
17. Yadollahi, Y.; Shamsaei, N. Additive Manufacturing of Fatigue Resistant Materials: Challenges and Opportunities. *Int. J. Fatigue* **2017**, *98*, 14–31. [CrossRef]

18. Leuders, S.; Thone, M.; Riemer, A.; Niendorf, T.; Troster, T.; Richard, H.A.; Maier, H.J. On the Mechanical Behavior of Titanium Alloy TiAl6V4 Manufactured by Selective Laser Melting: Fatigue Resistance and Crack Growth Performance. *Int. J. Fatigue* **2013**, *48*, 300–307. [CrossRef]

19. Reschetnik, W.; Bruggemann, J.-P.; Aydinoz, M.E.; Grydin, O.; Hoyer, K.-P.; Kullmer, G.; Richard, H.A. Fatigue Crack Growth Behavior of Additively Processed EN AW-7075 Aluminium Alloy. *Procedia Struct. Integr.* **2016**, *2*, 3040–3048. [CrossRef]

20. Riemer, A.; Richard, H.A. Crack Propagation in Additive Manufactured Materials and Structures. *Procedia Struct. Integr.* **2016**, *2*, 1229–1236. [CrossRef]

21. Strantza, M. Additive Manufacturing as a Tool for Structural Health Monitoring of Metallic Structures. Ph.D. Thesis, Vrije Universiteit Brussel, Brussels, Belgium, 2016.

22. Gunther, G.; Leuders, S.; Koppa, P.; Troster, T.; Henkel, S.; Biermann, H.; Niendorf, T. On the Effect of Internal Channels and Surface Roughness on the High-Cycle Fatigue Performance of Ti-6Al-4V Processed by SLM. *Mater. Des.* **2018**, *143*, 1–11. [CrossRef]

23. Ambrogio, G.; Gagliardi, F.; Muzzupappa, M.; Filice, L. Additive-Incremental Forming Hybrid Manufacturing Technique to Improve Customized Part Performance. *J. Manuf. Process.* **2019**, *37*, 386–391. [CrossRef]

24. Lauwers, B.; Klocke, F.; Klink, A.; Tekkaya, A.E.; Neugebauer, R.; Mcintosh, D. Hybrid Processes in Manufacturing. *CIRP Ann.* **2014**, *63*, 561–583. [CrossRef]

25. Lhuissier, P.; de Formanoir, C.; Martin, G.; Dendievel, R.; Godet, S. Geometrical Control of Lattice Structures Produced by EBM through Chemical Etching: Investigations at the Scale of Individual Struts. *Mater. Des.* **2016**, *110*, 485–493. [CrossRef]

26. Hinderdael, M.; De Baere, D.; Guillaume, P. Fatigue Performance of Powder Bed Fused Ti-6Al-4V Component with Integrated Chemically Etched Capillary for Structural Health Monitoring Application. *Proceedings* **2018**, *2*, 499. [CrossRef]

27. Hinderdael, M.; De Baere, D.; Moonens, M.; Vafadari, R.; Guillaume, P. Effect of Surface Roughness on Fatigue Crack Initiation in Additive Manufactured Components with Integrated Capillary for SHM Application. In Proceedings of the International Committee on Aeronautical Fatigue and Structural Integrity, Nagoya, Japan, 5–9 June 2017.

28. Schijve, J. *Fatigue of Structures and Materials*; Kluwer Academic Publishers: Dordrecht, The Netherlands, 2001.

29. Schijve, J.; Hoeymakers, A.H.W. Fatigue Crack Growth in Lugs. *Fatigue Eng. Mater. Struct.* **1979**, *1*, 185–201. [CrossRef]

30. Boljanovic, S.; Maksimovic, S. Fatigue Crack Growth Modelling of Attachment Lugs. *Int. J. Fatigue* **2014**, *58*, 66–74. [CrossRef]

31. Paris, P.C.; Gomez, M.P.; Anderson, W.E. A Rational Analytic Theory of Fatigue. *Trend Eng.* **1961**, *13*, 9–14.

32. Paris, P.C.; Erdogan, F. A Critical Analysis of Crack Propagation Laws. *J. Basic Eng.* **1963**, *85*, 528–533. [CrossRef]

33. Wyart, E. Three-Dimensional Crack Analysis in Aeronautical Structures using the Substructured Finite Element/Extended Finite Element Method. Ph.D. Thesis, Université Catholique de Louvain, Ottignies-Louvain-la-Neuve, Belgium, 2007.

34. Duflot, M. A Study of the Representation of Cracks with Level Sets. *Int. J. Numer. Methods Eng.* **2007**, *70*, 1261–1302. [CrossRef]

35. Moës, N.; Dolbow, J.; Belytschko, T. A Finite Element Method for Crack Growth without Re-meshing. *Int. J. Numer. Methods Eng.* **1999**, *46*, 131–150. [CrossRef]

36. Geuzaine, C.; Remacle, J.-F. Gmsh: A Three Dimensional Finite Element Mesh Generator with Built-in Pre- and Post-processing Facilities. *Int. J. Numer. Methods Eng.* **2009**, *79*, 1309–1331. [CrossRef]

Review

Evaluation of Cracking Patterns in Cement Composites—From Basics to Advances: A Review

Maciej Szeląg

Faculty of Civil Engineering and Architecture, Lublin University of Technology, 40 Nadbystrzycka Str., 20-618 Lublin, Poland; maciej.szelag@pollub.pl; Tel.: +48-81-538-4762

Received: 6 May 2020; Accepted: 27 May 2020; Published: 29 May 2020

Abstract: The structure and the development degree of a cracking pattern has a key impact on the durability of cement composites. This literature review focuses on the four most important aspects related to the evaluation of the surface cracking patterns, i.e., the process of formation, propagation and evolution of cracks into a branched system of cracks from the point of view of the fracture mechanics; the detection techniques of the cracking patterns on the surface of cement composites, where the tools of computer image analysis are the most used; parameters which can quantify the development degree and morphology of the cracks system; and also the influence of a cracking pattern on the functional features of cement composites. The studies described so far indicate the necessity of continuous development of this research area, because the knowledge of key relationships between the cracking patterns and functional properties of a cement composite is necessary to estimate the degree of material degradation. Researchers agree that the works carried out in the field of evaluation of the cracking patterns, to a large extent, contributes to the development of non-destructive testing methods in the field of cement composites technology.

Keywords: cracking pattern; cracks; cement composites; concrete; image analysis; non-destructive testing

---

1. Introduction

A cement composite is a conglomerate of aggregate grains combined into an artificial stone as a result of hardening of cement paste. This group consists of concretes—containing fine and coarse aggregate, mortars—containing fine aggregate and cement pastes, which in a sense can be considered as the micro-concretes. In this definition, the internal structure of a mature cement paste is composed of: relics of cement grains—as aggregate, gel-crystalline form of hydrated cement grains—as binder, free water remaining after the hydration reaction, as well as contractionary gel and capillary air pores of various moisture content [1]. The basic features of a fresh cement composite mix are its workability and potential strength. The latter feature manifests itself only after hardening but its value is pre-determined by the composition of the mixture and technology of its preparation. The mechanical strength first of all depends on the quality of the cement paste, i.e., the properties of the cement and the water/cement ratio (w/c). A fresh cement composite mix should contain enough cement and water to achieve the desired consistency on the one hand, and the hardened concrete on the other hand has the designed strength with filled aggregate cavities.

At the end of the 1920s, Brandtzaeg [2], who investigated longitudinal and transverse deformations of a concrete sample subjected to a uniaxial compression, found an increase in volume at close to destructive stress, which he attributed to an intensive increase in the number and size of the microcracks. Several years later the same phenomenon was observed by Yoshida [3]. For the next two decades, the phenomenon observed by the two aforementioned researchers was of little interest. It was only the growing problems related to safety and durability of reinforced concrete structures that drew the researchers' attention again to the changes occurring in the structure of concrete under a short- and long-term load. The increasingly improved measuring equipment made it possible to develop new research methods, both direct and indirect, allowing to conclude on the destruction of structural bonds at various stages of loading. The first property that has been observed is the fact that in the case of strong aggregate grains, the cracks run through the contact surface of the cement paste-aggregate, the so-called Interfacial Transition Zone (ITZ). On the other hand, in the case of grains weaker than hardened cement matrix, these grains may be broken up. Today, this feature of the cement composite is considered crucial in terms of its durability and strength.

The disadvantage of cement composites is their ability to brittle cracking. Nevertheless, they are characterized by low plasticity, because the destruction under a static load occurs with moderately low total deformation. The brittle behavior limit is assumed to be a deformation of the order of 0.001 to 0.005 (relative value) at destruction stage, with high strength concrete being more brittle than normal strength concrete [1,4]. A sudden release of an elastic energy causes the local tensile strength of the cement matrix to be exceeded, resulting in a crack formation [5–7]. A characteristic feature of cement matrix and cement composites is the propagation and development of cracks under increasing load or duration of impact. The cracks, by joining and intersecting, form a network of cracks in the material, which is also referred to as the cracking pattern [8–16]. An example of the cracking pattern on the surface of a cement matrix is presented in Figure 1. In the cement composite technology, one of the key research areas is the analysis of the process of crack formation and development, because the formation of the cracking pattern is accompanied by a decrease in durability of the material. This is manifested by the deterioration of virtually all properties of cement composites, ranging from a decrease in mechanical strength, through reduced frost resistance, water resistance or resistance to aggressive chemicals. In the final stage of development of the cracking pattern there is a total loss of material cohesion, which results in its destruction.

Figure 1. The cracking pattern on the surface of a cement matrix exposed to a temperature of 450 °C.

The aim of the article is to provide a current state-of-the-art in the field of the analysis of the cracking patterns in cement composites. This research issue has a very high scientific potential and a great practical significance, but due to its multidisciplinary character it is sometimes difficult to implement. The analysis of the cracking patterns combines many fields of science, including the fracture mechanics, image analysis and technical aspects of the materials engineering—in this case, the cement composites technology. The scope of the work includes mainly consideration of the surface cracks. This literature review is divided into 6 main parts. The first section is an introduction to the topic of the cracking patterns in cement composites. The second section discusses the basics related to the process of formation and propagation of cracks in brittle cement materials. The process of evolution of single cracks into a complete, branched system of cracks present in the structure of the material is discussed. The analysis of the cracking patterns requires their extraction from the material structure. Since computer image analysis is definitely the dominant method of an automatic crack detection, Section 3 discusses the most important techniques used to extract cracks for further quantitative analyses. Section 4 discusses the most important and most frequently used quantitative parameters, which are used to describe the morphology and complexity of the cracking patterns. Then, in Section 5, the results of the most important studies on the influence of the structure of the cracking patterns on the functional properties of a wide range of cement composites are discussed in a synthetic way. The last section is a summary of the literature review, which indicates the most important problems and difficulties in quantitative analysis of the cracking patterns, and indicates possible further directions of research and development in this area.

What distinguishes this review from others of similar subject matter is the fact that the paper presents in a comprehensive way the problem of analysis of the cracking patterns in the structure of cement composites. There are reviews on image analysis itself, in which only methods of the digital cracks extraction are discussed. In such works there is a lack of information about the influence of the cracking patterns on the functional properties of cement composites. So far, no work has been recorded in which this subject has been discussed in a collective way, i.e., starting from the presentation of the mechanism that causes cracks and their transformation into a cracking pattern, through the techniques used for analysis, as well as quantitative parameters that can characterize the cracking patterns, to the most important thing from the point of view of the cement composites technology—i.e., how the cracking patterns affect the functional properties of this group of materials.

2. Causes of Cracks and Cracking Patterns in Cement Composites

Destruction of the cement composite is a consequence of its cracking, so it is important to consider this problem in detail. The actual strength of a cement composite or other brittle material is much lower than the theoretical strength, determined on the basis of molecular cohesion (strength of ionic or atomic bonds), and calculated on the basis of surface energy of a perfectly homogenous solid body (without internal damage). Theoretically, the strength should be approximately equal to tenth of the Young's module, but practically it is much smaller. This state gave rise to the application of the Griffith's theory [17] to describe the physical and mechanical properties of cement composites. Griffith was a pioneer of the fracture mechanics and he assumed that crystals always have defects that cause stress concentration in a small area, which is enough to locally exceed the theoretical fracture strength of the material. Starting from the energy balance of the process, Griffith determined the so-called crack propagation criterion. It gives the stress value which causes further cracking of the brittle material:

$$\sigma_k = \sqrt{\frac{2EF}{\pi c}} \tag{1}$$

where:

E—the Young's modulus of elasticity [N/m^2],
F—the surface energy [N/m],
c—half the length of the crack [m].

Thus, for brittle materials that have a certain elasticity, the strength is dependent on the Young's modulus, surface energy and a crack length. However, the energy required to produce a new surface due to cracking is actually higher than σ_k. In order to take this into account, the Griffith formula has therefore introduced the fracture work W [J/m^2] = [N/m], which, in addition to the surface energy, includes the plastic deformation of the material that accompanies the crack propagation:

$$\sigma_k = \sqrt{\frac{2WE}{\pi c}} \tag{2}$$

The literature contains numerous studies on the applicability of Griffith's theory to hardened cement matrix [18]. According to Mindess [19], the W for a cement matrix is in a fairly wide range of 7–14 J/m^2. The determination of W itself is problematic, however, it is not necessary to determine σ_k. For this purpose, it is sufficient to experimentally determine the critical stress intensity factor (K_{IC}), which is the product of W and the Young's modulus of the material. During the determination of K_{IC}, the stress at which the material breaks is determined, which is equivalent to the moment when the crack begins to grow rapidly. The practical significance of the K_{IC} lies in the fact that by knowing its value it is possible to determine the value of destructive stress depending on the shape and dimension of the crack and vice versa, i.e., by knowing the value of an operational stress in the element, it is possible to calculate the size of the critical crack at which it will break. In the light of the tests carried out so far, it appears that the K_{IC} value for pastes made of Portland cement is in the range from 0.4–0.5 MN/m$^{3/2}$ [20–24]. The current guidelines for the K_{IC} testing of cement composites are contained in the RILEM Draft Recommendations [25].

It is known that the physical and mechanical properties of cement composites are mainly determined by such factors as: total porosity, pore size distribution, the presence of material defects and the degree of structure variation. In the literature there are many studies [26–29] which indicate a strong correlation between the porosity of a cement composite and its strength. In this aspect, however, the pore structure is also of more importance, where it has been shown that the strength increases with decreasing pore dimensions [30,31]. Thus, the application of the classical fracture mechanics to the cement matrix means that the factor determining the strength will not be the total porosity, but the dimension of the largest gap—in this case the air pore [32,33]. An experiment conducted by Birchall et al. [34] proved that the bending strength of classical cement pastes lies on the curve determined from the Griffith's equation (Figure 2a). The width of the gap (pore) was substituted for the dimension of the largest defect, naturally occurring in the material or artificially produced. On the other hand, the compliance of strength of cement paste with the overall porosity results from the fact that factors that reduce the overall porosity also reduce the dimension of the critical gap (pore) [35].

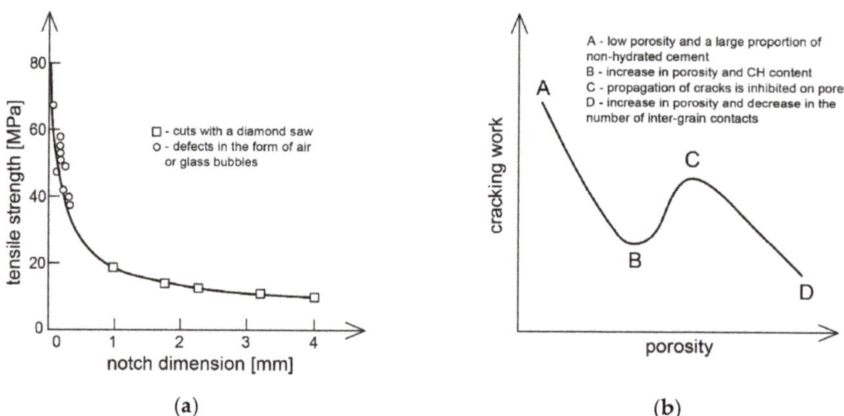

Figure 2. Dependency: (**a**) bending strength as a function of defect size (according to [34]); (**b**) cracking action as a function of porosity (according to [24]).

The dependence of W and K_{IC} on the porosity of the cement matrix is complicated. Beaudoin [24] developed a qualitative course of this dependence, as shown in Figure 2b. Both of the above parameters depend to a large extent on the way the material is dried, i.e., ultimately on the humidity of the environment in which the cement matrix matures, as the effect of water on the strength of the cement paste is commonly known. The results of [20,24] show that a decrease in relative humidity of the environment from 100% to 0% results in an increase in W from 7 to 14 J/m^2 and K_{IC} from 0.20 to 0.46 MN/m$^{3/2}$. Samples of cement paste soaked in water have a much lower tensile strength than dry ones. In the case of compressive strength, this feature increases slightly after drying the material [36,37]. Water, absorbing on the crack walls, shows the pushing pressure, which significantly facilitates the cracking process. On the other hand, the drying process removes the pushing water pressure on the crack planes near the crack limit. This was indirectly confirmed by an experiment conducted by Robertson and Mills [37]. They soaked the cement paste with organic solvents, which prevented significant reduction of tensile strength when moistened. Large molecules of organic solvents could not penetrate the microcracks, which prevented the pushing pressure.

The analysis of the effect of morphology of hydrated cement matrix on the cracking process was based on the hypothesis that portlandite crystals weaken its structure [38]. It was found that the cracks largely surround the areas where $Ca(OH)_2$ crystals occur. However, in the light of other studies [39] it was found that the strength of portlandite is similar to that of the CSH phase. On the other hand, the reason that cracks form relatively easily and run along these crystals is the morphology of the CH crystals themselves in the transition zone and high porosity in ordinary concretes. Ultimately, this results in a reduction in concrete strength. A radical change in this situation occurs when the *w/c* ratio is reduced or microsilica is applied. In his considerations, Mindess [18] believes that empirical formulas combining porosity with strength can be used, but only for the classical cement matrix. In the case of significant modification of the structure resulting, e.g., from the presence of an additive or reactive admixture, or as a result of a hydrothermal treatment, the classical empirical formulas cannot be used.

The properties of concrete are mostly determined by the properties of the cement matrix. It is known that the hydrated cement matrix contains numerous discontinuities in the form of pores, microcracks and voids, which do not necessarily act as defects themselves. A defect can be, e.g., a crack in the grain that is associated with the presence of such a void, or it can be caused by shrinkage or insufficient adhesion. This is a natural situation in a cement composite, taking into account the heterogeneous structure of the material and the way the different phases combine. It has been confirmed [40] that pores in the cement paste are not the only possible critical defects of the structure. The application of the Griffith's hypothesis assumes that first of all the voids in an unsegregated cement composite must be distributed at random, which is in line with the actual state of affairs [41]. Secondly, it is assumed that in the place where the defect exists there is a microscopic destruction, and the unit of material volume containing the weakest place determines the strength of the whole sample. Thus, each crack spreads over the entire section of the specimen subjected to a given stress or, in other words, a phenomenon occurring in an element is identified with the same phenomenon occurring in the entire volume. This occurs under the assumption of an even distribution of stresses and under the condition that the second consecutive weakest point in the material is not capable of carrying a stress *n/(n−1)* times greater than the stress at which the weakest point is destroyed, where n is the number of elements in the loaded section, each containing one defect. Thus, colloquially speaking, a cement composite is as strong as its weakest link.

Studies [42–44] indicate that in the ITZ and in the cement matrix itself there are very small cracks even before the load is applied to the cement composite. They are most probably the result of unavoidable differences in properties between the aggregate and hydrated cement matrix, combined with shrinkage and thermal stress. Microcracks are observed not only for cement composites of normal strength but also in unloaded concretes of *w/c* < 0.25, which mature in moist environment [45]. According to some studies [46], it is believed that the microcracks existing before loading are mainly responsible for low tensile strength of the cement composite. The microcracks are defined as cracks with a maximum width of up to 0.1 mm [1,46], which usually means the smallest dimension that can be followed by the naked eye.

The process of development of the microcracks under the compressive stress is already known. When a growing load is applied, the cracks remain stable to a load of about 30% of the limit load. Then the microcracks begin to propagate in terms of both length and width, and their number also increases. The stress at which the cracks develop depends on the *w/c* ratio of the cement matrix. This state is called as the stable crack propagation. As a result of further increase of the load up to a value between 70–90% of the maximum load, cracks are formed in the mortar. The cracks are then connected to the cracks caused by the breaking of adhesion and thus a certain system of continuous cracks is created, the so-called cracking pattern. This state is called as the state of rapid propagation of the cracks [42]. The stress level at the beginning of this state is higher in high strength concretes than in normal concretes. The increase in cumulative crack length is very large, while it is lower in high strength concretes [47]. The beginning of the state of rapid crack propagation corresponds to the point of discontinuity in volumetric deformations. If the load is fixed, the material can be destroyed as a function of time. However, as long as the cracks are stable, their presence is not harmful. This state is characteristic for the whole family of cement composites.

The structural heterogeneity of concrete is sometimes beneficial. For example, when the contact between coarse aggregate and cement paste is the site of local microcracks, it means that the presence of coarse aggregate grains prevents the opening of a single wide crack. These grains act as a blocking factor for the microcracks. The adhesion surfaces in the ITZ are formed at any angle to the direction of external force. As a result, local stresses differ significantly, upwards and downwards, from the nominal applied load. In the cement composite, cracks occur at each level of structural heterogeneity of the material [48–51]. Using an electron scanning microscope, the existence of a sub-microcracks was detected in the cement gel. However, there is no clear evidence that these sub-microcracks have a significant effect on concrete strength.

As shown above, the process of cracking and crack propagation is not a simple and unambiguous process. Very often, the formation of a crack may be initiated by one particular cause, and its development may be associated with a different cause [52]. Under certain physical conditions, the total width of cracks per unit length of the cement composite is usually constant. It is advantageous for the width of cracks to be as small as possible, which results in the fact that it is better if there are more cracks, but with a smaller widths than if there were fewer of them, but with a larger widths [12,13,15]. For example, the introduction of reinforcement controls shrinkage cracks by reducing the width of individual cracks, but does not change the total width of all cracks. The importance of the cracking process and the minimum width of the crack, which is considered significant, depend on the function of the structural element and on the exposure conditions of the cement composite. Table 1 summarizes the classification of cracks occurring in cement composites, together with the reasons for the formation and approximate time of occurrence of a given type of cracks.

Table 1. Types of cracks characteristic for structural elements made of cement-based composites (according to [52]).

Type of Cracks	Reasons	Occurence Time	Typical Location
plastic settlement	excessive secretion of cement milk—bleeding; rapid initial drying	10 min–3 h	sections of considerable depth; top of column, trough and coffered panels
plastic shrinkage	rapid initial drying; reinforcement at the surface; low speed of drained water discharge	30 min–6 h	surfaces; slabs
early thermal contraction	excessive heat generation; excessive temperature gradients; rapid cooling	1 day–(2–3 weeks)	thick walls; thick plates
prolonged shrinkage during drying	ineffective connections; excessive contraction; inadequate care	several weeks or months	thin panels; walls
map-cracking	excessive rubbing; rich mix; inadequate care	1–7 days	walls; plates
corrosion of the reinforcement	insufficient cover; poor concrete quality	over 2 years	columns; beams
alkaline reactions in the aggregate	reactive aggregate; cement with high alkali content	over 5 years	moistened elements
ankle cracks	aggregates damaged by frost	over 10 years	free ends of plates

3. Techniques for the Detection of Individual Cracks and the Whole Cracking Pattern

The first studies concerning the analysis of the morphology of cracks in cement composites were more qualitative than quantitative. Cracks were analyzed by means of a visual evaluation. This introduced great inaccuracies because the study was contaminated with human factor. Small portable measuring microscopes were started to be used for more accurate measurements, which were mainly used to determine the crack opening width in reinforced concrete structures [53]. It was only the development of digital technologies that initiated the development of research methods for the cracks analysis, as the computer image analysis was used for this purpose. The main problem of methods based on the digital image analysis is still the correct detection of cracks on the material surface. The reason for this is mainly a weak contrast of cracks to the remaining surface of the material. In the case of cement composites, cracks are visible as black or dark grey, and the natural shade of this group of materials is similar to grey. In recent years there has been a great development of digital testing methods for analyzing single cracks as well as the whole cracking patterns. This section discusses the most frequently used and at the same time most popular image analysis techniques, used in the analysis of cracks in cement composites, from the simplest in application to advanced segmentation algorithms.

3.1. Global Thresholding

The most basic and at the same time least complicated method of crack segmentation is the global thresholding [54–56]. It consists of determining for the whole analyzed image the global threshold value on the histogram, below which all pixels are interpreted as cracks and the rest are treated as background. This method is the least accurate of all methods of image analysis, because it can result in a very noisy image—cracks are very often integrated into the noise generated by the background. Depending on the complexity of the morphology of the analyzed surface, very often such an image is not suitable for analysis and drawing conclusions in terms of identification and evaluation of the cracking patterns. Currently, no analysis is performed using the global thresholding method, while the thresholding operation itself is popularly used at various stages of analysis, especially at an image pre-processing stage.

3.2. Locally Adaptive Thresholding

Using this group of algorithms, the threshold is calculated individually for each pixel depending on the local statistics of neighboring pixels. Such statistics can be, e.g., variance, range or surface-fitting parameters. Finally, the threshold value is a function dependent on the position of the pixel, i.e., $T(i,j)$, where: i, j—coordinates of the pixel [56]. This method is much more useful when assessing whether or not there is a crack in the analyzed, local area. Tang and Gu [57] used the locally adaptive thresholding on the histogram, which was initially smoothed with the Gaussian filter. They obtained a good quality image showing the cracking pattern on the surface of the concrete road pavement, however, it was contaminated with small objects with non-linear characteristics. They used morphology operators such as the dilation operator and erosion operator to remove them. Tong et al. [58] developed software for cracks inspections for concrete bridges using adaptive thresholding elements in their algorithm. Liu et al. [59] developed a multistage method for identifying cracks in concrete assets, without any focus on specific application. In order to solve the problem of low contrast between the cracks and background, they proposed an image enhancement algorithm that uses the multi-scale guided filters with gradient information. The result was a well-contrasted image that uses the adaptive threshold segmentation to create a binary image representing the cracking pattern. Tests carried out on various images showing the real surfaces of cement composites showed the high validity and robustness of this approach. An example of using the method is shown in Figure 3.

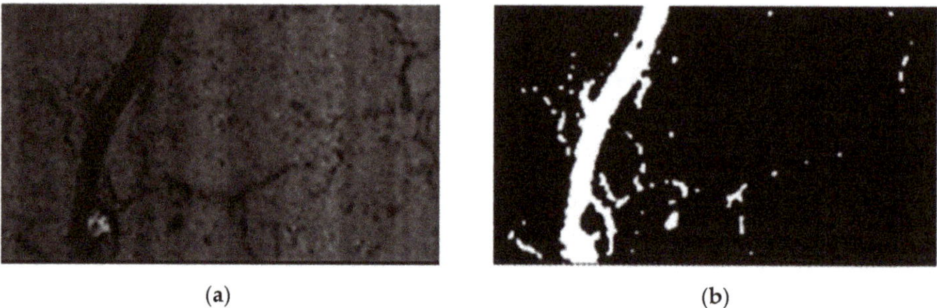

(**a**)   (**b**)

Figure 3. Example of the effect achieved by the locally adaptive thresholding: (**a**) original image; (**b**) after processing; based on [60].

3.3. Otsu Thresholding

The Otsu method is a very popular method, valued for its simplicity and efficiency. The algorithm was published in 1979 [61]. The aim of the algorithm is to binaryize the image, i.e., to convert the grayscale image to a binary image. The method is based on the histogram analysis and consists in minimizing the sum of the weighted variance of two classes (background and foreground objects), which is the same as maximizing the interclass variance. The method is particularly well suited for cases where the number of background and foreground pixels is similar. The idea of this method is shown in Figure 4.

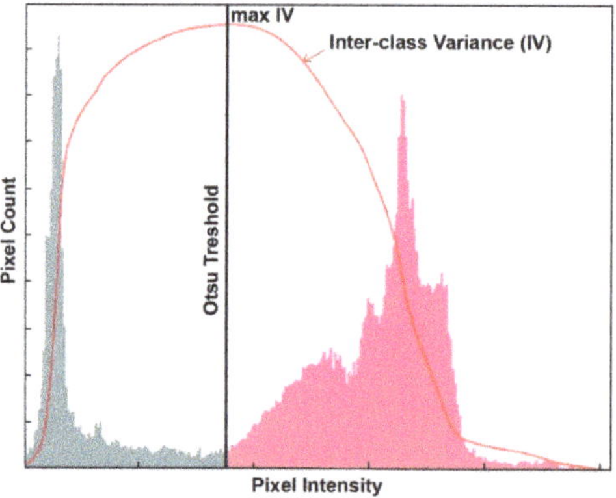

Figure 4. The Otsu method visualization.

Talab et al. [62] used the Otsu algorithm to detect major cracks on the surface of concrete structures. The original image was initially modified with a Sobel filter. The proposed method allowed for clear and accurate detection of cracks in images. Valenca et al. [63] developed an innovative method called the "MCRACK", which aimed at automatic identification of the cracking patterns on the surface of cement composites, using digital image processing techniques. The whole procedure consisted of 6 stages. In one of the stages it was necessary to create a binary image which reinforced the discontinuities on the surface of the material. For this purpose, several algorithms of image binearization were compared, including the Canny edge detector, the Otsu method and the manual threshold. Binearization with the Otsu algorithm proved to be the most stable and gave the most accurate results, so it was decided to implement it in the final method. Hoang [64] proposed a way to improve the Otsu method (Figure 5) by earlier implementation of an image enhancement algorithm called the Min-Max Grey Level Discrimination. The whole procedure allowed for positive identification of cracks on the surface of cement composites, additionally analyzing the morphology of individual cracks, i.e., determining perimeter, area, width, length and orientation.

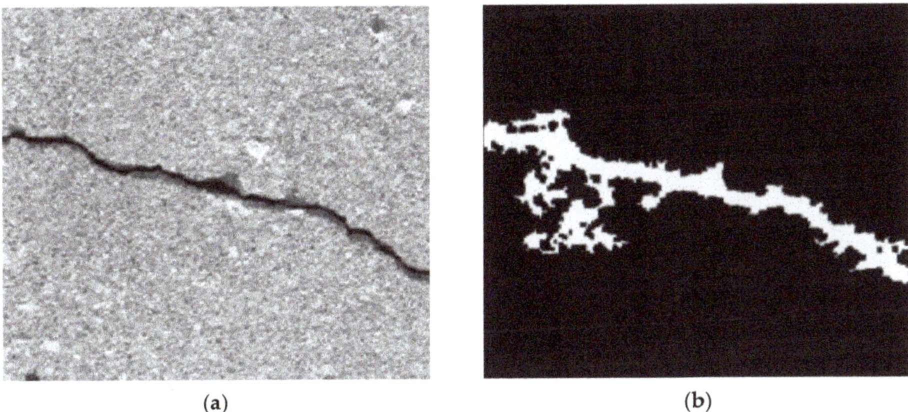

(a)   (b)

Figure 5. Extraction of cracks with Otsu thresholding: (**a**) original image; (**b**) after processing; based on [64].

3.4. Genetic Algorithms

Genetic algorithms are applicable to many optimization problems. An algorithm is a kind of heuristics that searches the space for alternative problem solutions to find the best solutions. By definition, the pattern of genetic algorithms resembles the phenomenon of biological evolution. Nishikawa et al. [65] designed an algorithm based on genetic programming which created an image filter for cracks detection on the surface of cement composites. The developed algorithm was an improved version of the filtering technique proposed by Aoki and Nagao [66], who suggested that the technique developed by them could be used to solve many engineering problems. The algorithm proposed by Nishikawa et al. [65] presented robust performance for removing of noises from the image and for cracks detection. In his next work Nishikawa et al. [67] expanded the capabilities of the previously developed genetic algorithm to measure the width of crack opening and determine the spatial orientation of the cracks on the analyzed concrete surface. The crack opening width measured by means of the modified algorithm was in high agreement with the width measured manually, which confirmed the possibility of practical application of the developed method for health monitoring of concrete structures. Genetic algorithms were also used in the detection of cracks on road surfaces, where the input material was a monochromatic surface image [68]. An example of using the genetic algorithm is shown in Figure 6.

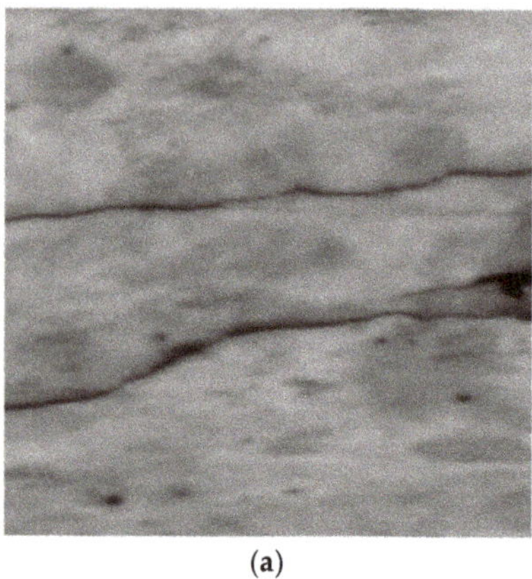 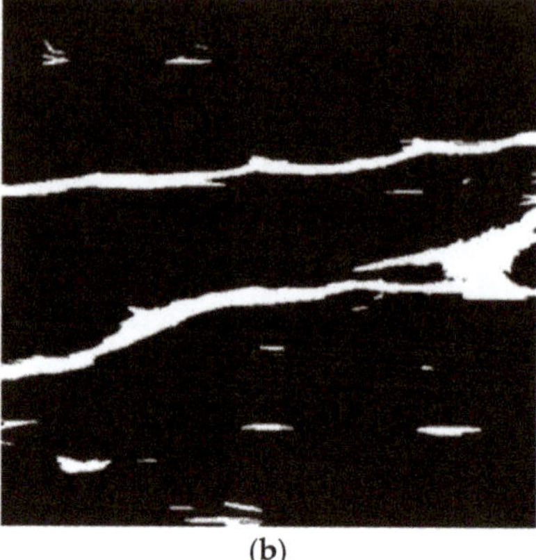

(a)   (b)

Figure 6. Extraction of cracks on a concrete surface using the genetic algorithm: (**a**) original image; (**b**) after processing; based on [69].

3.5. Fuzzy Logic Based Techniques

Fuzzy logic is one of the multi-value logic, i.e., one in which more than two logical values are adopted. This term is closely related to the fuzzy set theory. In the fuzzy logic, a series of intermediate values extend between state 0 (e.g., false) and state 1 (e.g., true), which determine the degree of belonging of an element to a set. The fuzzy logic is very useful in engineering applications, because the classical truth/false logic cannot effectively deal with many ambiguous problems [70]. The problem of this type in the cement composites technology is precisely the digital identification of cracks using an image analysis. The problem of ambiguous identification of cracks on the surface of a cement composite using the fuzzy logic has been shown in Figure 7. The fuzzy logic techniques are often used in combination with genetic algorithms and artificial neural networks, which may result in the creation of an intelligent system with the ability to generalize knowledge in a given area.

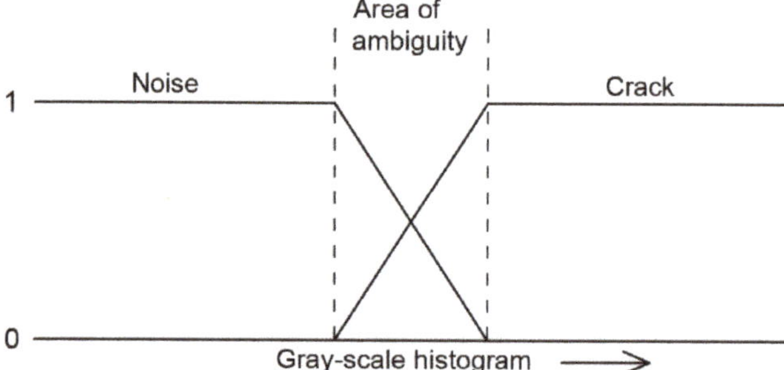

Figure 7. Model scheme of the fuzzy logic in an application to identify cracks on the surface of cement composites, where the input variable is the value on the gray-scale histogram.

Choudhary and Dey [71] used a fuzzy logic model in which they used "area" and "ratio" as input variables and the output variable was the "class" of the object. The class in this case was the pixel belonging to the crack or noise. Each variable consisted of two or more fuzzy subsets and corresponding trapezoidal membership functions. A total of 205 different images representing a cracked concrete surface were examined and it was found that the overall accuracy of the model was estimated at 90–94%. The fuzzy logic technique was also used to identify cracks on road pavements [72]. The whole developed method was based on the assumption that the crack is a continuous element and is darker on the histogram compared to its surroundings. The developed algorithm allowed positive identification of very thin cracks, even on very noisy pavement images. Yan et al. [73] developed an adaptive fuzzy image enhancement algorithm that effectively identified the cracking patterns on pavement surface.

3.6. Artificial Neural Networks

Neural networks are a very complex family of mathematical structures whose software or hardware models perform calculations or process signals through rows of processing elements called as the artificial neurons. The artificial neurons perform a basic operation on their input. There are different types of artificial neural networks, but their common feature is that their structure consists of neurons connected by synapses. Weights, or numerical values, are associated with synapses, whose interpretation depends on the model [74].

Lee and Lee [75] presented an integrated system consisting of three types of neural networks (image-based, histogram-based and proximity-based) to classify cracks on the concrete pavement surface. The system was validated on 124 actual pavement images. As a result of the analysis, it was concluded that the proximity-based neural network, despite the smallest computational requirements, is characterized by the highest accuracy in searching for the cracking patterns, which was estimated at about 95%. Bray et al. [76] developed a two-stage process of classifying cracks on the surface of concrete road surfaces using the artificial neural networks. In the first stage, based on the density and histogram, the artificial neural network divided into images with cracks and images without cracks. In the second stage, another neural network, after image segmentation, determined the type of cracks. The obtained results indicated 100% accuracy of the system in classifying images with cracks and 82% accuracy in classifying images without cracks. While the developed system gave good results in terms of cracks detection, it did not perform well in classifying the type of crack. Moon and Kim [77] developed an intelligent, automatic system for the cracks detection on concrete surfaces. The algorithm consisted of two main stages. In the first step, cracks were extracted from the digital image using filtering operations, the improved subtraction method and morphological operations. In the second stage, the cracks detection was performed by means of a backpropagation neural network. The whole algorithm was trained on 105 images of cracked concrete surfaces and then validated on another 120 images. The developed algorithm was characterized by a very high recognition rate, amounting to 90% for crack images and 92% for non-crack images, respectively. The authors emphasize the simplicity of the application of the algorithm, which makes it possible to assess the occurrence of cracks by non-expert inspectors. The versatility of programming artificial neural networks made it a very popular technique for detection of the cracking patterns, where further examples of application can be found in [78–81]. An example of the effect of identifying cracks on concrete surfaces with the implementation of artificial neural networks is shown in Figure 8.

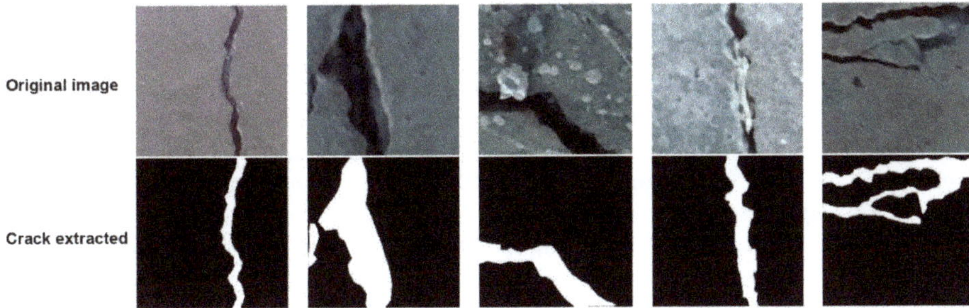

Figure 8. Results of crack detection by means of the artificial neural network, based on [79].

3.7. Dijkstra Algorithm

In theory, the algorithm is used to solve the problem of the shortest path, i.e., the problem of finding the shortest connection between given vertices in a weighted graph. The operation of the algorithm is based on the fact that when analyzing a given graph with a distinguished vertex there are distances from the source to all other vertices. In a very simple way, the algorithm can be modified to look only for the shortest path to one fixed vertex, interrupting the action when reaching the target vertex or transposing the graph's incident table. The feature of the Dijkstra algorithm is that it finds all the shortest paths between the selected vertex and all the others, while calculating the cost of passing each of these paths [82]. Using the Dijkstra algorithm is a very effective method to predict the occurrence of cracks in a given area, when there are cracks discontinuities due to a contamination of the digital image or local crack bridging is present.

Amhaz et al. [83] proposed an original method of automatic crack detection on digital pavement images, which main assumption is the location of minimal paths within each image. The developed minimal path selection algorithm consists of three main parts, i.e., endpoint selection, minimal path estimation using the Dijkstra algorithm and minimal path selection. The algorithm is supplemented by two post-processing steps, i.e., elimination of artifacts, and crack width detection. The developed algorithm was validated on real cracked pavement images and compared with five other crack extraction methods. The proposed algorithm automatically provided very robust and accurate results in a wide spectrum of situations, which is undoubtedly its advantage. Gunkel et al. [84] developed a procedure for automatic micro-crack detection that uses a shortest path algorithm. The method is very effective for the identification of microcracks that occur in a plastic deformation environment, where separation between cracks and plastic deformations is difficult. In the first stage of the method, cracks are detected as combined cluster systems, consisting of pixels with a gradient value below a certain threshold value. Then, the crack paths are identified with the Dijkstra algorithm as the longest shortest paths through the darkest areas of the cluster system. The developed method was applied to more than 2000 digital images of cracked surfaces. However, the procedure has some limitations because it assumes that the crack is only one path. The method is not suitable for identifying tree-like cracks.

Many researchers note [83–86] that using the Dijkstra algorithm to solve the problem of cracks detection on the surface of cement composites requires a very large amount of computational effort, which only highly specialized machines can cope with. It is necessary to further search and modify the algorithms that deal with the issue of the minimal path selection.

3.8. The Bayesian Classifier

One of the most effective methods of extracting the cracking patterns from a digital image of the cracked surface of a cement composite is the use of machine learning algorithms. One of the most popular algorithms used for this purpose is the Bayesian classifier, which is a simple probabilistic classifier. The classifier is based on the assumption of mutual independence of predictors (independent variables). They often have no relation to reality and therefore the classifier is often referred to as the Naive Bayesian classifier. In this case, the probability model can be derived using the Bayesian theorem, which binds the conditional probabilities of two mutually dependent events. Depending on the type of accuracy of the probability model, the Bayesian classifier can be effectively taught in a supervised learning mode. In many practical applications, the estimation of the parameter for the Bayesian classifier occurs using the maximum probability method "a posteriori", which is a kind of paradox because the classifier can be used without believing in the Bayesian theorem. A characteristic feature of the classifier is that the classification is valid as long as the correct class is more likely than others. Despite its naive design and many simplified assumptions, the Bayesian classifier is very effective in real situations [87].

Schmugge et al. [88] developed an automatic crack identification method for use in concrete and steel construction elements of nuclear power plants. In this case the Bayesian classifier was used to classify whether a pair of line segments belong to the same segment or not. These linear segments were a segment of the cracks. Then, certain geometrical features of the line were determined for each segment, e.g., distance, orientation and similarity. The tests carried out on real images using the developed method showed an improvement in the ability to identify cracks by 38% compared to the reference methods. Hutchinson and Chen [89] proposed an automatic system for the detection of various types of defects in concrete, not only cracks but also cavities. Their procedure is based on the Bayesian decision theory. The detection of a probable crack is based on the implementation of two algorithms, i.e., the Canny and the fast Haar transform. The Bayesian classifier, on the other hand, decided whether the identified area in the image, which may be a crack, is really a crack. The advantage of the developed method is a quite accurate localization of concrete surface damage, with a small calculation effort. Valenca et al. [90] went one step further and created a powerful tool called "SurfCrete" for analyzing concrete surfaces. With this method, it is possible to classify an area into different classes, e.g., crack, biological corrosion, exposed aggregate grain or repair mortar. In their work they used two types of classifiers, i.e., the Bayesian classifier and the Multi-Layer Perceptron. Each of the classifiers was taught on a fragment of the surface showing a given morphological change of the concrete surface. The effect of the method is the possibility to create a map of concrete surface damage. After validation of the method, the accuracy of approx. 94% was obtained on the real concrete surface images. The Bayesian classifier is also popular in cracks detection systems in road surfaces [91,92].

3.9. The AdaBoost Classifier

The AdaBoost classifier is another tool used by machine learning algorithms. It is one of the basic algorithms for boosting, which is a method by which a large number of weak classifiers can be used to get one better. In this case, a weak classifier is one that is relatively simple and can classify test data with an efficiency of more than 50%. The classifier works in such a way that in subsequent iterations it trains and then measures the error of all available weak classifiers. In each subsequent iteration the importance of badly classified observations is increased, so that the classifiers pay more attention to them. The algorithm can significantly improve the quality of classification, but this improvement is only observed when weak classifiers are used as components. When more complex classifiers are used, the use of the AdaBoost algorithm does not lead to a significant increase in effectiveness [93].

Cord and Chambon [94] used the AdaBoost classifier to classify defects on road surfaces. The whole method is based on an appropriate selection of linear and non-linear filters depending on the texture of the analyzed image. In this case, the task of the AdaBoost classifier is to choose the most optimal set of filters to extract cracks in a particular case. A number of textural descriptors are used to learn the classifier. The obtained results were compared with the results of the methods already described in the literature and high efficiency of the developed method was noticed. Prasanna et al. [95] developed an automatic method for the cracks detection on bridge deck surfaces, which they called as the STRUM. The use of machine learning algorithms allowed to eliminate the need for manual definition and tuning of threshold parameters. The developed method uses the robust curve fitting to locate potential cracked areas. Three types of classifiers were used within the algorithm, i.e., the AdaBoost, support vector machines and random forest. The procedure allows to develop the crack density maps in the analyzed area. The accuracy of the algorithm was estimated at about 90%.

3.10. Summary of Techniques for Identifying Cracks and Their Accuracy in the Literature Review Works

The authors of the works, which have been included in the literature review, stress the universality of individual methods, as they can be applied to virtually any engineering problem aimed at identifying surface linear structures. The presented crack identification techniques have been developed mainly to apply to the surface of cement composites, practically in every application, i.e., bridge structures, reinforced concrete tanks, floor and wall structures, structures in nuclear power plants, dams, etc. The second group of application that stands out is road surfaces, both concrete and asphalt.

Table 2 synthetically lists the types of techniques used to detect surface cracks in cement composites embedded in various types of structures, with references to literature items covering the scope of this literature review. In most cases the developed cracks detection algorithms are often a combination of many techniques, so the list also includes those methods which are not discussed in detail. In addition, the levels of accuracy of cracks identification obtained by individual researchers are presented, based on the literature survey.

Table 2. A summary of the cracks identification techniques and their accuracy, with literature references.

Digital Image Processing Technique	References	Accuracy Level (Only Where the Authors Have Given It)
Global thresholding	[54–56,84]	≤90% [54]
Locally adaptive thresholding	[55–60]	90–95% [58]
Otsu thresholding	[61–64,73]	≥95% [63]
Edge detection	[54,63,71]	≤90% [54]; 90–95% [71]; ≥95% [63]
Probabilistic approach	[55]	-
Median filtering	[55]	-
Sobel's filtering	[62]	-
Morphological approach	[57–59,77,81]	90–95% [58]; 90–95% [77]
Genetic algorithms	[65,67–69]	≤90% [67]
Fuzzy logic based techniques	[70–73]	90–95% [71]
Artificial neural networks	[71,75–81]	≤90% [76,78]; 90–95% [71,77]; ≥95% [75,79,80]
Wavelet approach	[76]	≤90% [76]
Dijkstra algorithm	[83–86]	-
The Bayesian classifier	[88–92]	90–95% [90]
The AdaBoost classifier	[94,95]	≤90% [95]
Statistical methods	[89]	-

Apart from methods closely related to the computer image analysis, there are also other methods of cracks detection, e.g., infrared thermography, x-ray tomography or acoustic emission.

4. Parameters to Describe the Morphology of the Cracking Patterns

The cracking pattern on the surface of a cement composite is a kind of surface structure. The morphology of this structure changes depending on many technological factors, such as the composition of the composite, the factor causing the cracking process, the duration of load, etc. Scientists have always wondered what reliable parameters from already existing ones can be used to quantify the cracking pattern. An alternative was to look for and define new quantitative parameters that would be used for this particular purpose. This section describes the most important and most frequently used quantitative parameters used to describe the morphology of the cracking patterns.

4.1. Opening Width of the Cracks

The crack opening width is one of the simplest parameters to measure, as measurements can be taken locally using portable optical microscopes. However, such a measurement is burdened by the error of the human eye. Greater measurement accuracy is achieved by using a digital measurement, on a previously prepared image of the tested surface. This parameter is also of great practical value, since practically every aspect of durability of cement composites depends on the opening width of cracks. Due to the simplicity of the measurement, it is very often used to measure the width of cracks opening as a supplementary test to the main research thread concerning the cracking patterns. It is also valuable to give the maximum crack width, because in the case of a cracked cement composite, its mechanical strength is largely dependent on this parameter.

Wagner et al. [96,97] proposed a classification of the measured crack width distributions as the crack width polygons. Cracks of particular widths are accumulated in percentage. The graph of the crack width polygons allows very quickly to determine the percentage of cracks with a width, which is below a certain limit. This parameter allows to characterize the whole of the cracking pattern, because it is physically very similar to the construction of the particle size distribution curves—in case of analyzing aggregates for cement composites.

4.2. Length and Orientation of the Cracks

The length of the cracks or the entire cracks system, as well as the orientation of the cracks, are one of the basic parameters for determining the properties of a cracks. It is relatively easy to measure when implementing computer image analysis into a research program. The combination of these two properties, which characterize the cracking pattern, can provide a lot of information, e.g., regarding the direction of shrinkage or swelling of the cement composite, the direction of the aggressive agent action, etc. The orientation of cracks on a degraded composite is very important for the prediction of its mechanical properties, as well as for predicting the direction of further development of the cracks network.

In his research, Szeląg [12–16,98] introduced a modification of the crack length measurement, by measuring the length of the perimeter line of the area which is limited by the cracks. This approach applies to the analysis of cracks that form closed areas called as the clusters, on the surface of the material. The clusters are formed only when there is a strong volumetric cracks propagation, e.g., in case of a thermal load.

4.3. The Crack Density

In the early 1990s, Mobasher et al. [99] defined a new parameter to describe the morphology of the cracking patterns, which is commonly used today, namely the crack density. This parameter was defined as the ratio of the average length of the crack to the area of the surface under analysis, according to the equation:

$$L_A = \frac{2\overline{L}}{a^2} \tag{3}$$

where:

L_A—the crack density [mm^{-1}],
$\overline{L}$—the average crack length [mm],
a^2—the area of the test section [mm^2].

A slightly modified version of the crack density, i.e., the parameter expressed as number of cracks per meter [13,96,100], is also encountered in the research. In this version, the crack density is more convenient to use, because in order to get the result it is necessary to count the number of cracks crossing the test line, and relate the result to the situation when the test line would be 1 m long. In this case, with the crack density it can be very easily calculated the crack spacing, which is a size that symbolizes the average distances between the cracks.

4.4. The Crack Area

The crack area is a surface parameter and determines which area is occupied by cracks. This parameter is used in two versions, i.e., as an absolute value expressed, e.g., in mm^2, and as a relative value in relation to the total area of the area under analysis, then often the crack area is given as a percentage value. Compared to the previously mentioned parameters, the crack area is more difficult to determine because it requires more complex digital image processing techniques.

Similarly as in the case of measuring the crack length, also in the case of the measurement of the crack area, Szeląg [12–16,98,101,102] applied a modification consisting in the measurement of the area which is limited by the cracks.

4.5. The Fractal Dimension

Very often such local parameters as its length or width of the opening are insufficient to describe the complexity of the cracking pattern. Thus, in recent years there has been a growing interest in using the concept of fractal dimension for this purpose. The use of fractal geometry to characterize the cracking pattern has many advantages compared to the use of Euclidean geometry. In order to use this dimension, it is necessary to assume that the structure created by the cracking pattern is in fact a fractal, i.e., it has certain characteristics such as [103,104]:

- formlessness—an unambiguous shape is impossible to determine,
- description by a recursive relationship, not a mathematical formula,
- self-similarity—an isolated fragment resembles its larger whole.

Looking at these assumptions, it is not difficult to resist the impression that the cracking pattern visible on the surface of cement composites fully meets these assumptions. In such a case, in the fractal geometry, it is the fractal dimension that determines how densely a fractal fills the metric space it is in. In the studies on the cracking patterns described so far, the fractal dimension is used as a stand-alone parameter, but often also serves as a component of an extensive index for evaluating the degree of material degradation [105].

4.6. A Summary of the Use of Individual Parameters to Describe the Morphology of the Cracking Patterns

Table 3 summarizes the parameters used to describe the morphology of the cracking patterns. The list includes works that have been included in this literature review.

Table 3. Analysis by parameter used, with references to the literature.

Parameter	References
Opening width of the cracks	[9,12–16,96,97,100–102,105–116]
Length and orientation of the cracks	[8,12–16,98,99,107,109,112,113]
The crack density	[8,9,13,96,99,100,110–112,114]
The crack area	[12–16,98,101,102,106,107,109]
The fractal dimension	[105,115–119]

5. Influence of Morphology of the Cracking Patterns on the Properties of Cement Composites

It is well known that cracks in cement composites are a defect in the material. Practically all functional properties of a cracked material deteriorate. The most important results of the research published so far concerning the influence of the structure of the cracking patterns on particular properties of cement composites are presented below in a synthetic way. The main focus is on performance at elevated temperatures, shrinkage and drying, water permeability, mechanical strength and chemical corrosion.

5.1. Performance at Elevated Temperatures

Mobasher et al. [99] examined the cracking process in cement composites, with different polypropylene fibers content, using an acoustic emission technique and quantitative image analysis. The morphology of the crack system was investigated for samples loaded to specific strain magnitudes. For this purpose, the crack density, length and spacing were determined. An average crack density of 1.27 mm^{-1} was obtained. It was observed that as the strain level increases, the formation of new cracks is accompanied by decrease in the crack spacing. Thus, the presence of polypropylene fibers in the cement composite resulted, first of all, in the necessity to supply more energy to the system so that the existing cracks could further propagate; secondly, an increase in the degree of cracks and microcracks dispersion in the material, which undoubtedly improved the mechanical properties of the composite.

Xu et al. [9] analyzed the morphology of the cracking patterns of concretes subjected to a high temperature. The subject of the study were concretes differing in the *w/b* ratio (water/binder) and fly ash content. A visual imaging technique was used to study, among other things, the crack density. It was noted that the crack density increased almost linearly up to 650 °C. There was a rapid increase in this magnitude between 650–800 °C. The degree of cracks dispersion increased as the fly ash content increased. Smaller *w/b* values were accompanied by a smaller number of cracks, however, the resulting cracks were characterized by a larger opening width and length, compared to concretes with a larger *w/b* ratio. It was found that a decrease in tensile strength with an increase in thermal load is identical with an increase in the crack density.

The studies carried out by Kim et al. [120] concerned the influence of thermal load (in the range 25–1000 °C) on the properties of cement paste. Among others, the 3D X-ray Computer Tomography was used to determine the change of pore structure and to evaluate the process of initiation and spatial configuration of thermal cracks, as a function of the temperature acting. It was observed that at temperatures higher than 900 °C a massive cracking network is formed in the edge zones of the material. This effect was considered to be the main reason for the occurrence of explosive spalling, which is one of the most dangerous phenomena accompanying cement composites exposed to the fire.

Magalhaes et al. [100] examined the mechanical response of a cement composite with PVA fibers. The samples were preheated at 90–250 °C. Next, uniaxial tensile, bending and compressive strength tests were carried out, during which the crack width and the crack density were measured. The studies showed that the crack density takes less and less values at the same level of deformation as the preheating temperature increases. In the case of the crack width, the inverse relationship was observed. The maximum crack widths were in the range 100–220 μm. The conducted research confirmed that in terms of mechanical performance, the more favorable situation is when there are more cracks in the analyzed area, but they are of smaller widths.

Szeląg evaluated the influence of microsilica [14,98,102], metacaolinite [16,101], polypropylene fibres [12] and multi-wall carbon nanotubes [13] on the process of formation of the cracking patterns in cement pastes, under thermal load, in the range 200–250 °C. The influence of the shape and size of the sample on the geometric characteristics of the cracking patterns was also investigated [15]. To describe the morphology of the cracking patterns, the following parameters were used: the area and perimeter of the cluster, the width of the crack opening and the crack density. It was found that modification of the cement matrix with the addition of microsilica causes densification of the crack network. There are more cracks, but they have a smaller opening widths, which has a positive effect on mechanical properties. In case of modification with metacaolinite, a decrease in the number of cracks formed with comparable widths of their opening in comparison to unmodified cement matrix was observed. The use of polypropylene fibers resulted in the reduction of area and perimeter of the clusters, with a negligible effect on the opening widths of the cracks. The polypropylene fibers melted due to thermal load and the cracks forming the cracking pattern were largely contained in cavities left by the fibers. In case of modification of the cement matrix with carbon nanotubes a drastic reduction of cracks system density was observed, with a simultaneous high increase in crack widths. It had a negative effect on mechanical performance. However, the effect observed was more related to the phenomenon of cement matrix foaming caused by the way carbon nanotubes were introduced than to their presence itself.

5.2. Shrinkage and Drying

One of the first preliminary observations on the influence of the drying process on the characteristics of the cracking pattern were made by Bazant et al. [110,111]. Concrete and cement pastes were the subject of research. They noticed that as a result of the drying process, a network of parallel cracks is formed, whose spacing and width of opening is proportional to the depth of penetration of drying. With typical test samples, most cracks are too thin to be observed with the naked eye. It was also found that the characteristics of the cracking pattern predominate in the direction of the development of discontinuous microcracking rather than in the direction of continuous macrocracks.

Wang et al. [106] studied plastic shrinkage behavior in cement composites with the addition of fly ashes and various types of dispersed fibers. The structure of air pores using a mercury-intrusion porosimetry and the cracking pattern using the image analysis were studied. The maximum crack opening width was measured as well as the crack area. It was observed that the addition of dispersed fibers generally increases the number of large pores in cement paste, which affects both bleeding behavior and crack distribution in the material. It was found that, depending on the type of fibers and their properties, the crack area can be reduced by 30–40%, with 0.1% (by volume) of fiber addition. It has a very big influence especially on the mechanical properties of the cement composite in the early stages of maturation.

Plastic shrinkage behavior of cement pastes with the addition of alkali-resistant glass fibers was studied by Bakhshi and Mobasher [107]. The addition of glass fibers in the amount of 0.06%, 0.11%, 0.17% and 0.23% by volume was studied. The tests were carried out in low-pressure conditions, and additionally the sample size, *w/c* ratio and conditions of initial care were also a variable factor. The morphology of the cracking pattern was analyzed using the image analysis. It was found that the drying process can be divided into two main stages, i.e., stage I—constant drying rate period, and stage II —falling drying rate period. The research confirmed that by using glass fibers it is possible to control the process of formation and propagation of cracks during the maturation process of the cement matrix. While it was not observed that the presence of fibers had an impact on the cracking pattern in the stage I, in stage II the presence of fibers reduced the diffusivity values. Scientists concluded that the reason for this was the bridging effect of the glass fibers. The results showed that in the case of 0.11% and 0.23% of fibers addition, the areal fraction of cracks was reduced by 22% and 61%, respectively. The maximum crack opening width was also reduced by 47% and 71%, respectively, compared to classical cement paste.

Bisschop and Wittel [8] conducted a study to evaluate the morphology of the cracking patterns in classical cement paste with the *w/c* = 0.5. The study determined the effect of sample thickness and methods of drying on the crack density, crack length and the crack penetration depth. The latter parameter was quantified using a fluorescent resin impregnation technique. It was found that for samples < 14 mm thick, the single-sided drying resulted in lower values of the crack penetration depth (15–20% of the sample thickness), compared to double-sided drying (20–35% of the sample thickness). In the case of samples that were dried in an Environmental Scanning Electron Microscope chamber at 25% humidity, 20 ± 0.2 °C, at 590 Pa, for 5 h, it was noticed that the cracking pattern was characterized by multiple crack cuts at an angle of about 120°. Many dead-end cracks were also found. In case of drying in a climatic chamber at normal atmospheric pressure it was found that the cracks intersections resemble more a 90°–180° system.

Lura et al. [113] studied the geometric features of the cracking pattern in cement paste. The observations were carried out on samples with a diameter of 10 mm, in which there was a centrally located steel rod with a diameter of 1.5 mm, 3 mm or 6 mm. The factor causing the cracking was the autogenous shrinkage of the cement paste. The crack detection was performed by impregnation with galium and analysis was performed by means of an optical and scanning electron microscopy. It was observed that as the thickness of the cement paste envelope around the steel rod increased (a rod diameter decreased), the samples showed the negative fracture geometry, which resulted in a stable crack growth. The length of the crack was greater the smaller the diameter of the rod was used.

Xuan et al. [114] examined the shrinkage cracking characteristics of concrete produced from the construction waste, mainly crushed concrete and masonry aggregates. The influence of four factors, i.e., masonry content, moisture content, degree of compaction and cement content, on the characteristics of the cracking pattern was examined. Two cracks characteristics, i.e., the crack width and the crack spacing, were studied. It was found that the crack width decreases with decreasing amount of cement, decreasing the compaction degree, increasing moisture and increasing the masonry content. In the case of the crack spacing it was found that this parameter decreases with decreasing the amount of cement, increasing the compaction degree, increasing moisture and decreasing the masonry content.

5.3. Water Permeability

Wagner et al. [96] made water permeability of strain-hardening cement-based composites dependent on quantitative parameters of the cracking patterns. The samples that were pre-cracked as a result of the uniaxial tension test were tested. To better describe the relationship between the cracking pattern and water permeability a new parameter was introduced, which was called the hydraulic crack pattern parameter R_{WP}. It was the product of chosen class width, the crack density and weighted crack width value. There was a strong correlation between R_{WP} and the water flow rate. In the course of the analyses, an equation binding the crack width with water permeability was also formulated for which the Hagen–Poiseuille equation was the starting point.

Lepech and Li [108] studied the effect of cracking degree of high performance fiber reinforced cementitious composites on water permeability. They limited the characteristics of the cracking pattern to determining the maximum crack width and to counting the number of cracks in the analyzed section. They found that with the increase of crack width, the value of the coefficient of permeability also increases. Moreover, normalized permeability per single crack was calculated. It was observed that this parameter also increases with the increase in crack opening width, although the number of cracks in the analyzed section is higher. The study also indicated a beneficial effect of PVA fibers in the context of reducing crack opening width and thus increasing water-tightness of the cement composite.

In the studies carried out by Aldea et al. [109], the relationships between cracks and water permeability of a normal strength concrete were observed. A controlled splitting tensile test was used to introduce cracks into the concrete structure and then the material was tested for water permeability. The results obtained indicated that water permeability increased significantly as the opening width of the cracks increases. Moreover, high repeatability of results for the same cracking levels was obtained. It was also found that the relationship between crack length and water flow is not linear, which was found for the relationship between the crack area and water flow.

An extensive research program was carried out by Torrijos et al. [112]. Two series of ordinary concretes were tested. Within the framework of the conducted works, the influence of the characteristics of the cracking patterns, expressed in the form of crack width and the crack density, was determined on a number of parameters related to permeability, i.e., the water absorption, water penetration, capillary absorption, water permeability. The cracking pattern was created by subjecting the samples to various degradation processes, including: low humidity drying, load at 150 °C and 500 °C and alkali-silica reaction. The highest crack density was found for concretes subjected to 500 °C load, cracks mainly occurred in the interface and to a lesser extent in the cement matrix. In case of samples subjected to drying in low humidity it was found that the velocity and capacity of capillary absorption increased with the crack density. This increase occurred to a certain maximum and then decreased, although the crack width and the crack density continued to increase. For concretes exposed to the alkali-silica reaction a high permeability was initially noticed, followed by a sharp drop after the first few hours. Rehydration of the gel at the crack sites was given as the reason, which resulted in sealing of the structure. The degree of internal destruction of the samples was evidenced by measuring the ultrasonic wave flow rate. A direct relationship between the pulse velocity and the crack density was found.

5.4. Mechanical Strength

Fahridzadeh et al. [105] proposed a new approach to assess the degree of degradation of reinforced concrete structures. They proposed to use the fractal dimension to describe the morphology of the cracking pattern on the surface of reinforced concrete. The effect was to develop the damage index, which was calculated according to the equation:

$$DI = \frac{D_i - D_l}{2 - D_l} \tag{4}$$

where:

DI—the damage index; $0 \leq DI \leq 1$,
D_i—the fractal dimension of the current status of the cracking pattern,
D_I—the fractal dimension of the cracking pattern computed during the first inspection.

The first inspection shall be considered to be the moment when the appearance of cracks on the surface of the component under test is observed as a result of the applied load. The tests carried out on the reversed cycled loaded reinforced concrete walls indicated high practical suitability of *DI*. It was found that the value of both the fractal dimension and *DI* increases with the next load cycle. It was also shown that *DI* can be used to estimate the remnant lateral stiffness of the wall.

The continuation of the above studies is presented in [115,117]. It was noted that in the literature there are practically no works that would correlate the morphology of the cracking patterns with the structural integrity of the studied material, in this case concrete. The multifractal analysis was used in the study and the considerations were carried out on synthetic crack patterns as well as on two real reinforced concrete walls subjected to cyclic mechanical loading. It was found that the development of cracks structure through an increase in their opening width, length, density, is accompanied by a correlated increase of the multifractal parameters. A rapid increase of these parameters was also observed when the concrete walls tested showed severe stiffness loss.

The subject of Ebrahimkhanlou et al. [118] was the analysis of the cracking patterns on the surface of pre-stressed concrete beams. The analysis was focused on the fractal and multifractal characteristics of cracks patterns, which were formed by a mechanical force. Once again, the high usefulness of the fractal dimension for quantitative description of the degree of development of the cracking patterns was confirmed. Moreover, it was proved that fractal analysis can be used to identify different cracking mechanisms.

Fooladi et al. [119] conducted research to determine the effect of the aggregate grading on the multifractal characteristics of the cracking patterns of concrete, with different compressive strengths. The study was carried out on two series of concretes for which the aggregate grading curve assumed the well-graded and gap-graded characteristics respectively. In the well-graded aggregate samples, it was observed that the value of the fractal dimension reached a saturation level, for which the increase in load did not cause any further change in the value of the fractal dimension. For samples with the gap-graded aggregate a continuous increase in the value of the fractal dimension was observed along with an increase in the value of mechanical load. In the study, the singularity spectrum analysis was also carried out, which, combined with the analysis of the fractal dimension, allowed to conclude that the development of cracks in both series of samples is characterized by multifractal character, but with different intensity. The study concluded that the multifractal approach allows to explain local irregularities of the cracking pattern.

5.5. Chemical Corrosion

Xu et al. [9] apart from studying the development of the cracking patterns as a function of the influence of elevated temperature of concrete with fly-ash, also conducted the rapid chloride diffusion test. The results obtained indicated that deterioration of concrete durability is correlated with the crack density.

Zhuang et al. [116] performed a fractal analysis of the crack patterns on the surface of reinforced concrete piles bonded with carbon fiber reinforced polymer. The factor causing the cracks was the corrosion of the reinforcement caused by a harsh marine environment. Within the framework of the research the damage index was developed to assess the degree of degradation of the piles, the method of calculation of which was very similar to the one presented in [105]. The assessment was made at different stages of reinforcement corrosion development. Together with the progressing corrosion, an increase in the value of fractal dimension of the cracking patterns was observed. Moreover, a study of maximum load of piles was carried out, which showed a very high correlation with the fractal dimension. The observed strength loss-fractal dimension dependence assumed a linear dependence. As a result of the conducted tests it was found that supplementing the measurements with the analysis of cracks width, a high quality information could be obtained regarding the degree of structure degradation in the situation when it is impossible to perform mechanical tests.

6. Summary

The paper provides a literature review of the development process, analysis and impact of the cracking patterns on the properties of cement composites. The review focuses on four main aspects related to the analysis of the cracking patterns:

- the process of cracks formation in brittle cement composites and their evolution into an extensive and complex system of cracks,
- methods and techniques of digital extraction of the cracking patterns for their further evaluation,
- quantitative parameters used to describe the complexity of the cracking patterns,
- evaluation of the influence of morphology of the cracking patterns on selected properties of cement composites.

Each of these four aspects requires a different scientific approach, making comprehensive analysis of the cracking patterns a difficult and multidisciplinary issue. The three main scientific areas that underpin this research area are the fracture mechanics, image analysis and cement composites technology.

The factors that determine the cracking process, i.e., the number and type of structure defects and the composition of the material, are discussed. It was found that critical stress intensity factor (K_{IC}) being the product of crack work and Young's modulus is the most commonly studied parameter which determines the ability of the material to propagate and develop cracks in the branched cracks system. The methodology of analysis of the cracking patterns requires the use of advanced tools for the detection and extraction of cracks. Research carried out so far in this field indicates that by far the most effective and popular group of tools is the computer image analysis. The simplest of them are threshold operations, e.g., the global thresholding, locally adaptive thresholding or the Otsu thresholding. However, such operations, despite the fact that they are simple to implement, are sometimes characterized by a very large detection error. More advanced techniques, such as the genetic algorithms, artificial neural networks or machine learning algorithms, are characterized by very good detection accuracy. Their disadvantage is the need for much more computational resources, which slows down the whole process and makes it sometimes impossible to analyze a large set of data. The literature review shows that the most accurate methods of cracks detection on cement composite surfaces are characterized by an accuracy of more than 95%.

Previous research indicates that the simplest parameter to measure, in terms of the complexity and morphology of the cracking patterns, is the width of the crack opening. However, this indicator does not provide information on the layout and organization of the cracks structure in the material. The parameter that provides this information is the crack density and the resultant the crack spacing. Recent studies postulate the use of fractal geometry to quantify the cracking patterns. In these studies, the concept of fractal dimension is used, which by definition determines the degree of complexity of the analyzed structure. Analyses carried out with the use of this parameter indicate a strong relationship with the mechanical features of a degraded cement composite. In the research results published so far, the greatest attention is paid to determining the influence of the cracking patterns on the performance of cement composites at elevated temperatures. The system of cracks formed due to the shrinkage and drying of the material as well as those formed due to chemical corrosion is also analyzed. The influence of the cracking patterns on water permeability and mechanical strength of cement composites is also evaluated. The presented results indicate that with the progressive development of the cracking patterns, practically every property of the hardened cement composite deteriorates.

The authors of many research point to the key importance of the degree of development of the cracking patterns for the functional properties and durability of cement composites. However, there are still few works in which the dependence of the cracking patterns on the physical and mechanical properties of the material is determined directly, in a numerical way. Knowledge in this area seems to be crucial in the aspect of designing durable cement composites, which would be resistant to cracking and crack development process in specific environments. From the methodological point of view, further development of works aimed at developing more and more accurate, automatic crack detection systems is indicated. Such systems could be used very successfully by building managers to effectively detect areas in need of remedial action. The search for new methods and the development of the relationship between the cracking patterns and material properties is crucial in the development of non-destructive testing methods in cement composite technology.

Funding: This research was funded by the Polish Ministry of Science and Higher Education with the statutory research number FN14/ILT/2019, the subsidy for the development of young scientists— FNM14.1/ILT/2019.

Acknowledgments: The author is supported by the Foundation for Polish Science (FNP).

Conflicts of Interest: The author declares no conflict of interest.

References

1. Neville, A.M. *Properties of Concrete*, 4th ed.; Arkady: Warsaw, Poland, 2000.
2. Brandtzaeg, A. *Failure of A Material Composed of Non-isotropic Elements*; Trondhjem Bruns i Komm.: Trondheim, Norway, 1927.
3. Yoshida, H. *Über Das Elastische Verhalten Von Beton: mit Besonderer Berücksichtigung Der Querdehnung*; Springer Verlag: Berlin, Germany, 1930.
4. VanMier, J. *Concrete Fracture: A Multiscale Approach*; CRC Press: Boca Raton, FL, USA, 2013.
5. Recho, N. *Fracture Mechanics and Crack Growth*; John Wiley & Sons: Hoboken, NJ, USA, 2012.
6. Kurumatani, M.; Terada, K.; Kato, J.; Kyoya, T.; Kashiyama, K. An isotropic damage model based on fracture mechanics for concrete. *Eng. Fract. Mech.* **2016**, *155*, 49–66. [CrossRef]
7. Hlobil, M.; Smilauer, V.; Chanvillard, G. Micromechanical multiscale fracture model for compressive strength of blended cement pastes. *Cem. Concr. Res.* **2016**, *83*, 188–202. [CrossRef]
8. Bisschop, J.; Wittel, F. Contraction gradient induced microcracking in hardened cement paste. *Cem. Concr. Compos.* **2011**, *33*, 466–473. [CrossRef]
9. Xu, Y.; Wong, Y.; Poon, C.; Anson, M. Influence of PFA on cracking of concrete and cement paste after exposure to high temperatures. *Cem. Concr. Res.* **2003**, *33*, 2009–2016. [CrossRef]
10. Tosun-Felekoglu, K.; Felekoglu, B. Effects of fibre hybridization on multiple cracking potential of cement-based composites under flexural loading. *Constr. Build. Mater.* **2013**, *41*, 15–20. [CrossRef]
11. Banthia, N.; Azzabi, M.; Pigeon, M. Restrained shrinkage cracking in fiber-reinforced cementitious composites. *Mater. Struct.* **1993**, *26*, 405–413. [CrossRef]
12. Szelag, M. Evaluation of cracking patterns of cement paste containing polypropylene fibers. *Compos. Struct.* **2019**, *220*, 402–411. [CrossRef]
13. Szelag, M. Properties of cracking patterns of multi-walled carbon nanotube-reinforced cement matrix. *Materials* **2019**, *12*, 2942. [CrossRef]
14. Szelag, M. Development of Cracking Patterns in Modified Cement Matrix with Microsilica. *Materials* **2018**, *11*, 1928. [CrossRef]
15. Szelag, M. Influence of specimen's shape and size on the thermal cracks' geometry of cement paste. *Constr. Build. Mater.* **2018**, *189*, 1155–1172. [CrossRef]
16. Szelag, M. The Influence of Metakaolinite on the Development of Thermal Cracks in a Cement Matrix. *Materials* **2018**, *11*, 520. [CrossRef]
17. Griffith, A. The phenomena of rupture and flow in solids. *Philos. Trans. R. Soc. Lond. Sereis A* **1921**, *221*, 163–198.
18. Mindess, S.; Barnes, P. *Structure and Performance of Cements*; Applied Science Publishers: London, UK, 1983.
19. Mindess, S. Fracture toughness testing of cement and concrete. In *Fracture Mechanics of Concrete: Material Characterization and Testing*; Springer: Dordrecht, Germany, 1984; pp. 67–110.
20. Powers, T.C. *Rheology of Freshly Mixed Concrete. The Properties of Fresh Concrete*; John Wiley: New York, NY, USA, 1968.

21. Oberholster, R.E. Pore structure, permeability and diffusivity of hardened cement paste and concrete in relation to durability: Status and prospects. In Proceedings of the 8th International congress on chemistry of cement, Rio de Janeiro, Brasil, 22–27 September 1986; pp. 323–335.

22. Brown, J.H.; Pomeroy, C.D. Fracture toughness of cement paste and mortars. *Cem. Concr. Res.* **1973**, *3*, 475–480. [CrossRef]

23. Hillemeier, B.; Hilsdorf, H. Fracture mechanics studies on concrete compounds. *Cem. Concr. Res.* **1977**, *7*, 523–535. [CrossRef]

24. Beaudoin, J. Effect of humiidty and porosity on fracture of hardened Portland-cement. *Cem. Concr. Res.* **1982**, *12*, 705–716. [CrossRef]

25. Shah, S. Determination of fracture parameters (KIcs and CTODc) of plain concrete using three-point bend tests. *Mater. Struct.* **1990**, *23*, 457–460. [CrossRef]

26. Chen, X.; Wu, S.; Zhou, J. Influence of porosity on compressive and tensile strength of cement mortar. *Constr. Build. Mater.* **2013**, *40*, 869–874. [CrossRef]

27. Kumar, R.; Bhattacharjee, B. Porosity, pore size distribution and in situ strength of concrete. *Cem. Concr. Res.* **2003**, *33*, 155–164. [CrossRef]

28. Lian, C.; Zhuge, Y.; Beecham, S. The relationship between porosity and strength for porous concrete. *Constr. Build. Mater.* **2011**, *25*, 4294–4298. [CrossRef]

29. Roy, D.; Gouda, G. Porosity-strength relation in cementitious materials with very high strengths. *J. Am. Ceram. Soc.* **1973**, *56*, 549–550. [CrossRef]

30. Taylor, H. Discussion of the paper "Microstructure and strength of hydrated cements" by RF Feldman and JJ Beaudoin. *Cem. Concr. Res.* **1977**, *7*, 465–468. [CrossRef]

31. Mindess, S. Relation between the compressive strength and porosity of autoclaved calcium silicate hydrates. *J. Am. Ceram. Soc.* **1970**, *53*, 621–624. [CrossRef]

32. Nadeau, J.; Mindess, S.; Hay, J. Slow crack growth in cement paste. *J. Am. Ceram. Soc.* **1974**, *57*, 51–54. [CrossRef]

33. Wittmann, F. Trends in research on creep and shrinkage of concrete. In *Cement Production an Use*; Engineering Foundation: New York, NY, USA, 1979; pp. 143–162.

34. Birchall, J.; Howard, A.; Kendall, K. Flexural strength and porosity of cements. *Nature* **1981**, *289*, 388–390. [CrossRef]

35. Alford, N.; Rahman, A. An assesment of porosity and pore sizes in hardened cement pastes. *J. Mater. Sci.* **1981**, *16*, 3105–3114. [CrossRef]

36. Wittmann, F. Surface tension skrinkage and strength of hardened cement paste. *Matériaux et Construction* **1968**, *1*, 547–552. [CrossRef]

37. Robertson, B.; Mills, R. Influence of sorbed fluids on compressive strength of cement paste. *Cem. Concr. Res.* **1985**, *15*, 225–232. [CrossRef]

38. Costa, U.; Massazza, F. Rheological properties of fly ash cement pastes. *Il Cemento* **1986**, *4*, 397–414.

39. Beaudoin, J. Calcium hydroxide in cement matrices: Physico-mechanical and physico-chemical contributions. In *Calcium Hydroxide in Concrete in Materials Science of Concrete*; American Ceramic Society: Westerville, OH, USA, 2000; pp. 131–132.

40. Alford, N.; Groves, G.; Double, D. Physical properties of high-strength cement pastes. *Cem. Concr. Res.* **1982**, *12*, 349–358. [CrossRef]

41. Neville, A. The influence of the direction of loading on the strength of concrete test cubes. *ASTM Bull.* **1959**, *239*, 63–65.

42. Hsu, T.; Slate, F.; Sturman, G.; Winter, G. Microcracking of plain concrete and the shape of the stress-strain curve. *J. Am. Concr. Inst.* **1963**, *60*, 209–224.

43. Akcaoglu, R.; Tokyay, M.; Celik, T. Assessing the ITZ microcracking via scanning electron microscope and its effect on the failure behavior of concrete. *Cem. Concr. Res.* **2005**, *35*, 358–363. [CrossRef]

44. Golewski, G. An assessment of microcracks in the Interfacial Transition Zone of durable concrete composites with fly ash additives. *Compos. Struct.* **2018**, *200*, 515–520. [CrossRef]

45. Jornet, A.; Guidali, E.; Muhlethaler, U. Microcracking in high-performance concrete. In Proceedings of the 4th euroseminar on microscopy applied to building materials, Boras, Sweden, 30 September 1993; Swedish National Testing and Research Institute: Stockholm, Sweden, 1993; p. 6.

46. Slate, F.; Hover, K. Microcracking in concrete. In *Fracture Mechanics of Concrete: Material Characterization and Testing*; Springer: Dordrecht, Germany, 1984; pp. 137–159.

47. Smadi, M.; Slate, F. Microcracking of high and normal strength concreters under short-term and long-term loadings. *Mater. J.* **1989**, *86*, 117–127.

48. Solomatov, V.I. Polystructural theory development of composite building materials. In *The Success of Modern Materials Science. Proceedings of the Anniversary Conference, RAASN*; MGUPS: Moscow, Russia, 2001; pp. 56–66.

49. Solomatov, V.I.; Vyrovoy, V.N.; Bobryshev, A.N. *Polystructural Theory of Composite Building Materials*; FAN: Tashkent, Uzbekistan, 1991.

50. Sukhanov, V.; Vyrovoy, V.; Dorofeev, V. Crack's role in structural development of the constructional composite materials. *Mod. Ind. Civ. Constr.* **2011**, *7*, 135–141.

51. Vyrovoy, V.N.; Dorofeev, V.S.; Sukhanov, V.G. *Composite Building Materials and Constructions. Structure, Self-Organization, Properties*; TES: Odess, Ukraine, 2010.

52. Report, C.S. *Non-Structural Cracks in Concrete. Technical Report no. 22*; Concrete Society: London, UK, 1992.

53. Henrichsen, A.; Laugesen, P.; Diamond, S.; Mindess, S.; Glasser, F.; Roberts, L.; Skalny, J.; Wakeley, L. Monitoring of concrete quality in high performance civil engineering constructions. *MRS Online Proceed. Library Arch.* **1995**, *370*, 49–56.

54. Acosta, J.A.; Figueroa, J.L.; Mullen, R.L. Low-cost video image processing system for evaluating pavement surface distress. *Transp. Res. Rec.* **1992**, *1348*, 63–72.

55. Fujita, Y.; Hamamoto, Y. A robust automatic crack detection method from noisy concrete surfaces. *Mach. Vis. Appl.* **2011**, *22*, 245–254. [CrossRef]

56. Sezgin, M.; Sankur, B. Survey over image thresholding techniques and quantitative performance evaluation. *J. Electron. Imag.* **2004**, *13*, 146–168. [CrossRef]

57. Tang, J.; Gu, Y. Automatic crack detection and segmentation using a hybrid algorithm for road distress analysis. In Proceedings of the 2013 IEEE International Conference on Systems, Man, and Cybernetics, Manchester, UK, 13–16 October 2013; pp. 3026–3030.

58. Tong, X.; Guo, J.; Ling, Y.; Yin, Z. A new image-based method for concrete bridge bottom crack detection. In Proceedings of the International Conference on Image Analysis and Signal Processing, Wuhan, China, 21–23 October. 2011; IEEE: Piscataway, NJ, USA, 2001; pp. 568–571.

59. Liu, X.; Ai, Y.; Scherer, S. Robust image-based crack detection in concrete structure using multi-scale enhancement and visual features. In Proceedings of the 2017 IEEE International Conference on Image Processing (ICIP), Beijing, China, 7–20 September 2017; IEEE: Piscataway, NJ, USA, 2017; pp. 2304–2308.

60. Gavilan, M.; Balcones, D.; Marcos, O.; Llorca, D.; Sotelo, M.; Parra, I.; Ocana, M.; Aliseda, P.; Yarza, P.; Amirola, A. Adaptive Road Crack Detection System by Pavement Classification. *Sensors* **2011**, *11*, 9628–9657. [CrossRef]

61. Otsu, N. A threshold selection method from gray-level histograms. *IEEE Trans. Syst. Man Cybern.* **1979**, *9*, 62–66. [CrossRef]

62. Talab, A.; Huang, Z.; Xi, F.; Liu, H. Detection crack in image using Otsu method and multiple filtering in image processing techniques. *Optik* **2016**, *127*, 1030–1033. [CrossRef]

63. Valenca, J.; Dias-da-Costa, D.; Julio, E. Characterisation of concrete cracking during laboratorial tests using image processing. *Constr. Build. Mater.* **2012**, *28*, 607–615. [CrossRef]

64. Hoang, D. Detection of Surface Crack in Building Structures Using Image Processing Technique with an Improved Otsu Method for Image Thresholding. *Adv. Civ. Eng.* **2018**, *2018*. [CrossRef]

65. Nishikawa, T.; Yoshida, J.; Sugiyama, T.; Saito, S.; Fujino, Y. Robust Image Procesing for detection of Concrete Cracks Using a Parallel Image-Filter. *J. JSCE* **2007**, *63*, 599–616.

66. Aoki, S.; Nagao, T. Automatic construction of tree-structural image transformations using genetic programming. In Proceedings of the 10th International Conference on Image Analysis and Processing, Kobe, Japan, 24–28 October 1999; IEEE: Piscataway, NJ, USA, 1999; pp. 136–141.

67. Nishikawa, T.; Yoshida, J.; Sugiyama, T.; Fujino, Y. Concrete Crack Detection by Multiple Sequential Image Filtering. *Comput. Aided Civ. Infrastruct. Eng.* **2012**, *27*, 29–47. [CrossRef]

68. Tomikawa, T. A study of road crack detection by the meta-genetic algorithm. In Proceedings of the 1999 IEEE Africon. 5th Africon Conference in Africa (Cat. No. 99CH36342), Cape Town, South Africa, 28 September–1 October 1999; IEEE: Piscataway, NJ, USA, 1999; pp. 543–548.

69. Medina, R.; Llamas, J.; Gomez-Garcia-Bermejo, J.; Zalama, E.; Segarra, M. Crack Detection in Concrete Tunnels Using a Gabor Filter Invariant to Rotation. *Sensors* **2017**, *17*. [CrossRef]

70. Zadeh, L. Fuzzy-logic. *Computer* **1988**, *21*, 83–93. [CrossRef]

71. Choudhary, G.; Dey, S. Crack Detection in Concrete Surfaces using Image Processing, Fuzzy Logic, and Neural Networks. In Proceedings of the 2012 IEEE Fifth International Conference on Advanced Computational Intelligence (ICACI), Nanjing, China, 18–20 October 2012; IEEE: Piscataway, NJ, USA, 2012; pp. 404–411.

72. Cheng, H.; Chen, J.; Glazier, C.; Hu, Y. Novel approach to pavement cracking detection based on fuzzy set theory. *J. Comput. Civ. Eng.* **1999**, *13*, 270–280. [CrossRef]

73. Maode, Y.; Shaobo, B.; Xue, L.; Yuyao, H. An adaptive fuzzy image enhancement algorithm for local regions. In Proceedings of the 2007 Chinese Control Conference, Hunan, China, 26–31 July 2007; IEEE: Piscataway, NJ, USA, 2007; Volume 5, pp. 308–311.

74. Haykin, S. *Neural Networks: A Comprehensive Foundation*; Prentice Hall PTR: New York, NY, USA, 1994.

75. Lee, B.; Lee, H. Position-invariant neural network for digital pavement crack analysis. *Comput. Aided Civ. Infrastruct. Eng.* **2004**, *19*, 105–118. [CrossRef]

76. Bray, J.; Verma, B.; Li, X.; He, W. A neural network based technique for automatic classification of road cracks. In Proceedings of the 2006 IEEE International Joint Conference on Neural Network Proceedings, Vancouver, BC, Canada, 16–21 July 2006; IEEE: Piscataway, NJ, USA, 2006; pp. 2161–4393.

77. Moon, H.; Kim, J. Intelligent crack detecting algorithm on the concrete crack image using neural network. In Proceedings of the 28th ISARC, Seoul, Korea, 9 June–2 July 2011; pp. 1461–1467.

78. Dung, C.; Anh, L. Autonomous concrete crack detection using deep fully convolutional neural network. *Autom. Constr.* **2019**, *99*, 52–58. [CrossRef]

79. Zhang, J.; Lu, C.; Wang, J.; Wang, L.; Yue, X. Concrete Cracks Detection Based on FCN with Dilated Convolution. *Appl. Sci.* **2019**, *9*. [CrossRef]

80. Cha, Y.; Choi, W.; Caicedo, J.; Pakzad, S. Vision-Based Concrete Crack Detection Using a Convolutional Neural Network. *Dyn. Civ. Struct.* **2017**, *2*, 71–73. [CrossRef]

81. Lee, B.; Kim, Y.; Yi, S.; Kim, J. Automated image processing technique for detecting and analysing concrete surface cracks. *Struct. Infrastruct. Eng.* **2013**, *9*, 567–577. [CrossRef]

82. Dijkstra, E. A note on two problems in connexion with graphs. *Numer. Math.* **1959**, *1*, 269–271. [CrossRef]

83. Amhaz, R.; Chambon, S.; Idier, J.; Baltazart, V. Automatic Crack Detection on Two-Dimensional Pavement Images: An Algorithm Based on Minimal Path Selection. *IEEE Trans. Intell. Transp. Syst.* **2016**, *17*, 2718–2729. [CrossRef]

84. Gunkel, C.; Stepper, A.; Muller, A.; Muller, C. Micro crack detection with Dijkstra's shortest path algorithm. *Mach. Vis. Appl.* **2012**, *23*, 589–601. [CrossRef]

85. Yu, S.; Jang, J.; Han, C. Auto inspection system using a mobile robot for detecting concrete cracks in a tunnel. *Autom. Constr.* **2007**, *16*, 255–261. [CrossRef]

86. Lee, S.Y.; Lee, S.; Shin, D.; Son, Y.; Han, C. Development of an inspection system for cracks in a concrete tunnel lining. *Can. J. Civ. Eng.* **2007**, *34*, 966–975. [CrossRef]

87. Lindley, D. Kendall's advanced theory of statistics, volume 2B, Bayesian inference, 2nd edition. *J. R. Stat. Soc. Ser. A Stat. Soc.* **2005**, *168*, 259–260. [CrossRef]

88. Schmugge, S.; Nguyen, N.; Thao, C.; Lindberg, J.; Grizzi, R.; Joffe, C.; Shin, M. Automatic detection of cracks during power plant inspection. In Proceedings of the 3rd International Conference on Applied Robotics for the Power Industry, Iguassu, Brazil, 14–16 October 2014; pp. 1–5.

89. Hutchinson, T.; Chen, Z. Improved image analysis for evaluating concrete damage. *J. Comput. Civ. Eng.* **2006**, *20*, 210–216. [CrossRef]

90. Valenca, J.; Goncalves, L.; Julio, E. Damage assessment on concrete surfaces using multi-spectral image analysis. *Constr. Build. Mater.* **2013**, *40*, 971–981. [CrossRef]

91. Oliveira, H.; Correia, P.L. Supervised strategies for cracks detection in images of road pavement flexible surfaces. In Proceedings of the 2008 16th European Signal Processing Conference, Lausanne, Switzerland, 25–29 August 2008; IEEE: Piscataway, NJ, USA, 2015; pp. 1–5.

92. Oliveira, H.; Correia, P. Automatic Crack Pavement Detection Using a Bayesian Stochastic Pattern Recognition System. In Proceedings of the RECPAD2007, Lisbon, Portugal, 31 Ocotber 2007.

93. Freund, Y.; Schapire, R.; Abe, N. A short introduction to boosting. *J. Jpn. Soc. Artif. Intell.* **1999**, *14*, 1612.

94. Cord, A.; Chambon, S. Automatic Road Defect Detection by Textural Pattern Recognition Based on AdaBoost. *Comput. Aided Civ. Infrastruct. Eng.* **2012**, *27*, 244–259. [CrossRef]

95. Prasanna, P.; Dana, K.; Gucunski, N.; Basily, B.; La, H.; Lim, R.; Parvardeh, H. Automated Crack Detection on Concrete Bridges. *IEEE Trans. Autom. Sci. Eng.* **2016**, *13*, 591–599. [CrossRef]

96. Wagner, C.; Villmann, B.; Slowik, V.; Mechtcherine, V. Water permeability of cracked strain-hardening cement-based composites. *Cem. Concr. Compos.* **2017**, *82*, 234–241. [CrossRef]

97. Wagner, C.; Dollase, A.; Slowik, V. Evaluation of crack patterns in SHCC with respect to water permeability and capillary suction. In Proceedings of the 3rd International Conference on Concrete Repair, Rehabilitation and Retrofitting (ICCRRR), Cape Town, South Africa, 3–5 September 2012; pp. 972–977.

98. Fic, S.; Szelag, M. Analysis of the development of cluster cracks caused by elevated temperatures in cement paste. *Constr. Build. Mater.* **2015**, *83*, 223–229. [CrossRef]

99. Mobasher, B.; Stang, H.; Shah, S. Microcracking in fiber reinforced-concrete. *Cem. Concr. Res.* **1990**, *20*, 665–676. [CrossRef]

100. Magalhaes, M.; Toledo, R.; Fairbairn, E. Thermal stability of PVA fiber strain hardening cement-based composites. *Constr. Build. Mater.* **2015**, *94*, 437–447. [CrossRef]

101. Szelag, M.; Szewczak, A. Evaluation of Dependencies between Physico-Mechanical Properties and the Thermal Cracks' Geometry of Cement Pastes Modified with Metakaolinite Using the LSM Method. 3rd World Multidisciplinary Civil Engineering, Architecture, Urban Planning Symposium (WMCAUS 2018). *IOP Conf. Ser. Mater. Sci. Eng.* **2019**, *471*. [CrossRef]

102. Szeląg, M.; Szewczak, A. Dependencies between Cracking Patterns and the Physico-Mechanical Properties of Microsilica Modified Cement Matrix. *In Proceedings of IOP Conference Series: Mater. Sci. Eng.* **2019**, *484*, 012015. [CrossRef]

103. Mandelbrot, B.B. *The Fractal Geometry of Nature*; WH freeman: New York, NY, USA, 1983; Volume 173.

104. Mandelbrot, B.B. *Fractals: Form, Chance, and Dimension*; WH freeman: New York, NY, USA, 1977; Volume 706.

105. Farhidzadeh, A.; Dehghan-Niri, E.; Moustafa, A.; Salamone, S.; Whittaker, A. Damage Assessment of Reinforced Concrete Structures Using Fractal Analysis of Residual Crack Patterns. *Exp. Mech.* **2013**, *53*, 1607–1619. [CrossRef]

106. Wang, K.; Shah, S.; Phuaksuk, P. Plastic shrinkage cracking in concrete materials–Influence of fly ash and fibers. *Mater. J.* **2001**, *98*, 458–464.

107. Bakhshi, M.; Mobasher, B. Experimental observations of early-age drying of Portland cement paste under low-pressure conditions. *Cem. Concr. Compos.* **2011**, *33*, 474–484. [CrossRef]

108. Lepech, M.; Li, V. Water permeability of engineered cementitious composites. *Cem. Concr. Compos.* **2009**, *31*, 744–753. [CrossRef]

109. Aldea, C.; Ghandehari, M.; Shah, S.; Karr, A. Combined effect of cracking and water permeability of concrete. In Proceedings of the 14th Engineering Mechanics Conference, Austin, TX, USA, 21–24 May 2000; pp. 21–24.

110. Bazant, Z.; Raftshol, W. Effect of cracking in drying and shrinkage specimens. *Cem. Concr. Res.* **1982**, *12*, 209–226. [CrossRef]

111. Bazant, Z.; Sener, S.; Kim, J. Effect of cracking on drying permeability and diffusivity of concrete. *Mater. J.* **1987**, *84*, 351–357.

112. Torrijos, M.; Giaccio, G.; Zerbino, R. Internal cracking and transport properties in damaged concretes. *Mater. Struct.* **2010**, *43*, 109–121. [CrossRef]

113. Lura, P.; Jensen, O.; Weiss, J. Cracking in cement paste induced by autogenous shrinkage. *Mater. Struct.* **2009**, *42*, 1089–1099. [CrossRef]

114. Xuan, D.; Molenaar, A.; Houben, L. Shrinkage cracking of cement treated demolition waste as a road base. *Mater. Struct.* **2016**, *49*, 631–640. [CrossRef]

115. Ebrahimkhanlou, A.; Farhidzadeh, A.; Salamone, S. Multifractal analysis of crack patterns in reinforced concrete shear walls. *Struct. Health Monit. Int. J.* **2016**, *15*, 81–92. [CrossRef]

116. Zhuang, N.; Dong, H.; Zhou, Y.; Chen, D. Cracking behavior of reinforced concrete piles externally bonded with carbon fiber reinforced polymer in a marine environment. *Constr. Build. Mater.* **2018**, *190*, 1154–1162. [CrossRef]

117. Ebrahimkhanlou, A.; Farhidzadeh, A.; Salamone, S.; Lynch, J.; Wang, K.; Sohn, H. Multifractal analysis of two-dimensional images for damage assessment of reinforced concrete structures. *Sens. Smart Struct. Techn. Civil Mech. Aeros. Syst.* **2015**, *9435*. [CrossRef]

118. Ebrahimkhanlou, A.; Athanasiou, A.; Hrynyk, T.; Bayrak, O.; Salamone, S. Fractal and Multifractal Analysis of Crack Patterns in Prestressed Concrete Girders. *J. Bridge Eng.* **2019**, *24*. [CrossRef]
119. Fooladi, A.; Banan, M. Multifractal Analysis of Crack Propagation in Concrete Specimens Considering the Influence of the Aggregates' Grading. *Iran. J. Sci. Technol.-Trans. Civ. Eng.* **2016**, *40*, 97–108. [CrossRef]
120. Kim, K.; Yun, T.; Park, K. Evaluation of pore structures and cracking in cement paste exposed to elevated temperatures by X-ray computed tomography. *Cem. Concr. Res.* **2013**, *50*, 34–40. [CrossRef]

MDPI

St. Alban-Anlage 66

4052 Basel

Switzerland

Tel. +41 61 683 77 34

Fax +41 61 302 89 18

www.mdpi.com

Materials Editorial Office

E-mail: materials@mdpi.com

www.mdpi.com/journal/materials